Soil Biology

Volume 61

Series Editor

Ajit Varma, Amity Institute of Microbial Technology,
Amity University, Noida, Uttar Pradesh, India

The Springer series *Soil Biology* publishes topical volumes in the fields of microbiology, environmental sciences, plant sciences, biotechnology, biochemistry, microbial ecology, mycology and agricultural sciences.

Special emphasis is placed on methodological chapters or volumes. This includes coverage of new molecular techniques relevant to soil biology research or to monitoring and assessing soil quality as well as advanced biotechnological applications. Leading international authorities with a background in academia, industry or government will contribute to the series as authors or editors.

Key Topics: microbial-plant interactions; microbial communities; root symbiosis, mycorrhiza; rhizosphere environment; soil fauna, e.g. termites etc.; biochemical processes, soil proteins, enzymes, nucleic acids; degradation of biomaterials, decomposition, nutrient cycles; soil genesis, mineralization; bioremediation of contaminated sites; biotechnological applications of soil microorganisms.

More information about this series at http://www.springer.com/series/5138

Chakravarthula Manoharachary •
Harikesh Bahadur Singh • Ajit Varma
Editors

Trichoderma: Agricultural Applications and Beyond

Springer

Editors
Chakravarthula Manoharachary
Department of Botany
Osmania University
Hyderabad, Telangana, India

Harikesh Bahadur Singh
Department of Biotechnology
GLA University
Mathura, Uttar Pradesh, India

Ajit Varma
Amity Institute of Microbial Technology
Amity University Uttar Pradesh
Noida, India

ISSN 1613-3382 ISSN 2196-4831 (electronic)
Soil Biology
ISBN 978-3-030-54757-8 ISBN 978-3-030-54758-5 (eBook)
https://doi.org/10.1007/978-3-030-54758-5

This Springer imprint is published by the registered company Springer Nature Switzerland AG.
The registered company address is: Gewerbestrasse 11, 6330 Cham, Switzerland

Preface

Agriculture plays an important role in a country's economy besides offering food security and nutritional security through increased crop production. In the last few years, agriculture has shown exceptional growth curve the world over due to the usage of various biotechnological approaches and innovation in the effective management of plant diseases and pests. Crop yields have increased in cereals, millets, vegetables, fruit crops, and other crops, thereby contributing toward food and nutritional security. However, in view of the arrival of invasive pathogens, shortage of monsoon, loss of soil fertility, global warming, injudicious use of chemical fertilizers and pesticides, and several other reasons, increased losses are going to be expected in the crops. Around 30–32% of crop losses are expected due to plant pathogens, loss of soil fertility, and change in seasons. Thus, the crop yields may get minimized in the future leading to hunger and poverty. It is difficult to eliminate hunger in developing countries in the near future. Therefore, there is an immediate need to reduce the crop losses not only by bringing better crop varieties but also by using biopesticides, biofertilizers, and growth-promoting microbes and application of other innovative methods.

Trichoderma Pers. is a commonly occurring fungus in soil and rhizosphere, and it has the ability to colonize and multiply on diversified habitats, which also include litter, air, mud, and seed, as endosymbiont in plant parts and in other habitats. *Trichoderma* is the anamorph discovered in 1801 by Persoon and its perfect stage falls under *Hypocrea* of Hypocreaceae of Ascomycota with more than 100 species. It is a complex genus posing problems in understanding its morphology, taxonomy, and identification.

In recent times, molecular approaches, biochemical parameters, secondary metabolites, electron microscopy, and other characters paved the way for better characterization and identification of different species and strains of *Trichoderma*. It also grows rapidly on different agar media.

The beneficial effects of *Trichoderma* include its ubiquitous occurrence and successful colonization of diversified substrates besides being a potential degrader of heterogeneous substrates. It is also an efficient biocontrol agent of plant diseases.

Large amounts of cellulase elaborated by *Trichoderma reesei* and other species have become important in the production of second-generation biofuels from cellulose waste. *Trichoderma* species have been employed in bioremediation and as bioprotectant to maintain plant health. Further, it has been used for the production of food additives and related products useful in food industry. Species of *Trichoderma* are also employed as growth promoters. In the last decade, this has emerged out as a potential bioprotectant for the efficient management of all kinds of soil-borne, root-borne, and foliar plant diseases caused by fungi, microbes, nematodes, and others. *Trichoderma* by its broad spectrum action against a number of plant pathogens has occupied a prime position among bioprotectants besides being easily cultivable and mass produced.

Trichoderma-based biopesticides have been proved successful in the plant disease management both under glasshouse conditions and in a large number of fields supporting cereals, millets, pulses, vegetables, fruit, and flowering plants. It is ecofriendly and its products are low cost when compared to chemical protectants. The mass production technology has been very widely used all over the world. The adverse impact of the injudicious use of chemical pesticides and fertilizers is of great concern, and therefore, development of alternate control strategies such as biological control is the need of the hour. In this context, the utilization of introduced resident microbes other than native host becomes really essential. Though biological control developed as an academic event initially, but now it has become an important branch of science, particularly in agriculture for application as a biopesticide/bioprotectant due to pressure from the public, environmental scientists, and agriculturists as the use of chemicals such as fungicides, pesticides, weedicides, insecticides, and others has become hazardous to nature, soil, and humans.

In the last 100 years, several research labs and researchers have contributed significantly not only in elaborating the mechanism of *Trichoderma* as a biocontrol agent but also in the production of biopesticides on a commercial basis. Thus, *Trichoderma*-based biopesticides have been produced commercially, marketed, and proved successful as biocontrol agents in the world. Most of the *Trichoderma* products are cost-effective and are adaptable to various soil and environmental conditions.

Trichoderma species as antagonist reduces the negative effects of plant pathogens and promote growth along with yield increase besides being adjusted to various adverse conditions. Biocontrol mechanism of *Trichoderma* has been envisaged through antibiosis, competition, and parasitism and also by inducing systemic resistance along with induction of plant growth through the production of growth hormones. *Trichoderma* and its bioformulations are in use for the control of soil-borne, root-borne, seed-borne, and foliar pathogens. *Trichoderma* species can be formulated as granules, pellets, dusts, wettable powders, and fluid drill gels for application on the crop plants. Granular or pellets and *Trichoderma*-enriched FYM have been used for soil application and have provided effective control of plant diseases. Talc, peat, lignite, kaolin-based formulations, seed biopriming, liquid formulations, and others have played an effective role in the control of plant diseases. There are several commercial formulations available in the markets the world over, but many local products of each country might have been adulterated. Therefore, it is necessary to establish the purity of the *Trichoderma* products under laboratory conditions before its application in the field. There are hundreds of commercial products based on *Trichoderma* on the global market. There is a need to enrich the rhizosphere with biocontrol agents such as *Trichoderma* after establishing their efficacy as potential biocontrol agents.

The book is organized into 16 chapters covering the following subjects: Systematic and taxonomy (Chap. 1), Biodiversity (Chap. 2), Beneficial effects on plant–pathogen interactions (Chap. 3), *Trichoderma*: Boon for agriculture (Chap. 4), Mass multiplication of *Trichoderma* in bioreactors (Chap. 5), *Trichoderma* species: A blessing for crop production (Chap. 6), Management of pests and pathogens of pulses (Chap. 7), Management of diseases of medicinal and aromatic (Chap. 8), *Trichoderma*: A globally dominant commercial biofungicide (Chap. 9), Modulation of microbiome through seed biopriming (Chap. 10), Opportunistic avirulent plant symbionts (Chap. 11), Biotechnological application of *Trichoderma* (Chap. 12), *Trichoderma* as biostimulant (Chap. 13), *Trichoderma* proteome (Chap. 14), Biotechnological innovations using *Trichoderma* (Chap. 15), and *Trichoderma* spp.: Expanding potential beyond agriculture (Chap. 16).

We are highly indebted to each and every person who helped us in preparing the exhaustive and informative volume dedicated to *Trichoderma*. Since the chapters were written independently by various authors, there might be some slight overlap or repetition in the content which was difficult to avoid under these conditions. We are also grateful to Dr. Sabine Schwartz, Springer Life Sciences Editorial Board, and Mr. Abdus Salam Mazumder, Project Coordinator (Books), Springer Nature, India, for their valuable help. It is our hope and belief that the valuable information shared in this volume will make an invaluable contribution to the scientific and teaching fraternity. We also believe that the exchange of knowledge shared in this volume will kindle further discussions and help in new knowledge generation in this vital subject area. Last but not least, we also anticipate that the book will be an asset for all the students, teachers, and researchers working in the area of fungal and microbial biotechnology, environmental science, microbiology, agriculture, forestry, and other allied science subjects.

Hyderabad, India — Chakravarthula Manoharachary
Mathura, India — Harikesh Bahadur Singh
Noida, India — Ajit Varma

Contents

Editors and Contributors

About the Editors

C. Manoharachary Professor C. Manoharachary has served Osmania University for 45 long years in different capacities including as Dean. He has also served as Vice-Chancellor of Oriental University, Indore. He has supervised 50 students for Ph.D., published >640 research papers, and written and edited 30 books. His contributions to fungal taxonomy are globally acclaimed. Prof. Manoharachary has discovered 20 new fungal genera and 82 new fungal species. He has immensely contributed to the advancement of teaching and research in mycology and plant pathology besides establishing excellent infrastructure at Osmania University, Hyderabad, India. He is the recipient of five National Awards including Dr. E. K. Janaki Ammal National Award by the Ministry of Environment and UGC J.C. Bose Award. Further, he has received five awards from state government including best teacher award and outstanding scientist award. He has been honored with six lifetime achievement awards, served as President of several Indian academies, viz., the Indian Phytopathological Society, Indian Botanical Society, Mycological Society of India, and Indian Science Congress Association—Botany Section. He served as chairman/expert member of UGC, DST, DBT, MOENF, ICAR, CSIR, and others. He did his postdoctoral work in the UK, the USA, and Germany. Chary is a fellow of the National Academy of Sciences, India, and fellow of the National Academy of Agricultural Sciences and several other academic societies.

Harikesh Bahadur Singh Prof. Harikesh Bahadur Singh Director, Somvanshi Research Foundation, Lucknow, and Former Professor and Head, Department of Mycology and Plant Pathology, Banaras Hindu University, India, has contributed significantly in developing formulations of biocontrol agents for management of soil-borne plant pathogens. The formulations have increased the yield of several crops and also controlled most of the soil-borne diseases. The products developed are useful as plant growth enhancers and biofungicides for seed, soil, and foliar applications and improve the soil health, crop productivity, and quality. The technology has been transferred to various industries for commercial production of biopesticides. Prof. Singh has received several national awards, namely, CSIR Technology Prize 1999, BRSI Industrial Medal Award 2007, CSIR Award for Science and Technology Innovations for Rural Development in 2011, Prof. P. Maheshwari Medel 2011, Mundkur Memorial Award 2015, M.S. Swaminathan Award 2016, and C.N.R. Rao Award 2017. During this journey, Dr. Singh has obtained 20 US patents and 4 PCTs and filed 7 Indian patents. Prof. Singh published 290 research papers, 42 review articles, 87 book chapters, and 20 edited books from CABI, Springer, and Elsevier; 4 authored books; and guided 23 Ph.D. students, 6 PDFs, and 2 DST Women scientists.

Ajit Varma Professor Ajit Varma completed his Ph.D. in 1964 from Allahabad University, Allahabad University, India. He is a former Professor at the School of Life Sciences, Jawaharlal Nehru University, India. Presently, he is Group Dy. Vice Chancellor at Amity University, Uttar Pradesh, India. He is Distinguished Scientist and Professor of Eminence at the Amity Institute of Microbial Technology and Vice Chairman of the Amity Science, Technology & Innovation Foundation at Amity University, Uttar Pradesh, India. He has visited many countries as Visiting Professor and Visiting Scientist. He has been awarded several national and international research fellowships. Prof. Varma has published more than 409 papers in respected journals and edited over 109 books. He has served as Editor in Chief of the Soil

Biology Series, Springer Verlag Germany. Prof. Varma has supervised >100 Ph.D. students and 1 D.Sc. student. Prof. Varma is a fellow of the Alexander-von-Humboldt Society, Germany, elected fellow of the National Academy of Agricultural Sciences, and fellow of the Microbiology Society of India and many more.

List of Contributors

Charles Oluwaseun Adetunji Applied Microbiology, Biotechnology and Nanotechnology Laboratory, Department of Microbiology, Edo University Iyamho, Edo State, Nigeria

Davoodbasha Mubarak Ali National Repository for Microalgae and Cyanobacteria – Freshwater (DBT, Govt. of India), Department of Microbiology, Bharathidasan University, Tiruchirappalli, India

P. Anbazhagan Department of Plant Pathology, College of Agriculture, Meerut, India

Udayashankar C. Arakere Department of Studies in Biotechnology, University of Mysore, Mysore, Karnataka, India

Raina Bajpai Department of Mycology and Plant Pathology, Institute of Agricultural Sciences, Banaras Hindu University, Varanasi, India

Arghya Chattopadhyay Department of Soil Science and Agricultural Chemistry, Institute of Agricultural Sciences, Banaras Hindu University, Varanasi, Uttar Pradesh, India

Srinivas Chowdappa Department of Microbiology and Biotechnology, Jnanabharathi Campus, Bangalore University, Bangalore, Karnataka, India

Sampa Das Division of Plant Biology, Bose Institute, Centenary Campus, Kolkata, West Bengal, India

O. Shiva Devika Department of Soil Science and Agricultural Chemistry, Institute of Agricultural Sciences, Banaras Hindu University, Varanasi, Uttar Pradesh, India

Nirmaladevi Dhamodaran Department of Microbiology, Ramaiah College of Arts, Science and Commerce, Bangalore, Karnataka, India

Nibha Gupta Plant Pathology and Microbiology Division, Regional Plant Resource Centre, Bhubaneswar, Odisha, India

Vijai Kumar Gupta AgroBioSciences (AgBS) and Chemical & Biochemical Sciences (CBS), Mohammed VI Polytechnic University (UM6P), Morocco

Anjana Hari Bioprocess Laboratory, Department of Microbial Biotechnology, Bharathiar University, Coimbatore, India

Akansha Jain Division of Plant Biology, Bose Institute, Centenary Campus, Kolkata, West Bengal, India

Yevgen Karpichev Department of Chemistry and Biotechnology, ERA Chair of Green Chemistry, Tallinn University of Technology, Tallinn, Estonia

Kandasamy Kathiresan Centre of Advanced Study in Marine Biology, Faculty of Marine Sciences, Annamalai University, Parangipettai, Tamil Nadu, India

Narasimhamurthy Konappa Department of Studies in Biotechnology, University of Mysore, Mysore, Karnataka, India

Soumya Krishnamurthy Department of Microbiology, Field Marshal K. M. Cariappa College, A Constituent College of Mangalore University, Madikeri, Karnataka, India

Krishna Kumar ICAR-Indian Institute of Pulses Research, Kanpur, India

C. Manoharachary Mycology and Molecular Plant Pathology Laboratory, Department of Botany, Osmania University, Hyderabad, Telangana, India

Deepak K. Maurya National Fungal Culture Collection of India (NFCCI), Biodiversity and Palaeobiology Group, MACS-Agharkar Research Institute, Gopal Ganesh Agarkar Road, Pune, India

Jhumishree Meher Department of Mycology and Plant Pathology, Institute of Agricultural Sciences, Banaras Hindu University, Varanasi, India

Monika Mishra ICAR-Indian Institute of Pulses Research, Kanpur, India

R. K. Mishra ICAR-Indian Institute of Pulses Research, Kanpur, India

D. Nagaraju Department of Botany, Govt. Degree College, Warangal, Telangana, India

K. Narasimha Murthy Department of Studies in Biotechnology, University of Mysore, Mysore, Karnataka, India

Subhadip Pal Department of Soil Science and Agricultural Chemistry, Institute of Agricultural Sciences, Banaras Hindu University, Varanasi, Uttar Pradesh, India

Sumita Pal Department of Mycology and Plant Pathology, Institute of Agricultural Sciences, Banaras Hindu University, Varanasi, India

Rakesh Pandey Department of Microbial Technology and Nematology, CSIR-Central Institute of Medicinal and Aromatic Plants, P.O. CIMAP, Lucknow, India

Sonika Pandey ICAR-Indian Institute of Pulses Research, Kanpur, India

Manoj Parihar Crop production division, ICAR-VPKAS, Almora, Uttarakhand, India

Vimala Prakash International Panaacea Limited, Haryana, India

Durga Prasad Department of Plant Pathology, Banda Agriculture University, Banda, India

Rahul Singh Rajput Department of Mycology and Plant Pathology, Institute of Agricultural Sciences, Banaras Hindu University, Varanasi, India

Amitava Rakshit Department of Soil Science and Agricultural Chemistry, Institute of Agricultural Sciences, Banaras Hindu University, Varanasi, Uttar Pradesh, India

Ratul Moni Ram Department of Mycology and Plant Pathology, Institute of Agricultural Sciences, Banaras Hindu University, Varanasi, India

Niranjana Siddapura Ramachandrapp Department of Studies in Biotechnology, University of Mysore, Mysore, Karnataka, India

Shiwali Rana National Fungal Culture Collection of India (NFCCI), Biodiversity and Palaeobiology Group, MACS-Agharkar Research Institute, Gopal Ganesh Agarkar Road, Pune, India

Utkarsh Singh Rathore ICAR-Indian Institute of Pulses Research, Kanpur, India

Susana Rodriguez-Couto Ikerbasque, Basque Foundation for Science, Bilbao, Spain

Kandasamy Saravanakumar Department of Medical Biotechnology, College of Biomedical Sciences, Kangwon National University, Chuncheon, Gangwon, Republic of Korea

Deepranjan Sarkar Department of Soil Science and Agricultural Chemistry, Institute of Agricultural Sciences, Banaras Hindu University, Varanasi, Uttar Pradesh, India

Birinchi Kumar Sarma Department of Mycology and Plant Pathology, Institute of Agricultural Sciences, Banaras Hindu University, Varanasi, India

Sabarathinam Shanmugam Bioprocess Laboratory, Department of Microbial Biotechnology, Bharathiar University, Coimbatore, India

Akanksha Singh Department of Microbial Technology and Nematology, CSIR-Central Institute of Medicinal and Aromatic Plants, P.O. CIMAP, Lucknow, India

Bhim P. Singh Department of Biotechnology, Mizoram University, Mizoram, India

Harikesh Bahadur Singh Department of Biotechnology, GLA University, Mathura, Uttar Pradesh, India

Paras Nath Singh National Fungal Culture Collection of India (NFCCI), Biodiversity and Palaeobiology Group, MACS-Agharkar Research Institute, Gopal Ganesh Agarkar Road, Pune, India

Ramji Singh Department of Plant Pathology, College of Agriculture, Meerut, India

Sanjay K. Singh National Fungal Culture Collection of India (NFCCI), Biodiversity and Palaeobiology Group, MACS-Agharkar Research Institute, Gopal Ganesh Agarkar Road, Pune, India

Sonam Singh Department of Soil Science and Agricultural Chemistry, Institute of Agricultural Sciences, Banaras Hindu University, Varanasi, Uttar Pradesh, India

C. Srinivas Department of Microbiology and Biotechnology, Jnanabharathi Campus, Bangalore University, Bangalore, Karnataka, India

Basavaraj Teli Department of Mycology and Plant Pathology, Institute of Agricultural Sciences, Banaras Hindu University, Varanasi, India

Ajay Tomer Department of Plant Pathology, Lovely Professional University, Jalandhar, Punjab, India

Kulbhusan Mani Tripathi ICAR-Indian Institute of Pulses Research, Kanpur, India

Zeba Usmani Department of Chemistry and Biotechnology, ERA Chair of Green Chemistry, Tallinn University of Technology, Tallinn, Estonia
Bioprocess Laboratory, Department of Microbial Biotechnology, Bharathiar University, Coimbatore, India

Anukool Vaishnav Department of Biotechnology, GLA University, Mathura, Uttar Pradesh, India

Ajit Varma Amity Institute of Microbial Technology Amity University Uttar Pradesh, Noida, India

H. S. Viswanath Department of Plant Pathology, College of Agriculture, Meerut, India

Myeong-Hyeon Wang Department of Medical Biotechnology, College of Biomedical Sciences, Kangwon National University, Chuncheon, Gangwon, Republic of Korea

Chapter 1
Advances in Systematics, Taxonomy, and Conservation of *Trichoderma* Species

Sanjay K. Singh, Paras Nath Singh, Deepak K. Maurya, and Shiwali Rana

Abstract *Trichoderma* is an important genus known for the past nearly 200 years. Till recently, *Trichoderma* and *Hypocrea* were treated as separate genera, with several species linked as asexual (anamorph) and sexual (teleomorph) morphs, respectively. As per the revised International Code of Nomenclature for Algae, Fungi and Plants (ICN) any fungi would no longer bear more than one name. Under this new provision of ICN, *Trichoderma* became valid and supersedes teleomorphic *Hypocrea*. Biotechnological applications of species of *Trichoderma* have seen tremendous changes in recent years, which has drawn serious attention toward fundamental taxonomy and systematics. The purpose of this chapter is to compile important information on current status of taxonomy, especially related to morphology, molecular and phylogeny of important species. Considering immense biotechnological importance of several species of this genus, it is pertinent to discuss importance of conservation of its species as it is largely ignored. Biological Resource Centres (BRCs)/Culture Collections play an important role in conserving the mycological resources. In order to reflect biodiversity, selected species of *Trichoderma* isolated from different natural sources and geographical locations, and already deposited at National Fungal Culture Collection of India (NFCCI) and a few newly isolated ones were reexamined morphologically as well as sequencing of recommended gene regions and their phylogenetic analysis were conducted.

Keywords Biodiversity · Fungi · Hypocreales · Systematics · Phylogeny · *Trichoderma* · India

---

S. K. Singh (✉) · P. N. Singh · D. K. Maurya · S. Rana
National Fungal Culture Collection of India (NFCCI), Biodiversity and Palaeobiology Group, MACS' Agharkar Research Institute, Pune, India
e-mail: sksingh@aripune.org

C. Manoharachary et al. (eds.), Trichoderma: *Agricultural Applications and Beyond*, Soil Biology 61, https://doi.org/10.1007/978-3-030-54758-5_1

1.1 Introduction

Trichoderma is a filamentous fungus belong to *Hypocreaceae* (Hypocreales) and is known since the centuries. Pers. Art. 59 of ICBN permitted the dual nomenclature, e.g., *Trichoderma* has been treated as anamorphic form of *Hypocrea*. However, Melbourne code in the year 2011 changed several age-old provisons including dual nomenclature, which has been abolished together with other regulatory changes. The ICBN has been changed to the International Code of Nomenclature (ICN). Among the most important changes which has taken place is one fungus one name (1F = 1 N) which has replaced the dual nomenclature provisons. This provision allows only one name. This nomeclatural change has been adopted in order to have unitary nomenclature, which simplifies the chaotic existence of pleomorphic fungi having more than one form (anamorph/teleomorph/synmorphs, etc.) in their life cycles. These forms compete with each other and as per rule of priority, the name established/published earlier will supersede the name published later and all the forms would be represented by single name (e.g., *Trichoderma* supersedes *Hypocrea*). The use of a single name is based either on priority or on commission-sanctioned decisions. The species of *Trichoderma* are found abundant in soil rich in organic matter mainly from tropics. Literature reveals that some species have been reported from other substrates including Mediterranean sponges (Brotman et al. 2010; Gal-Hemed et al. 2011; Druzhinina et al. 2011; Etschmann et al. 2015). This filamenous fungus has received world wide attention as biocontrol agents. Most of its species posses high rate of reproductive capacity, one of the important traits required for applications at large scale. Secondly, high survival, nutrients utilizing efficiency during adverse circumstances in rhizospheric condition and its high aggressiveness against many plant pathogens (Benítez et al. 2004; Harman 2006; Solanki et al. 2011; Keswani et al. 2013; Mukherjee et al. 2013a, b, c; Singh et al. 2013; Patil and Solanki 2016; Kumar et al. 2017; Deng et al. 2018; Topolovec-Pintarić 2019) make them especially interesting for their utilization in various sectors. *Trichoderma* species are widely used as biocontrol agents, and they can stimulate plant growth, and on the other hand, they can suppress plant diseases by direct and/or indirect mechanisms (Rai et al. 2019).

International Subcommission on *Trichoderma* and *Hypocrea* (ISTH) suggested a preference for adopting *Trichoderma* over *Hypocrea*. The generic name was proposed for acceptance by the Nomenclature Committee for Fungi (NCF) and the General Committee (GC) of the International Association for Plant Taxonomy (Rossman et al. 2013). The *Hypocrea* Fr., the type genus of the *Hypocreaceae* (Hypocreales) is now obsolete. Consequently, the name *Trichoderma* is now legitimate. Literature reveals the fact that much has been published about applications of this important genus and its species (Schuster and Schmoll 2010; Mukherjee et al. 2013a, b, c; Jaklitsch and Voglmayr 2014). From begining the taxonomy of *Trichoderma* was in the state of flux. It was described about 200 years ago, and mycologists mistook *Trichoderma* Pers.: Fr. for a Gasteromycete (Persoon 1794). Since then, several hundred epithets were added by the end of the nineteenth century which left scope to re-visit the generic circumscription defining the stable criteria. Some pioneering revisionary work during the twentieth century made the task easy where molecular phylogenetic study enabled rapid distinction of *Trichoderma* species (Jaklitsch 2009) irrespective of different morphs.

Species of *Trichoderma* were used as novel biological agents having immense properties and biological activities. Between 1992 and 1995, approximately 550 articles that cited the *Trichoderma* were cataloged in the USDA database AGRICOLA. Another global database, Index Fungorum, currently enlists 509 epithets in *Hypocrea*, including 55 names of varieties and forms (without having priority at species level). Similarly, 430 names are enlisted under *Trichoderma*. Many of these names do not represent to members of their respective genera and hence need to be corrected/synonymized, while the majority of entries require to be re-assessed especially after the execution of Melbourne code. One-stop history about some important work done on *Trichoderma* is enlisted in Table 1.1.

1.2 Biology and Species Concept

Species of *Trichoderma* is known to produce secondary metabolites and enzymes, and attack or restrict/kill the growth of other fungi. Because of these properties, it has attracted the attention of researchers in sectors like agriculture and industry. Biological control of plant diseases, industrially important enzyme production, and genetic control and manipulation in filamentous fungi are the major areas exploiting the potential of species of *Trichoderma*. In this scenario, when we address the issues relating to an increase in agriculture production for feeding the spiraling population and to enhance the economic wealth, *Trichoderma* serves as a valuable source. Whether and how this source can be utilized will depend upon understanding the biology of this taxon.

Generally, species of *Trichoderma* are defined based on morphology/morphotyping. The fundamental morphological features used in species recognition have been outlined from time to time by various researchers. Rifai (1969) and Bissett (1984, 1991a, b, c, 1992; Chaverri et al. 2003) provided comprehensive accounts about morphological characterization. It has been observed in hyphomycetous fungi that some morphological characters are not tenable hence difficult to consider as valid characters. These morphological characters are influenced by environmental conditions, hence are not good for species recognition. This situation is also found in *Trichoderma*. For example, conidial shape and size are considered as useful characters in other genera, have limited value in the identification of *Trichoderma*. Though variable conidial shapes like globose, subglobose, ellipsoidal, or oblong are useful in recognizing groups of species, they have limited value within the group.

Table 1.1 One-stop history of some important works done on *Trichoderma*

Author	Year	Details of contribution
Persoon	1794	Introduced *Trichoderma*
Bisby	1939	*Trichoderma viride* Pers. ex Fries, and notes on *Hypocrea*
Rifai	1969	Species aggregates
Montenecourt and Eveleigh	1979	Superior cellulase production by *T. reesei*, start of the genetic improvement
Chang et al.	1986	Mycoparasitism of *Trichoderma* strains first applications in biological control of plant pathogenic fungi
Bissett	1991a, b, c	Reported five sections and 27 morphological species
Kindermann et al.	1998	Studied molecular phylogeny based on ITS 1 & 2 (27 species)
Kullnig-Gradinger et al.	2002	Reported multigene phylogeny (46 species)
Chaverri and Samuels	2003	Reported details of *Hypocrea/Trichoderma* (Ascomycota, Hypocreales, Hypocreaceae): Species with green ascospores
Druzhinina et al.; Kopchinskiy et al.	2005; 2005	Reported oligonucleotide barcode for species identification and database for verified for type sequences (100 species)
Martinez et al.	2008	Release of *H. fecorina/T. reesei* genome
Druzhinina et al.	2012	Reported molecular phylogeny and species delimitation in the section Longibrachiatum of *Trichoderma*
JGI genome (https://genome.jgi.doe.gov/portal/)	2012	Reflects genomes of *T. harzianum, T. asperellum, T. longibrachiatum, T. virens, T. atroviride, T. asperellum*, and *T. citrinoviride* sequenced using NGS
JGI genome (https://genome.jgi.doe.gov/portal/)	2018	*Trichoderma helicum, T. koningiopsis, T. pleuroticola* and others under process
Baroncelli et al.	2015	Draft whole-genome sequence—*Trichoderma harzianum*
Li et al.	2017	Complete genome sequence, repeat-induced point mutation, and partitioning of CAZyme gene clusters—*Trichoderma reesei*
Fanelli et al.	2018	Draft whole-genome sequence—*Trichoderma harzianum* species complex
Kubicek et al.	2019	Evolution and comparative genomics of *Trichoderma* species

The differential color/shades of conidia may be taxonomically useful, but difficult to interpret and communicate. Similarly, conidial ornamentation (smooth/warted/tuberculate) is certainly a better and useful character. The other features like the shape of phialides has been considered as a strong character in differentiating the species.

1.3 Advances in Taxonomic History

Persoon (1794) proposed the genus with four species and then mycologists mistook *Trichoderma* Pers.: Fr. for a Gasteromycete (Persoon 1794). Its connection with *Hypocrea* (Teleomorph) was noticed by the Tulasne brothers in 1865. After the establishment, the taxonomy of the genus remained obscure and as a result, a few hundred epithets were included, and the genus became a dumping box of cryptic species, which attracted the attention of researchers. Later on, Bisby (1939) was in the opinion that morphological variations could be ascribed to a single species, *Trichderma viride*. Rifai (1969) made serious attempts and distinguished aggregates of nine taxa, which were not biological entities of single teleomorph species. Domsch et al. (1980) described a few additional species and keyed out the reported taxa. Teleomorph connections were established by raising isolates using ascospore (Dingley 1957; Rifai and Webster 1966; Webster and Rifai 1968). In a similar study, Doi (1969, 1972) studied cultural and anamorphic features of a plethora of teleomorphs, but no cultures were preserved as a record. Attempts were again made on teleomorph taxonomy with cultural studies initiated by Samuels and coworkers (Samuels et al. 1994; Samuels 1996). Besides, *Podostroma* Karst. (Doi 1967) and Sarawakus Boedijn (Rifai et al. 1985) were also associated with *Trichoderma*. Bissett (1984, 1991a, b, c, 1992), studied in detail and expanded to 21 taxa in sect. Pachybasium and 7 in sect. Longibrachiatum (Druzhinina et al. 2012), while the remaining sections have not yet been treated in a comparable way. All these studies reveals that the delimitation of species is extremely difficult on morphological basis alone. Other supplementary methods especially secondary metabolites has shown a distinct diversity in the genus *Trichoderma* (Okuda et al. 1982). Other physiological features detectable in microtiter plates and isoenzyme profiles have been used from time to time as a taxonomic tool (Leuchtmann et al. 1996; Samuels et al. 1994). Moreover, use of ITS rDNA region sequencing along with fingerprinting techniques have provided powerful tools for the comprehensive resolution of taxonomic entities (Meyer 1991; Meyer et al. 1992; Fujimori and Okuda 1994; Kuhls et al. 1995, 1996, 1997; Zimand et al. 1994; Kullnig-Gradinger et al. 2002; Hermosa et al. 2000; Druzhinina and Kubicek 2005; Samuels et al. 2006; Mukherjee et al. 2013a, b, c; Sriram et al. 2013; Jaklitsch et al. 2013; Chaverri et al. 2015; Jingade et al. 2018). Recently, a subcommission on Taxonomy of *Trichoderma* and *Gliocladium* along with an official website has been created which contains comprehensive informations. Latest information about different sections and the number of species placed under each section, details of typification, recommendation for genes to be sequenced, primers, etc. have been provided. It is therefore recommended to refer this website for having overall information (http://www.isth.info/).

1.4 Significance of Morphology

Rifai (1969) and Bissett (1991a) provide detailed accounts of morphological characters used in the differentiation of species of *Trichoderma*. Both authors pointed out the difficulties in defining morphological criteria. Samuels (1996) did detail observations and commented on the utility of morphological characters in defining species in *Trichoderma*. Certain characters used for characterization of other hyphomycetous genera are not useful for differentiating species of *Trichoderma*, probably due to a narrow range of morphological variations. However, critical observations of a few other characters may be sufficient for identifying species/strains of *Trichoderma*. As such, identification based on morphological characters remain the primary requirement for identifying species in *Trichoderma*.

Though colony characters sometimes serve as a distinctive characteristic of an individual species, it is difficult to describe with sufficient precision. To some extent other physiological characters like growth rate can be useful in differentiating similar species. Other noteworthy characters like conidiophores which are aggregated into fascicles or pustules have been considered as useful characteristics (Fig. 1.1). Diffusable pigments sometimes play an important role in differentiating closely related species. Strains/isolates belonging to sect. Longibrachiatum typically produce prominent bright greenish-yellow pigments, while dull yellowish pigments are common in several species, but they are not unique. Similarly, only *Trichoderma aureoviride* has been reported to produce crystals in the media. Besides, most of the strains of *Trichoderma* produces mouldy or musty odours; but aromatic odours similar to coconut are produced by *T. viride* and rarely by *T. atroviride*.

For the identification of strains, sections and species aggregates of *Trichodrma*, conidiophore branching pattern and their clustering into fascicles and pustules are very important features. Sec. Pachybasium as well as strains of other sections have been characterized by compact pustules. Verticillate conidiophores is the regular and common feature found in many species. Branches can be broad and relatively narrow in some species, while in certain species of sec. Pachybasium, conidophore apex terminates into a straight, coiled and undulate and sterile extension. Shape, size, and pattern of phialides also play an important role in the differentiation of species. Phialides are elongate, langeniform and often cylindrical in sect. Longibrachiatum; while in sect. Pachybasium they are short and plump. Terminal phialides in several species are found to be elongated and slender and frequently are more or less subulate. Sometimes intercalary phialides also known as aphanophialides (Gams 1971) are also found and reported in sect. Longibrachiatum. Shape of conidia has been varies from globose to obovoidal, ellipsoidal or short-cylindrical with tapering and truncated basal end. Even though conidial dimension in *Trichodrma* is not significantly different; still species can be distinguished by consistent and slight size. The ornamentation of the conidial surface can be considered as one of the stable criteria found in many species, which need SEM for precise observations. Conidia of *T. viride* aggregate are variousy roughned or verrucose (Qin and Zhuang 2016); whereas conidia of *T. saturnisporum* and *T. ghanense* have

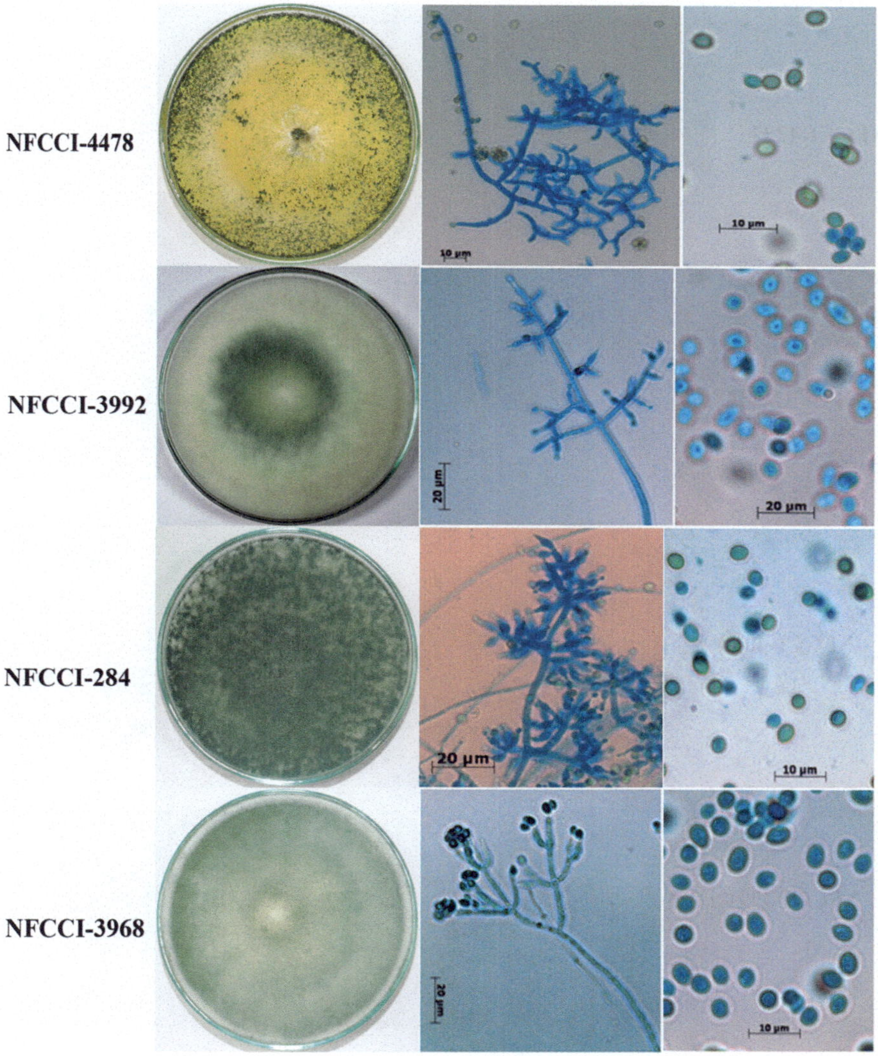

Fig. 1.1 Colony characteristics and morphology of representative *Trichoderma* species. Legends: NFCCI 4478—*Trichoderma longibrachiatum*, NFCCI 3992—*Trichoderma harzianum*, NFCCI 284—*Trichoderma asperellum*, NFCCI 3968—*Trichoderma virens*, NFCCI 4297—*Trichoderma koningii*, NFCCI 2745—*Trichoderma pleuroticola*, NFCCI 4068—*Trichoderma* sp., NFCCI 3991—*Trichoderma* sp.

winglike or bullate projection from the outer wall. Conidial pigments are also characteristic. In some species, mature conidia appear to be dark green in the microscopic mount, while in others appears only pale. In many species of *Trichodrma* chlamydospores are very common; but they are terminal or intercalary, globose to ellipsoidal, smooth walled and yellowish to colorless.

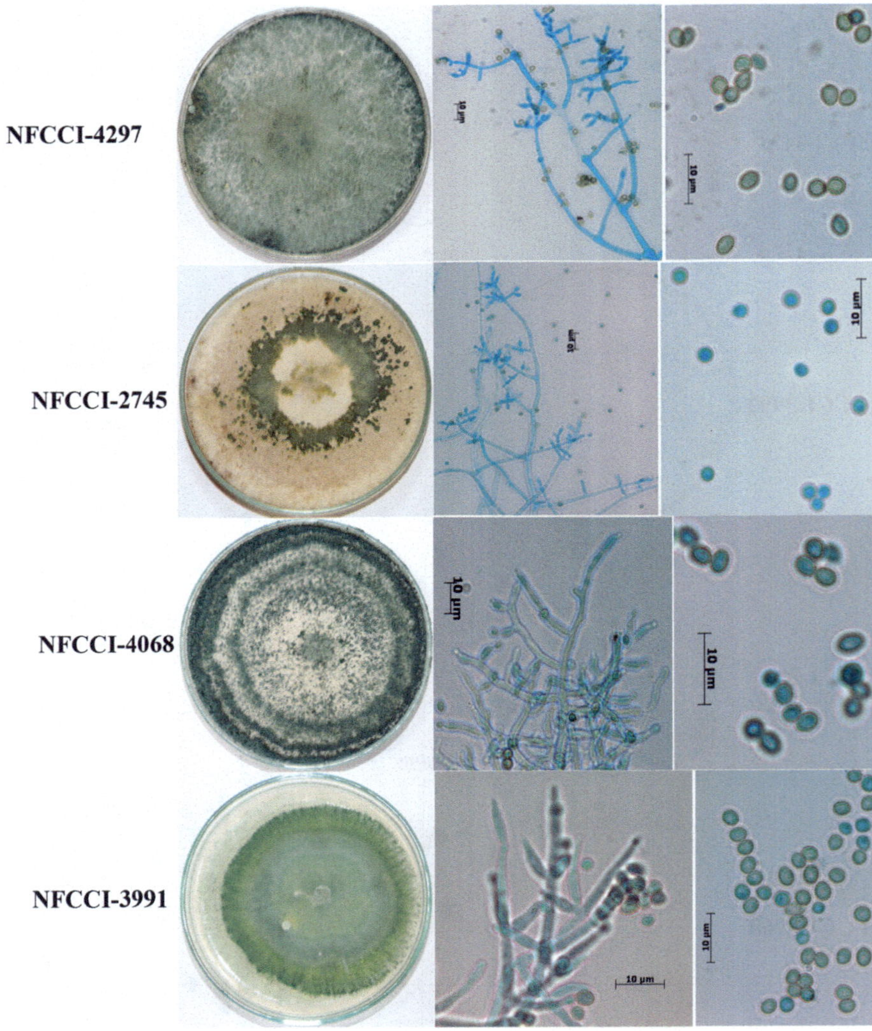

Fig. 1.1 (continued)

Trichoderma posses essential characters to become anamorphs of *Hypocrea* (Hypocreales) including lightly to brightly colored conidia produced from characteristic phialides developed from conidiophores (Samuels and Seifert 1987). Literature indicates that connection of *Trichoderma* with *Hypocrea* was noticed by Berkeley, way back in 1860, who suspected a link between *T. viride* and some unnamed ascomycete, which perhaps confirmed the connection between *T. viride* and *H. rufa* (Tulasne 1860; Smith 1902; Dodd et al. 2002). Tulasne and Tulasne (1865) confirmed the representation of the phialides and their disposition on the conidiophore. Brefeld and von Tavel (1891) proved the relationship between *H. rufa* and *T. viride* by single ascospore cultures. This study represented a major advancement in our understanding of the interrelationship between deuteromycetes and ascomycetes in general. Since then, several studies have successfully reported a number of cases of connection between *Trichoderma* and *Hypocrea*. This type of study continued further for more than a century as it was allowed under the provision of dual nomenclature (Art. 59 of ICBN).

Rules governing nomenclature was changed from 2013 (Melbourne code). As per the revised International Code of Nomenclature for Algae, Fungi and Plants (ICN) any fungi would no longer bear more than one name and abolished the long-existing Art. 49. Under this new provision of ICN, *Trichoderma* became valid and supersedes *Hypocrea* and therefore Rossman et al. (2013) proposed this generic name for acceptance by the Nomenclature Committee for Fungi (NCF) and the General Committee (GC) of the International Association for Plant Taxonomy (IAPT). In line with this proposal, Jaklitsch and Voglmayr (2014) re-combined 46 *Hypocrea* species in *Trichoderma*.

1.5 Molecular Taxonomy of *Trichoderma*

Earlier *Trichoderma* species were identified mainly based on distinctive morphological characters including rapid growth, conidial pigmentations, and branching pattern of conidiophores, shape and size of phialides. However, morphology-based analysis was questioned from time to time due to plasticity and unstable overlapping characteristic features. This has opened new avenues for refinement and consideration of valid and stable criteria for generic delimitation and to achieve the comprehensive species-level identification. The last two decades have seen dynamic changes in systematics and taxonomy in general. Taxonomy and phylogeny of the *Trichoderma* and its species have seen tremendous refinement including development and unveiling of the various phylogenetic markers which can be used for *Trichoderma*.

1.6 DNA Markers

It has been advocated that DNA data, and moreover protein data to some extent, gives a clearer idea of relationships. Analysis methods based on characterization of nucleic acid and/or proteins and their polymorphisms gave the number of useful markers for taxonomic studies; which finally reflect the phylogenetic relationship between the organisms. Physiological and phenotypic characters, isozyme, and molecular markers are used to identify *Trichoderma* (Okuda et al. 1982). The strategies to identify *Trichoderma* using DNA markers are sequence analysis of internal transcribed spacer (ITS) region, restriction fragment length polymorphisms

(RFLPs), random amplified polymorphic DNA (RAPD), and chromosome as well as karyotyping analysis (Muthumeenakshi et al. 1994). Several other markers like RAPD, SRAP, etc. have also been used for studying molecular diversity (Li and Quiros 2001). Zamir and Chet (1985) were the first to characterize *Trichoderma* species by isozyme patterns.

Studies of restriction fragment length polymorphisms (RFLP) as well as DNA fingerprinting Polymorphisms provide important information useful for taxonomic studies. Besides, for the evolutionary study of fungi, ribosomal and mitochondrial DNAs were widely used (Bruns et al. 1991).

The polymerase chain reaction (PCR) technique has revolutionized the molecular studies and opened new avenue of revealing DNA polymorphisms among closely related genotypes with high sensitivity protocols. In PCR-based fingerprinting, unkown DNA fragments were amplified by using GC-rich primers and artificially primed-PCR or microsatellite-complementary oligonucleotideds: (GACA)4, (GTG) 5 and M13 core sequence [Williams et al. 1990; Welsh and McClelland 1990; Lieckfeldt et al. 1993; Meyer et al. 1992; Rai et al. 2016]. As such, PCR-based fingerprinting has been widely used for characterizing *Trichoderma* species.

1.7 DNA Sequencing

As the sequences itself reflect the DNA structure, these data are considered to be phylogenetically more informative than other types of data. Moreover, data comparison based on cladistic and parsimony method will give phylogenetically more reliable results. which can not be compared with phenotypic data. For the sequence analysis, generally conserved genome region is selected. Ribosomal DNA containing conserved regions (18S rRNA, 28S rRNA and 5.8S rRNA genes) and highly variable regions such as internal transcribed spacer (ITS) and intergenic spacer (IGS) is generally used for the such studies. Most of the sequence studies in the case of fungi focuses on the rRNA genes (Bruns et al. 1991). Taylor et al. (1999) proposed phylogenetic species concepts based on the concordance of five or more gene regions. Consequently, combined sequence analysis of ITS1 and ITS2 with single-copy genes such as β-tubulin (Schardl et al. 1994; O'Donnell et al. 1998) or hydrophobin (Geiser et al. 1998) has been used with great success. Lieckfeldt et al. (2000) used endochitinase (ech42) gene sequence analysis for *Trichoderma* sect. *Trichoderma*, and showed that the results were concordant with those obtained from ITS1 and ITS2 sequence analysis.

Further, some studies have tried to establish a species phylogeny of the genus *Trichoderma*, based on the sequencing of multiple independent loci, by investigating all described species (Samuels et al. 1998, 1999, 2000, 2002). Multicopy loci ITS1, ITS2, 28S-rDNA; and the small mitochondrial rDNA subunit; and fragments from two single-copy gene loci, viz. translation elongation factor 1 (Berney et al. 2000), and endochitinase 42 have been considered to be informative. The 18S- and

28SrDNA sequence analyses have been used to delimit the genus and to estimate the chronology of its evolution (Lieckfeldt et al. 2000).

Later on, in several studies, it has been found that previously used markers of the ribosomal cluster, like ITS region, is of little use and subsequently rpb2 and tef1 exon (Chaverri et al. 2004; Overton et al. 2006a, b) or the tef1 intron 5 (Lu et al. 2004) were further used either exclusively or in combinatorial. It has been reported that tef1 intron 4, usually used in combination with intron 5, which provided the highest resolution for the species of the genus. It was particularly useful for the distinction of species in the sect. *Trichoderma* (Jaklitsch et al. 2006; Samuels et al. 2006). Taylor et al. (2000) suggested that less variable genes such as cal1 or chi18–5 can be used to fulfill the criteria of the genealogical concordance phylogenetic species recognition (GCPSR) concept (Druzhinina et al. 2012). As such, molecular studies have provided important tools for species-level identification of many complex genera of fungi including *Trichoderma*.

In order to determine the phylogenetic relationship of isolates selected in the present study with known taxa, ITS, TEF-1α and ech42 genomic regions were considered. The sequences of closely related strains have been retrieved from NCBI nucleotide database and for the phylogentic analysis a total 62 isolates were used. Genus *Nectria* was selected to be the outgroup taxon. Each gene region was alligned individually by using MAFFT v. 6.864b (Katoh and Standley 2013). The alignments were manually edited in Aliview (Larsson 2014) and then concatenated. Finally concatenated alignment was used for phylogentic analyis. With the help of ModelFinder (Kalyaanamoorthy et al. 2017), best substitution model was selected and phylogentic tree has been constructed using IQ-tree using v.1.6.11 and tree was visualized in FigTree v.1.4.4. Out of 286 models tested, TIM3+F+R7 was found to best-suitable model based on Bayesian Information Criterion (BIC). Tree was constructed by Maximum Liklihood method based on above mentioned model. Optimum log-likelhood valuse was obtained as −15847.815. Tree branches were tested using ultrafast bootstrap and SH-like approximate liklihood ratio test with 1000 replicates.

1.8 Ex Situ Conservation

Trichoderma are ubiquitous and their adaptability to varied conditions make them model organism for research in terms of biological and biotechnological perspectives. Conservation of these fungal strains is important due to the ongoing destruction of natural habitats, which necessitates the development of alternative strategies for conserving the natural fungal genetic resource. In this regard, ex situ conservation in laboratory condition plays an important role and can be achieved by preserving live cultures in germplasm banks and dried/exsiccate material in herbaria. Subsequent to the Convention on Biological Diversity (CBD), conservation strategies are being developed and laboratories in many member countries are progressively acting in conformity with the convention by respecting biological habitats and

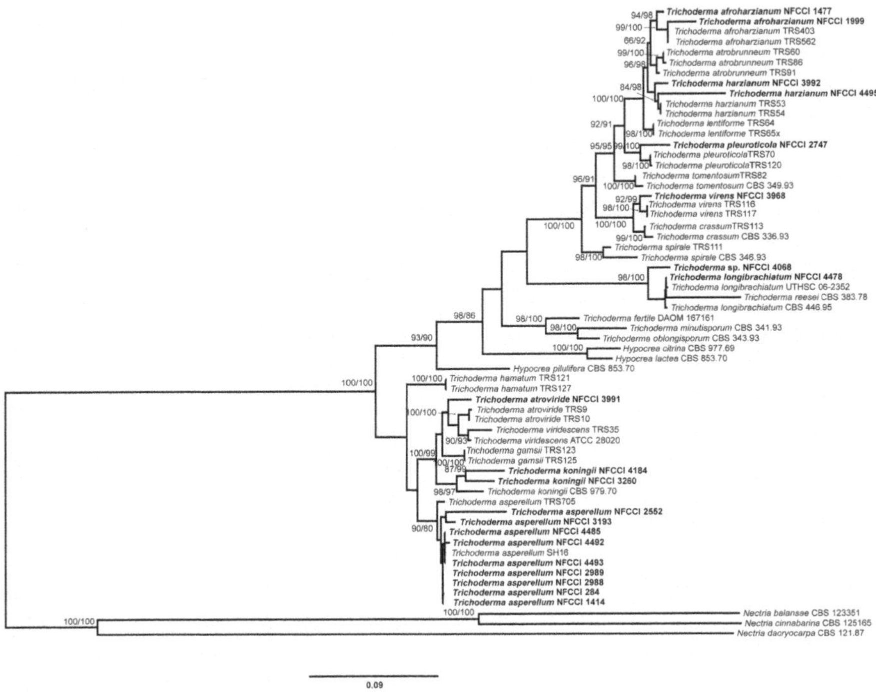

Fig. 1.2 Molecular phylogenetic analysis by maximum-likelihood (ML) method based on combined ITS, TEF-1α, and ech42 sequence data. Statistical supports are indicated next to each node, non-parametric maximum likelihood ultrafast-bootstrap (UFBS) values, and SH-aLRT obtained from 1000 replicates using IQ-TREE and the TIM3 + F + R7 model. Isolates used in the present study are highlighted in bold

the rights of local peoples. Apart from natural substrates, fungi grow best on media formulated from the natural materials from which they were originally isolated. Major factors affecting fungal growth are growth medium, temperature, light, aeration, pH, and water activity. As a general practice laboratory uses available standard media. However, optimization of the growth conditions of a particular fungus is an important step. Individual laboratories or specialized culture collections follow similar procedures in maintaining germplasm on a long-term basis as an extension of the taxonomic study. However, such fundamental aspects are getting less importance in a new era due to declining interest. Though several methods have been recommended for maintaining pure cultures, it is advised to take into account various considerations before applying a method to a particular group of fungi, viz. type and number of cultures to be maintained, manpower, time and facility available, climate and suitability of the method to a particular group of fungi. Several methods, briefly by Serial transfer, preservation in sterile water, spore suspension, wood chips/ toothpick sticks, on filter paper, oil overlay, skimmed milk and silica gel and by

Table 1.2 List of *Trichoderma* species used in the present study

S. No.	NFCCI No.	Fungus name	Geographical origin	Host	Target gene		
					ITS	*tef1α*	*Chi42*
1.	NFCCI 284	*Trichoderma asperellum*	Pune, Maharashtra	Soil	√	√	√
2.	NFCCI 1414	*Trichoderma asperellum*	Nasik, Maharashtra	Soil	√	√	√
3.	NFCCI 1477	*Trichoderma afroharzianum*	Pune, Maharashtra	Mushroom association	√	√	√
4.	NFCCI 1999	*Trichoderma afroharzianum*	Pune, Maharashtra	Sugarcane rhizosphere	√	√	√
5.	NFCCI 2552	*Trichoderma asperellum*	Pondicherry	Marine	√	√	√
6.	NFCCI 2745	*Trichoderma pleuroticola*	Kerala	Coconut husk	√	√	√
7.	NFCCI 2988	*Trichoderma asperellum*	Manipur	Rhizosphere soil (pea)	√	√	√
8.	NFCCI 2989	*Trichoderma asperellum*	Manipur	Rhizosphere soil (pea)	√	√	√
9.	NFCCI 3193	*Trichoderma asperellum*	Tamhinighat, Maharashtra	Rhizospheric soil(bamboo)	√	√	√
10.	NFCCI 3968	*Trichoderma virens*	Mokokchung, Nagaland	Forest soil	√	√	√
11.	NFCCI 3991	*Trichoderma* sp.	Jorhat, Assam	Bamboo	√	√	√
12.	NFCCI 3992	*Trichoderma harzianum*	Jorhat, Assam	Bamboo	√	√	√
13.	NFCCI 4068	*Trichoderma* sp.	Kanpur, Uttar Prades	Rhizospheric soil	√	√	√
14.	NFCCI 4184	*Trichoderma koningii*	Pune, Maharashtra	Grape leaf phylloplane	√	√	√
15.	NFCCI 4478	*Trichoderma longibrachiatum*	Haridwar, Uttarakhand	Ganga water	√	√	√
16.	NFCCI 4484	*Trichoderma asperellum*	Kusur, Pune, Maharashtra	Soil	√	√	√
17.	NFCCI 4490	*Trichoderma asperellum*	Mirzapur, Uttar Pradesh	Soil	√	√	√
18.	NFCCI 4493	*Trichoderma asperellum*	Kanpur, Uttar Pradesh	Soil	√	√	√
19.	NFCCI 3260	*Trichoderma koningii*	Banaskatha, Gujrat	Rhizospere of tomato	√	√	√
20.	NFCCI 1178	*Trichoderma harzianum*	Coimbatore, Tamilnadu	Sugarcane phylloplane	√	√	√

freeze-drying and cryopreservation (in LN_2) have been recommended to be used with their protocols, recovery steps, and important tips (Singh 2017; Singh and Baghela 2017; Singh et al. 2018).

1.9 Conclusions

Trichoderma is an important genus. Biotechnological applications of its species have seen tremendous changes in recent years. However, declining interests in fundamental taxonomy and systematics have drawn serious attention. The purpose of this study is to compile important information by applying morpho- and molecular analyses to select species available at the National Fungal Culture Collection of India (NFCCI) and a few newly isolated ones (Table 1.2). The identity of some of these species was ambiguous which has been re-confirmed by using a combination of basic and modern approaches of the sequencing of multigene regions and their phylogenetic analysis. The details of primers and methods used are briefly provided in Tables 1.3 and 1.4. Besides strategies required for ex-situ conservation of species of *Trichoderma* have been discussed

Table 1.3 Components of PCR master mixture used

PCR reaction mixture	Final volume of 50 μL
10× Taq DNA polymerase buffer	5 μL
Taq DNA polymerase (Sigma-Aldrich, USA)	1 μL
Deoxynucleotides (dNTP mixture-μM)	5 μL
Each primer (pmol)	2.5 μL each (5 μL)
Genomic DNA (5–20 ng)	1–2 μL
Sterile water	32 μL

Table 1.4 Details of primers used for PCR amplification

Target gene region	Primer name	Primer sequences	References
Internal transcribed spacer	ITS-4	5′-TCCTCCGCTTATTGATATGC-3′	White et al. (1990)
	ITS-5	5′-GGAAGTAAAAGTCGTAACAAGG-3′	
Translation Elongation factor 2	tef1fw	5′-GTGAGCGTGGTATCACCATCG-3′	Kullnig-Gradinger et al. (2002)
	tef1rev	5′-GCCATCCTTGGAGACCAGC-3′	
Endochitinase 42	Chit42-1a	5′-GCTYTCCATCGGTGGCTGGAC-3′	Kullnig-Gradinger et al. (2002)
	Chit42-2a	5′-GGAGTTGGGGTAGCTCAGC-3′	

as an important aspect needing further attention of the researchers. Biological Resource Centres (BRCs)/Culture Collections play an important role in conserving the mycological resources in general. In order to reflect biodiversity, selected species of *Trichoderma* isolated from different natural sources and geographical locations, and already deposited at the National Fungal Culture Collection of India (NFCCI), were reexamined morphologically as well as sequencing of recommended multigene and phylogenetic analysis was conducted.

Acknowledgments Authors thank Director, MACS' Agharkar Research Institute, Pune for providing research facility. Shiwali Rana thanks University Grant Commission, New Delhi for the award of Junior Research Fellowship.

References

Baroncelli R, Piaggeschi G, Fiorini L, Bertolini E, Zapparata A, Pè ME, Sarrocco S, Vannacci G (2015) Draft whole-genome sequence of the biocontrol agent *Trichoderma harzianum* T6776. Genome Announc 3(3):e00647–e00615

Benítez T, Rincón AM, Limón MC, Codon AC (2004) Biocontrol mechanisms of *Trichoderma* strains. Int Microbiol 7:249–260

Berney C, Pawlowski J, Zaninetti L (2000) Elongation factor 1-alpha sequences do not support an early divergence of the Acoela. Mol Biol Evol 17(7):1032–1039

Bisby GR (1939) *Trichoderma viride* Pers. ex Fries, and notes on *Hypocrea*. Trans Br Mycol Soc 23(2):149–168

Bissett J (1984) A revision of the genus *Trichoderma*. I. Section Longibrachiatum sect. nov. Can J Bot 62(5):924–931

Bissett J (1991a) A revision of the genus *Trichoderma*. II. Infrageneric classification. Can J Bot 69 (11):2357–2372

Bissett J (1991b) A revision of the genus *Trichoderma*. III. Section Pachybasium. Can J Bot 69 (11):2373–2417

Bissett J (1991c) A revision of the genus *Trichoderma*. IV. Additional notes on section Longibrachiatum. Can J Bot 69(11):2418–2420

Bissett J (1992) *Trichoderma atroviride*. Can J Bot 70(3):639–641

Brefeld O, von Tavel F (1891) Untersuchungen aus dem Gesammtgebiete der Mykologie: Ascomyceten II, 10, Heinrich Schöningh, Münster, pp 157–378

Brotman Y, Kapuganti JG, Viterbo A (2010) *Trichoderma*. Curr Biol 20(9):R390–R391

Bruns TD, White TJ, Taylor JW (1991) Fungal molecular systematics. Annu Rev Ecol Syst 22 (1):525–564

Chang Y-C, Chang Y-C, Baker R, Kleifeld O, Chet I (1986) Increased growth of plants in the presence of the biological control agent *Trichoderma harzianum*. Plant Dis 70:145–148

Chaverri P, Samuels GJ (2003) *Hypocrea/Trichoderma* (ascomycota, hypocreales, hypocreaceae): species with green ascospores. Stud Mycol 48:1–116

Chaverri P, Castlebury LA, Samuels GJ, Geiser DM (2003) Multilocus phylogenetic structure within the *Trichoderma harzianum/Hypocrea lixii* complex. Mol Phylogenet Evol 27 (2):302–313

Chaverri P, Candoussau F, Samuels GJ (2004) *Hypocrea phyllostachydis* and its *Trichoderma* anamorph, a new bambusicolous species from France. Mycol Prog 3(1):29–36

Chaverri P, Branco-Rocha F, Jaklitsch W, Gazis R, Degenkolb T, Samuels GJ (2015) Systematics of the *Trichoderma harzianum* species complex and the re-identification of commercial biocontrol strains. Mycologia 107(3):558–590

Deng JJ, Huang WQ, Li ZW, Lu DL, Zhang Y, Luo XC (2018) Biocontrol activity of recombinant aspartic protease from *Trichoderma harzianum* against pathogenic fungi. Enzym Microb Technol 112:35–42

Dingley JM (1957) Life history studies in the genus *Hypocrea* Fr. Trans R Soc N Z 84(4):689–693

Dodd SL, Lieckfeldt E, Chaverri P, Overton BE, Samuels GJ (2002) Taxonomy and phylogenetic relationships of two species of *Hypocrea* with *Trichoderma* anamorphs. Mycol Prog 1 (4):409–428

Doi Y (1967) Revision of the Hypocreales with cultural observations. III Three species of the genus *Podostroma* with *Trichoderma* or *Trichoderma*-like conidial states. Trans Mycol Soc Jpn 8:54–57

Doi Y (1969) Revision of the Hypocreales with cultural observations IV. The genus *Hypocrea* and its allies in Japan (1) general part. Bull Natl Sci Mus Tokyo 12:693–724

Doi Y (1972) Revision of the Hypocreales with cultural observations IV. The genus *Hypocrea* and its allies in Japan (2). Enumeration of the species. Bull Natl Sci Mus Tokyo 15:649–751

Domsch KH, Gams W, Anderson TH (1980) Compendium of soil fungi, vol 1. Academic Press, London

Druzhinina I, Kubicek CP (2005) Species concepts and biodiversity in *Trichoderma* and *Hypocrea*: from aggregate species to species clusters? J Zhejiang Univ Sci B 6(2):100

Druzhinina IS, Kopchinskiy AG, Komoń M, Bissett J, Szakacs G, Kubicek CP (2005) An oligonucleotide barcode for species identification in *Trichoderma* and *Hypocrea*. Fungal Genet Biol 42(10):813–828

Druzhinina IS, Seidl-Seiboth V, Herrera-Estrella A, Horwitz BA, Kenerley CM, Monte E, Prasun KM, Susanne Z, Igor VG, Kubicek CP (2011) *Trichoderma*: the genomics of opportunistic success. Nat Rev Microbiol 9(10):749

Druzhinina IS, Komoń-Zelazowska M, Ismaiel A, Jaklitsch W, Mullaw T, Samuels GJ, Kubicek CP (2012) Molecular phylogeny and species delimitation in the section Longibrachiatum of *Trichoderma*. Fungal Genet Biol 49(5):358–368

Etschmann MM, Huth I, Walisko R, Schuster J, Krull R, Holtmann D, Wiltmann C, Schrader J (2015) Improving 2- phenylethanol and 6- pentyl-α-pyrone production with fungi by microparticle-enhanced cultivation (MPEC). Yeast 32(1):145–157

Fanelli F, Liuzzi VC, Logrieco AF, Altomare C (2018) Genomic characterization of *Trichoderma atrobrunneum* (*T. harzianum* species complex) ITEM 908: insight into the genetic endowment of a multi-target biocontrol strain. BMC Genom 19(1):662

Fujimori F, Okuda T (1994) Application of the random amplified polymorphic DNA using the polymerase chain reaction for efficient elimination of duplicate strains in microbial screening. J Antibiot 47(2):173–182

Gal-Hemed I, Atanasova L, Komon-Zelazowska M, Druzhinina IS, Viterbo A, Yarden O (2011) Marine isolates of *Trichoderma* spp. as potential halotolerant agents of biological control for arid-zone agriculture. App. Environ Microbiol 77(15):5100–5109

Gams W (1971) *Cephalosporium*-artige schimmelpilze (Hyphomycetes)

Geiser DM, Frisvad JC, Taylor JW (1998) Evolutionary relationships in *Aspergillus* section Fumigati inferred from partial β-tubulin and hydrophobin DNA sequences. Mycologia 90 (5):831–845

Harman GE (2006) Overview of mechanisms and uses of *Trichoderma* spp. Phytopathology 96 (2):190–194

Hermosa MR, Grondona I, Iturriaga ET, Diaz-Minguez JM, Castro C, Monte E, Garcia-Acha I (2000) Molecular characterization and identification of biocontrol isolates of *Trichoderma* spp. Appl Environ Microbiol 66(5):1890–1898

Jaklitsch WM (2009) European species of *Hypocrea* Part I. The green-spored species. Stud Mycol 63:1–91

Jaklitsch WM, Voglmayr H (2014) New combinations in *Trichoderma* (Hypocreaceae, Hypocreales). Mycotaxon 126(1):143–156

Jaklitsch WM, Komon M, Kubicek CP, Druzhinina IS (2006) *Hypocrea crystalligena* sp. nov., a common European species with a white-spored *Trichoderma* anamorph. Mycologia 98 (3):499–513

Jaklitsch WM, Samuels GJ, Ismaiel A, Voglmayr H (2013) Disentangling the *Trichoderma viridescens* complex. Persoonia 31:112

Jingade P, Sannasi S, Jha CS, Mishra MK (2018) Molecular characterisation of *Trichoderma* species using SRAP markers. Arch Phytopathol Plant 51(3–4):128–138

Kalyaanamoorthy S, Minh BQ, Wong TKF, Haeseler A, Jermiin LS (2017) ModelFinder: fast model selection for accurate phylogenetic estimates. Nat Methods 14:587–589

Katoh K, Standley DM (2013) MAFFT multiple sequence alignment software version 7: improvements in performance and usability. Mol Biol Evol 30(4):772–780

Keswani C, Singh SP, Singh HB (2013) A superstar in biocontrol enterprise: *Trichoderma* spp. Biotech Today 3(2):27–30

Kindermann J, El-Ayouti Y, Samuels GJ, Kubicek CP (1998) Phylogeny of the genus *Trichoderma* based on sequence analysis of the internal transcribed spacer region 1 of the rDNA cluster. Fungal Genet Biol 24(3):298–309

Kopchinskiy A, Komon M, Kubicek CP, Druzhinina IS (2005) TrichoBLAST: a 525 multiloci database of phylogenetic markers for *Trichoderma* and *Hypocrea* powered by 526 sequence diagnosis and similarity search tools. Mycol Res 109:658–660

Kubicek CP, Steindorff AS, Chenthamara K, Manganiello G, Henrissat B, Zhang J, Cai F, Kopchinskiy AG, Kubicek EM, Kuo A, Baroncelli R, Sarrocco S, Noronha EF, Vannacci G, Shen Q, Grigoriev IV, Druzhinina IS (2019) Evolution and comparative genomics of the most common *Trichoderma* species. BMC Genomics 20(1):485

Kuhls K, Lieckfeldt E, Börner T (1995) PCR-fingerprinting used for comparison of ex type strains of *Trichoderma* species deposited in different culture collections. Microbiol Res 150 (4):363–371

Kuhls K, Lieckfeldt E, Samuels GJ, Kovacs W, Meyer W, Petrini O, Gams W, Börner T, Kubicek CP (1996) Molecular evidence that the asexual industrial fungus *Trichoderma reesei* is a clonal derivative of the ascomycete *Hypocrea jecorina*. Proc Natl Acad Sci USA 93(15):7755–7760

Kuhls K, Lieckfeldt E, Samuels GJ, Meyer W, Kubicek CP, Börner T (1997) Revision of *Trichoderma* sect. Longibrachiatum including related teleomorphs based on analysis of ribosomal DNA internal transcribed spacer sequences. Mycologia 89(3):442–460

Kullnig-Gradinger CM, Szakacs G, Kubicek CP (2002) Phylogeny and evolution of the genus *Trichoderma*: a multigene approach. Mycol Res 106(7):757–767

Kumar G, Maharshi A, Patel J, Mukherjee A, Singh HB, Sarma BK (2017) *Trichoderma*: a potential fungal antagonist to control plant diseases. SATSA Mukhapatra 21:206–218

Larsson A (2014) AliView: a fast and lightweight alignment viewer and editor for large data sets. Bioinformatics 30(22):3276–3278

Leuchtmann A, Petrini O, Samuels GJ (1996) Isozyme subgroups in *Trichoderma* section Longibrachiatum. Mycologia 88(3):384–394

Li G, Quiros CF (2001) Sequence-related amplified polymorphism (SRAP), a new marker system based on a simple PCR reaction: its application to mapping and gene tagging in *Brassica*. Theor Appl Genet 103(2–3):455–461

Li WC, Huang CH, Chen CL, Chuang YC, Tung SY, Wang TF (2017) *Trichoderma reesei* complete genome sequence, repeat-induced point mutation, and partitioning of CAZyme gene clusters. Biotechnol Biofuels 10(1):170

Lieckfeld E, Meyer W, Börner T (1993) Rapid identification and differentiation of yeasts by DNA and PCR fingerprinting. J Basic Microbiol 33(6):413–425

Lieckfeldt E, Yolaine C, Csaba F, Thomas B (2000) Endochitinase gene-based phylogenetic analysis of *Trichoderma*. Microbiol Res 155(1):7–15

Lu B, Druzhinina IS, Fallah P, Chaverri P, Gradinger C, Kubicek CP, Samuels GJ (2004) *Hypocrea/Trichoderma* species with pachybasium-like conidiophores: teleomorphs for *T. minutisporum* and *T. polysporum* and their newly discovered relatives. Mycologia 96 (2):310–342

Martinez D, Berka RM, Henrissat B, Saloheimo M, Arvas M, Baker SE, Chapman J, Chertkov O, Coutinho PM, Cullen D, Danchin EG, Danchin EG (2008) Genome sequencing and analysis of the biomass-degrading fungus *Trichoderma reesei* (syn. *Hypocrea jecorina*). Nat Biotech 26 (5):553

Meyer RJ (1991) Mitochondrial DNAs and plasmids as taxonomic characteristics in *Trichoderma viride*. Appl Environ Microbiol 57(8):2269–2276

Meyer W, Morawetz R, Börner T, Kubicek CP (1992) The use of DNA-fingerprint analysis in the classification of some species of the *Trichoderma* aggregate. Curr Genet 21(1):27–30

Montenecourt B, Eveleigh D (1979) Selective screening methods for the isolation of high yielding cellulase mutants of *Trichoderma reesei*. Adv Chem Ser 181:289–301

Mukherjee PK, Horwitz BA, Herrera-Estrella A, Schmoll M, Kenerley CM (2013a) *Trichoderma* research in the genome era. Annu Rev Phytopathol 51:105–129

Mukherjee PK, Horwitz BA, Singh US, Mukherjee M, Schmoll M (eds) (2013b) *Trichoderma*: biology and applications. CABI, Oxford

Mukherjee PK, Mukherjee AK, Kranthi S (2013c) Reclassification of *Trichoderma viride* (TNAU), the most widely used commercial biofungicide in India, as *Trichoderma asperelloides*. Open Biotechnol J 7:7–9

Muthumeenakshi S, Mills PR, Brownd AE, Seaby DA (1994) Intraspecific molecular variation among *Trichoderma harzianum* isolates colonizing mushroom compost in the British Isles. Microbiology 140(4):769–777

Nguyen L, Schmidt HA, von Haeseler A, Minh BQ (2015) IQ-TREE: a fast and effective stochastic algorithm for estimating maximum-likelihood phylogenies. Mol Biol Evol 32(1):268–274

O'Donnell K, Kistler HC, Cigelnik E, Ploetz RC (1998) Multiple evolutionary origins of the fungus causing Panama disease of banana: concordant evidence from nuclear and mitochondrial gene genealogies. Proc Natl Acad of Sci USA 95(5):2044–2049

Okuda T, Fujiwara A, Fujiwara M (1982) Correlation between species of *Trichoderma* and production patterns of isonitrile antibiotics. Agric Biol Chem 46(7):1811–1822

Overton BE, Stewart EL, Geiser DM (2006a) Taxonomy and phylogenetic relationships of nine species of *Hypocrea* with anamorphs assignable to *Trichoderma* section Hypocreanum. Stud Mycol 56:39–65

Overton BE, Stewart EL, Geiser DM, Wenner NG (2006b) Systematics of *Hypocrea citrina* and allied species. Stud Mycol 56:1–38

Patil HJ, Solanki MK (2016) Microbial inoculant: modern era of fertilizers and pesticides. In: Microbial inoculants in sustainable agricultural productivity. Springer, New Delhi, pp 319–343

Persoon CH (1794) Disposita methodica fungorum. Römer's Neues Mag Bot 1:81–128

Qin WT, Zhuang WY (2016) Seven wood-inhabiting new species of the genus *Trichoderma* (Fungi, Ascomycota) in *Viride* clade. Sci Rep 6:27074

Rai S, Kashyap PL, Kumar S, Srivastava AK, Ramteke PW (2016) Identification, characterization and phylogenetic analysis of antifungal *Trichoderma* from tomato rhizosphere. Springerplus 5 (1):1939

Rai S, Solanki MK, Solanki AC, Surapathrudu K (2019) Biocontrol potential of *Trichoderma* spp.: current understandings and future outlooks on molecular techniques. In: Plant health under biotic stress. Springer, Singapore, pp 129–160

Rifai MA (1969) A revision of the genus *Trichoderma*. Mycol Pap 116:1–56

Rifai MA, Webster J (1966) Culture studies on *Hypocrea* and *Trichoderma*: III. *H. lactea* (= *H. citrina*) and *H. pulvinata*. Trans Br Mycol Soc 49(2):297–310

Rifai MA, Kramadibat K, Basuki T (1985) The anamorph of *Sarawakus succisus*. Reinwardtia 10:265–270

Rossman AY, Seifert KA, Samuels GJ, Minnis AM, Schroers HJ, Lombard L, Crous PW, Põldmaa K, Cannon PF, Summerbell RC, Geiser DM, Zhuang W-Y, Hirooka Y, Herrera C, Salgado-Salazar C, Chaverri P (2013) Genera in Bionectriaceae, Hypocreaceae, and Nectriaceae (Hypocreales) proposed for acceptance or rejection. IMA Fungus 4(1):41–51

Samuels GJ (1996) *Trichoderma*: a review of biology and systematics of the genus. Mycol Res 100:923–935

Samuels GJ, Seifert KA (1987) Taxonomic implications of variation among hypocrealean anamorphs. Kodansha, Tokyo, pp 29–56

Samuels GJ, Petrini O, Manguin S (1994) Morphological and macromolecular characterization of *Hypocrea schweinitzii* and its *Trichoderma* anamorph. Mycologia 86(3):421–435

Samuels GJ, Petrini O, Kuhls K, Lieckfeldt E, Kubicek CP (1998) The *Hypocrea schweinitzii* complex and *Trichoderma* sect *longibrachiatum*. Stud Mycol 41:1–54

Samuels GJ, Lieckfeldt E, Nirenberg HI (1999) *Trichoderma asperellum*, a new species with warted conidia, and redescription of *T. viride*. Sydowia 51(1):71–88

Samuels GJ, Pardo-schultheiss R, Hebbar KP, Lumsden RD, Bastos CN, Costa JC, Bezerra JL (2000) *Trichoderma stromaticum* sp. nov., a parasite of the cacao witches broom pathogen. Mycol Res 104(6):760–764

Samuels GJ, Dodd SL, Gams W, Castlebury LA, Petrini O (2002) *Trichoderma* species associated with the green mold epidemic of commercially grown Agaricus bisporus. Mycologia 94 (1):146–170

Samuels GJ, Dodd SL, Lu BS, Petrini O, Schroers HJ, Druzhinina IS (2006) The *Trichoderma koningii* aggregate species. Stud Mycol 56:67–133

Schardl CL, Leuchtmann A, Tsai HF, Collett MA, Watt DM, Scott DB (1994) Origin of a fungal symbiont of perennial ryegrass by interspecific hybridization of a mutualist with the ryegrass choke pathogen, *Epichloe typhina*. Genetics 136(4):1307–1317

Schuster A, Schmoll M (2010) Biology and biotechnology of *Trichoderma*. Appl Microbiol Biotechnol 87(3):787–799

Singh SK (2017) *Ex situ* conservation of fungi: a review. In: Satyanarayana T, Deshmukh SK, Johari BN (eds) Developments in fungal biology and applied mycology. Springer Nature, Singapore, pp 543–562

Singh SK, Baghela A (2017) Cryopreservation of microorganisms. In: Varma A, Sharma AK (eds) Modern tools and techniques to understand microbes. Springer International Publishing AG, Cham, pp 321–333

Singh SP, Singh HB, Singh DK (2013) *Trichoderma harzianum* and *Pseudomonas* sp. mediated management of *Sclerotium rolfsii* rot in tomato (*Lycopersicon esculentum* Mill.). Life Sci 8 (3):801–804

Singh SK, Singh PN, Gaikwad SB, Maurya DK (2018) Conservation of Fungi: a review on conventional approaches. In: Sharma SK, Varma A (eds) Microbial resource conservation: conventional to modern approaches. Springer, Cham, pp 223–237

Smith AL (1902) The fungi of germinating farm seeds. Trans Br Mycol Soc 1:182–186

Solanki MK, Singh N, Singh RK, Singh P, Srivastava AK, Kumar S, Kashyap Prem L, Arora DK (2011) Plant defense activation and management of tomato root rot by a chitin-fortified *Trichoderma/Hypocrea* formulation. Phytoparasitica 39(5):471

Sriram S, Savitha MJ, Rohini HS, Jalali SK (2013) The most widely used fungal antagonist for plant disease management in India, *Trichoderma viride* is *Trichoderma asperellum* as confirmed by oligonucleotide barcode and morphological characters. Curr Sci 104(10):1332–1340

Taylor JW, Jacobson DJ, Fisher MC (1999) The evolution of asexual fungi: reproduction, speciation and classification. Annu Rev Phytopathol 37(1):197–246

Taylor JW, Jacobson DJ, Kroken S, Kasuga T, Geiser DM, Hibbett DS, Fisher MC (2000) Phylogenetic species recognition and species concepts in fungi. Fungal Genet Biol 31(1):21–32

Topolovec-Pintarić S (2019) *Trichoderma*: invisible partner for visible impact on agriculture. In: *Trichoderma*—the most widely used fungicide. IntechOpen, London

Tulasne L (1860) De quelques sphéries fungicoles, à propos d'un mémoire de M. Antoine de Bary sur les Nyctalis. Ann Sci Nat Bot 13:5–19

Tulasne LR, Tulasne C (1865) Selecta fungorum carpologia 3:27–35 (Eng transl by WB Grove)

Webster J, Rifai MA (1968) Culture studies on *Hypocrea* and *Trichoderma*: IV. *Hypocrea pilulifera* sp. nov. Trans Br Mycol Soc 51(3–4):511–514

Welsh J, McClelland M (1990) Fingerprinting genomes using PCR with arbitrary primers. Nucleic Acids Res 18(24):7213–7218

White TJ, Bruns T, Lee SJWT, Taylor J (1990) Amplification and direct sequencing of fungal ribosomal RNA genes for phylogenetics. PCR Protocol 18(1):315–322

Williams JG, Kubelik AR, Livak KJ, Rafalski JA, Tingey SV (1990) DNA polymorphisms amplified by arbitrary primers are useful as genetic markers. Nucleic Acids Res 18 (22):6531–6535

Zamir D, Chet I (1985) Application of enzyme electrophoresis for the identification of isolates in *Trichoderma harzianum*. Can J Microbiol 31(6):578–580

Zimand G, Valinsky L, Elad Y, Chet I, Manulis S (1994) Use of the RAPD procedure for the identification of *Trichoderma* strains. Mycol Res 98(5):531–534

Chapter 2
Biodiversity of *Trichoderma* Species in Different Agro-Ecological Habitats

Ramji Singh, Ajay Tomer, Durga Prasad, and H. S. Viswanath

Abstract *Trichoderma* is a fungal organism which is found worldwide in all climatic zones. Biodiversity of this fungus is quite useful for human beings, as they are successfully being used as antagonists of several plant pathogens, decomposer of woody and herbaceous plant residue, and they have an ability to survive along with the decaying plant materials and plant debris. Species of *Trichoderma* are characterized by their ability of rapid growth and also to assimilate a wide range of substrates. Rapid growth of *Trichoderma* is an added benefit for it to be a successful antagonist. Species of *Trichoderma* also have the ability to produce a large number of antimicrobial compounds with strong microbial inhibiting properties. *Trichoderma* spp. are commonly found as saprophytes in soil and root ecosystems of all types of plants and crops including field crops, pulses, oilseeds, vegetable crops, orchards, forests, and also on decaying woody materials. Some species of *Trichoderma* have also been detected and isolated from air, settled dust, and different water-related habitats, viz. marine environments and drinking water. Furthermore, certain species of this antagonist have been found as surface microflora and endophytes of plants, colonizers of mushroom-related natural and artificial substrata, and facultative pathogens of humans, thus demonstrating a high adaptability to various ecozones. *T. harzianum* are widely found in almost all agroecosystems, all types of forest areas, etc. *Trichoderma* from diverse habitat has been found to be equipped with several activities of very high benefit to agricultural crops in stress conditions. In addition to their use as antagonists of important plant pathogens, they are being used for several industrial purposes, viz., food industry, textile, paper and pulp industry, and also for the synthesis of nanoparticles for various specific uses. However, some of the *Trichoderma* spp. have also been found to be a human pathogen.

R. Singh (✉) · H. S. Viswanath
Department of Plant Pathology, College of Agriculture, SVPUAT, Meerut, India

A. Tomer
Department of Plant Pathology, Lovely Professional University, Jalandhar, Punjab, India

D. Prasad
Department of Plant Pathology, Banda Agriculture University, Banda, India

C. Manoharachary et al. (eds.), Trichoderma: *Agricultural Applications and Beyond*,
Soil Biology 61, https://doi.org/10.1007/978-3-030-54758-5_2

Keywords *Trichoderma* · Biodiversity · Agroecological zone

2.1 Prelude

Crop cultivation vastly depends upon the soil which is a substrate for several microbial and biochemical activities, thus it is a body containing millions of living microorganisms. These properties of soil support several biological functions related to ecology on the earth. Green revolution in India brought ample food production and self-sufficiency and India is a country now not only with sufficient food for domestic needs rather an exporter of food grain in the global market. Green revolution became a reality, mainly because of crop intensification and the use of fertilizer responsive varieties of cereals, especially wheat followed by rice. With advancing time after the green revolution scenario, cropping pattern and other agricultural operations in India have heavily affected the soil's physical, chemical, and biological health as well and ultimately the soil microbial wealth gets eroded. Soil biodiversity and physicochemical properties are reciprocal to each other with a positive correlation and very critical to sustainable cropping and sustainable farming as a whole. There is a continuous interaction and competition in the microbial ecosystem (pathogenic and antagonistic microbes) and as a result, the pathogen gets suppressed and in this way, the incidence and severity of several diseases get decreased. Creating a high level of microbial biodiversity in the soil may result in greater and sustainable crop productivity with comparatively less input. *Trichoderma* possesses a saprophytic ability and contributes greatly to the soil ecology and thus significantly helps to maintain soil physicochemical properties. Different other species of *Trichoderma* have been found to express a very high to moderate level of antagonistic activity against important soilborne microflora which are capable of causing plant diseases. This fungus is highly widespread, has the ability to survive and flourish over diverse ecosystems, be it temperate, tropical mangrove swamps, salt marshes, and estuarine. The majority of *Trichoderma* can be easily retrieved from the soil's top horizons. Two benchmark sites in Keny are presenting the land with intensive cropping, *T. harzianum* was in abundance whereas the frequency of other species of *Trichoderma* showed restricted recovery. The probable reason might be the difference in ecological components and also the type of crop being used for cultivation in that particular geographical area.

Trichoderma are free-living, soilborne, green spores containing ascomycetes, and ubiquitous in nature. These fungi are highly interactive in the rhizosphere and phyllosphere. *Trichoderma* spp. can be noticed in all types of soils representing almost every cropping system and also in each ecozone if some cultural, biological, or agricultural operations are followed. They are fast-growing, bearing bright green conidia on repetitively branched conidiophore. *Trichoderma* an imperfect fungi with its perfect state *Hypocrea* which also exhibit a very fast-growing habit in artificial culture and produce plenty of green conidiospores and sometime chlamydospores also, especially when the culture becomes older or with aging and growth medium geting exhausted. *Trichoderma* spp. are an integral component of eco-friendly, safe, and chemical-free disease management system, and thus have a greater importance in organic farming. Various species of *Trichoderma* are chiefly and widely being applied for management of plant diseases caused by fungi, and many a time also as a biofertilizer. The ability of *Trichoderma* to establish a mycorrhiza-like association with plants as endophyte is the main driving force; that is why it is being used as a bio-fertilizer. The key factor for the success of this genus in all ecological niches is because of its proactive mycoparasitic mechanisms along with effective inducer of plant defense and resistance to noxious chemicals. *Trichoderma*, in general acts as a plant growth promoter which is more pronounced when plants are under stress (Shoresh and Harman 2008).

In geographical areas with scarce moisture and hot weather, *T. harzianum* may be found in abundance whereas other species like *T. hamatum and T. koningii*can are chiefly isolated and retrieved in the areas of diverse climatic conditions (Denielson and Davey 1973). Continuance of dry conditions (Suboptimal moisture) in soil over a longer period may result in the declined population dynamics of *Trichoderma* and *Gliocladium*. These two genera are most dominating in the soil microbiota and can be noticed anywhere on the earth (Domsch et al. 1980).

Majority of *Trichoderma* species have a strong association with the rhizosphere of diverse groups of plants, and out of these several strains, some specific strains may often establish special endophytic relationship with specific compatible plants (Bailey et al. 2006; Evans et al. 2003; Hoyos-Carvajal et al. 2009; Sette et al. 2006; Viterbo and Chet 2006; Yedidia et al. 2000). As a beneficial plant-associated microflora, they get activated in the root zones and act as a successful antagonist to perform the function plant disease management following the practices of biological control. *Trichoderma* spp. mainly perform all these activities by secretion of certain biologically active compounds that are toxic to plant pathogenic microorganisms especially fungi-induced resistance (resistance inducer), and also by promoting growth and vigor in plants. Enhanced plant growth response occurs mainly due to solubilizing minerals present in the soil and making available the nutrients and plant growth regulators, which are earlier unavailable due to their insolubility (Vinale et al. 2006; Yedidia et al. 2000). Due to its characteristics of decomposition of biological wastes and agricultural residues, mycoparasitism and endophytic association, they are of utmost importance for sustainable agriculture and also for conserving the ecological balance (Harman et al. 2004). Because of incorrect and confusing names of several species in *Trichoderma*, its taxonomy has become unreliable, and this has made the comparisons of species little challenging and difficult (Kopchinskiy et al. 2005). There is greater variation among metabolic activities of different isolates of *Trichoderma* species which is sometimes as much as it is exhibited and observed among its species. Thus the study of variation in metabolic activity should be precise and careful so as to fully exploit the potential of this antagonist.

2.2 Study of Taxonomy: May Help to Explore Biodiversity

Genus *Trichoderma* for the first time was identified by Persoon (Bissett 1991a, b). In artificial cultures, *Trichoderma* grows quite faster on a diverse range of media containing some sources of nutrition like carbon, nitrogen, and growth factors. Very shortly they lose their ability to sporulate in culture media which can be restored after providing them the exposure to sun light or under NUV in the dark (Domsch et al. 1980; Bissett 1991a). *Trichoderma* can be successfully recovered from soil using a selective medium (Mukherjee and Dasgupta 2006).

Colonies of *Trichoderma* usually grow at a rapid rate, with smooth surface, translucent in color, subsequently becoming floccose, exhibiting various shades of greenish or pure white. Pigmentation can be noticed in the medium and backside of the colony does not show any change. The mycelium of this fungus is typically hyaline, septate, profusely branched with a smooth wall. Most of the species in *Trichoderma* produce chlamydospores after attaining maturity, and the substrate gets exhausted. The chlamydospores may be intercalary or some time terminal on short side branches of the mycelium also, ellipsoidal or globose, smooth-walled, and colorless. The sporophores of all species of this genus are profusely branched, look like a loose or compact bunch and in general, these conidiophores appear as a target board. The conidium is generally born on the hyphae erected in the air and very rarely on the surface hyphae. Branching of conidiophores are so profuse that the branching originated first may further produce secondary and tertiary branches. Branching may be either single or very often up to three. Side branches of the conidiophores arise at such an angle that conidiophores appear as conifer-like structures. In the majority of *Trichoderma* species, the terminal end of branching culminated into phialide but in the case of *T. hamatum* and *T. polysporum*, the main branches appear as straight or curved one or sometimes appear to be whip-like elongated sterile hyphae. Conidiophores are of such a shape that they appear to look like a flask of ninepin. In some of the cases, conidiophores appear to be as pear-shaped or most of the time ovoidal. The measurement of conidia is almost less than 15 μm in diameter. In very few instances conidia may form a short chain or their head may coalesce from each other to give an appearance that several conidia are joint together from the head. Their conidiphores terminate in phialide like structures which forms phialospores which are often used to denote conidia which are smooth or roughened at outer wall, colorless transparent or green having slight yellow tinchto fast green, semi-globose, semi obovoid or complete ovoid, or almost oblong. At the early stage, these conidia of *Trichoderma* may contain either single or several oil bodies and at maturity, these oil droplets inside the conidial body get disappeared (Bissett 1991c).

Taxonomical information and literature of *Trichoderma* have been thoroughly reviewed by Samuels (2006) for more intensive knowledge of species in *Trichoderma*. He has focused on the interaction between humans and this genus which belongs to the class hyphomycetes, and it has been noticed that *Trichoderma* is prevalent worldwide in every ecosystem and predominately in soils. The majority of species of *Trichoderma* have been utilized for several commercial purposes which are of direct beneficial use for human beings, viz. production of enzymes for commercial application, bioremediation, as a source of transgenes, and as a practice/input for plant disease management. Knowledge on the characterization of species in *Trichoderma* and species-to-species interaction and relationships and several other attributes related knowledge has not been properly coordinated with the new studies and findings related to the exploration of *Trichoderma* by molecular biologists and genetic engineering scientists, plant clinic experts and other scientists related to the field of medical science and public health.

With the advent of modern molecular tools, it has been possible to generate the data derived from nucleic acid analysis which has greatly helped the researchers to solve the confusion related to taxonomy in *Trichoderma*. DNA fingerprinting has been used to reclassify a vast number of species in *Trichoderma* (Meyer et al. 1992). Characterizing different strains of *Trichoderma*spp. has become more easier using PCR-based molecular tools (Lieckfeldt and siefert 2000).

2.3 Study at the Level of Species: Key to Biodiversity

Morphological Species Recognition (MSR) concept had been chiefly applied to define and recognize the species of *Trichoderma* in the past. MSR sometimes has also been applied in combination with other phonetic characters for species identification. Identifications however on the basis of morphological parameters/traits are highly prone to error, as some time definitive morphological characteristics are missing and the test culture exhibit a high level of variations. Hence, there is all possibility that more than 50% of *Trichoderma* samples have been inappropriately identified on the basis of morphology. Due to difficulty in making crosses to ascertain the reproductive behavior of *Trichoderma* strains, the identification of species on the basis of sexual or asexual reproduction which is also called as Biological Species Recognition (BSR) is also not a proper alternative for taxonomic purposes. Another attractive alternative for recognizing species of *Trichoderma* has been proposed which is known as Genealogical Concordance Phylogenetic Species Recognition (GCPSR). This technique follows on the concepts/principles of concordance of multiple gene phylogenies based on the Phylogenetic Species Concept (PSC) (Taylor et al. 2000).

2.4 Establishing Phylogenetic Relationship Among Species

Phylogenetic species concept is a technique to establish biodiversity which correlates the data pertaining to morphology, biogeography, biochemistry, ecology, and, most recently, phylogeny-related attributes. This approach/technique is adopted for differentiating and characterizing those *species* in *Trichoderma* which are of cryptic type (Rifai 1969). Seven phylogenetic lineages were determined by Chaverri et al. (2003), where he applied the PSC concept (Taylor et al. 2000), by examining the internal transcribed spacer regions of rDNA (ITS1 and ITS2), the large intron of the transcription elongation factor 1-(*tef1α)*, and short fragments of the actin (*act*1) and calmodulin (*cal1*) exon sequences in *H. lixii/T. harzianum*. However, due to the lack of reliability in the morphological distinction of these two species, they could not be recognized as closely related phylogenetic species. Similarly, applying GCPSR *T. koningii* can be considered as an evolutionary group of three distinguished subgroups of species. These three groups of species were recognized on the basis of phenotype of specimens, within these three lineages; further, there were 12 taxonomic species and one variety (Samuels 2006). Genetic diversity in *T. harzianum* has been revised after examining again, the three gene loci of 93 *Trichoderma*strains isolated from across the world (Druzhinina et al. 2010). These studies clearly indicate a complicated nature of species differentiation in *H. lixii* and *T. harzianum* and also for correlating the anamorph/teleomorph, thus a separate status for *H. lixii* and *T. harzianum* as two species could not be approved/resolved. According to Druzhinina et al. (2010), the isolates which were previously confirmed as *H. Jecorina* consist of four distinguishable species, including *H.jecorina/ T. Reesei*, *T.sensus*, and *T. tricot* representing major isolates in its perfect state and the wild-type strain of *T. reesei*. These isolates have been subsequently modified at genetic levels and employed in biofuel production adopting multiple genotyping and multiple methods of phenotype characterization. Finally, the strains isolated only capable of asexual reproduction were recognized as *T. parareesei.* Thus, before going for the selection of potential isolates for industrial use, the phylogenetic characters specially in the *T.* species are considered to be complicated ones and must be kept in mind. In the past, the name "*T. harzianum*" has been commonly denoted as abiocontrol agent, but now clear and ample evidences are there which is showing the application of several other genetically diverse species for the biocontrol of plant pathogens (Druzhinina et al. 2005).

2.5 How Species in *Trichoderma* genus Can Be Identified

2.5.1 Analysis of Morphological Characters

For morphological observations, *Trichoderma* needs to be grown first on a proper culture media after the screening of a variety of culture media. Out of several media, malt extract agar 2% (MEA) has been found to be most useful and relatively less complicated for the harvesting conidiospores along with studying its profusely branched conidiophores (macronematous). Pigments production by *Trichoderma* species can be easily observed by growing them on semisynthetic media like PDA. Fungal DNA can also be extracted from mycelium mat harvested from similar growth media. Four to seven days old culture of *Trichoderma* can be used for studying the morphological structure of conidiophores and the same can be collected from the edge or periphery of the fungal colony at the time of maturity of conidia (usually after 4–7 days of incubation). Conidial size and morphology can be observed at a little more advance stage, i.e., nearly after incubation for 14 days (Bissett 1984, 1991a, b, c, 1992). Species or species aggregates in *Trichoderma* are generally determined following characteristic morphological explanations in the available reports or studies and compilation related to taxonomic characteristics.

2.6 Molecular Analysis

Genome sequencing of *Trichoderma harzianum* has also been attempted. The most studied species with regard to genome sequencing are *T.reesei*, *T. atroviridae*, *T. virens*, *T. harzianum*, and *T. asperallum*. Out of these five, *T. reesei* is of saprophytic nature and can secret large amounts of cellulases and hemicellulases, and thus is of great industrial importance. However, the remaining four possesses mycoparasitism-specific genes and frequently live in association with plant foods and on dead fungal biomass (Kubicek et al. 2011; Druzhinina et al. 2011).

Morphological characters are many time variables and get changed with the effect of environment and other external factors; however, because of the fact that they can never be changed due to any other external factors except genetic one, molecular attributes, viz. DNA and RNA provide a solid base to determine existing diversity among the species of *Trichoderma.* Rifai (1969) initiated the first step toward a clear understanding of diversity in *Trichoderma*, where he introduced the principles of "species aggregates" in *Trichoderma* featuring nine of them, clarifying that these species aggregate could include multiple species which cannot be distinguished from each other by morphological characters. Later revision by Bissett and Gams (Bissett 1984; Bissett 1991a, b, c, 1992) resulted in the additional species which were distinguished on the basis of morphology, further on the basis of sexual and asexual stages; new members were added to *Trichoderma* although earlier they were species of the genus *Gliocladium*. Systematic and exhaustive observation clearly indicated serious overlaps in the morphology of *Hypocrea* genus *Trichoderma*, an anamorph (Chaverri and Samuels 2003; Jaklitsch 2009); this clearly concluded that sole morphology can never be a reliable trait for identification of species and also for establishing the genetic relationship between two species. Considering the morphology to be quite insufficient for establishing biodiversity, there have been more concentrated and focused efforts for more reliable information using molecular approaches and tools which can conclusively help in establishing *Trichoderma* biodiversity. Molecular markers as a tool for studying taxonomy have helped greatly in recognizing a large number of species (104 spp. in *Trichoderma*).

The molecular methods are presumed to be quite sufficient for the identification of species (e.g., Lieckfeldt and Seifert 2000), but subsequently, it was realized that some techniques like application of BLAST (Genbank) were found quite misleading. It is worthwhile to mention that scientists and technicians used these approaches for identifying fungal species using ITS sequence homology. Numerous identification errors among sequences deposited in Genebank has been encountered,and it was found that majority of the members of a particular genus which are identified at different times also could not be conclusively identified (Kopchinskiy et al. 2005) Thus, now it is quite apparent that ITS alone cannot be so sufficient and explanatory to identify phylogenetically closed species of *Trichoderma*. Determination of diversity in *Trichoderma*, based on ITS alone, can be misleading also (O'Donnell et al. 2000; Lieckfeldt and Seifert 2000; Chaverri et al. 2003; Hoyos-Carvajal et al. 2009). Determination of species in *Trichoderma* was done studying various genes possessed by it. Diverse types of gene governing different important functions, viz. translation elongation factor (TEF), RNA polymerization, chitinase activity, calmodulin 1, actin, −tubulin2, LAS1 nuclear protein, and ATP citrate lyase subunit A have been thoroughly studied. Genetic diversity of *Trichoderma* and *Hypocrea* in Manipur of Indo-Burma biodiversity hot spot region was investigated using ITS sequencing and morphological characteristics as well. On the basis of this analysis, 22 different species of *Trichoderma* and *Hypocrea* were identified with the dominance of *Trichoderma* species. According to phylogenetic studies, considerable variations were observed among the *Trichoderma* isolates collected from different districts of Manipur. Another study was also conducted to assess the genetic diversity of *Trichoderma* spp. from the Indo-Burma region of India. One hundred ninety-three soil samples representing cultivated soil of four agroclimatic zones across nine districts of Manipur (India) were used for isolating the fungus. Out of 193, 65 isolates of *Trichoderma* spp. showed interspecific variations on the basis of ITS-RFLP of rDNA region. Although initially isolates were categorized using the morphology of individual culture, 22 different types of representative *Trichoderma* spp. were reported on the basis of ITS sequencing. Four well-separated main clads were also visible after phylogenetic analysis and among all *Trichoderma* spp. was found to be predominant. Out of all the *Trichoderma* species, all the species were found to produce different metabolites like extrolites and enzymes playing a key role in biocontrol activities against fungal pathogens of important crops.

2.7 Diversity Due to Metabolic Traits

Fungal species exhibit a very vast difference in their metabolic activities which can provide a strong base for the establishment of fungal biodiversity. Metabolic tests can be conducted on the basis of a specific enzyme or many different enzymes capable of utilizing different substrates chitin or cellulose. However, other tests pertaining to metabolism can also be applied for the validation of new fungal species. These other techniques for metabolic tests can be helpful for the validation of new species and the data generated can also potentially be used to ascertain their roles in different activities related to ecology (Kubicek et al. 2003; Hoyos-Carvajal et al. 2009). Metabolic studies of different *Trichoderma* species helped to generate exhaustive quantitative data related to nutrient utilization, growth, and respiration on the different growth mediums. Data pertaining to metabolic traits may be very specific and highly correlated with particular species, and it may also be indicative of specific nutritional exploitation and also useful in ascertaining the role of a species in the ecological cycle. For example, the assimilation of polyols such as maltitol and adonitol may be the indicator of dehydrogenases activity which is relevant from the point of view of survival in drought conditions or dry environment/habitat in the arid regions.

2.8 Distribution and Biogeography of *Trichoderma:* Indicate the Status of Biodiversity

Some pioneer institutes situated in North America and some regions of South East Asia have been thoroughly engaged in conducting well-planned studies of taxonomical traits and diversity of *Trichoderma* (Bissett 1991a, b, c, 1992). These studies resulted in generating deep knowledge and proper understanding of the distribution for a particular taxonomic lineage of *Trichoderma* species. Across the world, some specific geographical areas have been thoroughly investigated, e.g., genetic diversity of *Trichoderma* in North Asia landscape near Vienna, and another hotspot of biodiversity, i.e., India have been exhaustively studied for analyzing the effect of weather and climatic parameters on the prevalence of *Trichoderma* species (Wuczkowski et al. 2003). The study conducted in the Indian hotspot of biodiversity resulted in identifying 1482 isolates of *Hypocrea/Trichoderma* representing undisturbed and disturbed ecology, and a fact was established that most of the strains were of Asian origin species. A diverse distribution pattern of *Trichoderma* prevalence was observed which were mainly affected by abiotic factors, soil types and management practices, independent of the crop variety used. The regions or habitats, previously untouched with the viewpoint of determining the biodiversity, almost resulted in the discovery of some new taxonomic group. In another study 76 isolates collected from Russia, Nepal, and North India were investigated, which resulted in the discovery of seven species (*T. asperellum*, *T. atroviride*, *T. ghanense*, *T. hamatum*, *T. harzianum*, *T. virens*, and *T. oblongisporum*) and five new taxonomic groups (Kulling et al. 2000). This study also confirmed that *T. harzianum* possesses a very high level of genetic diversity and also that majority of isolates were related to *T. Harzianum* complex. Kubicek et al. (2003) and Bissett et al. (2003) also conducted a similar study in Southeast Asia, where they tested 96 isolates of *Trichoderma* and reported *T. asperellum*, *T. atroviride*, *T. ghanense*, *T. hamatum*, *T. harzianum*, *T. koningii*, *T. spirale*, *T. virens*, *T. viride*, and *H. jecorina* (anam: *T. reesei*), along with seven new phylogenetic species. The high prevalence of *T. harzianum* complex clearly indicates a very high level of *Trichoderma* biodiversity related to metabolomic and morphological traits variability and this also justifies the wide distribution of *Trichoderma harzianum* species as cluster over a wide range of ecosystem (Kubicek et al. 2003). Variability of *Trichoderma* localized species and their relation with different climatic zones in Tunisia was also studied by Sadfi-Zouaoui et al. (2009). This study revealed that *T. harzianum* can be divided into six closely related groups with predominance. Predominance of *T. harzianum* and *T. longibrachiatum* was noticed in the soil of forest ecology in north Tunisia; forest soils in central Tunisia were predominant by *T. harzianum*, *T. saturnisporum*, and *Trichoderma* sp. indet.; fields under crops cultivation in northeast Tunisia were predominant by *T. atroviride* and *T. hamatum*; and oasis soils in south Tunisia were predominant by *T. harzianum* and *T. hamatum*. Diversity and geography of *Trichoderma* in China were assessed by Zhang et al. (2005). Soil samples collected from four regions: north (Hebei province), south-east (Zhejiang province), west (Himalayan, Tibet), and south-west (Yunnan province) resulted in the identification of 11 species, i.e., *T. asperellum*, *T. koningii*, *T. atroviride*, *T. viride*, *T. velutinum*, *T. cerinum*, *T. virens*, *T. harzianum*, *T. sinensis*, *T. citrinoviride*, and *T. longibrachiatum* along with two putative new species from the same soil samples. A north-south direction gradient in species distribution in eastern Asia was also established. Biodiversity of *Trichoderma* in some other Asian countries, viz. Mongolia, Japan, Vietnam, and Indonesia was also assessed (Tsurumi et al. 2010). Of the 332 strains of *Trichoderma* in most habitats, mainly four species, viz. *T. harzianum*, *T. hamatum*, *T. virens*, and *T. crissum* were found to be most prevalent. *T. koningiopsis*, *T. atroviridae*, and *T. stramineum* were also noticed; however, in the region representing cooler climate, the above-mentioned species were replaced by *T. polysporum* and *T. viridescens*. In areas witnessing tropical climate, *T. ghanense*, *T.brevicompactum*, and *T. erinaceum* were most prevalent.

2.9 Diversity of *Trichoderma* in Tropical Climate

Biodiversity of *Trichoderma* in central and south American regions has been explored but not up to the extent as in other climate and comprehensive studies undertaken for the purpose are comparatively very few. Most of the studies for the exploration of the biodiversity of *Trichoderma* in tropical climate has been focused mainly toward biocontrol of plant pathogens because of agricultural farming being the vital segment of local economies in these areas. Mostly plantation crops with high economic values are grown in the tropical climate; that is why biodiversity-related researches have been mostly focused to utilize them as biocontrol agents for important plant pathogens of economic importance for the important cash crops of such climate (Castro 1996; Carsolio et al. 1994; Hebbar et al. 1999; Hoyos et al. 2008; Rivas and Pavone 2010). The objectives of these studies were to manage the symbiotic fungus of the leaf-cutting ant *Atta cephalotes* (Lopez and Orduz 2003), and also to ascertain *Trichoderma*'s potential for utilizing them as plant growth promoter (Bae et al. 2009; Hoyos-Carvajal et al. 2009). Several studies on the taxonomy of the genus *Trichoderma* in the recent past are helping to augment the knowledge of the subject and still constantly increasing due to ever-increasing information on the distribution of *Trichoderma* species. Chances for better understanding the biogeography of *Trichoderma* species in the future is very high, as knowledge on *Trichoderma* is being generated, utilized, and analyzed very fast. Chances of resolving the complex species aggregates in *Trichoderma* are very bright with the fact that plenty of information being generated through several researches being pursued in new regions. The species of *Trichoderma* is usually mentioned as *T.koningii* in the scientific literature and texts; it is actually not common in its prevalence and also confined to temperate regions of the world (Samuels 2006). This scientific team also discovered and established numerous new species in *Trichoderma*, viz. *T. caribbaeum* var. *Aequatoriale*, *T. koningiopsis*, and *T. ovalisporum* as endophytes of *Theobroma* species in tropical America, and *T. ovalisporum* also from the woody liana *Banisteropsiscaapi* in Ecuador from within the *T. koningii* aggregate. *T. koningiopsis* (which was earlier known and identified as *T. koningii*) was very commonly observed in the tropical region of theAmerican continent, flourishing in natural habitat in the Eastern part of Africa, entire Europe, and near arctic region, i.e., Canada. The perfect state of this fungus was noticed in the eastern part of North America; its endophytic association with *Theobroma* was also noticed in Eastern North America. Another species, i.e., *T. stilbohypoxyli*, which parasitized on *Stilbohypoxylon* species in Central American country, i.e., Puerto Rico, have been commonly observed and recovered in the tropics. Several new species of *Longibrachiatum* section in *Trichoderma* have also been noticed from neotropical areas (Samuels et al. 1998). *T. viridescens* as a species was observed and recovered from high lands in Peru, and *T. neokoningii* was observed and recovered from the tropical part in Peru. These species were identified while revisiting the *T. viride* species complex by Jaklitsch (2009). Some new species, i.e., *T. scalesiae*, an endophyte of the *Scalesiapedunculata* trunk in the Galapagos Islands of Ecuador; *T. paucisporum*, a hyper-parasite of *Moniliophthoraroreri*, on pods of *Theobroma cacao* in Ecuador; and *T. gamsii*, an apparently cosmopolitan species that has been found in Italy, Rwanda, South Africa, and Romania and Guatemala have also been identified and isolated by Jaklitsch (2009). Some other new species of *Trichoderma* have also been determined through some recent studies in neotropical regions undertaken mainly to search some novel biocontrol agents in some selected crops, viz. cocoa (Samuels 2006).

2.10 The Soil-Inhabitant *Trichoderma* Species Are More Common in Tropical Regions

A scientifically well-planned survey for determining the biodiversity of *Trichoderma* species was carried out in seven countries situated in North, Central, and South America. Based on morphological traits, metabolic traits, and molecular attributes, a highly diverse group of 182 *Trichoderma* isolates belonging to a large number of species were identified (Hoyos-Carvajal et al. 2009). The number of isolates/species of *Trichoderma* identified through this study were 26 isolates of *T. asperellum*, 34 isolates of *T. asperelloides* (as *T. asperellum* "B"), 3 isolates of *T. atroviride*, 5 isolates of *T. brevicompactum*, 3 isolates of *T. crassum*, 3 isolates of *T. erinaceum*, 2 isolates of *T. gamsii*, 2 isolates of *T. hamatum*, 49 isolates of *T. harzianum*, 6 isolates of *T. koningiopsis* (6), 3 isolates of *T. longibrachiatum*, 1 isolate of *T. ovalisporum*, 2 isolates of *T. pubescens*, 4 isolates of *T. rossicum*, 1 isolate of *T. spirale*, 3 isolates of *T. tomentosum*, 8 isolates of *T. virens*, 7 isolates of *T. viridescens*, 3 isolates of *T. parareesei* (as *H. jecorina*), along with 11 presumptive new species that have not yet been described. On the basis of metabolic traits (assimilation of substrates), 12 Colombian species, viz. *T. asperellum*, *T. atroviride*, *brevicompactum*, *T. erinaceum*, *T. hamatum*, *T. harzianum*, *T. koningiopsis*, *T. longibrachiatum*, *T.virens*, *T. viridescens*, *T. parareesei*, and *Trichoderma* sp. 210 exhibited highly significant differences ($P < 0.0001$). *T. viridescens*, *T. asperellum*, *T. harzianum*, and *T. parareesei* exhibited the highest growth rate on 23 substrates. These four species, viz *T. viridescens*, *T. asperellum*, *T. harzianum*, and *T. parareesei* were isolated mostly from rhizosphere of Impatiens, a broad range of substrates, varied habitats, and from African palm on three nutritional bases, respectively. The nutritional sources where *T. viridescens* could grow quite faster were those substrates which were either not preferred or poorly preferred as sources of nutrition by any other species, indicating the ability of these isolates to grow on recalcitrant substrates; similar such growth habit, growth pattern, and substrate assimilation have been observed in other studies also. *Trichoderma* isolates representing areas of untouched forests were capable of multiplying on those substrates which are very hard to be assimilated. *Trichoderma viridescens* and *T. harzianum* exhibited very high multiplication performance on several substrates as the former could grow on 41 different substrates and the later on 34 different substrates. Growth of these two species of *Trichoderma* on a wide range of substrates indicates their capability of survival and adaptation over a broad range of habitats or niches which is also visible in their distributions over a wide climate. *T. erinaceum* isolated from maize rhizosphere exhibited relatively slower growth on 15 substrates, whereas, 210 isolates of *Trichoderma* sp. representing river sand were also found to be slow-growing on 11 substrates. Very few substrates (19–25 substrates) were assimilated by *T. longibrachiatum* and *Trichoderma* sp. 210 in section *Longibrachiatum*, along with *T. erinaceum*. Hoyos-Carvajal et al. (2009) also explored the biodiversity of *Trichoderma* in neotropical regions. This group identified 19 species out of total 182 isolations, and they also discovered 11 entirely new species from the rainforest's soil and other habitats, viz. sand of river, decomposed organic matters, and woody materials in Central and South American countries. Rivas and Pavone (2010) studied *Trichoderma* in Venezuelan soils, and they found that *T. harzianum*, followed by *T. virens*, *T. brevicompactum*, *T. theobromicola*, *T. koningiopsis*, *T. ovalisporum*, *T. asperellum*, *T. pleurotum*, and *T. koningiopsis* were most abundant species. In recent years, these observations have added some new species of *Trichoderma* from neotropics, mainly as endophytes of plants, and are evidence of significant biodiversity of *Trichoderma* in the tropical regions (Samuels 2006).

2.11 Diversity of Species and Secondary Metabolites

Production and secretion of secondary metabolites by *Trichoderma* species are well-known which are mainly utilized by producing species for performing a variety of biological activities. Research groups concentrating on biological control are well aware of the effective use of these secondary metabolites in biological control of plant pathogens. Molecular tools are now extensively being used and to date total of 1100 *Hypocrea* (perfect stage)/*Trichoderma* (imperfect stage) strains have been determined from 75 molecularly characterized species (Druzhinina et al. 2010). The species capable of mycoparasitism, viz. *T. atroviride* (*Hypocreaatroviridis*) and *T. virens* (formerly *Gliocladiumvirens*) and the species with saprophytic capacity, i.e., *T. reesei* (*Hypocreajecorina*) have been most frequently studied; a comparative genome analysis revealed important differences in their lifestyle (Kubicek et al. 2011). Studies pertaining to genome sequencing of *Trichoderma* have been concentrated on *T. harzianum*, *T. asperellum*, *T. longibrachiatum*, and *T. citrinoviride* (Mukherjee and Horwitz 2013). Scanning of scientific literature related to secondary metabolites, their structural type, bioassay, and their source of origin revealed that 20 different known *Trichoderma* species and various unidentified species produced approximately 390 non-volatile compounds (Li et al. 2019). The secondary metabolites produced and secreted by several *Trichoderma* spp. exhibit several characteristics such as siderophoric, antifungal, antibacterial, antialgal, antiviral, antitumor, antimicrobial, plant resistance inducers, enzyme inhibition, antibiotic, DPPH-radical-scavenging, cytotoxic, and anti-inflammatory. Major species of *Trichoderma* capable of producing strong secondary metabolites include *T. viride*, *T. virens*, *T. spirale*, *T. saturnisporum*, *T. reesei*, *T. polysporum*, *T. longibrachiatum*, and *T. koningiopsis.*

Genome sequencing of *Trichoderma* species has also been attempted with an aim to explore their capacity of producing secondary metabolites. The most studied species with regard to genome sequencing are *T.reesei*, *T. atroviridae*, *T.virens*, *T. harzianum*, and *T. asperallum*. Out of these five, *T. reesei* is of saprophytic nature and can secret large amounts of cellulases and hemicellulases, thus of great industrial importance. However, the remaining four possesses mycoparasitism specific genes and frequently live in association with plant foods and living or dead fungal biomass (Kubicek et al. 2011; Druzhinina et al. 2010). These secondary metabolites produced by several antagonists are mainly to support the attack and parasitism by antagonists on pathogenic microorganisms. Sometimes these secondary metabolites also act as communication molecules. In addition to their antagonistic activities, these metabolites also impart some beneficial effects to plants like plant growth promotion and defense inducer. According to their mode of action, they have been grouped into several categories, viz.:

(i) *Nonribosomal peptides*: They have been found associated with *T. attroviridae*, *T virens*, and *T. reesei* (Kubicek et al. 2011).

(ii) *Peptailbiotics*: These are produced by almost all species of *Trichoderma* but most prominently in *T. harzianum* (Neumann et al. 2015; Degenkolb et al. 2015).
(iii) *Epipolythiodioxypiperazines*: They are mainly produced by *T. virens* (Zeilinger et al. 2016).
(iv) *Siderophores*: These are produced by *T. virens, T. asperallum, and T. hamatum* (Mukherjee et al. 2012; Wallner et al. 2009; Oide et al. 2007; Shaw et al. 2016).
(v) *Polyketoides*: Mainly produced by *T. virens*, *T. attroviridae*, and also *T. ophioglossoides* (Kubicek et al. 2011).
(vi) *Terpenoids*: Produced by *T. attroviridae* and *T. reesei* (Bansal and Mukherjee 2016).

2.12 Root Colonization

Plant roots can be actively colonized by *Trichoderma* which can significantly alter several metabolic processes in plants viz. altering the production of growth regulators, soluble sugars, phenolic contents, amount of total amino acids, rate of photosynthesis, rate of transpiration, and relative water content in the plants (Yedidia et al. 2000; Bae et al. 2009). For effective colonization of broader host range by *Trichoderma*, it is assumed that the same has evolved in such a manner that it can overcome plant immunity and can create a proper environment for nutrient acquisition and reproduction in association with the host it has colonized (Khattabi et al. 2004). A compound namely hydrophobins play a key role in the recognition of host plant by *Trichoderma* and subsequent adhesion on the host surface during symbiotic associations (Viterbo and Chet 2006). A gene *qid74* of *T. harzianum* encoding for a cysteine-rich cell wall protein plays an important role in the adhesion of *Trichoderma* on tomato root surface. Appressorial formation in *Trichoderma* is governed by a class I hydrophobins which is encoded by the gene *TasHyd1*, secretion of cellulolytic and proteolytic enzymes facilitates root penetration by the degradation of cell wall of root surface made up of cellulose and protein (Viterbo and Chet 2006). An extensive exchange of molecular messages takes place during root colonization by *Trichoderma* spp. The deposition of fungal elicitors in the root cell apoplast also takes place during host colonization (Hermosa et al. 2012). During the process of host colonization by *Trichoderma* spp., the molecular and biochemical events which get activated in plants as a result or consequence are still not very clear. However, one thing is very clear that root colonization by *Trichoderma* spp. can be manifested in terms of plant disease control, increased plant growth promotion, and productivity of crop as well. Shoresh and Harman (2008) has reported that root colonization with *T. harzianum* Rifai strain 22 (T22) was found to induce strong changes in the proteome of shoots of corn seedlings, even though the antagonist applied was present only in roots. Cucumber plants inoculated with *T. asperellum* T34 also exhibited similar effects. During this study, four type proteins were identified, which play a very active role in providing tolerance to biotic and abiotic stress and imparting defense/resistance to diseases, generating energy and its metabolism, production of secondary metabolites, and synthesis of bioactive protein molecule (Segarra et al. 2007).

2.13 Substrates for Mass Multiplication

Level of ease with which a biocontrol agent can be mass multiplied on an easily available, less expensive, and suitable substrate is going to be a deciding factor for the commercial success of a biocontrol agent. In addition, its bio-efficacy and shelf life can never be overlooked. During the recent past, significant advancement has been made in the field of mass multiplication of biocontrol agents for their commercial production.

Solid and liquid fermentation technologies have been evaluated for various species of *Trichoderma*. Jayarajan may be considered a pioneer worker in this field who in the year 1990 successfully tested among 18 agricultural byproducts and wastes as suitable substrates for the mass multiplication of two *Trichoderma* species, i.e., *T. viride* and *T. harzianum*. Both of these species resulted in the highest population dynamics on the rind of cassava followed by its processing remains and well rotten cattle dung manure. Farm yard manure followed by wheat bran and rice bran were the appropriate rather best substrate for a very high population dynamics of *T. viride* and *T. harzianum*, whereas peat soil and rice straw were found to be unfavorable for growth and spore production of two species (Panicker and Jeyarajan 1993). Out of several cereal grains and agricultural wastes, a mixture of wheat bran–sawdust–tap water was found to be the best medium for mass multiplication of *T. harzianum* (Patel and Mishra 1994). Similarly wheat bran was found to be a more favorable and best-suited substrate for very high population dynamics which were significantly higher at 14 days as compared to 7 days of inoculation (Das et al. 1997).

Results of the study conducted by Jahagirdar et al. (1998) revealed wheat bran as the most suitable substrate for spore yield of *Trichodermaharzianum*. Deoiled cakes of *Pongemiapinnata* (Karanza) followed by *Azadirechtaindica* (neem) and *Arachis hypogaea* (groundnut) were most suitable among the cakes tested, whereas either neem cake alone or the neem cakes amended media was found to be favorable for *T. harzianum* (Karthikeyan and Bhaskaran 1998; Meena et al. 2001). A mixture of wheat straw + wheat bran (3:1) was found to be the most suitable material for faster multiplication of three isolates of *T. harzianum*. In this study, 11 semi-solid substrates were evaluated for mass multiplication of six *Trichoderma* isolates (Singh et al. 2001). Sugarcane waste + used tea leaves was found to be better for proper growth of *T. reesei* whereas used tea leaves + wheat bran also supported the highest population dynamics of *T. viride* and *T. koningii*. Among several solid substrates and incubation temperature tested, sporulation of *T. viride* and *T. harzianum* was optimal at 30–35 °C, with a highest on biogas slurry, followed by well rotten cattle dung manure, dried cow-dung, wheat bran, and sorghum grain (Sangle et al. 2003). While doing the mass multiplication work of *Trichoderma harzianum*, 50% moisture must be maintained in the solid substrates for appropriate sporulation. Sangle and Bambawale (2005) reported that population dynamics of *Trichoderma harzianum* and *T. viride* was quite high on several solid and liquid substrates with an inhibitory effect by neem cake on sporulation.

Mukherjee and Dasgupta (2006) reported effective mass multiplication of *T. harzianum* (isolate T_7). Multiplication was quite higher on bajra and wheat seeds with different materials as substrates: vermicompost, cow dung manure, and mustard oil cake and also concluded that cow dung may be used as a substrate for the mass multiplication of *Trichoderma* because of its cheaper availability. Wheat bran, rice bran, paddy straw, and neem cake for mass multiplication of *T. harzianum* has also been tested by Niranjana et al. (2009). Dutta and Das (2002) has standardized the load of inoculums for mass multiplication of *Trichoderma*. They noticed a gradual increase in the number of colonies up to first week, and thereafter the colonies gradually got declined in all the treatments except in treatments T2 (30 g inoculum/10 kg oilcake), Tr3 (40 g inoculum/10 kg oilcake), and Tr4 (50 g inoculum/10 kg oilcake); however, after the third week, cfu level enhanced and again got declined finally. Among all the treatments maximum colonies were noticed in the fourth week except in treatments Tr (20 g inoculum/10 kg oilcake) and Tr5 (60 g inoculum/10 kg oilcake).

2.14 Conclusions

Biodiversity of *Trichoderma* species represents a significant component, which plays an important and decisive role in augmenting soil biodiversity, maintaining its physicochemical properties along with plant health as well. With the advent of modern molecular and biochemical tools, it has now become quite easy to work out the population dynamics, biodiversity, specific roles, and interactions of *Trichoderma* species in the soil and plant ecosystem. These molecular and biochemical tools have greatly helped the scientists to resolve the taxonomy also. The development of latest molecular tools has overpowered other traditional methods by distinguishing the anamorphic forms which are most commonly encountered. Much of the novel biodiversity on *Trichoderma* spp. has been demonstrated from neotropics, although work to explore the diversity of *Trichoderma* has begun from individual regions, habitats, and substrates that exist in different climatic regions. The identification of *Trichoderma* species is now increasingly being done solely based on molecular data especially for economically important and species-rich genera, which proved to be highly reliant compared to the phenotypic traits having so many limitations. Many new species of *Trichoderma* having special characters and economic importance will undoubtedly be distinguished with the development of molecular tools for studying the ecology and environmental genomics. Crop cultivation is the major activity for livelihood security in neotropical regions. So, most of the investigations have been done on the line of application of *Trichoderma*, which is a major biological input for plant disease management without using chemicals for several plant pathogens of soilborne nature in these regions. Consequently, the discovery of new metabolites from different spp. of *Trichoderma* and their mechanisms of action, their growth on different substrates, and modes of colonization of roots of plants had been exclusively studied. We can now appreciate the importance of preserving the biodiversity of delicate ecosystems as reservoirs of metabolites and diverse and unique ecological niches for habitation by animals, plants, and microorganisms. Conservation is facilitated as we increase our knowledge regarding *Trichoderma* spp and their role in nutrient cycling and complex interactions within the soil biota.

References

Bae H, Sicher R, Kim M, Kim S, Strem M, Melnick R, Bailey B (2009) The beneficial endophyte *Trichoderma hamatum*isolate DIS 219b promotes growth and delays the onset of the drought response in *Theobroma cacao*. J Exp Bot 60(11):3279–3295

Bailey B, Bae H, Strem M, Roberts D, Thomas S, Crozier J, Samuels G, Choi I, Holmes K (2006) Fungal and plant gene expression during the colonization of cacao seedlings by endophytic isolates of four *Trichoderma* species. Planta 224(6):1449–1464

Bansal R, Mukherjee PK (2016) The terpenoid biosynthesis toolkit of *Trichoderma*. Nat Prod Commun 11:431–434

Bissett J (1984) *A* revision of the genus *Trichoderma*. I. Section Longibrachiatum sect. nov. Can J Bot 62(5):924–931

Bissett J (1991a) A revision of the genus *Trichoderma*. II. Infrageneric classification. Can J Bot 69 (11):2357–2372

Bissett J (1991b) A revision of the genus *Trichoderma*. III. Section Pachybasium. Can J Bot 69 (11):2373–2417

Bissett J (1991c) A revision of the genus Trichoderma. IV. Additional notes on section Longibrachiatum. Can J Bot 69(11):2418–2420

Bissett J (1992) *Trichoderma atroviride*. Can J Bot 70(3):639–641

Bissett J, Szacaks G, Nolan C, Druzinina I, Grandiger C, Kubicek C (2003) New infrageneric classification. Species of *Trichoderma* from Asia. Can J Bot 81(6):581

Carsolio C, Gutierrez A, Jiménez B, Van Montagu M, Herrera-Estrella A (1994) Characterization of ech-42, a *Trichoderma harzianum*Endochitinase gene expressed during mycoparasitism. Proc Natl Acad Sci USA 91(23):10903–10907

Castro B (1996) Antagonismo de algunos aislamientos de *Trichoderma koningii* originados en suelo colombiano contra *Roselliniabunodes, Sclerotinasclerotiorum*y *Pythiumultimum*. Fitopatol Colomb 19(2):7–18

Chaverri P, Samuels G (2003) *Hypocrea/Trichoderma* (Ascomycota, Hypocreales, Hypocreaceae): species with green ascospores. Stud Mycol 48(1):1–116

Chaverri P, Castlebury L, Samuels G, Geiser D (2003) Multilocus phylogenetic structure within the *Trichoderma harzianum*/Hypocrealixii complex. Mol Phylogenet Evol 27(2):302–313

Das BC, Roy SK, Bora LC (1997) Mass multiplication of *Trichoderma* species on different media. J Agric Sci Soc North East India 10(1):95–100

Degenkolb T, Fog Nielsen K, Dieckmann R, Branco-Rocha F, Chaverri P, Samuels GJ, Thrane U, von Döhren H, Vilcinskas A, Brückner H (2015) Peptaibol, secondary-metabolite, and hydrophobin pattern of commercial biocontrol agents formulated with species of the *Trichoderma harzianum* complex. Chem Biodivers 2015(12):662–684. https://doi.org/10.1002/cbdv.201400300

Denielson RM, Davey CB (1973) Non nutritional factor affecting the growth of *Trichoderma* in culture. Soil Biol Biochem 5(5):495–504

Domsch KH, Games W, Anderson TH (1980) Compendium of soil fungi, vol 1. Acedemic Press, London

Druzhinina I, Kopchinskiy A, Komon M, Bissett J, Szakacs G, Kubicek C (2005) An oligonucleotide barcode for species identification in *Trichoderma* and *Hypocrea*. Fungal Genet Biol 42:813–828

Druzhinina IS, Kubicek CP, Komoń-Zelazowska M, Mulaw TB, Bissett J (2010) The *Trichoderma harzianum* demon: complex specieation history resulting in coexistence of hypothetical biological species, recent agamospecies and numerous relict lineages. BMC Evol Biol 10:94, 14pp

Druzhinina IS, Seidl-Seiboth V, Herrera-Estrella A, Horwitz BA, Kenerley CM, Monte E, Mukherjee PK, Zeilinger S, Grigoriev IV, Kubicek CP (2011) Trichoderma—the genomics of opportunistic success. Nat Rev Microbiol 9:749–759

Dutta P, Das BC (2002) Management of collar rot of tomato by *Trichoderma* spp. and chemicals. Indian Phytopathol 55:235–237

Evans C, Holmes K, Thomas S (2003) Endophytes and mycoparasites associated with an indigenous forest tree, *Theobroma gileri*, in Ecuador and a preliminary assessment of their potential as biocontrol agents of cocoa diseases. Mycol Prog 2(2):149–160

Harman GE, Howell CR, Viterbo A, Chet I, Lorito M (2004) Trichoderma species—opportunistic, avirulent plant symbionts. Nat Rev Microbiol 2(1):43–56

Hebbar P, Lumsden R, Krauss U, Soberanis W, Lambert S, Machado R, Dessomoni C, Aitken M (1999) Biocontrol of cocoa diseases in Latin America—status of field trials. In: Hebbar P, Krauss U (eds) Workshop manual—research methodology for the biological control of plant diseases with special reference to fungal diseases of cocoa. CATIE, Turrialba

Hermosa R, Viterbo A, Chet I, Monte E (2012) Plant-beneficial effects of *Trichoderma* and of its genes. Microbiology 158(1):17–25

Hoyos L, Chaparro P, Abramsky M, Chet I, Orduz S (2008) Evaluation of *Trichoderma* spp. isolates against *Rhizoctonia solani* and *Sclerotium rolfsii* under *invitro* and greenhouse conditions. Agron Colomb 26(3):451–458

Hoyos-Carvajal L, Orduz S, Bissett J (2009) Genetic and metabolic biodiversity of *Trichoderma* from Colombia and adjacent neotropic regions. Fungal Geneti Biol 46(9):615–631

Jahagirdar S, Siddaramaiah AL, Narayanaswamy H (1998) Screening of substrates for mass multiplication of *Trichoderma viride*. Karnataka J Agric Sci 11:233–236

Jaklitsch WM (2009) European species of *Hypocrea* Part I. The green-spored species. Stud Mycol 63:1–91

Karthikeyan A, Bhaskaran R (1998) Effect of *Trichoderma harzianum*and neem cake in the control of basal stem rot of coconut. Neem News Lett 15:29–32

Khattabi N, Ezzahiri B, Louali L, Oihabi A (2004) Effect of nitrogen fertilizer and *Trichoderma harzianum* on *Sclerotium rolfsii*. Agronomie 24(5). https://doi.org/10.1051/agro:2004026

Kopchinskiy A, Komon M, Kubicek C, Druzhinina I (2005) Mycological research news. Mycol Res 109(6):657–660

Kubicek C, Bissett J, Druzhinina I, Kulling-Grandiger C, Szakacs G (2003) Genetic and metabolic diversity of *Trichoderma*: a case study on South East Asian isolates. Fungal Genet Biol 38 (3):310–319

Kubicek CP, Herrera-Estrella A, Seidl-Seiboth V, Martinez DA, Druzhinina IS, Thon M, Zeilinger S, Casas-Flores S, Horwitz BA, Mukherjee PK, Mukherjee M, Kredics L, Alcaraz LD, Aerts A, Antal Z, Atanasova L, Cervantes-Badillo MG, Challacombe J, Chertkov O, McCluskey K, Coulpier F, Deshpande N, von Döhren H, Ebbole DJ, Esquivel-Naranjo EU, Fekete E, Flipphi M, Glaser F, Gómez-Rodríguez EY, Gruber S, Han C, Henrissat B, Hermosa R, Hernández-Oñate M, Karaffa L, Kosti I, Le Crom S, Lindquist E, Lucas S, Lübeck M, Lübeck PS, Margeot A, Metz B, Misra M, Nevalainen H, Omann M, Packer N, Perrone G, Uresti-Rivera EE, Salamov A, Schmoll M, Seiboth B, Shapiro H, Sukno S, Tamayo-Ramos JA, Tisch D, Wiest A, Wilkinson HH, Zhang M, Coutinho PM, Kenerley CM, Monte E,

Baker SE, Grigoriev IV (2011) Comparative genome sequence analysis underscores mycoparasitism as the ancestral life style of *Trichoderma*. Genome Biol 12:R40. https://doi.org/10.1186/gb-2011-12-4-r40
Kulling C, Szakacs G, Kubicek C (2000) Molecular identification of *Trichoderma* species from Russia, Siberia and the Himalaya. Mycol Res 104(9):1117–1125
Li MF, Li GH, Zhang KQ (2019) Non-volatile metabolites from Trichoderma spp. Meta 9(3):58
Lieckfeldt E, Seifert K (2000) An evaluation of the use of ITS sequences in the taxonomy of the Hypocreales. Stud Mycol 45(1):35–44
Lopez E, Orduz S (2003) *Metarhiziumanisopliae* and *Trichoderma viride* for control nests of the fungus-growing and *Atta cephalotes*. Biol Control 27(2):194–200
Meena B, Ramamoorthy V, Muthusamy M (2001) Influence of organic amendments on the antagonistic population and their effect on sclerotial wilt of jasmine. Acta Phytopath Entomol Hung 36(3/4):299–309
Meyer W, Morawetz R, Börner T, Kubicek CP (1992) The use of DNA-fingerprint analysis in the classification of some species of the Trichoderma aggregate. Curr Genet 21(1):27–30
Mukherjee M, Dasgupta B (2006) Survival potential of *Trichoderma harzianum* in different mass multiplication media. Environ Ecol 24:737–741
Mukherjee PK, Horwitz BA, Singh US, Mukharjee M, Schmoll M (eds) (2013) *Trichoderma*: biology and applications. CABI, Oxfordshire
Mukherjee PK, Horwitz BA, Kenerley CM (2012) Secondary metabolism in *Trichoderma*—a genomic perspective. Microbiology 158:35–45. https://doi.org/10.1099/mic.0.053629-0
Neumann NKN, Stoppacher N, Zeilinger S, Degenkolb T, Brückner H, Schuhmacher R (2015) The peptaibiotics database—a comprehensive online resource. Chem Biodivers 12:743–751. https://doi.org/10.1002/cbdv.201400393
Niranjana SR, Lalitha S, Hariprasad P (2009) Mass multiplication and formulations of biocontrol agents for use against fusarium wilt of pigeonpea through seed treatment. Int J Pest Manag 55 (4):317–324
O'Donnell K, Kistler H, Tacke B, Casper H (2000) Gene genealogies reveal global phylogeographic structure and reproductive isolation among lineages of *Fusariumgraminearum*, the fungus causing wheat scab. Pro Natl Acad Sci USA 97 (14):7905–7910
Oide S, Krasnoff SB, Gibson DM, Turgeon BG (2007) Intracellular siderophores are essential for ascomycete sexual development in heterothallic *Cochliobolusheterostrophus*and homothallic *Gibberellazeae*. Eukaryot Cell 6:1339–1353. https://doi.org/10.1128/EC.00111-07
Panicker S, Jeyarajan R (1993) Mass multiplication of biocontrol agents *Trichoderma* Spp. Indian J Mycol Plant Pathol 23:328–330
Patel SI, Mishra A (1994) Some physiological studies on *Trichoderma harzianum* Rifai. Gujarat Agric Univ Res J 19(2):53–56
Rifai MA (1969) Revision of genus *Trichoderma*. Mycol Pap 116:1–56
Rivas M, Pavone D (2010) Diversidad de *Trichoderma* spp. enplantaciones de *Theobroma cacao* L. delestado Carabobo, Venezuela, y su capacidad biocontrol adora sobre *Crinipellis pernicosa* (Stahel) singer. Interciencia 35(10):777–783
Sadfi-Zouaoui N, Hannachi I, Rouaissi M, Hajlaoui M, Rubio M, Monte E, Boudabous A, Hermosa M (2009) Biodiversity of *Trichoderma* strains in Tunisia. Can J Microbiol 55(2):154–162
Samuels G (2006) *Trichoderma*: systematics, the sexual state, and ecology. Phytopathology 96 (2):195–206
Samuels G, Petrini O, Kulhs J, Lieckfeldt E, Kubicek C (1998) The *Hypocreaschweinitzii* complex and *Trichoderma* sect. Longibrachiatum. Stud Mycol 41(1):2–54
Sangle UR, Bambawale OM (2005) Evaluation of substrates for mass multiplication of *Trichoderma* spp. Indian J Plant Prot 33(2):298–300
Sangle UR, Bambawale OM, Ahmad N, Singh SK (2003) Substrate and temperature requirements for sporulation of sub-tropical isolates of *Trichoderma* spp. Ann Plant Prot Sci 11(2):192–195

Segarra G, Casanova E, Bellido D et al (2007) Proteome, salicylic acid, and jasmonic acid changes in cucumber plants in-oculated with *Trichoderma asperellum* strain T34. Proteomics 7:3943–3952

Sette L, Passarini M, Delarmelina C, Salati F, Duarte M (2006) Molecular characterization and antimicrobial activity of endophytic fungi from coffee plants. World J Microbiol Biotechnol 22 (11):1185–1195

Shaw S, LeCocq K, Paszkiewicz K, Moore K, Winsbury R, de Torres Zabala M, Studholme DJ, Salmon D, Thornton CR, Grant MR (2016) Transcriptional reprogramming underpins enhanced plant growth promotion by the biocontrol fungus *Trichoderma hamatum* GD12 during antagonistic interactions with *Sclerotiniasclerotiorum* in soil. Mol Plant Pathol 17:1425–1441. https://doi.org/10.1111/mpp.12429

Shoresh M, Harman G (2008) The molecular basis of shoot responses of maize seedlings to *Trichoderma harzianum*T22 inoculation of the root: a proteomic approach. Plant Physiol 147:2147–2163

Singh RS, Singh HV, Singh P, Kaur J (2001) A comparison of different substrates for the mass production of *Trichoderma*. Ann Plant Prot Sci 9(2):248–253

Taylor J, Jacobson D, Kroken S, Kasuga T, Geiser D, Hibbett D, Fisher M (2000) Phylogenetic species recognition and species concepts in fungi. Fugal Genet Biol 31(1):21–32

Tsurumi Y, Inaba S, Susuki S, Kamijo S, Widyastuti Y, Hop D, Balijinova T, Sukarno N, Nakagiri A, Susuki K, Ando K (2010) Distribution of *Trichoderma* species in four countries of Asia. 9th international Mycological congress. Edinburg, Scotland, Aug 2010

Vinale F, Marra R, Scala F, Ghisalberti E, Lorito M, Sivasithamparam K (2006) Major secondary metabolites produced by two commercial *Trichoderma* strains active against different phytopathogens. Lett Appl Microbiol 43(2):143–148

Viterbo A, Chet I (2006) TasHyd1, a new hydrophobin gene from the biocontrol agent *Trichoderma asperellum*, is involved in plant root colonization. Mol Plant Pathol 7(4):249–258

Wallner A, Blatzer M, Schrettl M, Sarg B, Lindner H, Haas H (2009) Ferricrocin, a siderophore involved in intra- and transcellular iron distribution in *Aspergillus fumigatus*. Appl Environ Microbiol 75:4194–4196. https://doi.org/10.1128/AEM.00479-09

Wuczkowski M, Druzhinina I, Gherbawy Y, Klug B, Prillinger H, Kubicek CP (2003) Species pattern and genetic diversity of *Trichoderma*in a mid-European, primeval floodplain-forest. Microbiol Res 158(2):125–133

Yedidia I, Benhamou N, Kapulnik Y, Chet I (2000) Induction and accumulation of PR proteins activity during early stages of root colonization by the mycoparasite *Trichoderma harzianum*strain T-203. Plant Physiol Biochem 38(11):863–873

Zeilinger S, Gruber S, Bansal R, Mukherjee PK (2016) Secondary metabolism in *Trichoderma*—chemistry meets genomics. Fungal Biol Rev 30:74–90. https://doi.org/10.1016/j.fbr.2016.05.001

Zhang C, Druzhinina I, Kubicek C, Xu T (2005) *Trichoderma* biodiversity in China: evidence for a North to South distribution of species in East Asia. FEMS Microbiol Lett 251(2):251–257

Chapter 3
Beneficial Effects of *Trichoderma* on Plant–Pathogen Interactions: Understanding Mechanisms Underlying Genes

Narasimhamurthy Konappa, Soumya Krishnamurthy, Nirmaladevi Dhamodaran, Udayashankar C. Arakere, Niranjana Siddapura Ramachandrappa, and Srinivas Chowdappa

Abstract *Trichoderma* is a genus of asexually reproducing filamentous fungi found in various ecosystems. It is among the utmost prevalent fungal genera commercially obtainable as a plant growth-promoting fungi (PGPF) and biocontrol agent. The biocontrol actions of *Trichoderma* are centered on the stimulation of various mechanisms such as competition for nutrients and space, mycoparasitism, alteration of the ecological conditions, antibiosis, and plant defensive mechanisms. Therefore, these fungi are commercially used in biocontrol of plant pathogens substituting the synthetic pesticides. The beneficial organism's genes and/or its products contain metabolites that reduce the harmful effects of plant pathogens and promote progressive responses in the plant. Certain genes have significant roles in the biocontrol process and are known as the biocontrol genes. These genes signal the secretion of enzymes and proteins that damage the plant pathogens. Some *Trichoderma* genes are also helpful in the control of different plant pathogens. In addition, *Trichoderma* produces plant growth-promoting molecules that stimulate growth and development of the plant. Within the rhizosphere, the conversation and recognition of signaling molecules by *Trichoderma* and plants may alter the physiological and biochemical characteristics of the plants as well as the biocontrol agent. A detailed realization of the molecular mechanisms underlying biocontrol would benefit from developing

N. Konappa · U. C. Arakere · N. S. Ramachandrappa
Department of Studies in Biotechnology, University of Mysore, Manasagangotri, Mysore, Karnataka, India

S. Krishnamurthy
Department of Microbiology, Field Marshal K. M. Cariappa College, A Constituent College of Mangalore University, Madikeri, Karnataka, India

N. Dhamodaran
Department of Microbiology, Ramaiah College of Arts, Science and Commerce, Bangalore, Karnataka, India

S. Chowdappa (✉)
Department of Microbiology and Biotechnology, Jnanabharathi Campus, Bangalore University, Bangalore, Karnataka, India

C. Manoharachary et al. (eds.), Trichoderma: *Agricultural Applications and Beyond*, Soil Biology 61, https://doi.org/10.1007/978-3-030-54758-5_3

Trichoderma strains with superior biocontrol properties. In this chapter, we summarize the interactions of *Trichoderma* with host plants and plant pathogens at the molecular level.

Keywords *Trichoderma* spp. · Biocontrol mechanisms · Antibiosis · Mycoparasitism · Induced systemic resistance · Secondary metabolites · Biocontrol genes

3.1 Introduction

The population of the world will reach around 9.1 billion people in 2050 which would need rising of total food production by some 70% (FAO 2009). The ever-increasing use of chemical inputs cause numerous harmful outcomes, development of resistance among pathogens, and their nontarget environmental effects (Sheikh et al. 2013). The pesticide consumption also increases year by year as 45.39 thousand tons of pesticides were consumed in the recent years (Krishijagran 2015). The number of biotic and abiotic stress causes yield losses up to a large extent. Biotic stress includes fungi, bacteria, viruses, nematodes, weeds, and insects which cause yield loss up to 42% and these pose the main danger to agriculture, food production, and supply (Agrios 2009; Kashyap et al. 2017; Sharma et al. 2017). Pesticide resistance and environment threat due to injudicious use of synthetic pesticides for disease control, hence, sustainable and ecofriendly approaches are new alternatives as a biological control in agriculture. The biological control, an eco-friendly approach, includes the use of particular microorganisms to control target phytopathogens and action on parasites, predators or pathogenic agents in controlling or maintaining the population density of another organism at a level lower than that would be present in their absence (Chernin and Chet 2002).

Plant-associated microorganisms are capable to stimulate plant growth by improving bio fertilization, bioremediation, production of phytohormones, and reducing biotic as well as abiotic stress (Mendes et al. 2011; Kumar et al. 2014; Babychan and Simon 2017) (Fig. 3.1). As a biocontrol agent, *Trichoderma* promotes ISR in plants, improves the uptake of nutrients by plants, improves growth and development of roots, promotes plant growth, and enhances crop productivity, increases biotic and abiotic stress resistance and soil remediation (Contreras-Cornejo et al. 2016; Waghunde et al. 2016; Kyriacou and Rouphael 2018). *Trichoderma* spp. is possibly the most commonly used microorganism for agricultural crop development (Rouphael et al. 2017). Root colonization by *Trichoderma* spp. leads to important metabolic variations in the plant and hormonal modifications, as well as phenolic compounds, soluble sugars, photosynthetic rate, amino acids, transpiration, and amount of water content (Zeilinger et al. 2016).

Trichoderma spp. and their metabolites secreted within the rhizosphere influence the growth rate and nutrition of the plant, ISR, and control the phytopathogens (Zeilinger et al. 2016). The mechanisms of biocontrol include competition for space,

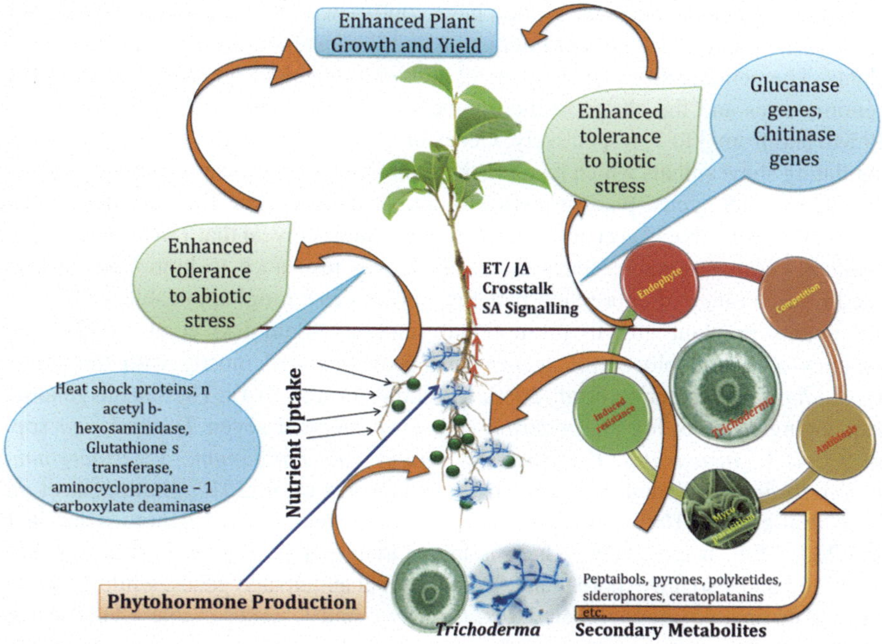

Fig. 3.1 Molecular mechanisms of *Trichoderma* species

resources, nutrients and synthesis, and production of antibiotics and extracellular degrading enzymes such as chitinase, β-1, 3-glucanase that target and break down cell wall of the pathogen resulting in its parasitization (Rai et al. 2016b). *Trichoderma* is an extensively studied genus that presently comprises more than 200 molecularly distinct species (Atanasova et al. 2013a). It is a free-living or saprophytic in soil, rhizosphere, and cellulosic materials; green spored ascomycete fungus with a worldwide distribution (Mukherjee et al. 2013; Waghunde et al. 2016). Members of the genus *Trichoderma* usually parasitize other fungi, saprophytically grow on wood, bark, and other substrates found in soil, interact with animals, plants, marine sponges and antagonistically kill other microbes (Kubicek et al. 2011; Holzlechner et al. 2016). Currently, *Trichoderma* spp. are the most effective biocontrol agents used with about 60% of the recorded bio fungicides all over the world being *Trichoderma* based (Verma et al. 2007) and used as formulations due to their unique plant protecting abilities (Sharma et al. 2015; Oros and Naár 2017). In India, only around 250 bio fungicide products are accessible for field use and have a very meager portion compared to chemical fungicide. Numerous species of *Trichoderma* such as *T. atroviride*, *T. asperellum*, *T. harzianum*, *T. virens*, *T. hamatum*, *T. asperelloides*, and *T. gamsii* are established as potential biological control agents in plant protection and many effective strains have been registered for commercial use in agriculture (Lorito et al. 2010).

In present years, enormous reports have contributed to unraveling the molecular basis of the plant–*Trichoderma* interaction and the resultant positive effects to host plants. The genome size is usually small and with a haploid nucleus. The expected genome sizes and the chromosome numbers of *Trichoderma* spp. array from 3 to 39 Mb and from 3 to 7, respectively. Genes involved in biocontrol play a key role in regulating some signals which result in the production of certain enzymes or proteins that inhibit pathogens, plant growth promotion and therefore they are designated biocontrol genes (Nicolás et al. 2014). Genomic studies reveal that *Trichoderma* spp. contains various valuable genes that help deliver resistance to biotic and abiotic conditions, a range of expression patterns, allows these fungi applicable as biocontrol agents in plant growth promotional activities (Samolski et al. 2012). The genetics of fungal biocontrol agents have been prepared mostly with the genus *Trichoderma* (Mukherjee et al. 2012a; Reithner et al. 2014). The recent genome sequencing projects for *Trichoderma* spp. have targeted seven *Trichoderma* spp. such as *T. atroviride*, *T. reesei*, *T. virens*, *T. harzianum*, *T. asperellum*, *T. longibrachiatum*, and *T. citrinoviride* (Srivastava et al. 2014; Baroncelli et al. 2016; Rai et al. 2016a). Interestingly, *T. atroviride* and *T. viren* genomes are 36.1 and 38.8 Mbp, respectively, which is larger than that of *T. reesei* with a size of 34.1 Mbp and also have more than 2000 additional anticipated genes, while *T. reesei* has 500 distinctive ones compared to *T. atroviride* and *T. virens* (Table 3.1). The aim of the present chapter focuses on the beneficial effects of *Trichoderma* in plant–pathogen interactions and an in-depth understanding of the molecular mechanisms involved.

3.2 Mycoparasitism

Mycoparasitism is a complex process involving a direct attack by fungal species on another (Harman 2000a, b). The consecutive events involved in this process comprise recognition, attack, penetration, and killing of the host fungus. Host recognition by the parasite leads to coiling and appressoria formation, secretion of hydrolytic enzymes aiding penetration of the hyphae, and killing of the host (Holzlechner et al. 2016). This process also includes the secretion of antimicrobial metabolites, finally the captivation and killing of the pathogen (Harman et al. 2004; Omann et al. 2012). Mycoparasitism of plant pathogens by *Trichoderma* spp. has been well investigated and extensively measured to be a main contributing feature to the biocontrol of a range of commercially significant diseases. It is mediated by physical penetration of the mycoparasite into the host hyphae with the aid of specialized structures called haustoria accompanied by the secretion of several degradative enzymes or bioactive metabolites crucial for the breakdown of host fungal structures and finally nutrient uptake from the host (Daguerre et al. 2014).

The remote detection is partly because of the consecutive expression of several fungi toxic pathogenesis-related proteins or hydrolytic enzymes or cell wall degrading enzymes (CWDEs), such as chitinases, glucanases, and proteases

Table 3.1 *Trichoderma* biocontrol genes and their mechanisms

Trichoderma species	Genes/elicitors	Pathogen	Mechanism	Reference(s)
T. arundinaceum	BcBOT	*Botrytis cinerea*	Biocontrol activity	Malmierca et al. (2016)
T. aggressivum	*zhd*	*F. graminearum, F. culmorum*	Zearalenone lactonohydrolase activity	Popiel et al. (2014)
T. arundinaceum	*tri4*	*B. cinerea, R. solani*	Biocontrol activity and induction of plant defense-related genes	Malmierca et al. (2012)
T. asperelloides	*chit36*	*Alternaria radicina, B. cinerea*, and *Alternaria dauci*	Enhanced tolerance	Baranski and Klocke (2008)
T. asperelloides	*chit36 + excy1*	*B. cinerea*	Enhanced tolerance to salinity and heavy-metal stresses	Brotman et al. (2012)
T. asperelloides	*TasSwo*	*B. cinerea and P. syringae*	Stimulating local defense responses in cucumber roots and leaves and affording local protection	Brotman et al. (2008)
T. asperellum	*TasHyd1*	*P. syringae*	Biocontrol activity	Viterbo and Chet (2006)
T. asperellum	*TaACCD*		Enhanced tolerance to salt stress	Zhang et al. (2016)
T. atrovide	*tmk1*	*B. cinerea, R. solani*	Mycoparasitism and plant protection	Reithner et al. (2007)
T. atrovide	*lae1*	*A. solani, B. cinerea Alternaria alternata*	Regulation of asexual development and mycoparasitism	Karimi-Aghcheh et al. (2013)
T. atrovide	*Xyr1*	*B. cinerea, Phytophthora capsici, R. solani*	Induction of systemic resistance in plants	Reithner et al. (2014)
T. atroviride	*chit42 (ech-42, ThEn-42, Tv-ech1, chi1, harchit)*	*Rhizoctonia solani, Alternaria solani, Botrytis cinerea and Alternaria alternata*	*Induced the resistance* and enhanced bio-control activity	Lorito et al. 1998

(continued)

Table 3.1 (continued)

Trichoderma species	Genes/elicitors	Pathogen	Mechanism	Reference(s)
T. atroviride	*chit42 (ech-42, ThEn-42, Tv-ech1, chi1, harchit)*	*Venturia inaequalis*	Increased resistance and reduced plant vigor	Bolar et al. (2001)
T. atroviride	*chit42 (ech-42, ThEn-42, Tv-ech1, chi1, harchit)*	*Penicillium digitatum*	Released endochitinase	Brants and Earle (2001)
T. atroviride	*chit42 (ech-42, ThEn-42, Tv-ech1, chi1, harchit)*	*Alternaria brassicicola*	Increased resistance	Mora and Earle (2001)
T. atroviride	*chit42 (ech-42, ThEn-42, Tv-ech1, chi1, harchit)*	*Phoma tracheiphila and B. cinerea*	Enhanced resistance	Gentile et al. (2007)
T. atroviride	*gluc78*	*Sclerospora graminicola*	Improved resistance	O'Kennedy et al. (2011)
T. atroviride	*chit42 + nag70*	*V. inaequalis*	Increased resistance to reduced plant vigor	Bolar et al. (2001)
T. atroviride	*chit42 + nag70*	*V. inaequalis*	Increased resistance	Schäfer et al. (2012)
T. atroviride	*chit42 + nag70 + gluc78*	*R. solani, Magnaporthe grisea.*	Overexpression of the glucanase alone provokes fatal influence on plant growth	Liu et al. (2004)
T. atroviride	*chit42 + (1,3-1,4)-β-glucanase*	*R. solani*	Mycorrhizal colonization not affected, enhanced to tolerance	Kogel et al. (2010)
T. atroviride	*prb l*	*R. solani, Pythium ultimum. Botrytis cinerea*	Glucose gene insertion, enhances the ISR activity	Brunner et al. (2005)
T. atroviride	*Gpr1*	*B. cinerea, R. solani, S. sclerotiorum*	Antagonistic interaction	Omann et al. (2012)
T. atroviride	*Pks4*	*Alternaria alternata, R. solani, Sclerotinia sclerotiorum*	Pigmentation and stress resistance and in protection against toxins	Atanasova et al. (2013a)
T. atroviride	*Taabc2*	*Beauveria bassiana, B. cinerea, Fusarium* spp., *P. ultimum, R. solani*	ABC transporter membrane pump in the interaction with different plant–pathogenic fungi	Ruocco et al. (2009)

T. atroviride	*nag1*	*B. cinerea*	Carbon starvation is antagonized via a BrlA-like cis-acting element	Brunner et al. (2003)
T. atroviride	*Tga3*	*B. cinerea*	Signal transduction	Zeilinger et al. (2005)
T. atroviride	Tanshinone I and Tanshinone IIA		Promoted growth and tanshinone biosynthesis	Ming et al. (2013)
T. atroviride P1	*Gluc78*	*Pythium and Phytophthora*	Cell wall degradation	Donzelli et al. (2001)
T. Brevicompactum T. arundinaceum	*tri14, tri12, tri11, tri10, tri3, tri4, and tri6*		Trichodermin biosynthesis with strong antifungal activity	Xuping Shentu et al. (2018)
T. brevicompactum IBT40841	*tri5*	*S. cerevisiae, Kluyveromyces marxianus, Candida albicans, C. glabrata, C. tropicalis and Aspergillus fumigates.*	Production of trichodermin and antifungal activity and increases biocontrol activity	Tijerino et al. (2011)
T. gamsii T6085		*Fusarium* spp.	Biocontrol activity	Baroncelli et al. (2016)
T. hamatum	*chit42 and prb1*	*Sclerotinia sclerotiorum*	Mycoparasitic activity	Steyaert et al. (2004)
T. hamatum LU593	Monooxygenase	*S. sclerotiorum, S. minor and S. cepivorum*	Antagonist activity against and enhanced biocontrol activity	Carpenter et al. (2008)
T. harzianum	*prb1 and ech42*	*Sclerotium rolfsii* and *Rhizoctonia solani*	Parasitic activity and regulation of hydrolytic enzymes	Cortes et al. (1998)
T. harzianum	*Ech42*	*Botrytis cinerea and R. solani*	Biocontrol activity	Woo et al. (1999)
T. harzianum	*Tri5*	*Fusarium* spp.	Increases the virulence	Gallo et al. (2004)
T. harzianum	*chit42 (ech-42, ThEn-42, Tv-ech1, chi1, harchit)*	*A. alternata*	Enhanced resistance	Saiprasad et al. (2009)
T. harzianum	*agn13.1*	*B. cinerea*	Significant resistance	Calo et al. (2006)

(continued)

Table 3.1 (continued)

Trichoderma species	Genes/elicitors	Pathogen	Mechanism	Reference(s)
T. harzianum	*bgn13.1*	*C. acutatum*	Enhanced tolerance	Mercado et al. (2007)
T. harzianum	*chit42 + chit33*	*R. solani, Pseudomonas syringae*	Enhanced tolerance to salt and heavy-metal stresses	Dana et al. (2006)
T. harzianum	*chit42 + harcho*	*C. sublineolum*	Increased resistance	Kosambo-Ayoo et al. (2011)
T. harzianum	*chit42 + harcho*	*Erysiphe graminis f.sp. tritici*	Enhanced resistance	Rana et al. (2012)
T. harzianum	*chit42 + StSy + Cu,Zn-SOD*	*Mycosphaerella fijiensis and B. cinerea*	Increased tolerance	Vishnevetsky et al. (2011)
T. harzianum	*Thmfs1*	*A. niger, B. cinerea, F. oxysporum, G. saubinetii, R. solani*	Biocontrol activity	Liu et al. (2012)
T. harzianum	*nox1*	*P. ultimum*	Production of reactive oxygen species to specific biocontrol activity	Montero-Barrientos et al. (2011)
T. harzianum	*bgn13.1*		Enhanced tolerance to crown rot diseases but interferes with plant growth	Mercado et al. (2015)
T. harzianum	*qid74*	*R. solani*	Increased plant biomass through an efficient use of NPK and micronutrients and mycoparasitic activity	Samolski et al. (2012)
T. harzianum	*Epl-1*	*S. sclerotiorum*	Involved in mycoparasitism, resistance induction, and self-cell wall protection	Gomes et al. (2015)
T. harzianum	Trichodiene	*Botrytis cinerea*	Induce systemic resistance in plants against stress	Malmierca et al. (2015)
T. harzianum	Harzianolide	*Sclerotinia sclerotiorum*	Plant growth regulator and systemic resistance elicitor	Cai et al. (2015)

T. harzianum	*Sm1*	*Biotic/abiotic stress*	Elicitor for triggering of plant defense	Freitas et al. (2014)
T. harzianum	*AOC3*, *PDF1.2* and *ERF2* genes	*Sclerotinia sclerotiarum*	Induced Systemic Resistance	Alkooranee et al. (2017)
T. harzianum	PAL1, *chit1*, β1,3-Glucanase, PR- 1, LOX 1 gene	*Fusarium oxysporum f. sp. radicis cucumerinum Botrytis cinerea*	Induced systemic resistance	Alizadeh et al. (2013)
T. harzianum CECT 2413	*ThPTR2*	*B. cinerea*	Enhances mycoparasitic activity and induces peptide transportation	Vizcaino et al. (2006)
T. harzianum CECT 2413	*Thctf1*	*R. solani, F. oxysporum and B. cinerea*	Antifungal activity and production of 6-pentyl-2H-pyran-2 and enhanced biocontrol activity	Rubio et al. (2009)
T. harzianum CECT 2413	*exc1 and exc2, chit42 and chit33 gene, prb1 and bgn13.1*	*F. oxysporum*	Mycoparasitic activity against and enhanced biocontrol activity and Expression of this gene helps in regulation of hydrolytic enzymes	Lopez-Mondejar et al. (2011)
T. harzianum Rifai	*qid74*	*R. solani*	Antagonism activity and mycoparasitic activity	Rosado et al. (2007)
T. harzianum T34	*ThPG1*	*R. solani and P. ultimum*	Secretion of plant cell wall degrading enzymes and enhanced biocontrol activity	Moran-Diez et al. (2009)
T. longibrachiatum	*Egl1*	*Pythium ultimum*	Enhanced biocontrol activity	Migheli et al. (1998)
T. longibrachiatum	*Hytlo1*		Established a mutually beneficial interaction with the colonized plant	Ruocco et al. (2015)
T. reesei	*gna3*	*P. ultimum*	Production of cell wall-degrading enzymes and mycoparasitism activity	Silva et al. (2009)
T. reesei	*pks4*	*A. alternata, R. solani, S. sclerotiorum*	pigmentation and stress resistance and in protection against toxins	Atanasova et al. (2013a)
T. virens	*TmkA*	*S. rolfsii and R. solani*	Shows increased biocontrol activity	Viterbo et al. (2005)

(continued)

Table 3.1 (continued)

Trichoderma species	Genes/elicitors	Pathogen	Mechanism	Reference(s)
T. virens	*TgaA*, *TgaB*	*S. rolfsii and R. solani*	Increases virulence in the plant–pathogenic interactions.	Mukherjee et al. (2004)
T. virens	*tvsp1*	*R. solani*	Biocontrol activity	Pozo et al. (2004)
T. virens	*Tac1*	*R .solani*, *S. rolfsii*, *Pythium* spp. *R. solani and P. ultimum*	Mycoparasitism and production of secondary metabolism	Mukherjee et al. (2007)
T. virens	*chit42 (ech-42, ThEn-42, Tv-ech1, chi1, harchit)*	*A. alternate*, *R. solani*	Increased resistance	Emani et al. (2003)
T. virens	*chit42 (ech-42, ThEn-42, Tv-ech1, chi1, harchit)*	*R. solani*	Enhanced resistance	Shah et al. (2009)
T. virens	*chit42 (ech-42, ThEn-42, Tv-ech1, chi1, harchit)*	*B. cinerea and Sclerotinia sclerotiorum*, *A. alternata*	Enhanced tolerance	Shah et al. (2009)
T. virens	*pacC*	*R. solani*, *S. rolfsii*	Antifungal activity	Trushina et al. (2013)
T. virens	*Vel1*	*P. ultimum*, *R. solani*	Morphogenesis and biocontrol properties	Mukherjee and Kenerley (2010)
T. virens	*pks4*	*A. alternata*, *R. solani*, *S. sclerotiorum*	Pigmentation and stress resistance and in protection against toxins	Atanasova et al. (2013a)
T. virens	*GliC gliI*, *gliF*	*P. ultimum*, *R. solani*	Mycoparasitism	Atanasova et al. (2013b)
T. virens	*gliK*	*P. ultimum*, *R. solani*	Mycoparasitism	Atanasova et al. (2013b)
T. virens	*gliM*	*P. ultimum*, *R. solani*	Mycoparasitism	Atanasova et al. (2013b)

T. virens	*ppt1*	*A. solani*, *B. cinerea*, *F. oxysporum*, *Fusarium* spp., *Phytophthora capsici*, *R. solani*, *S. cepivorum*, *S. rolfsii*	Secondary metabolism and induction of plant defense responses	Velázquez-Robledo et al. (2011)
T. virens	*Psy1*	*R. solani*	Biocontrol activity	Wilhite et al. (2001)
T. virens	4-phosphopantetheinyl transferase	*R. solani*	Biocontrol activity	Velázquez-Robledo et al. (2011)
T. virens	*ech42*	*Alternaria*	Enhanced tolerance	Kamble et al. (2016)
T. virens	*Sm2*	*Cochliobolus heterostrophus*	Important for plant protection	Gaderer et al. (2015)
T. virens	*tac1*	*R. solani and P. ultimum*	Mycoparasitic activity	Abbas et al. (2017)
T. virens IMI 304061	*TmkA*	*R. solani*	Induction of plant systemic resistance and biocontrol activity	Viterbo et al. (2005)
T. viride	*Xyn2/Eix*	*B. cinerea*	Elicits ET biosynthesis and hypersensitive response	Rotblat et al. (2002)

(Harman et al. 2004). Approximately 30% of the dry weight of the fungal cell is attributed to the presence of chitin, β-1, 3-glucans, and α-1, 3/1, 4-glucans. The biosynthesis of CWDEs is implicated in mycoparasitism which is regulated mainly at the transcriptional level and accountable genes are present as single-copy genes. Overall 20–30 genes, proteins, and other metabolites have a direct involvement in this communication. Morphological modifications and transformation to improve the copy number of these genes have been employed to overproduce these enzymes (Lu et al. 2004). The functions of different CWDEs in the course of mycoparasitism by *Trichoderma* spp. using a gene for gene approach and future studies will help in a better understanding of the process (Daguerre et al. 2014).

Since many of the lytic enzymes secreted by biocontrol agents are encoded by a single gene, it is considered to be a straightforward technique to isolate some of these genes and then transfer them to other biocontrol agents. The CWDEs are extracellular proteins with low molecular weight and high stability. Several forms or isozymes of a particular enzyme may be secreted that vary in size, regulation, and capacity to break down the cell walls of phytopathogens (Vos et al. 2015). Over 1100 *Trichoderma* spp. have been described containing 75 molecularly defined mycoparasitic against different plant pathogenic fungi (Druzhinina et al. 2011). Volatile secondary metabolites have also been implicated in mycoparasitism by *Trichoderma* spp. (Stoppacher et al. 2010).

3.3 Chitinases

Chitinases are among the most significant lytic enzymes produced by *Trichoderma*, which complete lysis of walls of fungal mycelia or conidia of phytopathogens. These chitinases are hydrolases that break down one of the major constituents of the fungal cell wall, chitin, a polymer composed of repeating units of *N*-acetyl-D-glucos-2-amine, linked by β-1, 4 glycosidic bonds (Bhattacharya et al. 2007). These are separated into β-*N*-Acetylhexosaminidases (GlcNAcases), endochitinases, and exochitinases. Endochitinases degrade chitin at interior sites releasing chitotriose, chitotetraose, and chitobiose. Exochitinases are further divided into chitobiosidases and *N*-acetyl-β-glucosaminidases (Prakash et al. 2010). Chitobiose, chitotriose, and chitotetraose are degraded into *N*-acetylglucosamine monomers by GlcNAcases in a similar manner to exochitinase. Genetic alteration of plant species with mycoparasitic genes from *Trichoderma* spp. signifies an advanced method of disease resistance. There are perhaps at least 30 chitinases alone, each with different genes and protein composition. Chitinase gene has been transmitted to several crops for developing fungal disease resistance. More specifically, De la Cruz et al. (1992) cloned *chit33* chitinase gene from *Sclerotinia sclerotiorum* and cloned *ech42* chitinase gene from *T. harzianum* (Garcia et al. 1994). A relative investigation of chitinases exposed that *Trichoderma* genomes harbor around 20 and 36 different genes encoding chitinases. The chitinolytic capability in *Trichoderma* is associated with varied chitinase genes including *ech42*, *chi33*, *nag1*, *chi18-13*, where these diverse enzymes could confer advanced mycoparasitic action against phytopathogens (Seidl et al. 2005).

Based on the previous investigations, the presence of fungal cell walls or colloidal chitin, as well as carbon starvation, induce the genes encoding endochitinase 42 (*ech42*), endochitinase 33 (*chit33*) and *N*-acetyl-β-D-glucosaminidase (*nag1*) (Peterbauer et al. 1996; Margolles-Clark et al. 1996). The expression of *ech42*, related to light-induced spore germination was suppressed by carbon catabolites (Lorito et al. 1996); whereas *N*-acetyl-β-D-glucosamine (GlcNAc) induced the transcription of *nag1* (exochitinase) (Peterbauer et al. 1996). Effective transformation and expression of several endochitinase genes, for example, *chit42* and *chit33* from *T. harzianum* have improved fungal tolerance in crops such as *Brassica juncea*, potato, apple, broccoli, rice, carrot, and lemon (Kamble et al. 2016; Lorito et al. 1998; Bolar et al. 2001; Mora and Earle 2001; Liu et al. 2004; Baranski et al. 2008; Distefano et al. 2008). Transgenic tomato plants overexpressing *chi194*, a wheat chitinase gene, under the control of maize ubiquitin 1 promoter have been reported (Girhepuje and Shinde 2011). Mishra et al. (2016) reported the transfer of a *Trichoderma* endochitinase gene into a guava plant (*Psidium guajava*). The gene, *ech42* in *T. harzianum*, encoding endochitinase was studied and cloned into pAN7-1 vector. The antifungal action was established against *B. cinerea* and *R. solani* pathogens using the wild type and disruptant strains (Woo et al. 1999). Genes *chit42* and *chit33* coding chitinase in *T. harizianum* play a key role in the mycoparasitic action against the phytopathogens, particularly *F. oxysporum* (Mondejar et al. 2011). The co-transformation of apple with nag70 (*nag1*) and *ech42* resulted in a synergistic rise in biocontrol activity against *Venturia inequalis* (Bolar et al. 2001). A dramatic increase in disease resistance of potato and tobacco against *A. alternata*, *A. solani*, *B. cinerea*, and *R. solani* was observed with the combination of *T. harzianum* and *T. atroviride* endochitinase *ech42*.

3.4 Glucanases

Glucanases are another class of cell wall degrading enzymes with a key role in mycoparasitism. Glucans are glucose polysaccharides that cross link chitin or chitosan polymers. Based on the chemical bonding among glucose subunits there are two types of glucans. β-glucans are defined by β-(1, 3) or β-(1, 6) bonds and afford rigidity to the cell wall. α-glucans are considered by α-(1, 3) and/or α-(1, 4) bonds and function as a part of the matrix. The second most plentiful polymer in fungal cell walls is β-1, 3-glucan (Latge 2007) with β-1, 6- branches, which are broken down by β-1, 3-glucanases. In the genomes of *Trichoderma* spp., genes encoding this class of enzymes are over represented when compared to other related fungi (Kubicek et al. 2011; Geraldine et al. 2013; Vos et al. 2015). β-1, 6-glucanases have been identified in the area of contact between *Trichoderma* spp. and its prey. In *T. harzianum* CECT 2413, the overexpression of *Bgn16.3* encoding β-1, 6-glucanase resulted in a more effective biocontrol agent with growth-inhibitory action on

B. cinerea, *R. solani*, and *Phytophthora citrophthora* (Montero et al. 2007). The *Bgn16.2* showed antifungal activities individually or in combination with other chitinases resulting in impairing the growth of *B. cinerea* and *Gibberella fujikuroi* (De la Cruz and Llobell 1999). Strains of *T. harzianum* and *T. virens* overproducing β-1, 6-glucanases were more effective in the biocontrol of *R. solani*, *B. cinerea* (Ihrmark et al. 2010), and *P. ultimum* (Djonovic et al. 2006).

Inhibition of spore germination or the growth of phytopathogens by β-1, 3-glucanases is in synergistic cooperation with chitinases (El-Katatny et al. 2001) as well as antibiotics (Harman et al. 2004). Numerous β-1, 3-glucanases have been identified, but only a few genes have been cloned; those are *lam1.3* (Cohen-Kupiec et al. 1999) from *T. harzianum*, *bgn13.1* (Benitez et al. 1998) and *glu78* (Donzelli et al. 2001) from *T. atroviride*, and *Tv-bgn1* and *Tv-bgn2* from *T. virens* (Kim et al. 2002). Increased biocontrol of *T. virens* against *R. solani*, *P. ultimum*, and *R. oryzae* was reported using co-overexpression of two β -glucanases *Bgn2* and *Bgn3* genes (Djonovic et al. 2007). Overexpression of *bgn13.1* in transformants has been described as inhibitory to the growth of *B. cinerea*, *R. solani*, and *P. citrophthora*. Transformant T28, with maximum *bgn13.1* glucanase activity in repressing as well as inducing situations, displayed strong suppression of pathogens. Expression and secretion of endo-β-1, 3-glucanase, *bgn13.1* in *T. harzianum* was noticed when grown on fungal plant pathogen cell walls (De la Cruz et al. 1995). The *Gluc78* from *T. atroviride* P1 revealed strong antimicrobial action against an array of fungi and oomycetes including *Pythium* and *Phytophthora*; the activity was in synergy with other enzymes. *Tv-bgn1* and *Tv-bgn2*, these glucanases have been identified and cloned (Donzelli et al. 2001). In *T. atroviride gluc78* gene coding for an antifungal glucan 1, 3-β-glucosidase was identified, cloned, and sequenced. The pGEM-T vector was used for cloning *gluc78* gene and the expression study carried out against the phytopathogens *R. solani* and *P. ultimum* (Donzelli et al. 2001).

T. asperellum α-1, 3-glucanase *agn13.2* and *T. harzianum* β-1, 6-glucanase *bgn16.2* have been reported with antifungal activity against *B. cinerea* (Sanz et al. 2005). Three α-1, 6-glucanases have been isolated from *T. harzianum* 2413 strain (Elad et al. 2000). *T. longibrachiatum* transformants exhibiting overexpression of β-1, 4-endoglucanase gene *egl1* showed biocontrol activity against *P. ultimum* in cucumber. Among 31 *T. harzianum* isolates, five of them *T*30, *T*31, *T*32, *T*57, and *T*78 encoded genes for *N*-acetyl-β-D-glucosaminidase (*exc1 and exc2*), chitinase (*chit42* and *chit33*), protease (*prb1*), and β-glucanase (*bgn 13.1*) which were cloned and expressed. These genes are critical in the mycoparasitic activity against the phytopathogenic fungi particularly *F. oxysporum* (Lopez-Mondejar et al. 2011). The adenalyte-cyclase encoding gene in *T. virens* termed as *tac1* gene was isolated and cloned and its role in mycoparasitic activity against *R. solani* and *P. ultimum* has been studied (Mukherjee et al. 2007). The *qid74*gene identified in *T. harzianum* CECT 2413 plays a significant role in cell protection and offers adherence to hydrophobic exteriors aiding the fungus in mycoparasitic activity against *R. solani* (Rosado et al. 2007). A gene *Taabc2* cloned from *T. atroviride* has a crucial role in ATP binding cassette (ABC) transporter in cell membrane pump that benefits in the mycoparasitic activity against *R. solani*, *B. cinerea*, and *P. ultimum* (Ruocco et al. 2009).

The *tag83* gene encoding exo-β-1, 3-glucanase enzyme was identified from *T. asperellum* and the expression of this gene exhibited parasitic activity against pathogens such as *R. solani* (Marcello et al. 2010). Two different types of β-1, 3 and β-1, 6 glucanase genes such as *TvBgn2* and *TvBgn3* transformants were expressed from *T. virens*. These genes secrete CWDEs that helps in the biocontrol activity. The glucose repressor gene *creI* from *T. harzianum* was isolated and characterized, and cloned using *pTZ57R/T* plasmid vector followed by transformation into *E. coli* DH10B and its role in the expression of cellulase and xylanase were studied (Saadia et al. 2008). Cellulase and xylanase are the major type of enzymes that involve in the cell wall degradation of the phytopathogens.

3.5 Proteases

Fungal proteases also play an important role in cell wall degradation and cleavage of peptide bonds in proteins (Haggag et al. 2006). Certain proteases secreted by *Trichoderma* spp. may be involved in the inactivation of extracellular enzymes produced by phytopathogenic fungi. Numerous studies substantiate the role of extracellular proteases in improved biocontrol efficiency of *T. virens*, *T. harzianum*, *T. asperellum*, *T. flavus* against pathogenic fungi and oomycetes such as *R. solani*, *F. oxysporum*, *B. cinerea*, *S. sclerotiorumor*, *P. ultimum*. The maximum mycoparasitic protease genes cloned so far is from *Trichoderma* spp. genes. The genes encode numerous serine proteases with subtilisin-like, chymotrypsin- or elastase-like, and trypsin-like activity and aspartic proteases. *T. virens Tvsp1* and *T. atroviride Prb1* are serine proteases (Pozo et al. 2004), while *T. asperellum TaAsp* and *T. harzianum* Sa76 are aspartic proteases (Yang et al. 2013). A novel serine protease gene from *T. harzianum* named *SL41* has been cloned and expressed effectively in *S. cerevisiae*. The cDNA of *Sl41* gene was sequenced and it was cloned in pMD18-T vector and was inserted into *E. coli* DH5-α (Liu et al. 2009). Numerous genes coding proteases and oligopeptide transporters are expressed earlier and during contact with the prey in different *Trichoderma* species (Seidl et al. 2009). A richness of genes encoding subtilisin-like serine proteases was also detected in a study of expressed sequence tags (ESTs) accumulated through the commencement of contact between *T. atroviridis* and its fungal preys *Rhizoctonia solani* and *S. sclerotiorum* (Seidl et al. 2009). The Protease *pra1* from *T. harzianum* isolate has an affinity for fungal cell walls (Elad et al. 2000) and this gene displays great potential in increasing biocontrol capacity, as serine proteases are active against oomycetes (Howell 2003). The alkaline protease *Prb1* from *T. harzianum* IMI 206040 strain has also been established to play a significant role in biological control efficiency (Benitez et al. 1998) and the *T. harzianum Prb1* gene transformants exhibited upto fivefold increase in the biocontrol effectiveness in the control of *R. solani*.

3.6 Mechanisms of Signal Transduction

Downstream transduction of signals, produced at the receptor sites, is necessary for further expression of genes in the host plants. Three significant signal transduction pathways are recognized in *Trichoderma* spp. that increase the expression of genes involved in biocontrol and mycoparasitism. Signal transduction pathways eliciting the genes involved in mycoparasitism have been deliberated in depth and contain heterotrimeric G-protein signaling, mitogen-activated protein kinase (MAPK) cascades, and the cAMP pathways (Zeilinger and Omann 2007). Adenylate cyclase and G-protein coupled receptors *Trichoderma* spp. are critical for the secretion of extracellular CWDE, production of antifungal metabolites, and development of infection Cyclic adenosine monophosphate (cAMP) is a significant regulator of structures. A positive trigger in the activity of adenylate cyclase by G-protein α-subunits *Tga3* and *Gna3* consequently improved mycoparasitism (Daguerre et al. 2014). Heterotrimeric G proteins contain a, b, and ɣ subunits are involved in transducing signals from transmembrane G protein-coupled receptors to a variability of intracellular targets. Depending on the system, Ga or Gbɣ transduces the signal by stimulating effectors such as adenylate cyclase or the MAPK cascade (Kaziro et al. 1991).

Cyclic adenosine monophosphate (cAMP) is a significant regulator of development, growth, and pathogenicity in filamentous fungi (Liebmann et al. 2003). The cAMP mediated signaling is a significant pathway in fungi in controlling the diversity, virulence, sexual development, nutritional status, stress, transcription, and cell cycle development (Kronstad et al. 1998). In most fungi, the adenylate cyclase activity is under the control of subunits of heterotrimeric G-proteins. The cAMP usually stimulates a cAMP-dependent protein kinase (PKA) that is composed of two regulatory and two catalytic subunits (Dickman and Yarden 1999), and the gene expression is regulated by means of phosphorylation of transcription factors. Lin et al. (2012) investigated the association of anthraquinone secondary metabolites emodin and pachybasin in the self regulation of coiling in *T. harzianum*. The addition of both of these *T. harzianum* derived metabolites improved the number of coils of the mycoparasite around hyphae of *R. solani* and resulted via stimulation of cAMP production. The detailed investigation of two genes in the heterotrimeric G protein signaling pathway such as the class I G-α subunits *Tga1* of *T. atroviride* and *TgaA* of *T. virens*, as well as the class III G-α subunits *Tga3* of *T. atroviride* and *Gna3* of *T. reesei*, have confirmed the functions of these genes are associated with biocontrol activity. The gene *Tga1* was reported crucial in the production of antifungal metabolites and regulation of coiling around the pathogenic hyphae (Rocha-Ramírez et al. 2002; Zeilinger et al. 2005). *TgaA* has a host-specific connection associated with the activity of MAP kinases while *Tga3* was found to be noteworthy for biocontrol activities.

Mitogen-activated protein kinase pathways transduce a great range of signals, containing those connected with pathogenesis. MAPK pathways signify one of the most prominent signal transduction systems in fungi. Numerous MAPKs convoluted in fungal mycoparasitism have been identified in *Trichoderma* spp. which harbor

MAPKKK, MAPKK, and MAPK signaling pathways, the three MAPK cascades which might act in mycoparasitism and biocontrol activity (Reithner et al. 2007; Kumar et al. 2010). The MAPKs in *Trichoderma* belong to the family of yeast and fungal extracellular related kinases (YERK1); other MAPKs include Pmk1 from *M. grisea*, *Fmk1* from *F. oxysporum*, *Bmp1* from *B. cinerea* or Ubc3/Kpp2 from *U. maydis*. The three MAPKs genes in the *Trichoderma* genome encode the so-called virulence MAPK (*TmkA/Tvk1*) ortholog of the pathogenicity related MAPKs of phytopathogens, the cell integrity kinase (*TmkB*), and the osmoregulatory MAPK (*Hog1*).

The expression levels of mycoparasitism-related genes (MGRs) in the MAP kinase encoding gene mutant of a *Trichoderma* strain raised during mycoparasitism when in direct contact with *R. solani*. The regulation of MGRs in *T. virens* is very complex; however, they share common elements including *Tvk1* like other fungi (Mendoza-Mendoza et al. 2003). The MAPK from *T. atroviride* (*Tmk1*) on characterization showed 98% similarity to *T. virens TmkA/Tvk1* (Reithner et al. 2007). *Δtmk1* mutants showed a reduction in radial growth and the conidiation was light-independent. The direct plate confrontation analyses against the pathogens *R. solani* and *B. cinerea* as hosts revealed that *T. atroviride Tmk1*—similar to *T. virens TmkA*—affected the host specificity as *Δtmk1* mutants had the ability to parasitize *R. solani* whereas they failed to attack *B. cinerea*. The *TmkA* mitogen-activated protein kinase from *T. Virens* is known to cause mycoparasitic activity to *R. solani* and *S. rolsfii* (Mukherjee et al. 2003). MAP kinase cascade connecting MPK4, MPK3, MPK11, and MPK6 and additional genes containing Ca^{2+} reliant proteinase kinases are triggered to found PTI (Bethke et al. 2012).

3.7 Competition

Starvation is a general cause of death of soilborne microorganisms (Benitez et al. 2004), so competition for limited nutrients is especially significant in the biocontrol of phytopathogens. Competition is the phenomenon in which the introduced biocontrol agent, i.e., *Trichoderma* and the pathogen compete for the obtainability of nutrients and space (Hjeljord et al. 2000). In most of the filamentous fungi, iron and carbon are two vital elements, essential for viability. This process could be connected also to the production of organic acids, such as gluconic, citric, and fumaric acids, which reduce soil pH and allow the solubilization of phosphates, micronutrients, and mineral cations like iron, manganese, and magnesium (Vinale et al. 2008a). The *Trichoderma* spp. displays natural resistance to fungicides, herbicides, and phenolic compounds and various toxic chemicals. *Trichoderma* spp. can, therefore, grow quickly and influence pathogens with the production of metabolic compounds that inhibit spore germination of the pathogen (fungistasis), cause death of the pathogen (antibiosis), or alter the conditions of the rhizosphere (Benitez et al. 2004). The disease inhibition activity of *Trichoderma* spp. is exerted either directly by obstructing growth and development of soilborne pathogens through competition for nutrients or excretion of antibiotics in the rhizosphere (Bakker et al. 2007; Sultana et al. 2009) or indirectly by stimulating a plant-mediated systemic resistance (van Wees et al. 2008). In their investigation, Lehner et al. (2013) describe the detection of around 12–14 siderophores in *T. atroviride*, *T. asperellum*, *T. gamsii*, *T. hamatum*, *T. virens*, *T. harzianum*, *T. polysporum*, and *T. reesei* by isotope-based screening using dimerum acid, coprogren, fusigen, fusarinine A, and the intracellular siderophore ferricrocin being produced by all species (Lehner et al. 2013).

3.8 Competition for Nutrients

Iron acts as a cofactor of several enzymes and an essential nutrient for the growth of plants and other microorganisms. Iron attainment is a significant component of microbial competition, particularly within the rhizosphere, where there is intense microbial activity. The biocontrol agent *Trichoderma* spp. may show rapid growth or utilize the available food source more efficiently in comparison to the phytopathogens, thereby suppressing the pathogen growth and taking over. This process is termed as competition for nutrients. The ability of *Trichoderma* spp. to scavenge iron from the environment makes it unavailable for the competing pathogens. Certain *Trichoderma* isolates produce highly efficient siderophores, iron-chelating compounds which bind with insoluble iron (FeIII) and converted to soluble form (FeII) for plant absorption and stop the growth of phytopathogens by depriving them of iron sources (Benitez et al. 2004). *Trichoderma* spp. are known to produce extracellular siderophores of the fusigen and coprogen family. Several *Trichoderma* spp., such as *T. viride*, *T. harzianum*, and *T. lignorum* are well-known siderophore producers better than the pathogenic strains of *Fusarium* such as *F. solani* and *F. oxysporum* (Dutta et al. 2006).

Competition for iron has been found to be among the critical factors in the antagonism of *T. asperellum* against *F. oxysporum* and may as well be beneficial for plants due to the iron solubilizing activity (Segarra et al. 2010). *T. virens* and *T. reesei* harbor an extra putative gene cluster for siderophore production (Mukherjee et al. 2012b). *T. virens* and *T. reesei* harbor two putative gene clusters covering an NRPS as the core member, whose orthologs (*SidD* and *NPS6*) are known to be intricate in siderophore production (Kubicek et al. 2011). During iron deprived situations, the synthesis of these iron scavenging siderophores are under the impact of HapX protein that specifically binds to CCAAT binding complex (CBC) (Thon et al. 2010). *T. harzianum* CECT 2413 strain encodes a high-affinity glucose transporter (Gtt1) and interestingly the Gtt1 gene is only expressed during very low glucose concentrations similar to the development of competence among microorganisms (Benitez et al. 2004). *Gtt1*, a high-affinity glucose transporter of the mycoparasitic fungus *T. harzianum*, has been characterized (Delgado-Jarana et al. 2003). Vargas et al. (2009) reported an intracellular invertase *TvInv* from *T. virens* that is involved in sucrose hydrolysis signifying the plant-derived sucrose as a vital nutritional resource to *Trichoderma*.

3.9 Competition for Root Colonization

From the standpoint of microbes, surfaces of living plants and soils are often nutrient-limited environments. Colonization of the root tissue is generally confined to the penetration of the first or second layers of cells and to the intercellular spaces (Brotman et al. 2008). Proteins of *Trichoderma* spp. involved in root colonization can act as MAMPs (Lorito et al. 2010; Ruocco et al. 2015). For example, swollenin, a protein encoded by *TasSwo* gene, induces defence responses in cucumber roots and leaves affording local defense against plant pathogens (Brotman et al. 2008) and endopolygalacturonases endoPGs produced by *Trichoderma* spp. aid in root penetration and constitute a preeliciting role in ISR (Baroncelli et al. 2015). Further, root penetration is accomplished via the secretion of cellulolytic, hemicellulolytic, and proteolytic enzymes (Viterbo et al. 2004).

Hydrophobins and expansin-like proteins (Brotman et al. 2008; Ruocco et al. 2015) are essential for the adherence to the root surface by *Trichoderma* spp. and in cell wall development, respectively. These are small secreted proteins that have a distinctive domain of eight cysteine residues at conserved positions. Hydrophobins were primarily separated into class I and class II hydrophobins according to their hydropathy patterns and solubility (Linder et al. 2005). *T. asperellum* harbors the *TasHyd1* hydrophobin gene, which has been revealed to support in plant root colonization, enabling the attachment of hyphal filaments to hydrophobic root surfaces (Viterbo and Chet 2006; Guzmán-Guzmán et al. 2017). Among the three hydrophobin genes *Hyd1*, *Hyd2*, and *Hyd3* recently identified in the fungus, only *Hyd3* is implicated in root colonization by *C. rosea* (Dubey et al. 2014). Hydrophobins in phytopathogenic fungi are necessary to anchor fungal cells to host plant surfaces and they could play a similar role in biocontrol agents such as *T. asperellum* and *C. rosea*.

Plant lytic enzymes involve actively in root colonization, similar to endopolygalacturonase *ThPG1* from *T. harzianum* and expansin-like proteins capable of recognizing cellulose swollenin *TasSwo* have also been revealed to be involved in plant root colonization (Moran-Diez et al. 2009). In *T. asperellum*, xylanases *Abf1* and *Abf2* along with proteases *PapA* and *PapB* are secreted in response to cucumber root attachment (Viterbo et al. 2004). The role of xylanases in plant root colonization by *Trichoderma* is not directly confirmed but these enzymes are upregulated during *Trichoderma*–plant interactions. Biochemically diverse microbe-associated molecular patterns (MAMPs) have been identified in *Trichoderma* (Shoresh et al. 2010), including the ceratoplatanin protein *SM1/Epl1* (Frischmann et al. 2013), ethylene-inducing xylanase (Ron and Avni 2004), and Swollenin protein from *T. asperellum*. *Epl1* has been defined with the ability to generate defense responses in plants (Salas-Marina et al. 2015; Ramada et al. 2016). The *SM1* is induced and expressed not just during plant interactions but also in the absence of a plant and further promotes the expression of genes related to pathogenesis and hypersensitive reactions (Djonovic et al. 2007).

3.10 Production of Antibiotics and Secondary Metabolites

Antibiosis occurs during microbial interactions and involves low molecular weight diffusible secondary metabolites (SMs) or antibiotics produced by *Trichoderma* strains that are detrimental for the growth of plant pathogen (Benitez et al. 2004; Viterbo et al. 2007). Fungal antibiosis is associated with the production of antibiotics and/or hydrolytic enzymes and secondary metabolites related to possible competition for nutrients in the rhizosphere and microbial antagonism (Harman et al. 2004). Antibiotics and secondary metabolites produced by *Trichoderma* spp. are crucial in their biocontrol activity (Ajitha and Lakshmedevi 2010). Secondary metabolites including antibiotics are not directly involved in the natural growth, development, or reproduction of the fungus. They are chemically dissimilar from natural compounds and may play important roles in the defense response, competition against other microorganisms, symbiosis, metal transport, differentiation, and stimulation or inhibition of spore formation and germination, etc. (Reino et al. 2008).

Based upon analytical studies, from the genus *Trichoderma* about 180 secondary metabolites (natural products) have been identified, representing various classes of chemical compounds and with the structures of more than 100 compounds described (Reino et al. 2008). Several molecules involved in the suppression of numerous soilborne plant pathogens have been described (Benitez et al. 2004). The communication between *Trichoderma* and their plant hosts is established by complex chemical interaction comprising volatile and diffusible secondary metabolites, small peptides, and/or antibiotics, which affect root growth, branching, and absorptive capacity (Lopez-Bucio et al. 2015). *Trichoderma* spp. produces several secondary metabolites, antibacterial and antifungal antibiotics which comprise volatile and nonvolatile toxic metabolites such as harzianic acid, alamethicins, tricholin, peptaibols, 6-*n*-pentyl-6H-pyran-2-one (6PP/6-PAP), formic aldehyde, acetaldehydes gliotoxin, viridian, Terpenoids, harzianopyridone, harziandione, massoilactone, viridin, gliovirin, glisoprenins, trichodermin, heptelidic acid, epipolythiodioxopi perazines (ETPs) (Gajera et al. 2013; Hermosa et al. 2014; Strakowska et al. 2014).

Various genes are components of large biosynthetic gene clusters harboring those encoding core enzymes such as polyketide synthases (PKSs), nonribosomal peptide synthetases (NRPSs), accessory enzymes and genes for transporters and transcription features (Bansal and Mukherjee 2016a). Genomes of some more mycotrophic species including *T. asperellum*, *T. parareesei*, *T. harzianum*, *T. gamsii*, and the opportunistic human pathogens *T. longibrachiatum* and *T. citrinoviride* were subsequently added to the public databases (Baroncelli et al. 2016). The hydrolytic enzymes along with antibiotics results in an advanced intensity of antagonism than that achieved by either mechanism singly (Monte 2001). Synergetic effects between an endochitinase from *T. harzianum* and gliotoxin and that of hydrolytic enzymes and peptaibols on conidial germination of *B. cinerea* have been reported (Howell 2003). A peptaibol synthetase from *T. virens* has recently been identified and the corresponding gene, which has been cloned, will facilitate studies of this compound and its contribution to biocontrol.

The genes involved in secondary metabolite biosynthesis in *Trichoderma* are present as clusters that can span more than 10 kb, with a few exceptions (Lo et al. 2012). These clusters encode the enzyme complexes such as the NRPS or PKS that comprise of various domains and modules with distinct activities (Strieker et al. 2010). The synthesis of the structural backbone of these unique secondary metabolites by PKS and NRPS utilizes building blocks such as malonyl groups and amino acids or their derivatives (Brakhage and Schroeckh 2011). The genes crucial in the biocontrol mechanisms of *Trichoderma* are of great value. The vast prospective of *Trichoderma* spp. to produce an array of diverse metabolites is reflected in the genomes of the species. Secondary metabolite genes of *Trichoderma* are organized just about the signature genes which encode NRPSs, PKSs, and terpene synthases, which define the biosynthetic pathways and clusters (Osbourn 2010).

3.11 Non-ribosomal Peptide Synthases (NRPSs)

The genome of *Trichoderma* is a repertoire for secondary metabolite production, including both beneficial and a few toxic compounds, which have been well characterized and few novel (Mukherjee et al. 2012b). Polyketide synthases and NRPSs are two major classes of secondary metabolites (Baker et al. 2012). NRPSs are large modular enzymes involved in the synthesis of Nonribosomal peptides (Mukherjee et al. 2012c). NRPS enzymes are composed of a series of modules that behave like an assembly line, each incorporating one monomer into the peptide (Strieker et al. 2010). The monomers may be peptaibols or even compounds that are non-amino acids. The peptides may be structurally linear or cyclic, and often go through large chemical modifications (Strieker et al. 2010). Peptaibols fit into the antifungal armory of *Trichoderma* and are now reported to trigger the apoptotic death of the host. *Trichoderma* spp. synthesize NRPSs, the large multifunctional enzyme domains that assemble various compounds using a diverse precursors such as non-proteinogenic amino acids and hydroxy or carboxyl acids (Mukherjee et al. 2011; Shi et al. 2012). Genes encoding hydrolytic enzymes like chitinases and glucanases and those for SMs like NRPSs are concurrently expressed to destroy the plant pathogens (Kubicek et al. 2011). Numerous NRPSs implicated in the synthesis of peptaibols in *Trichoderma* spp. have been recognized (Mukherjee et al. 2011). However, the characterization of NRPSs from additional biological control agents is still lacking.

3.12 Peptaibols

Peptaibols are short-chain linear polypeptides that generally exhibit strong antimicrobial effects against bacteria and fungi, and act in synergy with CWDEs inhibiting the growth of fungal pathogens and rendering the plant resistant to phytopathogens (Mukherjee et al. 2011). Peptaibols produced largely by members of *Trichoderma* are peptides composed of α-aminoisobutyric acid and a C-terminal 1, 2-amino alcohol constituting the major group which is characterized by an acylated N-terminus and an amide-bound amino alcohol at the C-terminus (Degenkolb et al. 2008). About 1000 various peptaibiotics that have been recognized and categorized into numerous groups on the basis of their chemical constructions and these include lipoaminopeptides, lipopeptaibols, peptaibols, and cyclic peptaibiotics (Neumann et al. 2015).

Trichoderma spp. are usually considered as the richest source of peptaibols and over 80% of the entries in the Comprehensive Peptaibiotics Database can be assigned to this fungal genus with *T. viride*, *T. brevicompactum*, *T. virens*, *T. parceramosum/T. ghanense*, and *T. harzianum* being the most extensively studied species (Stoppacher et al. 2013; Neumann et al. 2015). The biocontrol activity of peptaibols originates from their capacity of membrane altering properties, formation of pores in lipid membranes, as well as induction systemic resistance in plants against pathogens attack (Mukherjee et al. 2011). Numerous NRPSs involved in the synthesis of peptaibols in *Trichoderma* spp. have been studied (Mukherjee et al. 2011). There are two peptaibol synthetases such as of 18 and 14 modules in *Trichoderma* genomes and more than 700 peptaibol sequences are known, generally of *Trichoderma* origin (Degenkolb et al. 2008).

The genome of ITEM 908 harbors three loci with sequences encoding the homologs of potential peptaibol synthetases in *T. virens* (Mukherjee et al. 2012b). The three genes named *tex1, tex2, and tex3* have been identified as paptaibol synthetases. *Tex1* is a long chain peptide (18–25 remains) peptaibol synthetase and it is involved in the synthesis of 18 residue peptaibols (Wiest et al. 2002). *Tex1* accumulates an 18-residue peptaibol (trichovirin II) and by using *Dtex1* mutants trichovirin II type peptaibols revealed to activate induced resistance in hosts (Viterbo et al. 2007). Peptaibols of class 11, 14, and 18mer potentially inhibit pathogens including *A. solani*, *P. capsici*, *R. solani*, *S. rolfsii*, and *S. cepivorum* (Velázquez-Robledo et al. 2011). The three *Trichoderma* genomes discovered the presence of only 7, 14, and 18–20 module peptaibol synthetases (Degenkolb et al. 2012). Recently, the short peptaibol synthetase gene *tex2* has been delivered for the association of 11 and 14 modules peptaibols by a single NRPS *Tex2* of *T. virens* (Mukherjee et al. 2011; Reithner et al. 2011), later confirmed in *T. reesei* (Etxebeste et al. 2010). The *T. virens Tex2* was revealed to synthesize a total of 88 peptaibols belonging to 11 and 14-residue groups. The peptaibol trichokonin VI of *T. pseudokoningii* SMF2 was revealed to induce an extensive apoptotic programmed cell death in *F. oxysporum* (Shi et al. 2012). The *tex3*, homologous to *tex1* has seven complete modules arranged in a linear fashion (Mukherjee et al. 2012c) and homologs of all of these three genes in the genome of *T. atrobrunneum* ITEM908. Exogenous treatments of *Trichoderma* peptaibols in tobacco plants elicited a defense response by multiple defenses signaling pathways and resulting in increased resistance to the tobacco mosaic virus (Benitez et al. 2004; Luo et al. 2010; Holzlechner et al. 2016). The non-ribosomally synthesized peptaibols act as potential signature molecules forming the basis of mass spectrometry-based, species-specific monitoring approaches, as the peptaibiome of particular *Trichoderma* spp. is unique from that of closely related species (Marik et al. 2017).

3.13 Gliotoxin and Gliovirin

Gliotoxin and gliovirin are Epipolythiodioxopiperazines (ETPs), a class of peptides (Patron et al. 2007). The ETPs characterized by a diketopiperazine ring with a disulfide bridge derived from a cyclic peptide, produced by *Trichoderma* (Błaszczyk et al. 2014) and the genes for its biosynthesis in *T. virens* have been identified (Vargas et al. 2014). Gliotoxin belongs to the nonribosomal peptides (Patron et al. 2007). Gliotoxin derives from cyclic dipeptides that arise by the condensation of two α-amino acids and is produced biosynthetically from L-phenylalanine and L-serine via the cyclic dipeptide. The gliotoxin is produced by Q strains of *T. virens* whereas another ETP, gliovirin, is exclusively produced by the P strains of *T. virens*, both of which have potential antimicrobial activity (Scharf et al. 2016). Gliotoxin has attracted great attention for its function in the biocontrol of soilborne pathogens (Howell 2006). The *T. virens* veA ortholog *vel1* regulates gliotoxin biosynthesis, biocontrol activity, and many other secondary metabolism-related genes (Mukherjee and Kenerley 2010; Mukherjee et al. 2013). The gliotoxin genes clusters *gliZ*, *gliJ*, *gliA*, and *gliT* identified in the *T. virens* Q strain genome have a powerful role in the biocontrol of soilborne plant pathogens (Howell 2006).

3.14 Siderophores

The fungal siderophores, fusarinines, coprogens, and ferrichromes belong to the group of hydroxamate siderophores that share the structural unit N5-acyl-N5-hydroxyornithine (Renshaw et al. 2002; Lehner et al. 2013). Isotope assisted screening revealed an average 12-14 siderophores produced by *T. asperellum*, *T. atroviride*, *T. gamsii*, *T. harzianum*, *T. hamatum*, *T. virens*, *T. polysporum*, and *T. reesei* with dimerum acid, coprogren, fusarinine A, fusigen, and the intracellular siderophore ferricrocin (Lehner et al. 2013). Genome sequencing of *Trichoderma* spp. have revealed a single gene for ferricrocin synthesis, belonging to a secondary metabolism gene cluster (Kubicek et al. 2011). In *Trichoderma* spp. three NRPSs linked to siderophore biosynthesis have been known in different gene clusters (Mukherjee et al. 2013; Zeilinger et al. 2016). The genome of ITEM 908 harbors homologs of the aldehyde dehydrogenase (g626), the oxidoreductase (g625), the NRPS (g624), the ornithine monooxygenase (g623), and the transcription factor (g622). The second gene cluster comprises *NPS6*, a key enzyme that is accountable for extracellular siderophore production in *T. virens* (Mukherjee et al. 2013).

T. virens and *T. reesei* each contain two putative gene clusters having an NRPS as the core member, whose orthologs (*SidD* and *NPS6*) having potential in siderophore synthesis (Kubicek et al. 2011; Mukherjee et al. 2012c). *T. harzianum* produced the maximum number of siderophores, while, *T. reesei* biosynthesized one *cis*-fusarinine as the main siderophore and three others that were present only in *T. harzianum*.

3.15 6-Pentyl Pyrone (6-PP)/Pyrones

A volatile compound, 6-Pentyl pyrone (6-PP) with the unique coconut aroma, was produced by *Trichoderma* spp. (Vinale et al. 2008a, b). This compound fits into the chemically diverse class of low molecular weight metabolites with a high vapor pressure at room temperature and low water solubility grouped as volatile organic compounds (VOCs). Pyrones are derivative from fatty acids and the biosynthesis of 6-PP has been studied in *T. atroviride* by using [U-14C] and [1-14C] linoleic acid. *T. atroviride* exhibited an upregulation of the lipoxygenase gene thought to be involved in 6-PP biosynthesis and in *T. arundinaceum*, growth in co-culture with *B. cinerea* led to enhance expression levels of the "tri" biosynthetic genes (Malmierca et al. 2015). A lipoxygenase gene specific to *T. atroviride* may be involved in the biosynthetic pathway for the production of 6-PP but no useful characterization has yet been achieved (Kubicek et al. 2011).

A transcription factor gene called *Thctf1* was isolated from *T. harzianum* and involves in the synthesis of 6-pentyl-2H-pyran-2-one (6-PP) and displays antifungal activity against *R. solani*, *B. cinerea*, and *S. rolfsii*. The sequences were studied using the Laser gene package and cloned using pGEM-T vector (Rubio et al. 2009). Pyrones have been identified from numerous *T. harzianum* strains that are antagonistic to *G. graminis* var. *tritici* and *F. moniliforme*. The 6-PP secreted by *T. harzianum* potentially degrades mycotoxins including fusaric acid (FA) and additionally inhibits the mycelial growth of *F. moniliforme* (El-Hasan et al. 2008). Various *Trichoderma* spp. such as *T. viride*, *T. atroviride*, *T. harzianum*, *T. koningii* are able to produce the volatile antibiotic 6-PP which is antagonistic to *B. cinerea*, *R. solani* and *F. oxysporum* (Reino et al. 2008). The PR-1 gene was induced by 6-PP and harzianopyridone at 1 mg/l in canola cotyledons, indicating the initiation of an SA-dependent SAR response. At the same time, a chitinase PR-3 gene related to JA-dependent defense was induced by an equal amount of 6-PP, harzianopyridone or azaphilone (Viterbo et al. 2010). Recent studies revealed that *T. atroviride* produced 6-PP promoting plant growth and regulating the root architecture, preventing primary root growth and inducing lateral root formation (Garnica-Vergara et al. 2015).

3.16 Polyketides

The polyketides are a structurally diverse group of secondary metabolites, produced by numerous organisms, including filamentous fungi, with antibiotic activity such as (tetracyclines, polyenes, and macrolides), the mycotoxins (aflatoxin, fusaric acid, and fumonisin), the pigments (bikaverin and fusarubin) as well as the statins (lovastatin and compactin) (Zeilinger et al. 2016). These groups of molecules are that have carbon skeletons made up of polyenes, polyphenols, macrolides, enediynes, and polyethers. Polyketides are synthesized via pathways catalyzed by a collection of enzymes called PKSs, which are great multi-enzyme protein complexes that function with a coordinated group of active sites.

Genomes of *Trichoderma* spp. are rich in PKS encoding genes, suggesting the significance polyketides in the biology and activity of the fungus. There are several PKS genes involved in biosynthetic pathways and the genomes of *T. virens* and *T. atroviride* comprise 18 PKSs and the genome of *T. reesei* encodes 11 PKSs (Baker et al. 2012). The PKS genes are found usually as clusters along with genes coding cytochrome P450 monooxygenases, short-chain reductases or epimerases (Schmoll et al. 2016). Phylogenomic analysis of PKS genes of *T. reesei*, *T. virens*, and *T. atroviride* showed that most of the PKSs belonged to the lovastatin/citrinine or fumonisins clades that were present as orthologues in all three species studied (Baker et al. 2012). Two *T. atroviride* PKS genes were found to be expressed when confronted *R. solani*, indicating its possible role in mycoparasitism (Mukherjee et al. 2012b, c). Similar *gliP* and other SMs associated genes, PKSs in *T. virens* are regulated by the velvet complex protein *Vel1* (Mukherjee and Kenerley 2010).

There are numerous fungal SMs of interest produced by NRPS–PKS hybrid enzymes that consist of a PKS fused to a single, or in some cases truncated NRPS module (Fisch 2013). These hybrid enzymes are encoded in the genomes of *T. atroviride*, *T. reesei*, and *T. virens* (Kubicek et al. 2011). The first *Trichoderma* genome to be sequenced was from *T. reesei* and that contained 2 NRPS-PKS hybrid-encoding genes and the genes encoding terpenoid synthases (12 genes), NRPS (8 genes), and PKS (11 genes) (Martinez et al. 2008). The genome of *T. atroviride* harbors genes for 14 NRPSs, 18 PKSs, a single NRPS-PKS hybrid, and 14 terpenoid synthase domains (Kubicek et al. 2011). The efficient investigation of the *T. virens* showed that *Tex13*, a hybrid enzyme PKS/NRPS, was involved in inducing phenylalanine ammonialyase, the defense-related gene in maize seedlings; further the induction of *Tex13* is more than 40-fold during interactions of *T. virens* with maize roots (Mukherjee et al. 2012c).

3.17 Terpenoids/Steroids

Terpenoids are the most versatile natural products on earth and comprise a group of volatile and non-volatile secondary metabolites. The assembly of numerous activated forms of five carbon compounds isopentenyl/isoprene (C_5H_8) units depending on the number of carbon atoms. Each class contains molecules that are linear and cyclic; terpene cyclases generate the cyclization. Terpenoids recognized from *Trichoderma* spp. include volatile terpenes, the tetracyclic diterpene harziandione, sesquiterpenes such as the trichothecenes trichodermin and *harzianum* A and the triterpene viridin (Stoppacher et al. 2010; Cardoza et al. 2011). Compounds such as trichodermin isolated from *T. polysporum*, *T. sporulosum*, *T. virens*, and *T. reesei*, *Harzianum* A from *T. harzianum* and mycotoxin T2 detected in cultures of *T. lignorum* are examples of trichothecenes with antifungal activity. Trichothecene is synthesized by certain fungal genera such as *harzianum* A and trichodermin from *T. arundinaceum* and *T. brevicompactum*, respectively (Cardoza et al. 2011). The terpenoid Trichodermin is an extremely fungi toxic as well as phytotoxic, trichothecene type toxin produced by *T. brevicompactum* (Yuan et al. 2016). The production of trichodermin in *T. brevicompactum* involves the *tri5* gene which has a significant role such that its overexpression increases trichodermin production as well as the antimicrobial activity (Tijerino et al. 2011). A nonphytotoxic trichothecene, Harzianum A is antagonistic to fungal plant pathogens and triggers the genes responsible for plant defense. The *tri* gene cluster involved in harzianumA synthesis was characterized in *T. arundinaceum* (Malmierca et al. 2013). The triterpene biosynthetic pathway is catalyzed by enzymes encoded by the *erg1*, *erg7*, and *erg9* genes that are also capable of synthesis of viridin, a well-known antifungal molecule. In *T. harzianum*, the overexpression of *erg1* enhanced its antifungal effects against *B. cinerea* and reduced the lesion size. However, the induction of salicylate related plant defense genes and root colonization ability of *T. harzianum* was reduced (Cardoza et al. 2014).

The trichothecenes, sesquiterpenes are a huge group of toxic SMs produced by a few fungal species (Woloshuk and Shim 2013). The *tri* gene cluster for trichothecene biosynthesis has previously been defined in *T. arundinaceum* and *T. brevicompactum* and is made up of orthologues of seven genes present in the *Fusarium tri* cluster (Cardoza et al. 2011). Trichothecenes are sesquiterpenoid epoxides initially formed through isomerization–cyclization of farnesyl pyrophosphate from the parent compound trichodiene. Trichodiene synthase, encoded by *tri5* gene is the key enzyme catalyzing this reaction. The genes involved in trichothecene biosynthesis including *Tri5* are all organized in a coordinately regulated gene cluster.

Terpenes were isolated from *T. lignorum* HKI 0257, a new sesquiterpenoid named lignoren. This compound has a santalene-like structure and displays a sensible antimicrobial activity against *B. subtilis*, *M. smegmatis*, *P. aeruginosa*, *S. salmonicolor*, and *Rhodotorula rubra* (Berg et al. 2004). A recent study reported that the *T. reesei* genome encodes 6 terpene synthases or cyclases, 7 in *T. atroviride*, and 11 in *T. virens*, of which two, three, and six are part of biosynthetic gene clusters (Bansal and Mukherjee 2016b). Harzianic acid (HA), a nitrogen heterocyclic compound produced by *T. harzianum* has growth-promoting effect (Vinale et al. 2009) which acts as an antagonistic effect on fungal pathogens as reported in canola seedlings (Vinale et al. 2009). Also, they promote nutrients uptake and growth of plants by their ability to produce siderophores (Vinale et al. 2013).

3.18 Induced Systemic Resistance by *Trichoderma*

Induced systemic resistance is one of the most important mechanisms of biocontrol effects of *Trichoderma* (Harman 2006; Vinale et al. 2008a). Induction of metabolic changes in plants is brought about by several strains of *T. virens*, *T. asperellum*, *T. harzianum*, and *T. atroviride* that result in increased resistance to a wide range of plant pathogenic microorganisms. The colonization and induction of plant resistance by *Trichoderma* with some species is related to that elicited by rhizobacteria, which enhance the defense system but do not involve the production of pathogenesis-related proteins (PR proteins) (Harman et al. 2004). Induced resistance conferred to host plants by microorganisms are of two different kinds named induced systemic resistance (ISR) and systemic acquired resistance (SAR), which differ by the biochemical pathways involved (Birkenbihl et al. 2017). The SAR is triggered by previous exposure and infections by avirulent pathogens, whereas ISR is triggered by previous colonization of the rhizosphere by *Trichoderma* spp. SAR is a salicylic acid-dependent pathway, whereas ISR is salicylic acid independent (Hermosa et al. 2013; You et al. 2016; Birkenbihl et al. 2017). These defense pathways involve the evolution of pattern recognition receptors that specifically recognize microbe-based signals referred to as pathogen or microbe-associated molecular patterns (PAMPs or MAMPs) (Hermosa et al. 2012). The ability of *Trichoderma* spp. hyphae to release a variety of MAMPs for molecular recognition may contribute to signal cascade by signaling molecule within the plant such as salicylic acid (SA), jasmonic acid (JA), and ethylene (ET) (Lorito et al. 2010).

In the interactions of *Trichoderma* with plants, different classes of metabolites may act as elicitors or so-called resistance inducers (Woo and Lorito 2007). These metabolites are usually proteins including enzymes (serine proteases, xylanases, chitinases, phenylalanine ammonia lyase, peroxidase, polyphenol oxidase, lipoxygenase, cellulases, and glucanases) (Shoresh et al. 2005), proteins as PR (pathogenesis-related protein), gene products resembling proteins encoded by avirulent genes, low molecular compounds released from cell walls of fungi or plants by fungal hydrolytic enzymes and phytoalexin accumulation in host plants (Tuão Gava and Pinto 2016). *Trichoderma* endochitinase can also increase defense, probably via induction of plant defense-related proteins. Expression of *T. atroviride* endochitinase *Ech42* displayed enhanced resistance toward *Fusarium* sp. infection (McIntyre et al. 2004). Expression of *T. harzianum* chitinase *Chit42* in tobacco and potato plants resulted in improved resistance to the foliar pathogens *A. alternata*,

A. solani, *B. cinerea*, and to the soilborne pathogen *R. solani* (Howell 2003). Similarly, effects were seen on the heterologous expression of *Chit42* in strawberry infected with *Colletotrichum* and with *Chit42* and a β-1, 6 glucanase in melon and tomato plants. *T. harzianum* efficiently increased the SA and JA contents in melon thus altering the plant responses against *F. oxysporum* (Martínez-Medina et al. 2010). Expression of fungal chitinases in plants with CBDs, such as *Chit42CBD*, which already has increased antifungal activity, may result in greater resistance against phytopathogens (Limon et al. 2004). *Eix* also acts as a fungal elicitor that regulates phytoalexin production and defense gene expression through calcineurin B-like protein-interacting protein kinases, OsCIPK14/15, in rice (Kurusu et al. 2010). *T. longibrachiatum* cellulases, *T. viride* xylanase *Xyn2/Eix*, *T. harzianum* endopolygalacturonase *ThPG1* generates a response in *Arabidopsis* (Moran-Diez et al. 2009).

The *T. asperellum* swollenin *TasSwo* stimulates defense responses in cucumber roots and leaves and affords local protection against phytopathogens (Moran-Diez et al. 2009). *T. asperellum* produces the class I hydrophobin *TasHyd1*, which aids in root surface colonization, possibly by improving its attachment to the root surface and protecting the hyphal tips from plant defense compounds (Viterbo and Chet 2006). In oil palm plants the expression of defense gene chitinase was increased in plants treated with *T. harzianum* and *Ganoderma boninense* compared to those treated with *G. boninense* alone (Naher et al. 2012). In a study on cucumber plants, *T. asperellum* induced a systemic response of two defense-related genes encoding phenylalanine and hydroperoxidase lyase and systemic accumulation of phytoalexins against *P. syringae* pv. *lachrymans* (Yedidia et al. 2003). *T. asperellum* T203 modulated the expression of the genes *Lox1* (Lipoxygenase 1), a constituent of the JA biosynthetic pathway; PAL1, an element of the biosynthetic pathway for SA; and *Etr1* and *Ctr1*, both components of ET signaling (Shoresh et al. 2005). Contreras-Cornejo et al. (2011) who recommended that JA as an important factor in boosting plant immunity involved in defense responses elicited by *Trichoderma* in *Arabidopsis* against *B. cinerea*. Similar soil application of *T. viride* to tomato plants with *F. oxysporum* or *R. solani* resulted in an increase in the expression of JA-related *PDF1* and *PDF2* genes (Hafez et al. 2013).

Molecular confirmation exhibited that *A. thaliana* root colonization by *T. asperelloides* T203 activates a quick rise in the expression of transcription factors (WRKY18, WRKY40, WRKY60, and WRKY33) activating JA pathway responses and represses SA signaling. WRKY18, WRKY40, and WRKY60 are pathogen-induced and encode three structurally related WRKY proteins that exert a positive role in JA-mediated defense (Brotman et al. 2013). Mathys et al. (2012) reported that induced resistance in *Arabidopsis* roots treated with *T. hamatum* is regulated by JA and ET related genes. Additionally, the JA inducible genes lipoxygenase *(Lox1*) and phenylalanine ammonialyase (*Pal1*) and the ET-inducible genes ethylene receptor (*ETR1*) and constitutive triple response 1 (*CTR1*) were found to be induced both locally and systemically on treatment with *T. asperellum* T-203 spores alone. Tucci et al. (2011) observed that *Trichoderma* CF treatment triggered ISR through SA-dependent gene expression

Several SMs and proteins involved in mycoparasitism and antibiosis have been identified as ISR elicitors. Secondary metabolites like alamethicin and trichokinin (20mer peptaibol), 18mer peptaibol, 6-pentyl-a-pyrone, harzianolide, and harzianopyridone at high doses have antimicrobial effects but at low concentrations are ISR inducers (Vinale et al. 2014). Peptaibols produced by *Trichoderma* may act as elicitors of plant defense mechanisms against pathogens (Wiest et al. 2002). A peptaibol synthetase from *T. virens* was purified and the achieved cloning of the corresponding gene will facilitate an understanding of the role of this class of compounds in plant defense response. Application of alamethicin, a long sequence peptaibol with a 20-residue produced by *T. viride*, elicits JA and SA biosynthesis in lima bean in *Phaseolus lunatus* (lima bean) (Maischak et al. 2010) and *A. thaliana* hypersensitive reaction to pathogen attack (Rippa et al. 2010). The 18mer peptaibols from *T. virens* elicit systemic induced defense responses in cucumber against the leaf pathogen *P. syringae* (Viterbo et al. 2007; Luo et al. 2011).

Early defense responses triggered by SMs from *T. atroviride* induced intracellular Ca^{2+} variations in soybean cells (Navazio et al. 2007). The defense mechanisms plants and their developmental responses to *Trichoderma* share common components. This was evident when 1 ppm of 6-pentyl-a-pyrone, harzianolide, and harzianopyridone activated plant defense mechanisms and regulated plant growth in pea, tomato, and canola (Vinale et al. 2008b), suggesting that plants' *Epl-1* has been described as being able to trigger defense reactions in plants (Gomes et al. 2015; Ramada et al. 2016; Salas-Marina et al. 2015). Fernanda Blauth de Lima (2017) reported that, when challenged by the *Guignardia citricarpain* citrus black, *T. harzianum* T1A there was a decrease in the total amount of secreted proteins, particularly those involved in primary metabolism while the secretion of proteins related to fungal interaction increased. *T. harzianum* T1A secretes proteins related to the control of *G. citricarpa* and induction of plant resistance, even in the absence of pathogen challenge.

A PKS/NRPS hybrid enzyme involved in defense responses in maize was identified (Mukherjee et al. 2012c). Non-enzymatic proteins such as small cysteine-rich hydrophobin-like protein of the cerato-platanin (CP) family *Sm1* secreted by *T. virens* and *Epl1* secreted by *T. atroviride* trigger the activation of plant defense mechanisms and the induction of systemic resistance in cotton and maize (Seidl et al. 2006). In response to invasion by a pathogen, the *Sm1* of *T. virens* acts as an elicitor inducing the expression of CAD1-C gene encoding (+)-δ-cadinene synthase in cotton petioles which is the primary precursor for phytoalexin production (Djonovic et al. 2006; Yoshikuni et al. 2006). Induction of defense mechanisms in plants is also brought about by another group of proteins that are the products of avirulence-like (*Avr*) genes (Woo et al. 2006). The hydrophobin-like protein produced by T22 was identified to induce both enhanced root development and disease resistance (Ruocco et al. 2007). Early defense reactive oxygen species (ROS) such as H_2O_2 and nitric oxide also are associated in *Trichoderma*-mediated plant immunity in cotton, rice, and *A. thaliana* (Gupta et al. 2014; Contreras-Cornejo et al. 2014).

3.19 Stress Tolerances

The genus *Trichoderma* is able to inhabit and colonize diverse niches due to its metabolic versatility and tolerance to stress conditions. Among fungal biocontrol agents, *Trichoderma* spp. have gained much interest due to their high reproductive capacity, prolific producers of secondary metabolites, survived under unfavorable conditions, and ability to resist against plant pathogenic fungi (Contreras-Cornejo et al. 2016). *Trichoderma* spp. colonize plants and produce certain compounds (gibberellins, ethylene, auxins, plant enzymes, antioxidants) and phytoalexins and phenols that confer abiotic stresses tolerance and enhance the branching capacity of the root system (Brotman et al. 2013; Lopez-Bucio et al. 2015). Several recent studies report that *Trichoderma* induces tolerance against abiotic stresses and improves plant growth (Zeilinger et al. 2016; Yasmeen and Siddiqui 2017). *Trichoderma* spp. can also promote growth and induce resistance to a variety of abiotic stresses, including water deficit, temperature, salinity, and osmotic stress (Zelicourt et al. 2016).

Trichoderma spp. are significant for regulating numerous genes involved in plant defense against biotic and abiotic stresses and improving the plant basal metabolism (Domínguez et al. 2016). The genes responsible for resistance to salt or other stresses in *T. harzianum*, *ThHog1* (Delgado-Jarana et al. 2006), *Hsp70* (Montero-Barrientos et al. 2010) and Thkel1 (Hermosa et al. 2011) have been successively characterized. In an HSP24-carrying transgenic mutant of *S. cerevisiae*, the small heat shock protein *Hsp24* of *T. harzianum* was shown to enhance salt, heat, and drought tolerances (Liming et al. 2008). Cloning of *hsp70* gene in *pGEM-T* vector and its expression in different isolates of *T. harzianum* enhanced fungal resistance to heat and other stresses such as oxidative tolerances, osmotic and salt tolerance (Montero-Barrientos et al. 2010). The sequences were analyzed using DNA star package and aligned using CLUSTAL X algorithm. The genome of *T. reesei* revealed three genes for potential small heat shock proteins; in *T. atroviride* there were four genes and in *T. virens* five genes were present. All of them are homologs to *N. crassa Hsp30* (Plesofsky-Vig and Brambl 1995). *Hsp30* of *N. crassa* was found essential for carbon utilization at high temperatures (Plesofsky-Vig and Brambl 1995).

Montero-Barrientos et al. (2007) studied the response of the small heat shock protein *Hsp23* of *T. virens* T59 to high and low temperatures and reported the expression of *Hsp23* was improved on ethanol addition. The *Hsp23* gene when transferred to the biocontrol strain *T. harzianum* T34 resulted in higher biomass production in the mutant strains than in the wild type T34 strain along with improved thermotolerance (Bonaccorsi et al. 2006). The *Thkel1* gene encodes putative kelch-repeat proteins which modulate glucosidase activity and confer salt tolerance, enhance seed germination, and osmotic stress in *Arabidopsis* plants, probably due to the glucosidase activity and abscisic acid (ABA) level modulations (Hermosa et al. 2011). The vector used for cloning was pSIL-KEL and was transformed into *T. harzianum*. The *Thkel1*gene expression was studied by growing the fungus under various biotic and abiotic stress conditions (Hermosa et al. 2011).

Rana et al. (2012) reported that genes encoding an endochitinase (*chit42*) and a chitosanase (*harcho*) from *T. harzianum*, if co-transformed in wheat plants resulted in an increased tolerance to the powdery mildew pathogen (*Blumeria graminis* f.sp. *tritici*). Under conditions of water scarcity, *T. harzianum* T22 modulated the expression of genes that encoding enzymes that scavenge ROS, such as SOD, catalase, and ascorbate peroxidase, in both root and shoots of tomato plants (Shoresh et al. 2010; Mastouri et al. 2012). The highly conserved ribosomal protein subunits like *Rpl44* and *Rps3ae* are also promising candidates for enhanced tolerance in crop plants (Liang et al. 2015) and these genes are generally found downstream to those resistant pathways likely having a direct contribution to stress tolerance. Systemic induction of about 40 genes by *T. harzianum* 382 in tomato plants with functions related to biotic or abiotic stress, as well as RNA, DNA, and protein metabolism (Shoresh et al. 2010). About 205 differentially expressed proteins were identified, in roots and shoots of maize plants inoculated by *T. harzianum* T226. From *T. virens* glutathione transferase gene *TvGST* was cloned. The expression of this gene in transgenic plants showed tolerance to cadmium accumulation in plants thus acting as a cadmium tolerance gene (Dixit et al. 2011).

3.20 Hyphal Growth

In *T. reesei* the *TrCCD1* gene helps in hyphal growth, development of conidiospores, and production of carotenoid pigment, therefore improving biocontrol activity (Zhong et al. 2009). Chitinase degrade chitin, the linear homopolymer of β-1, 4-*N*-acetyl-D-glucosamine, which is the main cell wall constituent of plant pathogenic fungi thus inhibiting the in vitro germination and hyphal growth (Lorito et al. 1996). These genes find application in improving plant defense against fungal pathogens. Bae and Knudsen (2000) reported that the to monitor hyphal growth, activities, and existence of a *T. harzianum* strain, transformed strain ThzID1 with plasmids carrying the *gfp* (pTEFEGFP), *Gus* (pNOM102), and *hygB* (pAN7-2) genes. The mitotic stability of the cotransformants and their ability to colonize the inactive sclerotia of the plant pathogen *S. sclerotiorum* in soil were studied.

References

Abbas A, Jiang D, Fu Y (2017) *Trichoderma* Spp. as antagonist of *Rhizoctonia solani*. J Plant Pathol 8:402

Agrios GN (2009) Plant pathogens and disease: general introduction. Elsevier, University of Florida, Gainesville, FL

Ajitha PS, Lakshmedevi N (2010) Effect of volatile and von-volatile compounds from *Trichoderma* spp. against *Colletotrichum capsici* incitant of anthracnose on Bell peppers. Nat Sci 8:265–296

Alizadeh H, Behboudi K, Ahmadzadeh M, Javan-Nikkhah M, Zamioudis C, Pieterse CMJ, Bakker PAHM (2013) Induced systemic resistance in cucumber and *Arabidopsis thaliana* by the combination of *Trichoderma harzianum Tr6* and *Pseudomonas* sp. *Ps14*. Biol Control 65:14–23

Alkooranee JT, Aledan TR, Ali AK, Lu G, Zhang X, Wu J, Fu C, Li M (2017) Detecting the hormonal pathways in oilseed rape behind induced systemic resistance by *Trichoderma harzianum* TH12 to *Sclerotinia sclerotiorum*. PLos One 12:e0168850

Atanasova L, Le Crom S, Gruber S, Coulpier F, Seidl-Seiboth V, Kubicek CP et al (2013a) Comparative transcriptomics reveals different strategies of *Trichoderma* mycoparasitism. BMC Genomics 14:121

Atanasova L, Knox BP, Kubicek CP, Druzhinina IS, Baker SE (2013b) The polyketide synthase gene pks4 of *Trichoderma reesei* provides pigmentation and stress resistance. Eukaryot Cell 12:1499–1508

Babychan M, Simon S (2017) Efficacy of *Trichoderma* spp. against *Fusarium oxysporum* f. sp. *lycopersici*. (FOL) infecting pre- and post-seedling of tomato. J Pharmacogn Phytochem 6:616–619

Bae Y, Knudsen GR (2000) Cotransformation of *Trichoderma harzianum* with beta-glucuronidase and green fluorescent protein genes provides a useful tool for monitoring fungal growth and activity in natural soils. Appl Environ Microbiol 66:810–815

Baker SE, Perrone G, Richardson NM, Gallo A, Kubicek CP (2012) Phylogenetic analysis and evolution of polyketide synthase- encoding genes in *Trichoderma*. Microbiology 158:147–154

Bakker PAHM, Pieterse CMJ, Van Loon LC (2007) Induced systemic resistance by *Fluorescent Pseudomonas* spp. Phytopathology 97:239–243

Bansal R, Mukherjee P (2016a) Identification of novel gene clusters for secondary metabolism in *Trichoderma* genomes. Microbiology 85(2):185–190

Bansal R, Mukherjee P (2016b) The terpenoid biosynthesis toolkit of *Trichoderma*. Nat Prod Commun 11:431–434

Baranski R, Klocke E (2008) Chitinase CHIT36 from *Trichoderma harzianum* enhances resistance of transgenic carrot to fungal pathogens. J Phytopathol 156:513–521

Baranski R, Klocke E, Nothnagel T (2008) Chitinase CHIT36 from *Trichoderma harzianum* enhances resistance of transgenic carrot to fungal pathogens. J Phytopathol 156:513–521

Baroncelli R, Piaggeschi G, Fiorini L, Bertolini E, Zapparata A, Pè ME et al (2015) Draft whole-genome sequence of the biocontrol agent *Trichoderma harzianum* T6776. Genome Announc 3: e00647–e00615

Baroncelli R, Zapparata A, Piaggeschi G, Sarrocco S, Vannacci G (2016) Draft whole-genome sequence of *Trichoderma gamsii* T6085, a promising biocontrol agent of Fusarium head blight on wheat. Genome Announc 4:e01747–e01715

Benitez T, Limón C, Delgado-Jarana J, Rey M (1998) Glucanolytic and other enzymes and their genes. In: Harman GE, Kubicek CP (eds) *Trichoderma* and *Gliocladium*. Vol. 2: enzymes, biological control and commercial application. London, Taylor and Francis, pp 101–128

Benitez T, Rincon AM, Limon MC, Codon AC (2004) Biocontrol mechanisms of *Trichoderma* strains. Int Microbiol 7:249–260

Berg A, Wangun HVK, Nkengfack AE, Schlegel B (2004) Lignoren, a new sesquiterpenoid metabolite from *Trichoderma lignorum* HKI 0257. J Basic Microbiol 44:317–319

Bethke G, Pecher P, Eschen-Lippold L, Tsuda K, Katagiri F, Glazebrook J et al (2012) Activation of the *Arabidopsis thaliana* mitogen-activated protein kinase MPK11 by the flagellin-derived elicitor peptide, flg22. Mol Plant Microbe Interact 25:471–480

Bhattacharya D, Nagpure A, Gupta RK (2007) Bacterial chitinases: properties and potential. Crit Rev Biotechnol 27:21–28

Birkenbihl RP, Liu S, Somssich IE (2017) Transcriptional events defining plant immune responses. Curr Opin Plant Biol 38:1–9

Błaszczyk L, Siwulski M, Sobieralski K, Lisiecka J, Jędryczka M (2014) *Trichoderma* spp. application and prospects for use in organic farming and industry. J Plant Protect Res 54:309–317

Bolar JP, Norelli J, Harman GE, Brown SK, Aldwinckle HS (2001) Synergistic activity of endochitinase and exochitinase from *Trichoderma harzianum* against the pathogenic fungus Venturia inaequalis in transgenic plants. Transgenic Res 10:533–543

Bonaccorsi ED, Ferreira AJ, Chambergo FS, Ramos AS, Mantuani MC, Farah JP, Sorio CS, Gombert AK, Tonso A, El-Dorry H (2006) Transcriptional response of the obligatory aerobe *Trichoderma* reesei to hypoxia and transient anoxia: implications for energy production and survival in the absence of oxygen. Biochemistry 45:3912–3924

Brakhage AA, Schroeckh V (2011) Fungal secondary metabolites e strategies to activate silent gene clusters. Fungal Genet Biol 48:15–22

Brants A, Earle ED (2001) Transgenic tobacco cell cultures expressing a *Trichoderma harzianum* endochitinase gene release the enzyme into the medium. Plant Cell Rep 20:73–78

Brotman Y, Briff E, Viterbo A, Chet I (2008) Role of swollenin, an expansin-like protein from *Trichoderma*, in plant root coloniza- tion. Plant Physiol 147:779–789

Brotman Y, Landau U, Pninic S, Lisec J, Balazadeh S et al (2012) The LysM receptor-like kinase LysMRLK1 is required to activate defense and abiotic-stress responses induced by overexpression of fungal chitinases in Arabidopsis plants. Mol Plant 5:1113–1124

Brotman Y, Landau U, Cuadros-Inostroza Á, Takayuki T et al (2013) *Trichoderma*-plant root colonization: escaping early plant defense responses and activation of the antioxidant machinery for saline stress tolerance. PLoS Pathog 9:e1003221

Brunner K, Montero M, Mach RL, Peterbauer CK, Kubicek CP (2003) Expression of the ech42 (endochitinase) gene of *Trichoderma atroviride* under carbon starvation is antagonized via a BrlA-like cis-acting element. FEMS Microbiol Lett 218:259–264

Brunner K, Zeilinger S, Ciliento R, Woo SL, Lorito M, Kubicek CP, Mach RL (2005) Genetic improvement of a fungal biocontrol agent to enhance both antagonism and induction of plant systemic disease resistance. Appl Environ Microbiol 71:3959–3965

Cai F, Chen W, Wei Z, Pang G, Li R, Ran W, Shen Q (2015) Colonization of *Trichoderma harzianum* strain SQR-T037 on tomato roots and its relationship to growth, nutrient availability and soil microflora. Plant Soil 388:337–350

Calo L, García I, Gotor C, Romero LC (2006) Leaf hairs influence phytopathogenic fungus infection and conferred an increased resistance when expressing a *Trichoderma* 1,3-glucanase. J Exp Bot 56:3911–3920

Cardoza RE, Malmierca MG, Hermosa MR, Alexander NJ, McCormick SP, Proctor RH et al (2011) Identification of loci and functional characterization of trichothecene biosynthesis genes in filamentous fungi of the genus *Trichoderma.* Appl Environ Microbiol 77:4867–4877

Cardoza RE, Malmierca MG, Gutierrez S (2014) Overexpression of *erg1* gene in *Trichoderma harzianum* CECT 2413: effect on the induction of tomato defence-related genes. J Appl Microbiol 117:812–823

Carpenter MA, Ridgway HJ, Stringer AM, Hay AJ, Stewart A (2008) Characterization of a *Trichoderma hamatum* monooxygenase gene involved in antagonistic activity against fungal plant pathogens. Curr Genet 53:193–205

Chernin L, Chet I (2002) Microbial enzymes in biocontrol of plant pathogens and pests. In: Burns R, Dick R (eds) Enzymes in the environment: activity, ecology, and applications. Marcel Dekker, New York, pp 171–225

Cohen-Kupiec R, Broglie KE, Friesem D, Broglie RM, Chet I (1999) Molecular characterization of a novel β-1,3-exoglucanase related to mycoparasitism of *Trichoderma harzianum.* Gene 226:147–154

Contreras-Cornejo HA, Macías-Rodríguez L, Beltrán-Peña E, Herrera-Estrella A, López-Bucio J (2011) *Trichoderma*-induced plant immunity likely involves both hormonal and camalexin-dependent mechanisms in *Arabidopsis thaliana* and confers resistance against necrotrophic fungus *Botrytis cinerea*. Plant Signal Behav 6:1554–1563

Contreras-Cornejo HA, Macıas-Rodrıguez L, Lopez-Bucio J (2014) Enhanced plant immunity using *Trichoderma.* In: Gupta VK (ed) Biotechnology and biology of *Trichoderma.* Elsevier, Oxford, pp 495–504

Contreras-Cornejo HA, Macías-Rodríguez L, del-Val E, Larsen J (2016) Ecological functions of *Trichoderma* spp. and their secondary metabolites in the rhizosphere: interactions with plants. FEMS Microbiol Ecol 92:fiw03

Cortes C, Gutierrez A, Olmedo V, Inbar J, Chet I, Herrera Estrella A (1998) The expression of genes involved in parasitism by *Trichoderma harzianum* is triggered by a diffusible factor. Mol Gen Genet 260:218–225

Daguerre Y, Siegel K, Edel-Hermann V, Steinberg C (2014) Fungal proteins and genes associated with biocontrol mechanisms of soil-borne pathogens: a review. Fungal Biol Rev 28:97–125

Dana MM, Pintor-Toro JA, Cubero B (2006) Transgenic tobacco plants overexpressing chitinases of fungal origin show enhanced resistance to biotic and abiotic stress agents. Plant Physiol 142:722–730

de la Cruz J, Llobell A (1999) Purification and properties of a basic endo-β-1,6-glucanase (BGN16.1) from the antagonistic fungus *Trichoderma harzianum*. Eur J Biochem 265:145–151

de la Cruz J, Hidalgo-Gallego A, Lora JM, Benitez T, Pintor-Toro JA, Llobell A (1992) Isolation and characterization of three chitinases from *Trichoderma harzianum*. Eur J Biochem 206:859–867

de la Cruz J, Pintor-Toro JA, Benitez T, Llobell A, Romero LC (1995) A novel endo-β-1,3-glucanase, BGN13.1, involved in the Mycoparasitism of *Trichoderma harzianum*. J Bacteriol 177:6937–6945

de Lima FB, Félix C, Osório N, Alves A, Vitorino R, Domingues P, Ribeiro R, Esteves A (2017) *Trichoderma harzianum* T1A constitutively secretes proteins involved in the biological control of *Guignardia citricarpa*. Biol Control 106:99–109

Degenkolb T, Döhren HV, Nielsen KF, Samuels GJ, Brückner H (2008) Recent advances and future prospects in peptaibiotics and mycotoxin research and their importance for chemotaxonomy of *Trichoderma* and *Hypocrea*. Chem Biodivers 5:671–680

Degenkolb T, Karimi Aghcheh R, Dieckmann R, Neuhof T, Baker SE, Druzhinina IS et al (2012) The production of multiple small peptaibol families by single 14-module peptide synthetases in *Trichoderma/Hypocrea*. Chem Biodivers 9:499–535

Delgado-Jarana J, Moreno-Mateos MA, Benítez T (2003) Glucose uptake in *Trichoderma harzianum:* role of gtt1. Eukaryot Cell 2:708–717

Delgado-Jarana J, Sousa S, González F, Rey M, Llobell A (2006) *ThHog1* controls the hyperosmotic response in *Trichoderma harzianum*. Microbiology 162:1687–1700

Dickman MB, Yarden O (1999) Serine/threonine protein kinases and phosphatases in filamentous fungi. Fungal Genet Biol 26:99–117

Distefano G, La Malfa S, Vitale A, Lorito M, Deng Z (2008) Defence related gene expression in transgenic lemon plants producing an antimicrobial *Trichoderma harzianum* endochitinase during fungal infection. Transgenic Res 17:873–879

Dixit P, Mukherjee PK, Ramachandran V, Eapen S (2011) Glutathione transferase from *Trichoderma virens* enhances cadmium tolerance without enhancing its accumulation in transgenic *Nicotiana tabacum*. PLoS One 6(1):1–15

Djonovic S, Pozo MJ, Dangott LJ, Howell CR, Kenerley CM (2006) Sm1, a proteinaceous elicitor secreted by the biocontrol fungus *Trichoderma virens* induces plant defense responses and systemic resistance. Mol Plant-Microbe Interact 19:838–853

Djonovic S, Vittone G, MendozaHerrera A, Kenerley CM (2007) Enhanced biocontrol activity of *Trichoderma virens* transformants constitutively coexpressing β-1, 3- and β-1, 6-glucanase genes. Mol Plant Pathol 8(4):469–480

Domínguez S, Rubio MR, Cardoza RE, Gutiérrez S, Nicolás C, Bettiol W et al (2016) Nitrogen metabolism and growth enhancement in tomato plants challenged with *Trichoderma harzianum* expressing the *Aspergillus nidulans* acetamidase amdS gene. Front Microbiol 7:1182

Donzelli BG, Lorito M, Scala F, Harman GE (2001) Cloning, sequence and structure of a gene encoding an antifungal glucan 1, 3-beta-glucosidase from *Trichoderma atroviride* (T. *harzianum*). Gene 277:199–208

Druzhinina IS, Seidl-Seiboth V, Herrera-Estrella A, Horwitz BA, Kenerley CM, Monte E, Mukherjee PK, Zeilinger S, Grigoriev IV, Kubicek CP (2011) *Trichoderma*: the genomics of opportunistic success. Nat Rev Microbiol 9:749–759

Dubey MK, Jensen DF, Karlsson M (2014) Hydrophobins are required for conidial hydrophobicity and plant root colonization in the fungal biocontrol agent *Clonostachys rosea*. BMC Microbiol 14:18

Dutta S, Kundu A, Chakraborty M, Ojha S, Chakrabarti J, Chatterejee N (2006) Production and optimization of Fe(III) specific ligand, the siderophore of soil inhabiting and wood rotting fungi as deterrent to plant pathogens. Acta Phytopathol Entomol Hung 41:237–248

Elad Y, Freeman S, Monte E (eds) (2000) Biocontrol agents: mode of action and interaction with other means of control. IOBC wprs Bulletin, vol 24. Sevilla, España

El-Hasan A, Walker F, Buchenauer H (2008) *Trichoderma harzianum* and its metabolite 6-pentyl-alpha-pyrone suppress fusaric acid produced by *Fusarium moniliforme*. J Phytopathol 156:79–87

El-Katatny MH, Gudelj M, Robra KH, Elnaghy MA, Gübitz GM (2001) Characterization of a chitinase and an endo-ß-1,3- glucanase from *T. harzianum* Rifai T-24 involved in control of the phytopathogen *Sclerotium rolfsii*. Appl Microbiol Biotechnol 56:137–143

Emani C, Garcıa JM, Lopata-Finch E, Pozo MJ, Uribe P, Kim DJ, Sunilkumar G, Cook DR, Kenerley CM, Rathore KS (2003) Enhanced fungal resistance in transgenic cotton expressing an endochitinase gene from *Trichoderma virens*. Plant Biotechnol J 1:321–336

Etxebeste O, Garzia A, Espeso EA, Ugalde U (2010) *Aspergillus nidulans* asexual development: making the most of cellular modules. Trends Microbiol 18:569–576

FAO (2009). http://www.fao.org/fileadmin/templates/wsfs/docs/issues_papers/hlef 2050global_ agr culture.pdf. Retrieved on 20 Jan 2018

Fisch KM (2013) Biosynthesis of natural products by microbial iterative hybrid PKS-NRPS. RSC Adv 3:18228–18247

Freitas RS, Steindorff AS, Ramada MHS, de Siqueira SJL, Noronha EF, Ulhoa CJ (2014) Cloning and characterization of a protein elicitor Sm1 gene from *Trichoderma harzianum*. Biotechnol Lett 36:783–788

Frischmann A, Neudl S, Gaderer R, Bonazza K, Zach S, Gruber S, Spadiut O, Friedbacher G, Grothe H, Seidl-Seiboth V (2013) Self-assembly at air/water interfaces and carbohydrate binding properties of the small secreted protein EPL1 from the fungus *Trichoderma atroviride*. J Biol Chem 288:4278–4287

Gaderer R, Lamdan NL, Frischmann A, Sulyok M et al (2015) *Sm2*, a paralog of the *Trichoderma ceratoplatanin* elicitor *Sm1*, is also highly important for plant protection conferred by the fungal- root interaction of *Trichoderma* with maize. BMC Microbiol 15:2

Gajera H, Domadiya R, Patel S, Kapopara M, Golakiya B (2013) Molecular mechanism of *Trichoderma* as bio-control agents against phytopathogen system—a review. Curr Res Microbiol Biotechnol 1:133–142

Gallo A, Mule G, Favilla M, Altomare C (2004) Isolation and characterization of a trichodiene synthase homologous gene in *Trichoderma harzianum*. Physiol Mol Plant Pathol 65:11–20

Garcia I, Lora JM, De la Cruz J, Benitez T, Llobell A, Pintor-Toro JA (1994) Cloning and characterization of a chitinase (CHIT42) cDNA from the mycoparastic fungus *T. harzianum*. Curr Genet 27:83–89

Garnica-Vergara A, Barrera-Ortiz S, Munoz-Parra E (2015) The volatile 6-pentyl-2H-pyran-2-one from *Trichoderma atroviride* regulates *Arabidopsis thaliana* root morphogenesis via auxin signaling and ETHYLENE INSENSITIVE 2 functioning. New Phytol 209:1496–1512

Gentile A, Deng Z, LaMalfa S, Distefano G, Domina F, Vitale A, Polizzi G, Lorito M, Tribulato E (2007) Enhanced resistance to *Phoma tracheiphila* and *Botrytis cinerea* in transgenic lemon plants expressing a *Trichoderma harzianum* chitinase gene. Plant Breed 126:146–151

Geraldine AM, Cardoso Lopes FA, Costa Carvalho DD, Barbosa ET, Rodrigues AR, Brandão RS, Ulhoa CJ, Junior ML (2013) Cell wall-degrading enzymes and parasitism of sclerotia are key factors on field biocontrol of white mold by *Trichoderma* spp. Biol Control 67:308–316

Girhepuje PV, Shinde GV (2011) Transgenic tomato plants expressing a wheat endochitinase gene demonstrate enhanced resistance to *Fusarium oxysporum* f. sp. *lycopersici*. Plant Cell Tissue Organ Cult 105:243–251

Gomes EV, Nascimento CM, Graciano PR, Azevedo RR et al (2015) The Cerato-Platanin protein Epl-1 from *Trichoderma harzianum* is involved in mycoparasitism, plant resistance induction and self-cell wall protection. Sci Rep 5:17998

Gupta KJ, Mur LA, Brotman Y (2014) *Trichoderma asperelloides* suppresses nitric oxide generation elicited by *Fusarium oxysporum* in Arabidopsis roots. Mol Plant-Microbe Interact 27:307–314

Guzmán-Guzmán P, Alemán-Duarte MI, Delaye L, Herrera-Estrella A, Olmedo-Monfi V (2017) Identification of effector-like proteins in *Trichoderma* spp. and role of a hydrophobin in the plant-fungus interaction and mycoparasitism. BMC Genet 18:16

Hafez EE, Meghad A, Elsalam HAA, Ahmed SA (2013) *Trichoderma viride*-Plant pathogenic fungi interactions. World Appl Sci J 21:1821–1828

Haggag WM, Kansoh AL, Aly AM (2006) Proteases from *Talaromyces flavus* and *Trichoderma harzianum*: purification, characterization and antifungal activity against brown spot disease on faba bean. Plant Pathol Bull 15:231–239

Harman GE (2000a) Myths and dogmas of biocontrol. Plant Dis 84:377–391

Harman GE (2000b) Myths and dogmas of biocontrol: changes in perceptions derived from research on *Trichoderma harzianum* T-22. Plant Dis 84:377–393

Harman GE (2006) Overview of mechanisms and uses of *Trichoderma* spp. Phytopathology 96:190–194

Harman GE, Howell CR, Viterbo A, Chet I, Lorito M (2004) *Trichoderma* species–opportunistic, avirulent plant symbionts. Nat Rev Microbiol 2:43–56

Hermosa R, Botella L, Keck E, Jiménez JA, Montero-Barrientos M, Arbona V, Gómez-Cadenas A, Monte E, Nicolás C (2011) The overexpression in *Arabidopsis thaliana* of a *Trichoderma harzianum* gene that modulates glucosidase activity, and enhances tolerance to salt and osmotic stresses. J Plant Physiol 168:1295–1302

Hermosa R, Viterbo A, Chet I, Monte E (2012) Plant-beneficial effects of *Trichoderma* and of its genes. Microbiology 158:17–25

Hermosa R, Rubio ME, Cardoza MB, Nicolás E, Monte E, Gutiérrez S (2013) The contribution of *Trichoderma* to balancing the costs of plant growth and defense. Int Microbiol 16:69–80

Hermosa R, Cardoza MB, Rubio ME, Gutiérrez S, Monte E (2014) Secondary metabolism and antimicrobial metabolites of *Trichoderma*. In: Gupta VK, Schmoll M, Herrera-Estrella A, Upadhyay RS, Druzhinina I, Tuohy M (eds) Biotechnology and biology of *Trichoderma*. Elsevier, Netherlands, pp 125–137

Hjeljord LG, Stensvand A, Tronsmo A (2000) Effect of temperature and nutrient stress on the capacity of commercial *Trichoderma* products to control *Botrytis cinerea* and *Mucor piriformis* in greenhouse strawberries. Biol Control 19:149–160

Holzlechner M, Reitschmidt S, Gruber S, Zeilinger S, Marchetti-Deschmann M (2016) Visualizing fungal metabolites during mycoparasitic interaction by MALDI mass spectrometry imaging. Proteomics 16(11–12):1742–1746

Howell CR (2003) Mechanisms employed by *Trichoderma* species in the biological control of plant diseases: the history and evolution of current concepts. Plant Dis 87:4–10

Howell CR (2006) Understanding the mechanisms employed by *Trichoderma virens* to effect biological control of cotton diseases. Phytopathology 96:178–180

Ihrmark K, Asmail N, Ubhayasekera W, Melin P, Stenlid J, Karlsson M (2010) Comparative molecular evolution of *Trichoderma* chitinases in response to mycoparasitic interactions. Evol Bioinform 6:1–26

Kamble S, Mukherjee PK, Eapen S (2016) Expression of an endochitinase gene from *Trichoderma virens* confers enhanced tolerance to Alternaria blight in transgenic *Brassica juncea* (L.) czern and coss lines. Physiol Mol Biol Plants 22:69–76

Karimi-Aghcheh R, Bok JW, Phatale PA, Smith KM, Baker SE, Lichius A, Omann M, Zeilinger S, Seiboth B, Rhee C, Keller NP, Freitag M, Kubicek CP (2013) Functional analyses of *Trichoderma reesei* LAE1 reveal conserved and contrasting roles of this regulator. G3 (Bethesda) 3:369–378

Kashyap PL, Kumar S, Srivastava AK (2017) Nanodiagnostics for plant pathogens. Environ Chem Lett 15:7–13

Kaziro Y, Itoh H, Kozasa T, Satoh T (1991) Structure and function of signal-transducing GTP-binding proteins. Annu Rev Biochem 60:349–400

Kim DJ, Baek JM, Uribe P, Kenerley CM, Cook DR (2002) Cloning and characterization of multiple glycosyl hydrolase genes from *Trichoderma virens*. Curr Genet 40:374–384

Kogel KH, Voll LM, Schäfer P, Jansen C, Wu Y et al (2010) Transcriptome and metabolome profiling of field grown transgenic barley lack induced differences but show cultivar-specific variances. Proc Natl Acad Sci USA 107:6198–6203

Kosambo-Ayoo LM, Bader M, Loerz H, Becker D (2011) Transgenic sorghum (*Sorghum bicolor* L. *Moench*) developed by transformation with chitinase and chitosanase genes from *Trichoderma harzianum* expresses tolerance to anthracnose. Afr J Biotechnol 10:3659–3670

Krishijagran (2015) Outlook of pesticide consumption in India–Krishi Jagran. Available online: http://www.krishijagran.com/corporate-watch/Industry-Profile/2014/11/Outlook of Pesticide-Consumption-in-India. Retrieved on 29 July 2016

Kronstad J, De Maria AD, Funnell D et al (1998) Signalling via cAMP in fungi: interconnections with mitogen-activated protein kinase pathways. Arch Microbiol 170(6):395–404

Kubicek CP, Herrera-Estrella A, Seidl-Seiboth V, Martinez DA, Druzhinina IS, Thon M et al (2011) Genome Biol. Comparative genome sequence analysis underscores mycoparasitism as the ancestral life style of *Trichoderma*. Genome Biol 12:R40

Kumar A, Scher K, Mukherjee M, Pardovitz-Kedmi E, Sible GV, Singh US, Kale SP, Mukherjee PK, Horwitz BA (2010) Overlapping and distinct functions of two *Trichoderma virens* MAP kinases in cell- wall integrity, antagonistic properties and repression of conidiation. Biochem Biophys Res Commun 398(4):765–770

Kumar V, Shahid M, Singh A, Srivastava M, Mishra A, Srivastava YK, Pandey S, Shrarma A (2014) Effect of Biopriming with Biocontrol Agents *Trichoderma harzianum* (Th.Azad) and *Trichoderma viride* on Chickpea Genotype (Radhey). J Plant Pathol Microbiol 5:1–4

Kurusu T, Hamada J, Nokajima H et al (2010) Regulation of microbe associated molecular pattern-induced hypersensitive cell death, phytoalexin production, and defense gene expression by calcineurin B-like protein-interacting protein kinases, OsCIPK14/15, in rice cultured cells. Plant Physiol 153:678–692

Kyriacou MC, Rouphael Y (2018) Towards a new definition of quality for fresh fruits and vegetables. Sci Hortic 234:463–469

Latge JP (2007) The cell wall: a carbohydrate armour for the fungal cell. Mol Microbiol 66:279–290

Lehner SM, Atanasova L, Neumann NK, Krska R, Lemmens M, Druzhinina IS, Schuhmacher R (2013) Isotope-assisted screening for iron-containing metabolites reveals a high degree of diversity among known and unknown siderophores produced by *Trichoderma* spp. Appl Environ Microbiol 79:18–31

Liang X, Liu Y, Xie L, Liu X, Wei Y, Zhou X et al (2015) A ribosomal protein AgRPS3aE from halophilic *Aspergillus glaucus* confers salt tolerance in heterologous organisms. Int J Mol Sci 16 (2):3058–3070

Liebmann B, Gattung S, Jahn B, Brakhage AA (2003) cAMP signaling in *Aspergillus fumigatus* is involved in the regulation of the virulence gene *pksP* and in defense against killing by macrophages. Mol Gen Genomics 269:420–435

Liming Y, Qian Y, Pigang L, Sen L (2008) Expression of the HSP24 gene from *Trichoderma harzianum* in Saccharomyces cerevisiae. J Thermal Biol 33:1–6

Limon MC, Chacón MR, Mejías R, Delgado-Jarana J, Rincón AM, Codón AC, Benítez T (2004) Increased antifungal and chitinase specific activities of *Trichoderma harzianum* CECT 2413 by addition of a cellulose binding domain. Appl Microbiol Biotechnol 64:675–685

Lin YR, Lo CT, Liu SY, Peng KC (2012) Involvement of pachybasin and emodin in self-regulation of *Trichoderma harzianum* mycoparasitic coiling. J Agric Food Chem 60:2123–2128

Linder MB, Szilvay GR, Nakari-Setala T, Penttilala ME (2005) Hydrophobins: the protein-amphiphiles of filamentous fungi. FEMS Microbiol Rev 29:877–896

Liu M, Sun ZX, Zhu J, Xu T, Harman GE, Lorito M (2004) Enhancing rice resistance to fungal pathogens by transformation with cell wall degrading enzyme genes from *Trichoderma atroviride*. J Zhejiang Univ Sci 5:133–136

Liu Z, Faris JD, Oliver RP, Tan KC, Solomon PS, McDonald MC, McDonald BA, Nunez A, Lu S et al (2009) *SnTox3* acts in effector triggered susceptibility to induce disease on wheat carrying the Snn3 gene. PLoS Pathog 5:e1000581

Liu M, Liu J, Wang WM (2012) Isolation and functional analysis of *Thmfs1*, the first major facilitator superfamily transporter from the biocontrol fungus *Trichoderma harzianum*. Biotechnol Lett 34:1857–1862

Lo HC, Entwistle R, Guo CJ, Ahuja M, Szewczyk E, Hung JH et al (2012) Two separate gene clusters encode the biosynthetic pathway for the meroterpenoids austinol and dehydroaustinol in *Aspergillus nidulans*. J Am Chem Soc 134:4709–4720

Lopez-Bucio J, Pelagio-Flores R, Herrera-Estrella A (2015) *Trichoderma* as biostimulant: exploiting the multilevel properties of a plant beneficial fungus. Sci Hortic 196:109–123

Lopez-Mondejar R, Ros M, Pascual JA (2011) Mycoparasitism-related genes expression of *Trichoderma harzianum* isolates to evaluate their efficacy as biological control agent. Biol Control 56:59–66

Lorito M, Farkas V, Rebuffat S, Bodo B, Kubicek CP (1996) Cell wall synthesis is a major target of mycoparasitic antagonism by *Trichoderma harzianum*. J Bacteriol 178:6382–6385

Lorito M, Woo SL, Fernandez Garcia I, Colucci G et al (1998) Genes from mycoparasitic fungi as a source for improving plant resistance to fungal pathogens. Proc Natl Acad Sci USA 95:7860–7865

Lorito M, Woo SL, Harman GE, Monte E (2010) Translational research on *Trichoderma*: from 'omics to the field. Annu Rev Phytopathol 48:395–417

Lu Z, Tombolini R, Woo S, Zeilinger S, Lorito M, Jansson JK (2004) *In vivo* study of *Trichoderma*-pathogen-plant interactions, using constitutive and inducible green fluorescent protein reporter systems. Appl Environ Microbiol 70:3073–3081

Luo Y, Zhang DD, Dong XW, Zhao PB, Chen LL, Song XY et al (2010) Antimicrobial peptaibols induce defense responses and systemic resistance in tobacco against tobacco mosaic virus. FEMS Micro Lett 313:120–126

Luo Y, Ruan LF, Zhao CM, Wang CX, Peng DH, Sun M (2011) Validation of the intact zwittermicin A biosynthetic gene cluster and discovery of a complementary resistance mechanism in *Bacillus thuringiensis*. Antimicrob Agent Chemother 55:4161–4169

Maischak H, Zimmermann MR, Felle HH, Boland W, Mithofer A (2010) Alamethicin induced electrical long distance signaling in plants. Plant Signal Behav 5:988–990

Malmierca MG, Cardoza RE, Alexander NJ, McCormick SP, Hermosa R, Monte E, Gutierrez S (2012) Involvement of *Trichoderma trichothecenes* in the biocontrol activity and induction of plant defense-related genes. Appl Environ Microbiol 78:4856–4868

Malmierca MG, Cardoza RE, Alexander NJ, McCormick SP, Collado IG, Hermosa R, Monte E, Gutierrez S (2013) Relevance of trichothecenes in fungal physiology: disruption of tri5 in *Trichoderma arundinaceum*. Fungal Genet Biol 53:22–33

Malmierca MG, McCormick SP, Cardoza RE, Alexander NJ, Monte E, Gutierrez S (2015) Production of trichodiene by *Trichoderma harzianum* alters the perception of this biocontrol strain by plants and antagonized fungi. Environ Microbiol 17:2628–2646

Malmierca MG, Izquierdo-Bueno I, McCormick SP, Cardoza RE, Alexander NJ, Barua J, Lindo L, Casquero PA, Collado IG, Monte E, Gutiérrez S (2016) Trichothecenes and aspinolides produced by *Trichoderma arundinaceum* regulate expression of *Botrytis cinerea* genes involved in virulence and growth. Environ Microbiol 18(11):3991–4004

Marcello CM, Steindorff AS, Silva SP, Silva RN, Bataus LAM (2010) Expression analysis of the exo-β-1, 3-glucanase from the mycoparasitic fungus *Trichoderma asperellum*. Microbiol Res 165:75–81

Margolles-Clark E, Harman GE, Penttila M (1996) Enhanced expression of endochitinase in *Trichoderma harzianum* with the cbh1 promoter of *Trichoderma* reesei. Appl Environ Microbiol 62:2152–2155

Marik T, Urbán P, Tyagi C, Szekeres A, Leitgeb B, Vágvölgyi M et al (2017) Diversity profile and dynamics of peptaibols produced by green mould *Trichoderma* species in interactions with their hosts *Agaricus bisporus* and *Pleurotus ostreatus*. Chem Biodivers 14:e1700033

Martinez D, Berka RM, Henrissat B, Saloheimo M, Arvas M, Baker SE, Chapman J, Chertkov O, Coutinho PM et al (2008) Genome sequencing and analysis of the biomass-degrading fungus *Trichoderma reesei* (syn. *Hypocrea jecorina*). Nat Biotechnol 26:553–560

Martínez-Medina A, Pascual J, Pérez-Alfocea F, Albacete A, Roldán A (2010) *Trichoderma harzianum* and *Glomus intraradices* modify the hormone disruption induced by *Fusarium oxysporum* infection in melon plants. Phytopathology 100(7):682–688

Mastouri F, Bjorkman T, Harman GE (2012) *Trichoderma harzianum* enhances antioxidant defense of tomato seedlings and resistance to water deficit. Mol Plant-Microbe Interact 25:1264–1271

Mathys J, De Cremer K, Timmermans P, Van Kerckhove S, Lievens B, Vanhaecke M, Cammue BP, De Coninck B (2012) Genome-wide characterization of ISR induced in *Arabidopsis thaliana* by *Trichoderma hamatum* T382 against *Botrytis cinerea* infection. Front Plant Sci 3:108

Mcintyre M, Nielsen J, Arnau J, Brink H, Hansen K, Madrid S (2004) Proceedings of the 7th European conference on fungal genetics. Copenhagen, Denmark, pp 125–130

Mendes R, Kruijt M, de Bruijn I, Dekkers E, van der Voort M, Schneider JHM, Piceno YM, DeSantis TZ, Andersen GL, Bakker PAHM, Raaijmakers JM (2011) Deciphering the rhizosphere microbiome for disease-suppressive bacteria. Science 332:1097–1100

Mendoza-Mendoza A, Pozo MJ, Grzegorski D, Martinez P, Garcia JM, Olmedo-Monfil V, Cortes C, Kenerley C, Herrera-Estrella A (2003) Enhanced biocontrol activity of *Trichoderma* through inactivation of a mitogen-activated protein kinase. Proc Natl Acad Sci U S A 100:15965–15970

Mercado JA, Martín-Pizarro CL, Pascual MA, Quesada F et al (2007) Evaluation of tolerance to *Colletotrichum acutatum* in straw-berry plants transformed with *Trichoderma* derived genes. Acta Hortic 738:383–388

Mercado JA, Barcelo M, Pliego C, Rey M, Caballero JL, Munoz-Blanco J, Ruano-Rosa D, Lopez-Herrera C, Santos B, Romero-Munoz F, Pliego-Alfaro F (2015) Expression of the b-1, 3-glucanase gene *bgn13*,1 from *Trichoderma harzianum* in strawberry increases tolerance to crown rot diseases but interferes with plant growth. Transgenic Res 24:979–989

Migheli Q, Gonzales-Candelas L, Dealessi L, Camponogara A, Ramon Vidal D (1998) Transformants of *Trichoderma langibrachaitum* overexpressing the beta-1-4-endoglucanase gene agl1 show enhanced biocontrol of *Pythium ultimum* on cucumber. Phytopathology 88:673–677

Ming Q, Su C, Zheng C, Jia M, Zhang Q, Zhang H et al (2013) Elicitors from the endophytic fungus *Trichoderma atroviride* promote *Salvia miltiorrhiza* hairy root growth and tanshinone biosynthesis. J Exp Bot 4:5687–5694

Mishra M, Jalil SU, Mishra RK, Kumari S, Pandey BK (2016) *In vitro* screening of guava plantlets transformed with endochitinase gene against *Fusarium oxysporum* f. sp. *psidii*. Czech J Genet Plant Breed 52:6–13

Monte E (2001) Understanding *Trichoderma*: between biotechnology and microbial ecology. Int Microbiol 4:1–41

Montero M, Sanz L, Rey M, Llobell A, Monte E (2007) Cloning and characterization of bgn16·3, coding for a β-1, 6-glucanase expressed during *Trichoderma harzianum* mycoparasitism. J Appl Microbiol 103:1291–1300

Montero-Barrientos M, Cardoza R, Gutiérrez S, Monte E, Hermosa R (2007) The heterologous overexpression of hsp23, a small heat-shock protein gene from *Trichoderma virens*, confers thermotolerance to T. *harzianum*. Curr Genet 52:45–53

Montero-Barrientos M, Hermosa R, Cardoza RE, Gutierrez S, Nicolás C, Monte E (2010) Transgenic expression of the *Trichoderma harzianum* HSP70 gene increases Arabidopsis resistance to heat and other abiotic stresses. J Plant Physiol 167:659–665

Montero-Barrientos M, Hermosa R, Cardoza RE, Gutierrez S, Monte E (2011) Functional analysis of the *Trichoderma harzianum nox1* gene, encoding an NADPH oxidase, relates production of reactive oxygen species to specific biocontrol activity against *Pythium ultimum*. Appl Environ Microbiol 77:3009–3016

Mora A, Earle ED (2001) Resistance to Alternaria brassicicola in transgenic broccoli expressing a *Trichoderma harzianum* endochitinase gene. Mol Breed 8:1–9

Moran-Diez E, Hermosa R, Ambrosino P, Cardoza RE, Gutierrez S, Lorito M, Monte E (2009) The *ThPG1* endopolygalacturonase is required for the *Trichoderma harzianum*-plant beneficial interaction. Mol Plant-Microbe Interact 22:1021–1031

Mukherjee PK, Kenerley CM (2010) Regulation of morphogenesis and biocontrol properties in *Trichoderma virens* by a VEL-VET protein, Vel1. Appl Environ Microbiol 76:2345–2352

Mukherjee PM, Latha J, Hardar R, Horwitz BA (2003) TmkA, Mitogen activated Protein Kinase of *Trichoderma virens* is involved in biocontrol properties and repression of conidiation in the dark. Eukaryot Cell 2:446–455

Mukherjee PM, Latha J, Hadar R, Horwitz BA (2004) Role of two G- protein alpha subunits, *TgaA* and *TgaB*, in the antagonism of plant pathogens by *Trichoderma virens*. Appl Environ Microbiol 70(1):542–549

Mukherjee M, Mukherjee PK, Kale PS (2007) cAMP signalling is involved in growth, germination, mycoparasitism and secondary metabolism in *Trichoderma virens*. Microbiology 153:1734–1742

Mukherjee PK, Wiest A, Ruiz N, Keightley A, Moran-Diez ME, McCluskey K, Pouchus YF, Kenerley CM (2011) Two classes of new peptaibols are synthesized by a single non-ribosomal peptide synthetase of *Trichoderma virens*. J Biol Chem 286:4544–4554

Mukherjee M, Mukherjee PK, Horwitz BA, Zachow C, Berg G, Zeilinger S (2012a) *Trichoderma*-plant-pathogen interactions: advances in genetics of biological control. Indian J Microbiol 52:522–529

Mukherjee PK, Horwitz BA, Kenerley CM (2012b) Secondary metabolism in *Trichoderma*—a genomic perspective. Microbiology 158:35–45

Mukherjee PK, Buensanteai N, Moran-Diez ME, Druzhinina IS, Kenerley CM (2012c) Functional analysis of non-ribosomal peptide synthetases (NRPSs) in *Trichoderma* virens reveals a polyketide synthase (PKS)/NRPS hybrid enzyme involved in the induced systemic resistance response in maize. Microbiology 158:155–165

Mukherjee PK, Horwitz BA, Herrera-Estrella A, Schmoll M, Kenerley CM (2013) *Trichoderma* research in the genome era. Annu Rev Phytopathol 51:105–129

Naher L, Yusuf UK, Siddiquee S, Ferdous J, Rahman MA (2012) Effect of media on growth and antagonistic activity of selected *Trichoderma* strains against Ganoderma. Afr J Microbiol Res 6:7449–7453

Navazio L, Baldan B, Moscatiello R, Zuppini A, Woo SL, Mariani P, Lorito M (2007) Calcium-mediated perception and defense responses activated in plant cells by metabolite mixtures secreted by the biocontrol fungus *Trichoderma atroviride*. BMC Plant Biol 7:41

Neumann NK, Stoppacher N, Zeilinger S, Degenkolb T, Bruckner H, Schuhmacher R (2015) The peptaibiotics data- baseea comprehensive online resource. Chem Biodivers 12:743–751

Nicolás C, Hermosa R, Rubio B, Mukherjee PK, Monte E (2014) *Trichoderma* genes in plants for stress tolerance- status and prospects. Plant Sci 228:71–78

O'Kennedy MM, Crampton BG, Lorito M, Chakauya E, Breese WA, Burger JT, Botha FC (2011) Expression of β -1, 3-glucanase from a biocontrol fungus in transgenic pearl millet. S Afr J Bot 77:335–345

Omann MR, Lehner S, Escobar Rodrıguez C, Brunner K, Zeilinger S (2012) The seven-transmembrane receptor Gpr1 governs processes relevant for the antagonistic interaction of *Trichoderma atroviride* with its host. Microbiology 158:107–118

Oros G, Naár Z (2017) Mycofungicide: *Trichoderma* based preparation for foliar applications. Am J Plant Sci 8(02):113–125

Osbourn A (2010) Secondary metabolic gene clusters: evolutionary toolkits for chemical innovation. Trends Genet 26:449–457

Patron NJ, Waller RF, Cozijnsen AJ, Straney DC, Gardiner DM, Nierman WC, Howlett BJ (2007) Origin and distribution of epipolythiodioxopiperazine (ETP) gene clusters in filamentous ascomycetes. BMC Evol Biol 7:174

Peterbauer CK, Lorito M, Hayes CK, Harman GE, Kubicek CP (1996) Molecular cloning and expression of the nag1 gene (N-acetyl-b-D-glucosaminidase-encoding gene) from *Trichoderma harzianum* P1. Curr Genet 30(4):325–331

Plesofsky-Vig N, Brambl R (1995) Disruption of the gene for hsp30, an alpha-crystallin-related heat shock protein of *Neurospora crassa*, causes defects in thermotolerance. Proc Natl Acad Sci USA 92:5032–5036

Popiel D, Koczyk G, Dawidziuk A, Gromadzka K, Blaszczyk L, Chelkowski J (2014) *Zearalenone lactonohydrolase* activity in Hypocreales and its evolutionary relationships within the epoxide hydrolase subset of a/b-hydrolases. BMC Microbiol 14:82

Pozo MJ, JongMin B, Garcia JM, Kenerley CM (2004) Functional analysis of *tvsp1*, a serine protease encoding gene in the biocontrol agent *Trichoderma virens*. Fungal Genet Biol 41:336–348

Prakash NU, Jayanthi M, Sabarinathan R, Kangueane P, Mathew L, Sekar K (2010) Evolution, homology conservation, and identification of unique sequence signatures in GH19 family chitinases. J Mol Evol 70:466–478

Rai S, Kashyap PL, Kumar S et al (2016a) Identification, characterization and phylogenetic analysis of antifungal *Trichoderma* from tomato rhizosphere. Springer Plus 5:1939

Rai S, Kashyap PL, Kumar S et al (2016b) Comparative analysis of microsatellites in five different antagonistic *Trichoderma* species for diversity assessment. World J Microbiol Biotechnol 32:8

Ramada MH, Steindorff AS, Bloch C Jr, Ulhoa CJ (2016) Secretome analysis of the mycoparasitic fungus *Trichoderma harzianum* ALL 42 cultivated in different media supplemented with *Fusarium solani* cell wall or glucose. Proteomics 16:477–490

Rana IA, Loerz H, Schaefer W, Becker D (2012) Overexpression of chitinase and chitosanase genes from *Trichoderma harzianum* under constitutive and inducible promoters in order to increase disease resistance in wheat (*Triticum aestivum* L). Mol Plant Breed 3:37–49

Reino JL, Guerrero RF, Hernandez-Galan R, Collado IG (2008) Secondary metabolites from species of the biocontrol agent *Trichoderma*. Phytochem Rev 7:89–123

Reithner B, Schuhmacher R, Stoppacher N, Pucher M, Brunner K, Zeilinger S (2007) Signaling via the *Trichoderma atroviride* mitogen-activated protein kinase Tmk 1 differentially affects mycoparasitism and plant protection. Fungal Genet Biol 44:1123–1133

Reithner B, Ibarra-Laclette E, Mach RL, Herrera-Estrella A (2011) Identification of mycoparasitism-related genes in *Trichoderma atroviride*. Appl Environ Microbiol 77:4361–4370

Reithner B, Mach-Aigner AR, Herrera-Estrella A, Mach RL (2014) The transcriptional regulator *Xyr1* of *Trichoderma atroviride* supports the induction of systemic resistance in plants. Appl Environ Microbiol 80:5274–5281

Renshaw JC, Robson GD, Trinci APJ, Wiebe MG, Livens FR, Collison D, Taylor RJ (2002) Fungal siderophores: structures, functions and applications. Mycol Res 106:1123–1142

Rippa S, Eid M, Formaggio F, Toniolo C, Béven L (2010) Hypersensitive-like response to the pore-former peptaibol alamethicin in *Arabidopsis thaliana*. Chem Biol Chem 11:2042–2049

Rocha-Ramírez V, Omero C, Chet I, Horwitz BA, Herrera-Estrella A (2002) *Trichoderma atroviride* G- protein α-subunit gene tag1 isinvolved in mycoparasitic coiling and conidiation. Eukaryot Cell 1:594–605

Ron M, Avni A (2004) The receptor for the fungal elicitor ethylene-inducing xylanase is a member of a resistance-like gene family in tomato. Plant Cell 16:1604–1615

Rosado IV, Rey M, Codon AC, Govantes J, MorenoMateos MA, Benitez T (2007) QID74 Cell wall protein of *Trichoderma harzianum* is involved in cell protection and adherence to hydrophobic surfaces. Fungal Genet Biol 44:950–964

Rotblat B, Enshell-Seijffers D, Gershoni JM, Schuster S, Avni A (2002) Identification of an essential component of the elicitation active site of the EIX protein elicitor. Plant J 32:1049–1055

Rouphael Y, Cardarelli M, Bonini P, Colla G (2017) Synergistic action of a microbial-based biostimulant and a plant derived-protein hydrolysate enhances lettuce tolerance to alkalinity and salinity. Front Plant Sci 8:131

Rubio MB, Hermosa R, Reino JL, Collado IG, Monte E (2009) Thctf1 transcription factor of *Trichoderma harzianum* is involved in 6-pentyl- 2H-pyran-2-one production and antifungal activity. Fungal Genet Biol 46:17–27

Ruocco M, Lanzuise S, Woo SL, Lorito M (2007) The novel hydrophobin HYTRA1 from *Trichoderma harzianum* T22 plays a role in *Trichoderma*-plant interactions. XIII Inernational Congress. Mol Plant Microbe Interact:394

Ruocco M, Lanzuise S, Vinale F, Marra R, Turra D, LoisWoo S, Lorito M (2009) Identification of a new biocontrol gene in *Trichoderma atroviride*: the role of an ABC transporter membrane pump in the interaction with different plant pathogenic fungi. Am Phytopathol Soc 22(3):291–301

Ruocco M, Lanzuise S, Lombardi N, Woo SL, Francesco Vinale F, Marra R, Varlese R, Manganiello G, Pascale A, Scala V, Turra D, Scala F, Lorito M (2015) Multiple roles and effects of a novel *Trichoderma* hydrophobin. Mol Plant-Microbe Interact 28:167–179

Saadia M, Ahmed S, Jamil A (2008) Isolation and cloning of cre1 gene from a filamentous fungus *Trichoderma harzianum*. Pak J Bot 40(1):421–426

Saiprasad GVS, Mythili JB, Anand L, Suneetha C, Rashmi HJ, NaveenaC GG (2009) Development of *Trichoderma harzianum* gene construct conferring antifungal activity in transgenic tobacco. Indian J Biotechnol 8:199–206

Salas-Marina MA, Isordia-Jasso M, Islas-Osuna MA, Delgado-Sánchez P, Jiménez-Bremont JF et al (2015) The Epl1 and Sm1 proteins from *Trichoderma atroviride* and *Trichoderma virens* differentially modulate systemic disease resistance against different life style pathogens in *Solanum lycopersicum*. Front Plant Sci 23:77

Samolski I, Rincon AM, Pinzón LM, Viterbo A, Monte E (2012) The *qid74* gene from *Trichoderma harzianum* has a role in root architecture and plant biofertilization. Microbiology 158:129–138

Sanz L, Montero M, Redondo J, Llobell A, Monte E (2005) Expression of an alpha-1,3-glucanase during mycoparasitic interaction of *Trichoderma asperellum*. FEBS J 272:493–499

Schäfer T, Hanke MV, Flachowsky HS, König A, Peil M et al (2012) Chitinase activities, scab resistance, mycorrhization rates and biomass of own rooted and grafted transgenic apple. Genet Mol Biol 35:466–473

Scharf DH, Brakhage AA, Mukherjee PK (2016) Gliotoxin e bane or boon? Environ Microbiol 18:1096–1109

Schmoll M, Dattenbock C, Carreras-Villasenor N, Mendoza-Mendoza A, Tisch D, Aleman MI, Baker SE, Brown C, Cervantes-Badillo MG, Cetz-Chel J et al (2016) The Genomes of three uneven siblings: footprints of the lifestyles of three *Trichoderma* Species. Microbiol Mol Biol Rev 80:205–327

Segarra G, Casanova E, Aviles M, Trillas I (2010) *Trichoderma asperellum* strain T34 controls Fusarium wilt disease in tomato plants in soilless culture through competition for iron. Microb Ecol 59:141–149

Seidl V, Huemer B, Seiboth B, Kubicek CP (2005) A complete survey of *Trichoderma* chitinases reveals three distinct subgroups of family 18 chitinases. FEBS J 272:5923–5939

Seidl V, Marchetti M, Schandl R, Allmaier G, Kubicek CP (2006) Epl1, the major secreted protein of *Hypocrea atroviridis* on glucose, is a member of a strongly conserved protein family comprising plant defense response elicitors. FEBS J 273:4346–4359

Seidl V, Song L, Lindquist E et al (2009) Transcriptomic response of the mycoparasitic fungus *Trichoderma atroviride* to the presence of a fungal prey. BMC Genomics 10:567

Shah JM, Raghupathy V, Veluthambi K (2009) Enhanced sheath blight resistance in transgenic rice expressing an endochitinase gene from *Trichoderma virens*. Biotechnol Lett 31:239–244

Sharma KK, Singh US, Sharma P, Kumar A, Sharma L (2015) Seed treatments for sustainable agriculture-a review. J Appl Nat Sci 7(1):521–539

Sharma S, Rai P, Rai S, Srivastava M et al (2017) Genomic revolution in crop disease diagnosis: a review. In: Singh SS (ed) Plants and microbes in an ever changing environment. Nova Science Publishers, Hauppauge, pp 257–293

Sheikh M, Safi-uddin A, Khan Z, Rizvi R, Mahmood I (2013) Antibacterial and antifungal potential of some medicinal plants against certain phytopathogenic microorganisms. Arch Phytopathol Plant Protect 46(9):1070–1080

Shentu X, Yao J, Yuan X, He L, Sun F, Ochi K, Yu X (2018) *Tri11*, *tri3*, and *tri4* genes are required for trichodermin biosynthesis of *Trichoderma brevicompactum*. AMB Express 8:58

Shi M, Chen L, Wang XW, Zhang T, Zhao PB, Song XY, Sun CY, Chen XL, Zhou BC, Zhang YZ (2012) Antimicrobial peptaibols from *Trichoderma pseudokoningii* induce programmed cell death in plant fungal pathogens. Microbiology 158:166–175

Shoresh M, Yedidia I, Chet I (2005) Involvement of jasmonic acid/ethylene signaling pathway in the systemic resistance induced in cucumber by *Trichoderma asperellum* T203. Phytopathology 95:76–84

Shoresh M, Harman GE, Mastouri F (2010) Induced systemic resistance and plant responses to fungal biocontrol agents. Annu Rev Phytopathol 48:21–43

Silva RDN, Steindorff AS, Ulhoa CJ, Felix CR (2009) Involvement of G-alpha protein GNA3 in production of cell wall-degrading enzymes by *Trichoderma* reesei (*Hypocrea jecorina*) during mycoparasitism against *Pythium ultimum*. Biotechnol Lett 31:531–536

Srivastava M, Shahid M, Pandey S, Singh A, Kumar V, Gupta S, Maurya M (2014) *Trichoderma* genome to genomics: a review. J Data Min Genomics Proteom 5:162

Steyaert JM, Stewart A, Jaspers MV, Carpenter M, Ridgway HJ (2004) Co-expression of two genes, a chitinase (*chit42*) and proteinase (*prb1*), implicated in mycoparasitism by *Trichoderma hamatum*. Mycologia 96(6):1245–1252

Stoppacher N, Kluger B, Zeilinger S, Krska R, Schuhmacher R (2010) Identification and profiling of volatile metabolites of the biocontrol fungus *Trichoderma atroviride* by HS-SPME-GC-MS. J Microbiol Methods 81:187–193

Stoppacher N, Neumann NKN, Burgstaller L, Zeilinger S, Degenkolb T, Breuckner H, Schuhmacher R (2013) The comprehensive peptaibiotics database. Chem Biodivers 10:734–743

Strakowska J, Błaszczyk L, Chełkowski J (2014) The significance of cellulolytic enzymes produced by *Trichoderma* in opportunistic lifestyle of this Fungus. J Basic Microbiol 54:1–12

Strieker M, Tanovic A, Marahiel MA (2010) Nonribosomal peptide synthetases: structures and dynamics. Curr Opin Struct Biol 20:234–240

Sultana F, Hossain MM, Kubota M, Hyakumachi M (2009) Induction of systemic resistance in *Arabidopsis thaliana* in response to a culture filtrate from a plant growth-promoting fungus, Phoma sp. GS8e3. Plant Biol 11:97–104

Thon M, Al Abdallah Q, Hortschansky P, Scharf DH, Eisendle M, Haas H et al (2010) The CCAAT-binding complex coordinates the oxidative stress response in eukaryotes. Nucleic Acids Res 38:1098–1113

Tijerino A, Cardoza RE, Moraga J, Malmierca MG, Vicente F, Aleu J, Collado IG, Gutiérrez S, Monte E, Hermosa R (2011) Overexpression of the trichodiene synthase gene tri5 increases trichodermin production and antimicrobial activity in *Trichoderma brevicompactum*. Fungal Genet Biol 48:285–296

Trushina N, Levin M, Mukherjee PK, Horwitz BA (2013) PacC and pH dependent transcriptome of the mycotrophic fungus *Trichoderma virens*. BMC Genomics 14:1–21

Tuão Gava CA, Pinto JM (2016) Biocontrol of melon wilt caused by *Fusarium oxysporum* Schlect f. sp. *melonis* using seed treatment with *Trichoderma* spp. and liquid compost. Biol Control 97:13–20

Tucci M, Ruocco M, De Masi L, De Palma M, Lorito M (2011) The beneficial effect of *Trichoderma* spp. on tomato is modulated by the plant genotype. Mol Plant Pathol 12:341–354

Van Wees SCM, Van der Ent S, Pieterse CMJ (2008) Plant immune responses triggered by beneficial microbes. Curr Opin Plant Biol 11:443–448

Vargas W, Mandawe J, Kenerley C (2009) Plant-derived sucrose is a key element in the symbiotic association between *Trichoderma virens* and maize plants. Plant Physiol 151(2):792–808

Vargas WA, Mukherjee PK, Laughlin D, Wiest A, Moran-diez ME, Kenerley CM (2014) Role of gliotoxin in the symbiotic and pathogenic interactions of *Trichoderma virens*. Microbiol 4:2319–2330

Velázquez-Robledo R, Contreras-Cornejo HA, Macías-Rodríguez L, Hernández-Morales A, Aguirre J, Casas-Flores S, López-Bucio J, Herrera-Estrella A (2011) Role of the 4-phosphopantetheinyl transferase of *Trichoderma virens* in secondary metabolism, and induction of plant defense responses. Mol Plant-Microbe Interact 24:1459–1471

Verma M, Brar SK, Tyagi RD, Surampalli RY, Valero JR (2007) Antagonistic fungi, *Trichoderma* spp.: panoply of biological control. Biochem Eng J 37:1–20

Vinale F, Sivasithamparam K, Ghisalberti E, Marra R, Woo S, Lorito M (2008a) *Trichoderma*–plant–pathogen interactions. Soil Biol Biochem 40:1–10

Vinale F, Sivasithamparam K, Ghisalberti EL, Marra R, Barbetti MJ, Li H et al (2008b) A novel role for *Trichoderma* secondary metabolites in the interactions with plants. Physiol Mol Plant Pathol 72:80–86

Vinale F, Ghisalberti EL, Sivasithamparam K, Marra R, Ritieni A, Ferracane R, Woo S, Lorito M (2009) Factors affecting the production of *Trichoderma harzianum* secondary metabolites during the interaction with different plant pathogens. Lett Appl Microbiol 48:705–711

Vinale F, Nigro M, Sivasithamparam K, Flematti G, Ghisalberti EL, Ruocco M, Varlese R, Marra R, Lanzuise S, Eid A, Woo SL, Lorito M (2013) Harzianic acid: a novel siderophore from *Trichoderma harzianum*. FEMS Microbiol Lett 347:123–129

Vinale F, Sivasithamparam K, Ghisalberti EL, Woo SL, Nigro M, Marra R, Lombardi N et al (2014) *Trichoderma* secondary metabolites active on plants and fungal pathogens. Open Mycol J 8:127–139

Vishnevetsky J, White TL Jr, Palmateer AJ, Flaishman M, Cohen Y, Elad Y, Velcheva M, Hanania U, Sahar N, Dgani O, Perl A (2011) Improved tolerance toward fungal diseases in transgenic *Cavendish banana* (Musa spp. AAA group) cv. Grand Nain. Transgenic Res 20:61–72

Viterbo A, Chet I (2006) *TasHyd1*, a new hydrophobin gene from the biocontrol agent *Trichoderma asperellum*, is involved in plant root colonization. Mol Plant Pathol 7:249–258

Viterbo A, Harel M, Chet I (2004) Isolation of two aspartyl proteases from *Trichoderma asperellum* expressed during colonization of cucumber roots. FEMS Microbiol Lett 238:151–158

Viterbo M, Harel B, Horwitz A, Chet I, Mukherjee PK (2005) *Trichoderma* mitogen-activated protein kinase signaling is involved in induction of plant systemic resistance. Appl Environ Microbiol 71:6241–6246

Viterbo A, Wiest A, Brotman Y, Chet I, Kenerley CM (2007) The 18mer peptaibols from *Trichoderma virens* elicit plant defence responses. Mol Plant Pathol 8:737–746

Viterbo A, Landau U, Kim S, Chernin L, Chet I (2010) Characterization of ACC deaminase from the biocontrol and plant growth-promoting agent *Trichoderma asperellum* T203. FEMS Microbiol Lett 305:42–48

Vizcaino JA, Cardoza RA, Hauser M, Hermosa R, Rey M, Lobell A, Becker JM, Gutierrez S, Monte E (2006) *ThPTR2*, a di/tri-peptide transporter gene from *Trichoderma harzianum*. Fungal Genet Biol 43:234–246

Vos CM, De Cremer K, Cammue BPA, De Coninck B (2015) The toolbox of *Trichoderma* spp. in the biocontrol of *Botrytis cinerea* disease. Mol Plant Pathol 16:400–412

Waghunde RR, Shelake MR, Sabalpara NA (2016) *Trichoderma*: a significant fungus for agriculture and environment. Afr J Agric Res 11:1952–1965
Wiest A, Grzegorski D, Xu B et al (2002) Identification of peptaibols from *Trichoderma virens* and cloning of a peptaibol synthetase. J Biol Chem 277:20862–20868
Wilhite SE, Lumsden RD, Straney DC (2001) Peptide synthetase gene in *Trichoderma virens*. Appl Environ Microbiol 67:5055–5062
Woloshuk CP, Shim WB (2013) Aflatoxins, fumonisins, and trichothecenes: a convergence of knowledge. FEMS Microbiol Rev 37:94–109
Woo S, Lorito M (2007) Exploiting the interactions between fungal antagonists, pathogens and the plant for biocontro. In: Vurro M, Gressel J (eds) Novel biotechnologies for biocontrol agent enhancement and management. Springer, Netherlands, pp 107–130
Woo S, Donzelli B, Scala FRM, Harman G, Kubicek C, Del Sorbo G, Lorito M (1999) Disruption of the *ech42* (endochitinase-encoding) gene affects biocontrol activity in *Trichoderma harzianum* P1. Mol Plant Microbiol Interact 12:419–429
Woo SL, Scala F, Ruocco M, Lorito M (2006) The molecular biology of the interactions between *Trichoderma* spp., phytopathogenic fungi, and plants. Phytopathology 96:181–185
Yang Y, De Coninck B, Cammue BPA, Vos C (2013) Induced systemic resistance (ISR) signaling pathways involved in the *Trichoderma hamatum* Tomato *Botrytis cinerea* tripartite system. IOBC Bull 89:263–266
Yasmeen R, Siddiqui ZS (2017) Physiological responses of crop plants against *Trichoderma harzianum* in saline environment. Acta Bot Croat 76(2):154–162
Yedidia I, Shoresh M, Kerem Z et al (2003) Concomitant induction of systemic resistance to *Pseudomonas syringae* pv. *lachrymans* in cucumber by *Trichoderma asperellum* (T-203) and accumulation of phytoalexins. Appl Environ Microbiol 69:7343–7353
Yoshikuni Y, Martin VJ, Ferrin TE et al (2006) Engineering cotton (+)- delta-cadinene synthase to an altered function: germacrene D-4-ol synthase. Chem Biol 13:91–98
You J, Zhang J, Wu M, Yang L, Chen W, Li G (2016) Multiple criteria-based screening of *Trichoderma* isolates for biological control of *Botrytis cinerea* on tomato. Biol Control 101:31–38
Yuan XF, Shentu XP, Yu XP (2016) Cloning of tri cluster and analysis of tri genes expressions under different Trichodermin-producing conditions in *Trichoderma brevicompactum*. Chin J Biol Control 32(1):93–100
Zeilinger S, Omann M (2007) *Trichoderma* biocontrol: signal transduction pathways involved in host sensing and mycoparasitism. Gene Regul Syst Biol 1:227–234
Zeilinger S, Reithner B, Scala V, Peissl I, Lorito M, Mach RL (2005) Signal transduction by *Tga3*, a novel G protein and subunit of *Trichoderma atroviride*. Appl Environ Microbiol 71:1591–1597
Zeilinger S, Gruber S, Bansal R, Mukherjee PK (2016) Secondary metabolism in *Trichoderma*—chemistry meets genomics. Fungal Biol Rev 30:74–90
Zelicourt AD, Colcombet J, Hirt H (2016) The role of MAPK modules and aba during abiotic stress signaling. Trends Plant Sci 21:677–685
Zhang F, Liu Z, Gulijimila M, Wang Y, Fan H, Wang Z (2016) Functional analysis of the 1-aminocyclopropane-1-carboxylate deaminase gene of the biocontrol fungus *Trichoderma asperellum* ACCC30536. Can J Plant Sci 96:265–275
Zhong YH, Wang TH, Wang XL, Zhang GT, Yu HN (2009) Identification and characterization of a novel gene, TrCCD1, and its possible function in hyphal growth and conidiospore development of *Trichoderma reesei*. Fungal Genet Biol 46:255–263

Chapter 4
Trichoderma: Boon for Agriculture

C. Manoharachary and D. Nagaraju

Abstract *Trichoderma* Pers. is one of the important soil fungi growing on diversified habitats. It is represented by more than 100 species. Various species have been identified using morpho-taxonomy and molecular methods. *Trichoderma* spp. have been reported to grow luxuriantly on different media. The beneficial effects of *Trichoderma* include its utility as a biocontrol agent of soil-borne, root-borne, foliar, and aerial fungal/bacterial pathogens, and also it has been considered as a plant growth promoter. *Trichoderma* is also employed as a bioremedial agent and also used in the industry. *Trichoderma* as a potential biocontrol agent has attracted the attention of researchers all over the world. The present paper reviews diversity, taxonomy, conservation, growth, its utility as a biocontrol agent, plant growth promoter, and other related aspects. In recent times it has also been used as a seed primer to control several diseases. In order to maintain soil health and plant health, *Trichoderma* seems to be a promising fungus as nature's gift to boost agriculture besides maintaining soil and plant health.

Keywords Agriculture · Antagonism · Biocontrol · Fungus · Plant growth · Root-borne · Soil-borne · Soil health · *Trichoderma*

4.1 Diversity, Taxonomy, and Conservation

Trichoderma Pers. is a commonly known soil inhabitant and has the ability to colonize diversified habitats. It is the most common culturable fungus. *Trichoderma* Pers. is the name given for its conidial state in the year 1801 by Persoon. Its perfect stage mostly falls under Ascomycota, Class Sordariomycetes, Order Hypocreales,

---

C. Manoharachary (✉)
Mycology and Molecular Plant Pathology Laboratory, Department of Botany, Osmania University, Hyderabad, Telangana, India

D. Nagaraju
Department of Botany, Government Degree College, Warangal, Telangana, India

C. Manoharachary et al. (eds.), Trichoderma: *Agricultural Applications and Beyond*, Soil Biology 61, https://doi.org/10.1007/978-3-030-54758-5_4

and Family Hypocreaceae. The type species of *Trichoderma* is *Trichoderma fuliginoides*. However, its perfect stage belongs to mostly the genus *Hypocrea*. If the genus *Trichoderma* is examined critically, it was originally described in 1794 by Christian Henrik Persoon. For a longer time the taxonomy of this genus remained a complicated subject, and it was difficult to resolve for many years. *Trichoderma* has been considered to have only one species that is *Trichoderma viride* producing green colored conidia, hence named as green mold. The genus *Trichoderma* has been divided by Bissett (1991, 1992) into five sections after Rifai. Due to the entry of molecular tools from 1995 onwards, some changes were made in Bissett's scheme. Druzhinina and Kubicek (2005) concluded that the genus *Trichoderma* was considered as a complete fungus due to the discovery of its perfect stage as *Hypocrea*. The above scientists have also identified 88 species.

The genus *Trichoderma* remained as a monotypic genus till Rifai (1969) recognized nine species. According to Index Fungorum (2019), it seems to be as many as 427 records of *Trichoderma* are known by specific names. Recent contributions on the genus *Trichoderma* include those of Gams and Bissett (1998), Nagamani et al. (2002), and Samuels (2006). These works include a record of 135 species.

The genus *Trichoderma* Pers. is a ubiquitous fungus distributed widely in soils besides colonizing the root, litter, and other habitats. *Trichoderma* with more than 100 species is a complex genus posing problems in understanding its morphology, taxonomy, and identification. The authors have isolated around 17 species of *Trichoderma* from various soils of Telangana and Andhra Pradesh, India, respectively (Table 4.1).

The fungus *Trichoderma* can be grown on a wide variety of growth media and this has helped in easy culturing for identification. In recent times molecular approaches like RFLP, PCR, RAPD, ITS, DNA Barcoding, LSU, and other tools along with electron microscopy paved the way for better characterization of different species and strains. Bissett (1992), Rifai (1969), and Samuels et al. (1998) elevated Rifai species aggregates to species level and recognized several species within five sections of the genus (*Thachybasidium, Longibrachiatum, Trichoderma, Satumispermum, Hypocreanum*). Several research articles have been published on the genus *Trichoderma* to decide whether these artificial approaches have a molecular or physiological basis. Kiffer and Morelet (2000) and Samuels et al. (1998) have recognized 36 species based on molecular methods.

Molecular methods have added additional strength to the taxonomy of *Trichoderma* and its phylogenetic classification (Druzhinina and Kubicek 2005; Druzhinina et al. 2006; Kullnig et al. 2000; Kullnig-Gradinger et al. 2002; Jaklitsch 2009; Lieckfeldt and Seifert 2000; Lübeck et al. 2000; Overton et al. 2006; Taylor et al. 2000). The image analysis of HPLC and chromatograms were found useful for the separation of *Trichoderma* strains (Thrane et al. 2001). Several researchers have followed morphotaxonomic, biochemical, and molecular methodologies not only in the identification of species but also in strain differentiation. Mostly the species of *Trichoderma* were typified only with its asexual stage as many species were not linked with the sexual stage. Some species of *Trichoderma* have been identified as

Table 4.1 List of *Trichoderma* spp. recorded by the author(s)

S. No	Name	Substrate
1.	*Trichoderma asperellum*	Herbicide treated soil, Hyderabad
2.	*T. atroviride*	Polluted soils, Patancheru
3.	*T. aureoviride*	Forest soil, Vikarabad
4.	*T.citrinoviride*	Forest soil, Vikarabad
5.	*T. fasciculatum*	Cultivated soil of ground nut
6.	*T. fertile*	Cultivated soil of ground nut
7.	*T. harzianum*	Pigeon pea cultivated soil
8.	*T. konilangbra*	Polluted soil
9.	*T. koningii*	Forest soil, Vikarabad
10.	*T. longibrachiatum*	Forest soil of Vikarabad
11.	*T. piluliferum*	Pigeon pea cultivated soil
12.	*T. polysporum*	Cultivated soil, Chittoor
13.	*T. pseudokoningii*	Forest soil, Vikarabad
14.	*T. reesei*	Polluted soil, Hyderabad
15.	*T. strictipilis*	Polluted soil, Hyderabad
16.	*T. virens*	Forest soil, Vikarabad
17.	*T. viride*	Rhizosphere soil of *Ocimum* sp., Hyderabad

Source: Nagarmani A, Kunwar IK, Manoharachary C (2006) IK International Pub., New Delhi., India, p. 477

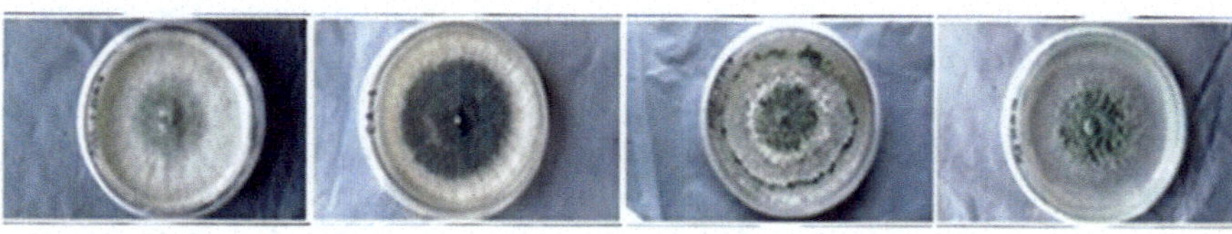

Fig. 4.1 Growth of *T. harzianum* on different agar media (Author's data)

mycoparasites and many of the species found to be antagonistic to pathogens. In recent times species of *Trichoderma* have also been reported to be endophytic.

Trichoderma spp. grow rapidly on different agar media. Colony color changes from white to various shades of green or remains light green. Yellow amber or brownish color pigments may be found on the reverse side of the colony. Exudates may be hyaline to yellow in color. Very few species have coconut odor (Fig. 4.1).

Mycelium is hyaline, profusely branched, smooth-walled, chlamydospores intercalary or terminal, globose, ellipsoidal, colorless, and smooth. Conidiophores are highly ramified, loosely or compactly tufted, they form into distinct concentric rings bearing conidia. Conidiophores may be single or irregularly branched on the aerial hyphae or submerged hyphae. Conidiophores are broad, flexuous, or with straight areas with numerous side branches. Side branches arise at right angles or at a wide angle to the axis in pairs or verticel. The apex of the conidiophores is terminated by sterile elongation. Phialides usually are in divergent verticals produced terminally on

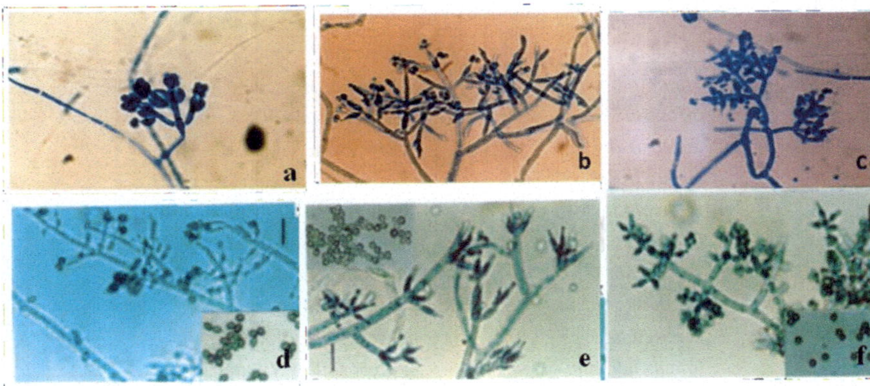

Fig. 4.2 (**a**) *Trichoderma virens*, (**b**) *T. pseudokoningii*, (**c**) *T. harzianum*, (**d**) *T. aureoviride*, (**e**) *T. koningii*, (**f**) *T. pilulifoerum* (all 400×) (Author's data)

branches. These are flask shaped or fine pin shaped to ovoid or ampulliform. Phialides arise in irregular whorls, beneath the terminal phialides. Philospores, the conidia are colorless or grayish, brown, yellowish-green to dark green. The spores are produced as single or successively accumulate at the tip of phailide to form heads. Conidia are smooth or minutely ornamented, sinuate, some with bullet-like or wing-like projection, globose, subglobose, obovoid, elliptic-cylindrical to almost oblong. Sometimes angular, base is occasionally distinctly truncate. The lectotype species is T*richoderma viride* Pers. ex S.F Gray. Some species studied by authors are shown in Fig. 4.2a–f.

The teleomorph of the genus *Trichoderma* falls under the genus *Hypocrea* which is an ascomycete. *Hypocrea* is characterized by ascocarps, which are fleshy, stromatic with shades of light or dark brown, yellow, or orange. Pulvinate stromata may be effused in some or differently organized. The perithecia are completely immersed in the substratum. Ascospores are hyaline or green, typically spinulose. The genus *Hypocrea* has got 200 species so far described (Kirk et al. 2008). However, very few have been grown in pure culture. *Trichoderma* spp. have been found to grow luxuriantly on different growth media (Fig. 4.1).

4.2 Beneficial Effects of *Trichoderma*

The beneficial effects of *Trichoderma* include its ubiquitous occurrence and successful colonization of diversified habitats. It efficiently fights their competitors on the establishment in a habitat as it launches its potential degradable mechanism of enzymes for the decomposition of the heterogeneous substrates. Efficient biocontrol stages of this genus are being developed as a biological fungicide(s) to control several plant diseases using its metabolites and novel antibiotics elaborated. The cellulase, elaborated by *Trichoderma reesei* has become important in the production of second-generation biofuels from cellulosic waste. The *Trichoderma* has been considered as a protector of plant health. *Trichoderma* spp. have been extensively used for the production of food additives and related products useful in the food industry. *T. harzianum* and *T. aggressivum* have been found to cause green mold disease on button mushrooms all over the world.

T. harzianum has also been reported as a biocontrol agent of foliar pathogens. In this twenty-first century, the environment is dumped with a number of xenobiotic compounds in particular the pesticides, fungicides, insecticides, and herbicides which ultimately reach a destination other than target species. Noticeable amounts of above compounds reach the soil, thus potentially affecting soil fertility and plant health. In this regard, *Trichoderma* as a potential soil-borne fungus degrades many such toxic compounds. Species of *Trichoderma* not only act as biocontrol agents but also acclaimed as effective in nullifying the ill effects of pesticides and other xenobiotic compounds. *Trichoderma,* because of its metabolic regulation and functional genomics, its utility has gained importance for commercial purposes. It is also employed as a bioremedial agent. Another important aspect is its application as a plant growth promoter.

Trichoderma also serves the purpose as a source of industrial enzymes, protein producers and also few species find their application in the wine and beer industry. *Trichoderma*-based products have found widespread use in agriculture. Thus, the genus *Trichoderma* is considered as an important fungus from the point of plant disease control, resistance induction, abiotic stress tolerance, plant growth promotion, biodegradation, bioremediation, pharmaceuticals, industry, medicine, and others. Very rarely *Trichoderma* has been considered as toxic domestic mold. *T. longibrachiatum* produces *trilongines* which affects neurons, heart, and anemocytes. A calcineurin inhibitor is also known to be produced by *T. polysporum.*

4.3 *Trichoderma*: As a Potential Biocontrol Agent

Biocontrol means reduction of inoculums(s) density or disease-producing activities of a pathogen. This is accomplished by fungi, bacteria, actinomycetes, and viruses. Biocontrol has attracted the attention of the researcher's world over for the last 70 years. This is because of the fact that the biological control agents are considered eco-friendly, i.e., the management of plant diseases in the absence of various agricultural practices including fungicide and pesticide application. The most commonly used biocontrol agents include *Trichoderma, Gliocladium, Aspergillus, Penicillium, Chaetomium, Glomus, Dactylella, Paecilomyces,* and others. Biocontrol can be achieved by following various steps namely competition, hyper-parasitism, induced resistance, and hypho-virulence. The mycoparasitism along with the production of volatile and non-volatile antibiotics has been considered as an important mechanism for operating *Trichoderma* as a commercial biocontrol agent. Probably, the production of these biocontrol agents through mass multiplication and also by using genetic engineering mechanisms may provide better application and improvements in the efficacy of wild strains as biocontrol agent(s).

In recent times chemical disease control which includes fungicides, pesticides, herbicides, and others have created the occurrence of toxic residues in food, soil, river, groundwater, and also in various crop produce, besides the development of resistance of plant pathogens to these chemicals. In developing and developed countries, the application of modern methods of crop protection, but for the production of disease-resistant varieties, there has been an overall negative impact on the environment, soil fertility, and society. Effective biocontrol takes advantage of naturally occurring competition of living organisms for limited ecological niches. Further two organisms at the same time cannot occupy the same space.

4.4 Mechanisms of Biological Control

Biological control means suppression or elimination of the disease-causing pathogen by introducing or using another biological entity. The term "Biological Control" was introduced by von Tubeuf (1914) and later it was elaborated by Hartley (1921) by introducing another microorganism to control some root-borne diseases. Many diseases are caused by crop plants and forest plants by pathogenic microbes and fungi besides the insects. However, just the presence or introduction of an organism to suppress or eliminate the disease-causing organism needs to build up its mass or inoculum. This process involves only biological action by one or more organisms other than man (Cook and Baker 1983). This also involves decreasing pathogen's activity accomplished by one or more biological entities along with host plants other than man (Baker 1987). Managining plant diseases using biological entities is called "Biological Control, (Harman 2000)". Mechanism of biological control involves the following aspects:

1. Antibiosis denotes the suppression of disease-causing pathogen by using another fungus/microbe through the elaboration of antibiotic substances. For example, the Take All disease of wheat has been controlled by elaborating phenazin produced by *Pseudomonsas fluorescens*. *Trichoderma virens* controls damping-off of cotton caused by *Pythium ultimum* which produces gliovirin.
2. In the mechanism of competition, many disease-causing agents and biological control agents compete with each other for nutrients to achieve growth, multiplication, and survival. The biocontrol agents which occur in soil or introduced into the soil limit the growth of pathogens as the antagonist posses a more efficient uptake system of nutrients than the pathogen.
3. Many fungi are parasitic on disease-causing pathogens called mycoparasites. Four steps namely: Chemotropism, Recognition, Attachment, Cell wall degradation, and Penetration are the steps involved in mycoparasitism. For example, *Trichoderma* spp. parasitize *Rhizoctonia* spp., the soil and the root pathogen and

kills it. *Dasturella* has been found to be parasitic on rust pathogens. Similarly, *Ampelomyces* grows as mycoparasite on powdery mildews.

4. Another mechanism of biological control is the acquisition of systemic resistance. SAR can be induced into the plants by inoculating the plants with a necessary pathogen/non-pathogen or synthetic chemical compounds which trigger defense against many disease-causing pathogens. This has been demonstrated by inoculating *Trichoderma* spp. or plant growth-promoting rhizobacteria into the same host plant which were found resistant to many disease-causing pathogens. This is because of the induction of plant systemic resistance through plant chemicals such as polyacrylic acid, ethylene, salicylic acid, jasmonic acid, and others that offer resistance toward disease-causing pathogen.
5. Many predators are used in the control of many diseases. For example, Lady Beetles are voracious predators of aphids, mites, scale insects, and caterpillars which cause diseases. The eggs of the insects which act as a host are used as food by the insect larvae. Most insect parasitoids play an important role in pest and disease control.
6. Another mechanism called Antagonism involves using of "X" organism to kill "Y" pathogen. "X" secretes a substance that is of antibiotic nature and kills the "Y" pathogen.

   The microbes, fungi, and others which occur abundantly in nature are of great importance. These organisms are of great use as biological control agents which act as an alternate mechanism employed against chemical control for better control of plant diseases and pests. Biological control reduces the excessive usage of agrochemicals. Thus, biological control can be achieved through antagonism of soil-borne and root-borne pathogens by eliciting a plant-mediated resistance response and through a mechanism that involves antibiosis, parasitism, competition, cell wall degradation, and other such activities.

4.5 Antagonism by *Trichoderma* spp.

The authors have tested *Trichoderma* isolates on soil-borne and root-borne pathogens like *Macrophomina phaseolina* and *Rhizoctonia solani* (isolated from diseased roots of green gram and chickpea.). The data indicates that the *Trichoderma* has shown maximum inhibition to the above pathogens (Table 4.2). The effect of culture filtrates of *Trichoderma* spp. and respective pathogens as mentioned above were tested on seed germination percentage (Table 4.3). The results clearly indicated the

Table 4.2 Antagonistic activity by *Trichoderma harzianum* on *Macrophomina phaseolina* and *Rhizoctonia solani*

S. No	Pathogen	Control	% Inhibition
1	*M. phaseolina*	26	70.00
2	*R. solani*	39	64.00

Author's data

Table 4.3 Effect of pathogen and *Trichoderma* spp. culture filtrates on seed germination (%)

S. No	*Trichoderma*	Green gram	Chickpea
1	Control	98	100
2	*F. oxysporum*	60	50
3	*M. phaseolina*	65	48
4	*R. solani*	50	60
5	*T. harzianum*	95	98
6	*T. viride*	85	90

Author's data

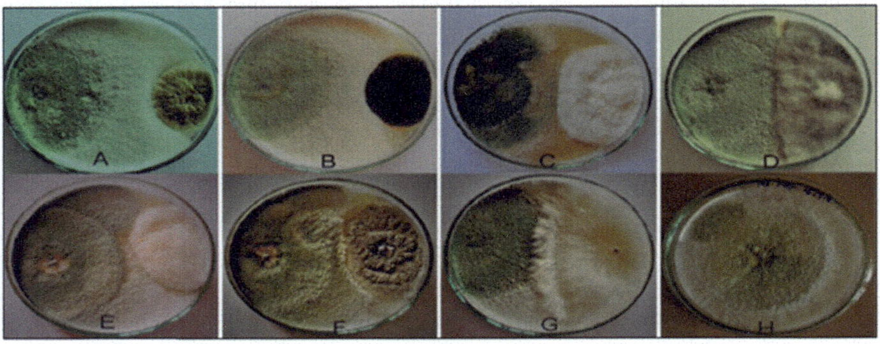

Fig. 4.3 Antagonistic activity of *Trichoderma harzianum* against (**a**) *Aspergillus flavus*, (**b**) *A. niger*, (**c**) *Fusarium roseum*, (**d**) *Macrophomina phaseolina*, (**e**) *Phythium myriotylum*, (**f**) *Rhizoctonia solani*, (**g**) *Sclerotium rolfsii*, (**h**) *Trichoderma harzianum*

role of *Trichoderma harzianum* as a biocontrol agent and also as a plant growth promoter.

The antagonistic activity of *T. harzianum* and *T. virens* was studied against *Aspergillus flavus, Fusarium oxysporum, Botrytis cinerea, Macrophomina phaseolina,* and *Rhizoctonia solani* isolated from Tomato and Brinjal, respectively. The data clearly indicated the inhibition of the above pathogens by *T. harzianum* and *T. virens* (Fig. 4.3a–j), respectively.

In vitro antagonism of *Trichoderma viride* was tested against *Cylindrocladium parvum* infecting *Eucalyptus*. Mycoparasitic interaction was assessed by following Bell's ranking (Bell et al. 1982). It was observed that after 5 days the pathogen was significantly suppressed recording to Bell's ranking No.4. Results also indicated the antagonism of *T. viride* against *Cylindrocladium parvum* (Tables 4.4 and 4.5). Similar observations were made by Mandels (1975) and Ngueko and Xu (2002) on root pathogens. Further, the authors have also observed antagonism of *T. viride* on *C. purvum* (Fig. 4.4a, b).

T1—Application of fungicide Carbendazim (Bavistin) to growing media @0.2%; T2—Application of *Cylindrocladium* culture filtrate having spore concentration of 2×10^6 CFU/ml; T3—Application of *Trichoderma* culture @ 100 ml of PDA liquid dissolved in 1 l of distilled water and later after 7 days application of

Table 4.4 In vitro antagonistic activity of *Trichoderma viride* isolate against *Cylindrocladium parvum*

Treatment	Linear growth of *Trichoderma* (cm)	Linear growth of *Cylindrocladium parvum* (cm)	Bell's ranking
Dual inoculation	20.44	3.8	R4
Single inoculation	22	19	

Author's data

R1—Complete overgrowth, R2—75% overgrowth, R3—50% overgrowth, R4—Locked at the point of contact

Table 4.5 In vivo antagonism of *Trichoderma viride* isolate against *Cylindrocladium parvum*

Treatment	% Mortality			Mean
	R1	R2	R3	
T1	3	4	2	3
T2	100	100	100	100
T3	2	3	3	3
T4	5	6	5	5

Author's data

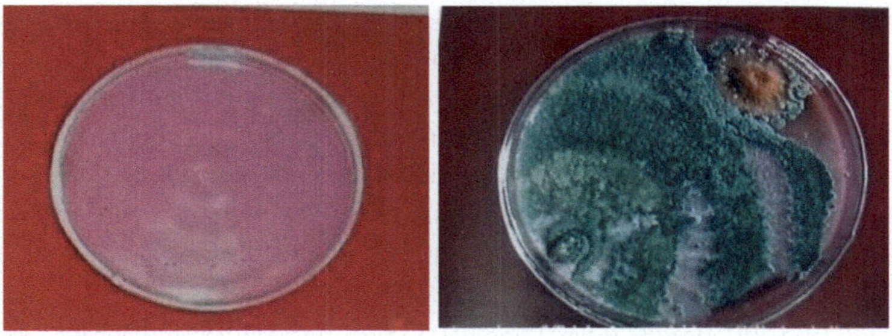

Fig. 4.4 (**a**). Colonies of *Trichoderma viride* on 2% water agar (**b**). Antagonism of *Trichoderma viride* on *Cylindrocladium parvum* (Author's data)

Cylindrocladium culture filtrate having the spore concentration of 2×10^6 CFU/ml; T4—Control (without any application). It has been noticed that the *Eucalyptus* seedling was suffering from damping-off disease caused by *C. parvum*. Interestingly the above disease was controlled by applying *Trichoderma viride* in rooting medium having the plantlets (Fig. 4.5a, b). Sanjeev Kumar et al. (2014) have conducted antagonism tests of *Trichoderma* spp. against *Helminthosporium maydis* as a soil application and also under laboratory condition which resulted in effective control of *Maydis* leaf blight.

Fig. 4.5 (a) Damping-off of *Eucalyptus* seedling caused by *C. parvum* control (b) Plants having rooting media inoculated with *Trichoderma viride* (Author's data)

4.6 Application of *Trichoderma* as Biocontrol Agent for Disease Control

Trichoderma spp. are free-living fungi present in soil and rhizosphere. They release a variety of compounds that induce localized or systemic resistance responses in plants. *Trichoderma* strains have long been recognized as biological agents, for the control of plant disease besides their role in plant growth (Kumar et al. 2012). Biocontrol of air-borne diseases has been made possible through the study of phyllosphere fungi. Its resistance to fungicide to Benomyl is achieved by substituting single amino acid in one of the β-tubulines of *Trichoderma viride* which in turn offers biocontrol. Some of the foliar pathogens could also be controlled by this method.

Trichoderma spp. has been used extensively for the control of diseases caused by *Fusarium, Phytophthora, Sclerotium,* and other soil-borne plant pathogens. Various crops namely cereals, millets, pulses, oilseeds, vegetables, horticultural crops, cash crops, spices, etc. are cultivated by farmers both in India and in the world. These crops suffer because of many diseases caused by phytoplasma, viruses, bacteria, fungi, nematodes, and insects and incur heavy losses. More than 40% of crop yield losses are reported due to different diseases and pests. Some of the pathogens are not manageable due to the resistance developed by the pathogens against fungicides and other chemicals. It has been observed that rice crop suffers from sheath blight caused by *Rhizoctonia solani* and it has been controlled by *Trichoderma harzianum.* Root rot of wheat caused by *Sclerotium rolfsii* and *Fusarium oxysporum* has been controlled by using *Trichoderma harzianum* in several countries. Charcoal rot of maize caused by *Macrophomina phasceolina* can be managed with *Trichoderma* spp. by soil application or seed priming. Species of *Fursarium, Rhizoctonia, Macrophomina,* and other soil-borne and root-borne pathogens have been controlled by *Trichoderma* spp. under field application. Vegetables like Chillies, Cucurbits, Brinjal, Pea, Potato, and others suffer from seedling blight, root rot, seed rot, charcoal rot, damping-off, wilt diseases and others. Major pathogens causing soil- and root-borne diseases can be controlled by applying *Trichoderma* spp. in the form of formulations, as soil inoculants or using commercial products both under glasshouse and field conditions (Singh 2014).

Trichoderma strains are known to induce resistance in plants and such plants are known to produce ethylene, hypersensitive responses, and other defense-related reactions in plant cultivars. Introduction of the endochitinase gene from *Trichoderma* into plants such as tobacco and potato resulted in increased resistance to fungal pathogens. Selected transgenic lines are highly tolerant to foliar pathogens such as *Alternaria alternata, A. solani, Botrytis cirerea,* and *Rhizoctonia* spp. as well as to other soil-borne plant pathogens.

Biological control agents like *Trichoderma* colonize the rhizosphere without releasing any toxic residues and offer effective control to soil-borne, root-borne, and foliar disease. Biological control of some soil-borne fungal diseases has been correlated with the chitinase production by *Trichoderma* (Cumagun 2014; Harman and Kubicek 1998; Mukherjee et al. 2012; Mukhopadhyay 1996; Kumar et al. 2012).

4.7 *Trichoderma*: As a Biocontrol Agent and Plant Growth Promoter

Trichoderma is the most common genus present in all types of habitats. It comprises around 3% of the total forest fungal population and 1.5% of other soils. It exhibits competition toward soil-borne and plant pathogenic fungi for key exudates that stimulate the germination of plant pathogenic fungi in soil. Various species of *Trichoderma* act against important foliar, soil-, and root-borne plant pathogens. (Weindling 1934). The antagonistic nature of *Trichoderma* was demonstrated more than 70 years ago and many studies approved that *Trichoderma* spp. form excellent biological control. In recent times, Herrera-Estrella and Chet (2004) discussed the role of *Trichoderma* as a biocontrol agent which is due to mycoparasitism-related genes. Antibiosis has been recognized as the role of mycoparasitism due to its related genes. Strain improvement and related activities of *Trichoderma* spp. are made possible due to the induction of resistance and plant growth promotion, *Trichoderma* spp. posses genes for biological control action and also for crop improvement. It has been observed that the siderophore production and induction of systemic resistance has been made possible by *Trichoderma* (Chet 1987). In the mycoparasitism which is the main mechanism involved in the biocontrol of plant diseases by *Trichoderma* includes (1) Chemotropic growth of *Trichoderma*, (2) Recognition of the host by *Trichoderma*, (3) Secretion of extracellular enzymes, (4) Penetration of hyphae, and (5) Lysis of the host. In this regard, *T. harzianum* not only acted as a biocontrol agent but also acted as a growth promoter which has been verified both in greenhouse and hydroponic systems. *Trichoderma* produces proteins, Avr homologs, oligosaccharides, and low molecular weight compounds that exert induced resistance in plants and also antagonistic gene expression along with phytoalexin biosynthesis.

Trichoderma spp. are known to induce the growth of various crops and increasing crop yield following seed priming (Mukhopadhyay 1996). This is due to (1) Suppression of harmful microbes/pathogens, (2) Production or activation of growth-stimulating factors, and (3) Increased nutrient uptake and sequestering of nutrients.

4.8 Commercial Formulations and Products of *Trichoderma*

Several species of *Trichoderma* possess the innate capacity and natural resistance against fungicides, pesticides, etc. Mass multiplication of *Trichoderma* is an important issue for commercialization. Papavizas (1985) and Singh et al. (2002) reported that an effective means of disease management by the application of biological control agents has to be understood deeply. They have suggested proper culturing methods, delivery systems, and augmentation of the soil directly using biocontrol agents like *Trichoderma* spp.

Nelson and Hoitink (1982) tried composted hardwood bark as a substrate for large-scale production of *Trichoderma* as a fungal antagonist for commercialization. Jin et al. (1996) used shakers and liquid fermenters for mass production of *Trichoderma.* Backman and Rodriguez-Kabana (1975) used diatomaceous earth granules and molasses for *Trichoderma* formulation. Hadar et al. (1979) used wheat bran and Sivan et al. (1984) developed wheat bran and peat as formulations, Mukhopadhyay et al. (1986) used sorghum grains as powdered formulation. Several workers have used cow dung, rice bran, sugarcane bagasses, organic manure, paddy straw, maize cob, chickpea husk, oat seeds, ground nutshell, cocoa hulls and several other substrates for the multiplication of *Trichoderma.* Kumar and Marimuthu (1997) used decomposed coconut coir and Lewis et al. (1998) have employed BIODAC cellulose granules for mass multiplication. The main problem in the commercial use of *Trichoderma* being mass multiplication, delivery methods, viability, and shelf life of *Trichoderma* at various temperatures. Commercial use of *Trichoderma* is also dependent on effective biocontrol strain and compatibility with other disease management systems.

Trichoderma spp. can be formulated as granules, pellets, dusts, and wettable powders and fluid drill gels. Granular or pellets preparations and *Trichoderma* enriched FYM have been used for soil application directly and have provided effective control of diseases both in nurseries and field conditions. Talc-based formulations of *Trichoderma* by TNAU have become quite popular in India for the management of several soil-borne diseases of various crops through seed treatment at 4–5g/kg seed (Jeyarajan et al. 1994). The talc formulations of *Trichoderma* have a shelf life of 3–4 months. Seedling roots can be treated with spore or cell suspension of antagonists either by drenching the bioagent in nursery beds or by

Table 4.6 List of commercial formulations of *Trichoderma* spp. and their target pathogens (Available in India)

Commercial product	Source	Target pathogen/disease
Root shield	*Trichoderma harzianum* strain-KRL—AG2 (T-22)	*Pythium, Rhizoctonia, Fusarium*
BINAB T	*Trichoderma harzianum/ Trichoderma polysporium*	Wood decay fungi
Promote	*Trichoderma harzianum* and *Trichoderma viride*	*Pythium, Rhizoctonia, Fusarium* (transplanted trees)
Trichodex	*Trichoderma harzianum*	*Plasmopara, Colletotrichum, Monilin* (various plants)
F-stop	*Trichoderma harzianum*	*Rhizoctonia, Pythium* (ornamental and food crops)
Soil Gard (Glio Gard)	*Gliocladium* (*Trichoderma virens* GL-21)	*Rhizoctonia solani, Pythium* (ornamental and food crops)
Monitor SD	*Trichoderma* spp.	Soil-borne plant pathogens
Monitor WP	*Trichoderma* spp.	Soil-borne plant pathogens
Root pro	*Trichoderma harzianum*	*Rhizoctonia solani, Pythium* spp., *Fusarium* spp., and *Sclerotium rolfsii*
Supresivit	*Trichoderma harzianum*	Various fungi
Trieco	*Trichoderma viride*	*Rhizoctonia* spp., *Pythium* spp., *Fusarium* spp., graymold
Trichoderma 2000	*Trichoderma* spp.	*Rhizoctonia solani, Sclerotium rolfsii, Pythium* spp., *Fusarium* spp.
T-22, T-22 planter box	*Trichoderma harzianum* strain KRL AG2	*Pythium* spp., *Rhizoctonia solani, Fusarium* spp. and *Sclerotinia* spp.

Source: Woo et al. (2014) *Trichoderma*-based products and their widespread use in agriculture

dipping roots in bioagent suspension before transplanting. Soil applications are ideally suited for greenhouse and nursery. *Trichoderma* is capable of colonizing FYM and therefore application of colonized FYM to the soil is more appropriate and beneficial. Liquid formulation of *Trichodema* performed better than powdered formulation in case of foliar application. It can be applied at 5–10 ml/l depending upon its concentration and application against a particular disease. Some commonly used commercial formulations are given in Table 4.6.

Effective management of plant pathogens and pests by means of biocontrol agents have become important in view of environment and health issues raised by the use of fungicides and pesticides in crops. The effective biological control system is dependent on:

1. Effectiveness of biocontrol agent(s)
2. Mass production and viability
3. Healthy storage and delivery system
4. Shelf life

Mainly all the above aspects are dependent upon isolation, identification, purification, conservation, and formulation of biocontrol agents such as *Trichoderma* spp. The ideal property of an antagonist includes:

1. Cost-effective production
2. Liquid formulations
3. Preservation in purity
4. Long Shelf Life

For the application of such antagonists to control plant diseases so as to be economically feasible to produce and adapt by the farmers (Cumagun 2014).

To achieve the above we need feasible mass production, healthy and effective formulation technology along with quality control, an effective delivery system, and with a large scope for commercialization for the management of various plant diseases. Among many antagonists, *Trichoderma* has been exploited and success has been achieved (Kumar et al. 2017). Among the *Trichoderma* spp., *T viride (T. asperllium)*, and *T harzianum* have been exploited to a larger extent and successful commercial products of different species of *Trichoderma* are made available in the world market including in India. In order to get *Trichoderma* successful formation, the species in question has to satisfy the rhizosphere competence, CSA, plant growth initiation, mass multiplication, mass spectrum action, safety to environmental ecosystem, compatibility with other biocontrol agents, must tolerate biotic and abiotic stress and UV radiation (Jeyarajan and Nakkeeran 2000; Kumar et al. 2014). Some of the formulations of *Trichoderma* which are in use for crop health and crop growth are given in Table 4.6.

Talc-based formulation of *Trichoderma viride* was developed at TNAU Coimbatore, India for seed treatment. The yearly requirement for farmers has been estimated to be 5000 tons to cover 50% area in India (Jeyarajan 2006). Wheat bran formulation-vermiculate (Lewis 1991) of *Trichoderma* multiplied in molasses yeast medium proved good (Connick et al. 1991). Pesta granule-based formulation and alginate pills-based formulations were developed by Fravel et al. (1999). Press Mud is made available for mass multiplication of *Trichoderma* as a formulation (Sabalpara 2014; Sawant and Sawant 1989, 1996). *Trichoderma* formulations based on coffee husk which is effective against *Phytophthora* foot rot of black pepper was developed in Karnataka and Kerala. Project Directorate of Biological Control (PDBC) in India has used oil-based formulation in the form of emulsion to control soil-borne diseases of groundnut and others. Balasubramanian et al. (2008) have produced *Trichoderma* sp. as Banana waste-based formulation. Besides the above microencapsulation has been developed to prolong shelf life further. Dried conidial pellets of *Trichoderma harzianum* were found more effective as antagonist formulation than liquid formulation. In order to achieve compatibility with other biological systems, there is a need to use highly disease-specific and consortial strains of *Trichoderma* or other biological control agents. The success story of biological control agent of different plant pathogens by *Trichoderma* also depends upon the modes of delivery and application which include seed treatment, seed priming (Pusa biogranule, PUSA bio pellet), seed priming (Fig. 4.6), liquid coating on seed

Fig. 4.6 Seed treatment with *Trichoderma* on (**a**) pea, (**b**) black gram, (**c**) moong. 1 = Normal seeds, 2 = Seeds treated with *Trichoderma* (Author's data)

(agro-lig) and double coating which is applied directly on the seed coat followed by the particulate formation in the second layer (Cook 1993; Cumagun 2014; Dubey et al. 2011; Harman 1991; Mathre et al. 1999; Mukhopadhyay et al. 1992; Nagaraju et al. 2012; Taylor et al. 1991).

Trichoderma has been used as a biocontrol agent in the last 60–100 years for managing particularly soil-borne, root-borne, and foliar diseases and also as a plant growth stimulant. In order to achieve success in this area, there is a need for accurate strain identification by molecular approach, use of consortia of biocontrol agents and as plant growth promoter, physiological and genetic enhancement of biocontrol mechanism, and manipulation of formulations. In recent times *Trichoderma* has been reported as an opportunistic, virulent, and plant symbiotic fungus as they compete and survive in the soil and also in other ecosystems. It has also been reported as an endophyte in different plant tissues. One of the advantages which is believed to be one of the important mechanisms of biocontrol effects of *Trichoderma* spp. is to induce metabolic changes in plants that offer induced resistance. Mutation is also employed to generate variability among *Trichoderma* populations. Protoplast fusion is another mechanism for strain improvement as *Trichoderma* spp. are known to release glucanases, chitinases, cellulases, and various other enzymes (Lalithakumari and Mathivanan 2003). One of the impediments to mass produce *Trichoderma* as a biocontrol agent include non-avialbility of effective methods for mass culturing and delivery of biocontrol agents under healthy and viable condition. Both liquid formulation and solid substrate fermentation products/formulations are made available besides the others. The most important aspect is the shelf life which has to be studied in depth with scientific accuracy. There are several commercial products of *Trichoderma* spp. that are available all over the world (Table 4.7). These

Table 4.7 Commercial products of *Trichoderma* available in various countries and their target pathogens/pest

S. No	Species name	Product	Country	Target uses/biological activity	Target pathogen/pest
1	*T. viride*	Agrigold Trichogold	India	Biofertilizers; cotton, pulses, vegetables, oilseeds, fruit plants and flower-bearing plants; stimulates seed germination	Effective application on root and stem rot diseases, wilts, blights/leaf spots
2	*T. viride*	ANOKA	India	Biocontrol agent; damping-off	Used for the control of seed root, rot, collar rot, nematodes wilt
3	*T. harzianum* DSM 14944	Agroguard WG	Colombia	Diverse crops	*Pythium, Rhizoctonia, Sclerotinia, Sclerotium, Phoma*
4	*T. viride*	BASDERMA	India	Protects the crops from diseases throughout their growth period; cauliflower, cotton, tobacco, soybean, sugarcane, redgram, bengalgram, banana, tomato, chillies	Parasitizes and kills the pathogenic fungi, exudes certain toxins like gliotoxin, viridin, and trichodermin that are harmful for the growth of the pathogenic fungi
5	*T. viride*	Biocure F	EU	Diverse crops; NPOP, NOP	IMO OVP; *Pythium* sp., *Rhizoctonia solani*, *Fusarium* spp., *Botrytis cinerea*, *Sclerotium* rolfsii, *Sclerotinia homoeocarpae, Ustilago tritici*
6	*T. harzianum*	Bioderma H	India	Biofungicide; bacterial and fungal diseases of cotton, cereals, pulses, vegetables, oilseeds, fruit plants	Root and foliar pathogens, *Phytophthora, Fusarium* and bacteria; damping-off *Pythium;* foliar *Alternaria, Macrophomina, Myrothecium, Ralstonia*
7	*T. atroviride* IMI 206040	Binab T Vector	USA, EU	A powder for treating flowers against fungal pathogens through vector transmission; Strawberries	*Botrytis cinerea*; controls fungal pathogens such as *Botrytis, Verticillium, Pythium, Fusarium, Phytophthora, Rhizoctonia,*
8	*T. viride*	Bio-shield	India	Cotton, groundnut, sunflower, *Sesamum*, urad, moong, arhar, gram (Chana), soyabean, tomato, chillies, tea, coffee	Used against seed-borne plant pathogenic fungi, e.g., *Fusarium, Pythium, Phytophthora, Rhizoctonia, Sclerotium* etc.

9	*T. harzianum* strain SF	Bio-Tricho	South Africa, Brazil	Can be applied to all crops	Control *Botrytis* and root diseases such as *Rhizoctonia, Phytophthora, Pythium, Fusarium*, etc.
10	*T. asperellum* ICC 012	Bioten	Spain	Flowers, ornamentals in vase (*Chrysanthemums, cyclamens*, poinsettia, primula etc.); Horticulture (tomato, peppers, salads and aromatic herbs	*Rhizoctonia solani, Sclerotinia Sclerotiorum, Verticillium dahliae, Thielaviopsis basicola, Phytophthora capsici*
11	*T. viride*	Bio-Tricure	India	Biopesticide; fungal diseases of cotton, tobacco, cereals, Pulses, vegetables, oilseeds, fruit plants and floriculture	Root and stem rots by *Sclerotinia* and *Rhizoctonia*, wilts by *Fusarium* and *Verticillium*, blights or leaf spots caused by *Alternaria, Ascochyta, Cercospora, Macrophomina, Myrothecium, Ramularia*
12	*T. viride*	Bioveer	India	Biofungicide, phosphate biofertilizer and also produces plant growth-promoting substances; patchouli, coleus, root crops diseases	Active on root rot, foot rot, collar rot, stem rot, damping-off, wilt, blight/leaf spot; sheath rot, sheath blight, and bacterial leaf blight (BLB) of rice
13	*T. harzianum* ATCC 52443	BioFungo WP	Colombia, Equador	Biofungicide	*Botrytis cinerea, Sphaerotheca pannosa* on roses
14	*T. harzianum* e *T. virens*	Bio Traz	Chile	Protecting the plant from pathogens, nutrient competition	Recommended for the control of foliar/aerial diseases caused by *Botrytis* in grape, *Monilinia* on stone fruits and cherries
15	*T. asperellum*	Bioprotection TR	Costa Rica	Antagonistic fungus, stimulates resistance and plant growth promoter	*Rhizoctonia Pythium. Phytophthora, Fusarium, Rhizopus, Mucor,* Botrytis, *Colletotrichum*
16	*T. viride*	Coimbatore	India	Biofertilizer; plant growth promoter, biocontrol of soil-borne plant- pathogenic fungi; sugarcane, pulses, oilseeds, cotton	Prevents the crops from diseases such as root rots, wilts, brown rot, damping-off, charcoal rot and other soil-borne diseases in crops

(continued)

Table 4.7 (continued)

S. No	Species name	Product	Country	Target uses/biological activity	Target pathogen/pest
17	*T. harzianum*	Commander Fungicide	India	Bio-control agent, protects the root system against diseases caused by soil-borne pathogens	Soil nematode, *Fusarium* wilt and blister blight
18	*T. asperellum* (=*T. atroviride*) SKT-1	Ecohope, Ecohope-dry	Japan	Seed and root diseases	*Giberella fujikuroi, Burkholderia glumae e Acidovorax avenae*
19	*T. viride* (TNAU strain)	Ecosom-TV	India	Biofungicide; seed/soil treatment of root rot of pulses, damping-off of chilli seedling, wilting and other root rot diseases; cereals, millets, pulses, oilseeds	*Rhizoctonia* spp., *Pythium* spp., *Fusarium* spp., and *Alternaria* spp.
20	*T. harzianum*	Ecotrich ES	Brazil	Lettuce, cotton, onion, ginger, carrot, sunflower, beans, tobacco, corn, tomatoes, wheat	*Rhizoctonia solani* in beans, strawberry and soya, *Sclerotinia* spp. in beans and soya, *Pythium* in lettuce
21	*T. viride*	Tricho shield combat	India	Biofungicide for bio management of seed and soil-borne plant pathogenic fungi	Biomanagement of soil-borne fungal infections of crops
22	*T. harzianum*	Trichosoil	Uraguay	Prevention and control of root pathogens	*Fusarium, Sclerotinia, Pythium*
23	*T. harzianum* strain kd	Eco-T	South Africa, Kenya, and Zambia, UK, and India	Control of crop root diseases and for enhanced plant growth	*Rhizoctonia, Pythium, Fusarium, Phytophthora*
24	*T. asperellum* T34	T34 biocontrol	European Union, UK	In greenhouse, carnations	*Fusarium oxysporum*
25	*T. harzianum* strain T-22	TRIANUM-P	EU, New Zealand, Australia	Tomato, vegetables, soft fruit, herbs, bulbs, ornamentals, perennials,	*Rhizoctonia* spp., *Fusarium* sp., *Pythium* sp.
26	*T. asperellum*	Trifender	Hungary	Prevention and control of root pathogens; diverse crops	*Pythium, Phytophthora, Fusarium, Sclerotinia, Rhizoctonia*
27	*T. asperellum* TV1	VIRISAN	European Union, USA	Tomato	*Rhizoctonia* spp., *Fusarium* sp., *Pythium* sp.

28	*T. harzianum*	Supresivit	Czech Republic	Biological control; strawberry	Soil-borne fungal pathogens such as *Verticillium dahlia*, Kleb, *Pythium* spp., *Phytophthora* spp., *Rhizoctonia* spp.; other pathogens *Botrytis Cinerea*
29	*T. harzianum* T-35 + *T. harzianum* T-315	Root-pro	Israel	Nursery and field soil amendment; soil-fungicide for use on greenhouse and nursery crops	Control of soil-borne diseases *Pythium* spp., *Sclerotium rolfsi*, *Fusarium* spp., *Rhizoctonia solani*
30	*T. harzianum* Rifai strain T-22 (KRL-AG2)	RootShield PLUS WP,	USA, Canada, EU	Horticulture and agriculture. Pathogen control, promotes a healthier root system, increasing root mass potential	Root disease control *Fusarium*, *Pythium*, *Rhizoctonia*, *Thielaviopsis,* and *Cylindrocladium*; PLUS *Phytophthora*, *Pythium* (P. *aphanidermatum*)
31	*Trichoderma* spp.	Tricho plus	South Africa	Control of soil-borne diseases. Wide range of crops including: maize, potatoes, beans, cucumbers, tomatoes and flowers	*Rhizoctonia, Pythium* and *Sclerotinia* spp.
32	*T. harzianum* T-22	GROW BOOST plant strengthener	UK	Wide range of plants including vegetables and salads, vegetables of the brassica family	Healthier and stronger plants, suppression of soil-borne diseases
33	*T. harzianum*, *T. koningii*	Promot WP	Germany, Kenya	Horticultural and ornamental crops	Control of damping-off and root rot caused by *Pythium*
34	*T. harzianum* Rifai strain T-39	Trichodex	South Africa, Australia, USA	Tomato, horticulture crops	*Botrytis cinerea*, *Collectotrichum* spp., *Plasmopara viticola*, *Pseudoperonospora cubensis*, *Rhizopus stolonifera*, *Sclerotinia sclerotiorum*

Source: Woo et al. (2014) *Trichoderma*-based products and their widespread use in agriculture

are used in field applications to control several soil-borne, root-borne, and foliar diseases to boost agricultural production.

4.9 Management of Soil-Borne Diseases by *Trichoderma viride*

Soil-borne pathogenic fungi attack most of the vegetable crops resulting in heavy losses. Presently, the most widely used control measures for suppressing the pathogens is the use of fungicides which not only cause environmental pollution but also lead to the development of resistant strains. Biocontrol of fungal pathogens is a positive preposition to decrease dependence on costly chemicals that are in vogue. The genus *Trichoderma* by virtue of its broad-spectrum action against a number of plant diseases caused by fungi, bacteria, and even nematodes has occupied the top position among the bio-protectants developed for plant disease management. *Trichoderma* has become an important biocontrol agent offering protection to fungal diseases of crops caused by plant pathogens.

Different crops viz. pulses, oilseeds, vegetables, cereals, horticultural crops, spices, floriculture crops, cash crops, etc. are cultivated by farmers who suffer from many diseases caused by fungi, bacteria, viruses, phytoplasma, nematodes, etc. Among these pathogens, some of the fungal pathogens viz. *Fusarium* spp., *Macrophomina phaseolina, Rhizoctonia solani, Pythium* spp., *Phytophthora* spp. etc., and most of the nematodes as soil inhabitants in nature may cause >70% crop yield losses. Often these pathogens are unable to manage with fungicides due to many reasons. In the last 25 years, researchers in the management of the soil-borne diseases of these crops through biological agents have made fruitful recommendations to the farmers. *Trichoderma* spp. viz,. *T. asperellum, T. asperelloides, T. harzianum, T. koningii, T. viride. T.virens* etc. have been proven quite beneficial for the management of biotic stresses, i.e., seed and soil-borne diseases of the crops particularly wilt (*Fusarium* spp), root rot (*M. phaseolina, R. solani*), Collar rot (*A. niger*), stem rot (*S. rolfsii*), etc. which cause huge qualitative and quantitative crop yield losses. *Trichoderma* spp. not only act as biocontrol agents but also stimulate plant resistance to biotic and abiotic stresses (temperature, drought, salinity, etc.) and also help in plant growth and development resulting in an increase in crop production. The mechanism of biocontrol which involves mycoparasitism, antibiosis, and competition for nutrients, may also induce defense or systemic resistance in plants. Several mechanisms by which *Trichoderma* may influence plant development have been proposed, such as the production of phytohormones, the solubilization of sparingly soluble minerals, reduction of pollutant toxicity (organic or heavy metal), and the regulation of rhizospheric microflora. *Trichoderma* spp. are found distributed all over the world in different eco-systems as decomposers of waste organic matter as they produce cell wall degrading enzymes viz. cellulase, protease, chitinase, xylanase, endoglucanase, etc. A number of commercial formulations of *Trichoderma* spp. are available for the use of the farmers through the technologies of seed treatment, seed biopriming, enrichment of organic manures, cakes, etc. (Kumar et al. 2014).

4.10 Genetics of *Trichoderma*

The correct identification of *Trichoderma* spp. is a difficult task. Though morpho-taxonomic criteria have helped in the identification but failed to segregate strains, races, ecotypes, biotypes, etc. However, the modern classificatory system is evolved based on multigene phylogeny (Atanasova et al. 2013) which enabled the scientists to recognize around 200 species of *Trichoderma.* Most of the strains of *Trichoderma viride* have been assigned to *T. asperellum* and *T. harzianum* is now considered as *T. afroharzianum.* Both the species are now in use as biocontrol agents (Chaverri et al. 2015). Genes involved in signal transduction and responsible for G-protein, CAMP, and MAP kinase pathways have been studied at length in species like *T. atroviride* (Reithner et al. 2005) due to their involvement in mycoparasitism. Degradation of cell wall components of a pathogen by secreting hydrolytic enzymes such as chitinases, glucanases, proteases, and laccases play an important role in mycoparasitism and as a biocontrol agent. Genes responsible for cell wall degradation of various soil-borne pathogens by *Trichoderma* spp. have been identified (Dubey et al. 2012; Woo et al. 1999). The role of *Trichoderma* spp. as a biocontrol agent and plant growth promoter needs to be able to cope up with abiotic and biotic stresses. Genes responsible for such tolerance have been identified in *T. harzianum* and *T. atroviride* (Montero-Barrientos et al. 2007). It has been established that *Trichodrma* spp. elaborate a variety of secondary metabolites both of high and low molecular weight (terpenes, peptide synthetase, trichothecenes, PKS metabolite, etc.) have been responsible as plant growth regulators in inducing resistance and biocontrol activity. It has been observed that a network of genes seems to be influencing secondary metabolism. Harman et al. (2004) and Mukherjee et al. (2012) observed that *Trichoderma*–root interaction involves entry into the root by secreting hydrolytic enzymes, transfer of carbohydrate to biocontrol agent, and induction of resistance in plants through elicitors. *Trichoderma* spp. have been considered as opportunistic fungi. This indicates the role of genes of *Trichoderma* which is in direct interaction with plants. So far seven species of *Trichoderma* have been sequenced (Mukherjee 2011). Gene expression studies in *T. virens* while interacting with maize or tomato root revealed that there is host specificity at transcriptome level (Moran-Diez et al. 2015). All the above information documents that perhaps, a blend of classical methods in collaboration with genomics in *Trichoderma* will be utilizable as beneficial plant microbe.

4.11 *Trichoderma*: Agriculture

Population is increasing around the world and it has become a challenge to India having 1.3 crore population. There is a need to grow more food in order to establish security for food, nutrition, environment, and also economic stability. Natural resources are depleting rapidly throughout the world because of greater human activity, industrialization, environmental degradation, natural calamities, and others. Therefore, it is essential that the dynamic equilibrium of natural resources be established by imposing hard and fast rules and regulations as there is no end to human selfishness. Keeping in view the changes that are occurring in the environment due to various factors along with global warming, it is expected that there will be a huge disturbance in the environment due to climate change. Hence, there is an urgent need for establishing sustainable agriculture, protection, and replenishing natural resources for the benefit of mankind. The greater need is there to step up agricultural production due to overgrowth of population in various countries of the world and in particular of developing countries. Sustainable agriculture can be achieved if the major diseases of crop plants and others caused by microbes, pathogenic fungi, nematodes, and by insects are controlled to a greater extent. Soil-borne, root-borne, and some foliar diseases can be controlled through the intervention of some potential *Trichoderma* spp. as a biocontrol agent. For field application, researchers–industry partnership is essential so as to produce *Trichoderma* as a biocontrol agent on a commercialized basis. In our studies it has been possible to control fusarium wilt disease in tomato, chickpea and soybean under green house conditions besides showing an increase in plant growth and root development on inoculation with *Trichoderma* spp. (Fig. 4.7a–c). Modern agriculture is mostly dependent on the extensive use of fungicides, pesticides, chemical fertilizer, herbicides, and others. The nonjudicious use of the above by the farmer led to the loss of soil fertility, soil health, crop yield, and other related disturbances besides health hazards to the farmer. The crucial factor to achieve sustainable agriculture and food security is to maintain soil and plant health. Soil stabilization and soil fertility status affect crop productivity. Application of *Trichoderma* spp, PGPR, beneficial microbes, cow dung, and other decomposed organic wastes are

Fig. 4.7 Effect of *Trichoderma* spp. on plant growth and root development in tomato (**a**), chickpea (**b**), and Soybean (**c**). (Left is untreated and right is treated) (Author's data)

known to boost crop productivity besides maintaining soil and plant health (Authors unpublished data). The ill-effects of fungicides, pesticides, and agrichemicals due to their unscientific application and usage of them in larger quantities need to be managed by biological methods. Such ill effects will be brought down by the existing natural soil microflora and also by the application of biofertilizers. One of the innovations in nature has been the use of *Trichoderma* as a biocontrol agent and plant growth promoter to achieve the much needed sustainable agriculture followed by food security, nutritional security besides maintaining soil and plant health. Therefore, organisms like *Trichoderma*, PGPR, and others are considered as nature's gift to boost agriculture.

References

Atanasova L, Druzhinina IS, Jaklitsch WM (2013) Two hundred *Trichoderma* species recognized based on molecular phylogeny. In: Mukherjee PK, Singh US, Horwitz BA, Schmoll M, Mukherjee M (eds) *Trichoderma*: biology and applications. CABI, Nosworthy Way

Backman PA, Rodriguez-Kabana R (1975) A system for growth and delivery of biological control agents to the soil. Phytopathology 65:819–821

Baker KF (1987) Evolving concept of biological control of plant pathogens. Annu Rev Phytopathol 26:67–85

Balasubramanian C, Udaysoorian P, Prabhu C, Kumar GS (2008) Enriched compost for yield and quality enhancement in sugarcane. J Ecobiol 22:173–176

Bell DK, Wells HD, Markham CR (1982) *In vitro* antagonism of *Trichoderma* species against six fungal pathogens. Phytopathology 72:379–382

Bissett J (1991) A revision of the genus *Trichoderma*. II. Infrageneric classification. Can J Bot 69:2357–2372

Bissett J (1992) A revision of the genus *Trichoderma*. III. Section Pachybasium. Can J Bot 69:2373–2417

Chaverri P, Branco-Rocha F, Jaklitsch W, Gazis R, Degenkolb T, Samuels GJ (2015) Systematics of the *Trichoderma harzianum* species complex and the re-identification of commercial biocontrol strains. Mycologia 107(3):558–590

Chet I (1987) *Trichoderma*—application, mode of action and potential as biocontrol agent of soil-borne plant pathogenic fungi. In: Chet I (ed) Innovative approaches to plant disease control. Wiley, New York, pp 137–160

Connick W, Daigle D, Quimby P (1991) An improved invert emulsion with high water retention for mycoherbicide delivery. Weed Technol 5:442–444

Cook RJ (1993) The role of biological control in the 21st century. In: Lumsden RD, Vaughn JL (eds) Pest management. Biologically based technologies. American Chemical Society, Washington, DC, pp 1–20

Cook RJ, Baker KF (1983) The nature and practices of biological control of plant pathogens. APS Books, St Paul, MN, 539pp

Cumagun CJR (2014) Advances in formulation of *Trichoderma* for biocontrol, vol 31. Elsevier, Dordrecht, pp 1–5

Druzhinina IS, Kopchinskiy AG, Komon M, Bissett J, Szakacs G, Kubicek CP (2005) An oligonucleotide barcode for species identification in *Trichoderma* and *Hypocrea*. Fungal Genet Biol 42:813–828

Druzhinina IS, Kopchinskiy AG, Kubicek CP (2006) The first 100 *Trichoderma* species characterized by molecular data. Mycoscience 47:55–64

Dubey SC, Bhavani R, Singh B (2011) Integration of soil application and seed treatment formulations of *Trichoderma* species for management of wet root rot of mungbean caused by *Rhizoctonia solani*. Pest Manag Sci 67:1163–1168

Dubey SC, Tripathi A, Singh B (2012) Combination of soil application and seed treatment formulations of *Trichoderma* species for integrated management of wet root rot caused by *Rhizoctonia solani* in chickpea. Indian J Agric Sci 82:357–364

Fravel DR, Rhodes DJ, Larkin RP (1999) *Production and commercialization of biocontrol products*. In: Albajes R, Lodovica Gullino M, Van Lenteren JC, Elad Y (eds) Integrated pest and disease management in greenhouse crops. Kluwer Academic, Boston, pp 365–376

Gams W, Bissett J (1998) Morphology and identification of *Trichoderma*. In: Harmann GE, Kubicek CP (eds) *Trichoderma* and *Gliocladium*. Taylor and Francis, London, pp 3–34

Hadar Y, Chet I, Henis Y (1979) Biological control of *Rhizoctonia solani* damping-off with wheat bran culture of *Trichoderma harzianum*. Phytopathology 69:64–68

Harman GE (1991) Seed treatments for biological control of plant disease. Crop Prot 10:166–171

Harman GE (2000) Myths and dogmas of biocontrol. Changes in perceptions derived from research on *Trichoderma harzianum* T22. Plant Dis 84:377–393

Harman GE, Kubicek CP (1998) *Trichoderma* and *Gliocladium* enzymes, biological control and commercial applications, vol 2. Taylor and Francis, London, p 393

Harman GE, Howell CR, Vitarbo A, Chet I, Lorito M (2004) *Trichoderma* species—a opportunistic, avirulent plant symbionts. Nat Rev Microbiol 2:43–56

Hartley C (1921) Damping-off in forest nurseries. US Depart Agric Bull 934:1–99

Herrera-Estrella A, Chet I (2004) The biological control agent *Trichoderma* from fundamentals to applications. In: Arora DK (ed) Fungal biotechnology in agricultural, food and environmental applications. Marcel Dekker, New York, pp 147–156

Index Fungorum (2019). http://www.indexfungorum.org/names/Names.asp

Jaklitsch WM (2009) European species of *Hypocrea*. Part I. The green-spored species. Stud Mycol 63:1–91

Jeyarajan R (2006) Prospects of indigenous mass production and formulation of *Trichroderma*. In: Rabindra RJ, Ramanujam B (eds) Current status of biological control of plant diseases using antagonistic organisms in India. Project Directorate of Biological Control, Bangalore, pp 74–80, 445

Jeyarajan R, Nakkeeran S (2000) Exploitation of microorganisms and viruses as biocontrol agents for crop disease mangement. In: Upadhyay RK et al (eds) Biocontrol potential and their exploitation in sustainable agriculture. Kluwer Academic/Plenum, Boston, pp 95–116

Jeyarajan R, Ramakrishnan G, Dinakaran D, Sriela R (1994) Development of product of *Trichoderma viride* and *Bacillus subtilis* for root rot disease of pulses and oil seeds. J Biol Control **7**(1):58–62

Jin X, Taylor AG, Harman GE (1996) Development of Media and automated liquid fermentation methods to produce desicacation-tolerant propagules of *Trichoderma harzianum*. Biol Control 7:267–274

Kiffer E, Morelet M (2000) The deuteromycetes: mitosporic fungi classification and generic keys. Science, Enfield, NH, p 273

Kirk PM, Cannon PF, Stalpers JA (2008) Dictionary of the fungi, 10th edn. CABI, Wallingford, p 960

Kullnig CM, Szakacs G, Kubicek CP (2000) Molecular identification of *Trichoderma* species from Russia, Siberia and the Himalaya. Mycol Res 104:1117–1125

Kullnig-Gradinger C, Szakacs G, Kubicek CP (2002) Phylogeny and evolution of the genus *Trichoderma*: a multigene approach. Mycol Res 106:757–767

Kumar A, Marimuthu T (1997) Decomposed coconut coir pith—a conducive medium for colonization of *Trichoderma viride*. Acta Phytopathol Entomol Hung 32(1–2):51–58

Kumar S, Lal M, Singh V (2012) Exploitation of *Trichoderma* spp. as biocontrol agent for plant disease management. Rashtriya Krishi 7(2):71–73

Kumar S, Thakur M, Rani A (2014) *Trichoderma*: mass production, formulation, quality control, delivery and its scope in commercialization in India for the management of plant diseases. Afr J Agric Res 9(53):3838–3852

Kumar G, Maharshi A, Patel J, Mukherjee A, Singh HB, Sarma BK (2017) *Trichoderma*: a potential fungal antagonist to control plant diseases. SATSA Mukhapatra 21:206–218

Lalithakumari D, Mathivanan N (2003) Strain improvement in filamentous fungi by protoplast fusion. In: Mathivanan N, Prabavathy VR, Gomathinayagam S (eds) Innovative methods and techniques for integrated pest and disease management. Centre for Advanced Studies in Botany, University of Madras, Chennai, pp 76–97

Lewis JA (1991) Formulation and delivery system of biocontrol agents with emphasis on fungi *Beltsville symposia.* In: Keister DL, Cregan PB (eds) The rhizosphere and plant growth. Agric Res 14:279–287

Lewis JA, Larkin RP, Rogers DL (1998) A formulation of *Trichoderma* and *Gliocladium* to reduce damping-off caused by *Rhizoctonia solani* and saprophytic growth of the pathogen in soilless mix. Plant Dis 82(5):501–506

Lieckfeldt E, Seifert KA (2000) An evaluation of the use of ITS sequences in the taxonomy of the Hypocreales. Stud Mycol 45:35–44

Lübeck M, Poulsen SK, Lübeck PS, Jensen DF, Thrane U (2000) Identification of *Trichoderma* strains from building materials by ITS1 ribotyping, UP-PCR fingerprinting and UP-PCR cross hybridization. FEMS Microbiol Lett 185:129–134

Mandels M (1975) Microbial sources of cellulase. Biotechnol Bioeng Symp 5:81–105

Mathre DE, Cook RJ, Callan NW (1999) From discovery to use: traversing the world of commercializing biocontrol agents for plant disease control. Plant Dis 83:972–983

Montero-Barrientos M, Cardoza RE, Gutierrez S, Monte E, Hermosa R (2007) The heterologous overexpression of hsp23, a small heat-shock protein gene from *Trichoderma viriens*, confers thermotolerance to *T.harzianum*. Curr Genet 52:45–53

Moran-Diez ME, Trushina N, Lamdan NL, Rosenfelder L, Mukherjee PK, Kenerley CM, Horwitz BA (2015) Host-specific transcriptomic pattern of *Trichoderma viriens* during interaction with maize or tomato roots. BMC Genomics 16:8

Mukherjee PK (2011) Genomics of biological control—whole genome sequencing of two mycoparasitic *Trichoderma* spp. Curr Sci 101:268

Mukherjee PK, Horwitz BA, Kenerley CM (2012) Secondary metabolism in Trichoderma—a genomic perspective. Microbiology 158:35–45

Mukhopadhyay AN (1996) Recent innovations in plant disease control by ecofriendly biopesticides. In: 83rd Annual Meeting of Indian Science Congress, Patiala, January, pp 1–8

Mukhopadhyay AN, Patel GJ, Brahbhatt A (1986) *Trichoderma harzianum*: a potential biocontrol agent for tobacoo damping-off. Tobacoo Res 12:26–35

Mukhopadhyay AN, Shrestha SM, Mukherjee PK (1992) Biological seed treatment for control of soil-borne plant pathogens. FAO Plant Prot Bull 40:21–30

Nagamani A, Manoharachary C, Agarwal DK, Chowdhry PN (2002) Monographic contribution on *Trichoderma*. Elegent Printers, New Delhi, p 47

Nagaraju A, Sudhisha J, Murthy SM, Ito S (2012) Seed priming with *Trichoderma harzianum* isolates enhances plant growth and induces resistance against *Plasmopora halstedii*, an incitant of sunflower downy mildew disease. Australas Plant Pathol 1:609–620

Nelson EB, Hoitink HAJ (1982) Factors affecting suppression of Rhizoctonia solani in container media. Phytopathology 72:275–279

Ngueko RB, Xu T (2002) Anatogonism in vitro of *Trichoderma harzianum* C184 against the root pathogens of banana and plantain in Cameroon. J Zhejiang Uni (Agril Life Sci) 28:407–410

Overton BE, Stewart EL, Geiser DM (2006) Taxonomy and phylogenetic relationships of nine species of *Hypocrea* with anamorphs assignable to *Trichoderma* section *Hypocreanum*. Stud Mycol 56:39–65

Papavizas GC (1985) *Trichoderma* and *Gliocladium*: biology, ecology and potential for biocontrol. Annu Rev Phytopathol 23:23–54

Persoon CH (1794) Dispositio methodica fungorum. Römer's Neues Mag Bot 1:81–128

Persoon CH (1801) Synopsis methodica fungorum, pp 1–706

Reithner B, Brunner K, Schuhmacher R, Peissl I, Seidl V, Krska R, Zeilinger S (2005) The G protein alpha subunit Tga1 of *Trichoderma atroviride* is involved in chitinase formation and differential production of antifungal metabolites. Fungal Genet Biol 42:749–760

Rifai MA (1969) A revision of the genus *Trichoderma*. Mycol Pap 116:1–56

Sabalpara AN (2014) Mass multiplication of biopesticides at farm level. J Mycol Plant Pathol 44 (1):1–5

Samuels G (2006) *Trichoderma*: systematics, the sexual state, and ecology. Phytopathology 96 (2):195–206

Samuels GJ, Petrini O, Kuhls K (1998) The *Hypocrea schweinitzii* complex and *Trichoderma* sect. *Longibrachiatum*. Stud Mycol 41:1–54

Sawant IS, Sawant SD (1989) Coffee fruit skin and cherry husk as substrates for mass multiplication of *Trichoderma harzianum* as antagonist to citrus *Phytophthora*. Indian Phytopathol 42:336

Sawant IS, Sawant SD (1996) A simple method for achieving high cfu of *Trichoderma harzianum* on organic wastes for field applications. Indian Phytopathol 9:185–187

Singh HB (2014) Management of Plant Pathogens with Microorganisms. Proc Indian Natl Sci Acad 80(2):443–454

Singh HB, Singh A, Nautiyal CS (2002) Commercialization of biocontrol agents: problem and prospects. In: Rao GP, Manoharachari C, Bhat DJ, Rajak RC, Lakhanpal TN (eds) Frontiers of fungal diversity in India. International Book Distributing Company, India, pp 847–861

Sivan A, Elad Y, Chet I (1984) Biological control effects of new isolate of *Trichoderma harzianum* on *Phythium aphanidermatum*. Phytopathology 74:498–501

Taylor AG, Min T, Harman GE, Jin X (1991) Liquid coating formulation for the application of biological seed treatments of *Trichoderma harzianum*. Biol Control 1:16–22

Taylor JW, Jacobson DJ, Kroken S, Kasuga T, Geiser DM, Hibbett DS, Fisher MC (2000) Phylogenetic species recognition and species concepts in fungi. Fungal Genet Biol 31:21–32

Thrane U, Poulsen SB, Nirenberg HI, Lieckfeldt E (2001) Identification of *Trichoderma* strains by image analysis of HPLC chromatograms. FEMS Microbiol Lett 203:249–255

von Tubeuf CF (1914) Biologische Bekampfung von Pilzkrankheiten der Pflanzen. Naturwiss Z Forst Landwirtsch 12:11–19

Weindling R (1934) Studies on a lethal principle effective in the parasitic action of *Trichoderma lignorum* on *Rhizoctonia solani* and other soil fungi. Phytopathology 24:1153–1179

Woo SL, Donzelli B, Scala F, Mach R, Harman GE, Kubicek CP, Del Sorbo G, Lorito M (1999) Disruption of the *ech42* (endochitinase-encoding) gene affects biocontrol activity in *Trichoderma harzianum* P1. Mol Plant-Microbe Interact 12:419–429

Woo SL, Ruocco M, Vinale F, Nigro M, Marra R, Lombardi N, Pascale A, Lanzuise S, Manganiello G, Lorito M (2014) *Trichoderma*-based products and their widespread use in agriculture. Open Mycol J 8(Suppl-1, M4):71–126

Chapter 5
Mass Multiplication of *Trichoderma* in Bioreactors

Vimala Prakash and Kausik Basu

Abstract Global statistics show that the biopesticides market is far behind as compared to their chemical counterparts. Therefore, in order to increase their share in the global market, low-cost mass production technologies need to be developed. The most exploited genus in the biopesticide industry is *Trichoderma* whose formulation is developed using various organic and inorganic carriers either through solid or liquid fermentation technologies. However, the major trouble faced by both farmers and manufacturers regarding the bioproduct is the instability of the developed product under different environmental conditions. Therefore, through this chapter, we have tried to summarize the various production technologies with necessary precautions to be taken for making a successful *Trichoderma* product for the end users.

Keywords *Trichoderma* · Propagules · Fermentation · Quality assurance · Shelf life

5.1 Introduction

Industrial production of effective biocontrol agents for commercial use requires well-defined criteria of selection and also thorough knowledge of the process for the development of microbial products for disease control in plants. This chapter discusses the selection of novel microbial control agents and their production process for commercial use. An important factor in the success of *Trichoderma* is the adoption of cost-effective means of production and development of a product that remains stable in the environment and multiplies and performs the desired functions as a biocontrol agent (Singh et al. 2013; Fraceto et al. 2018).

V. Prakash (✉) · K. Basu
International Panaacea Limited, Gurugram, Haryana, India
e-mail: vimala.prakash@iplbiologicals.com

C. Manoharachary et al. (eds.), Trichoderma: *Agricultural Applications and Beyond*, Soil Biology 61, https://doi.org/10.1007/978-3-030-54758-5_5

In the process of mass manufacturing or production of *Trichoderma* spp., the factors to be considered are the types of production, i.e., submerged fermentation, solid-state fermentation, and formulation of the products, i.e., liquid, powder, and granular etc. The company or the manufacturer has to make a choice on the type of product the customer requires and accordingly the infrastructure of the plant has to be made.

5.2 *Trichoderma* spp. Commercially Important Propagules: Mycelia, Chlamydospores, and Spores

Trichoderma spp.is a fungus that produces different structures throughout their life cycle. The main structural organization and components are mycelia, spores, and chlamydospores. Each is considered as propagules which have capabilities of multiplying and growing into a new colony if suitable substrate and other favorable conditions are available (Singh et al. 2014b).

Solid-state fermentation (SSF) produces mostly aerial spores also known as conidia and the other kind of spores known as blastospores are produced under submerged fermentation (SmF). Spores produced by SSF, i.e., aerial spores have longer viability compared to submerged spores. Other advantages of SSF aerial spores are that they have more virulence resistance to stress conditions, higher UV resistance over SmF. Blastospores also have few advantages for commercial uses as its infection is much faster and also germination is faster than SSF spore.

5.2.1 Mass Production

There are various research going on across the country and throughout the world to identify, isolate, and characterize the most efficient *Trichoderma* as a biocontrol agent. But there is a huge gap when the efficient strain has to be mass multiplied and supplied as a product. A lot of emphasis is to be given to this area of research where such knowledge is very much required (Keswani et al. 2016). The mass production of the *Trichoderma* in a bioreactor is a very important step in the process of commercialization of the *Trichoderma* and having a potential strain does not serve the purpose unless and until the art of mass multiplication is known (Kumar et al. 2014; Prasad and Rangeshwaran 2000). Hence, as a futuristic approach, a lot of research has to be undertaken in generating as much knowledge as possible in this front. Any typical mass production unit of *Trichoderma* comprises of the following:

1. Production Lab—Production lab comprises basic infrastructure like laminar airflow, BOD incubators, orbital shaker, etc. The production lab is the starting point of the activities of commercialization. Activities done in production lab are as follows:

   (a) Culture maintenance at 4, −20, −80 °C as glycerol stock, agar slant, or Lyophilized form

(b) Seed culture preparation
(c) In-process sample checking
(d) Harvest sample analysis
(e) Finished good analysis

5.2.2 Production Area (Upstream and Downstream)

The production area is the next step in the mass production of *Trichoderma*. It comprises of fermenters depending upon the type of fermenters (SmF/SSF), utilities like boilers, compressors, water supply etc. The fermentation process of *Trichoderma* is conducted by the adoption of standard operating procedures and documentation of the process. The in-process samples during fermentations are drawn and analyzed in the production lab. The samples are analyzed for various parameters like pH, microscopy, spore count, detection of contaminants (i.e., other than the desired microbes) and substrate residual analysis. These parameters are very much required to be studied to assess the quality of the product (Al-Taweil et al. 2009).

The output of the fermentation is formulated into different formulations. The formulation can be of anything ranging from liquid, powder, granules, water dispersable granules, aqueous suspension, suspension concentrate, etc. The formulation is an important step in the product development. Intense research is required to optimize the formulation steps to ensure the spore survivability in different forms. If the formulation and the formulants are not compatible with *Trichoderma*, this may lead to a very poor shelf life of the product.

5.2.3 Quality Assurance Unit

In any manufacturing unit, the role of the quality assurance unit is of utmost importance as it plays a major role in ensuring product quality (Keswani et al. 2016). The activities of the quality assurance unit are as below:

1. Seed inoculum purity checking.
2. Pre-inoculum and inoculum purity checking.
3. Fermenter in-process sample checking. The samples are checked for growth, pH contamination, microbial count, and spore count.
4. Harvest samples checking. The quality assurance department checks the harvest samples, assesses the quality of the product, and approves the batch to the harvested.
5. Formulated sample checking. The samples after formulations are also checked by the quality assurance unit. The parameters checked are pH, moisture, CFU count, total viable count, and presence of contamination in any formulation should not exceed 10^4 per gram.

6. The formulated product should be tested negative for human pathogens like *Salmonella*, *Shigella*, and *Vibrio*. The quality assurance unit conducts the entire above tests to ensure the quality of the product.

5.2.4 Packing and Dispatch of the Final Product

Once the product is formulated, necessary quality checks are done. The product is either stored as semifinished goods or as packed units.

5.3 Product Stability Study

Any formulated product of *Trichoderma* should remain viable during the period of storage as claimed in the label. The quality assurance unit conducts periodical tests for viability of the *Trichoderma*, moisture, pH, and presence of any contaminants (Fig. 5.1) if any (Bhat et al. 2009). Selection of compatible polymers are again an

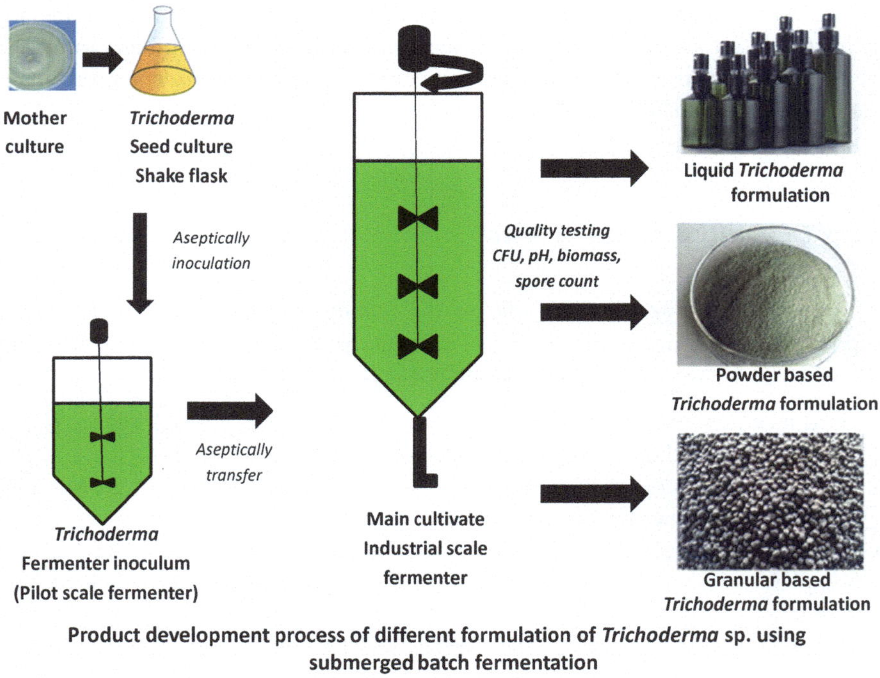

Fig. 5.1 Product development process flow in submerged fermentation

important factor for ideal product development. Different statistical approaches are employed for the optimization study.

Bioinoculants normally has to be produced at large scale with utmost care and good manufacturing practices. Good manufacturing practices play a major role in maintaining the quality of the bio inoculants. Mostly bio inoculants/bio fungicides like *Trichoderma* spp. are being manufactured across the country as small scale industry practices (Ramanujam et al. 2010).

5.4 Different Types of Fermentation and Bioreactor Designing

Trichoderma spp. as the commercial formulation is manufactured using fermentation technology which is through SmF and SSF procedures.

5.4.1 Submerged Fermentation

Industrial process using SmF can be carried out by three major types of fermentation:

1. Batch fermentation
2. Fed-batch fermentation
3. Continuous fermentation

Batch fermentation is also known as closed-system fermentation. In this approach, the fermentation medium is prepared, inoculated and the process runs till the harvest of the products. There is no intermediate addition or removal of broth or ingredient from the vessel (Fig. 5.2). This type of fermentation is mainly followed in the case of mass multiplication or whole-cell fermentation.

Fed-batch fermentation is also known as semi-closed system fermentation. In this process, the medium is prepared, inoculated and the process runs for a stipulated time. The new ingredients or substrate are added at regular intervals. This process is known as feeding strategy. There is no removal of cells or ingredients from the system/vessels (Fig. 5.3). This type of fermentation is generally followed in a case where secretory secondary metabolic substances are products (intercellular, intracellular, or extracellular).

Continuous fermentation is also known as open system fermentation where a sterile medium is added and inoculated and after a certain period of time, and cell and medium containing metabolites are removed and replenished with new medium (Fig. 5.4). The log phase is prolonged. This type of fermentation is usually followed in pharmaceutical industries where metabolites are products. This type of fermentation process runs for several weeks to months.

Fig. 5.2 Typical schematic of batch fermentation

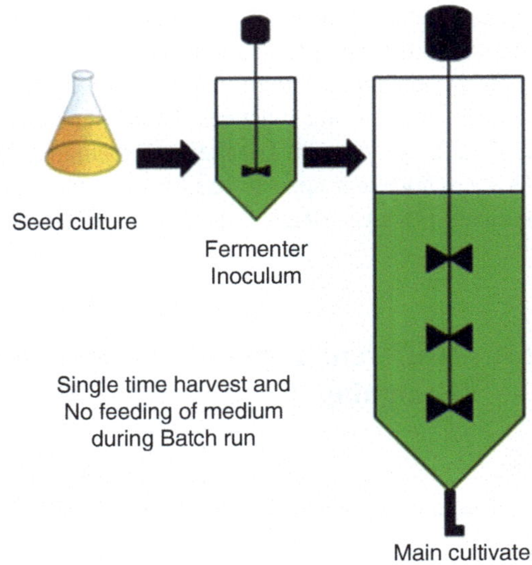

Fig. 5.3 Typical schematic of fed fermentation

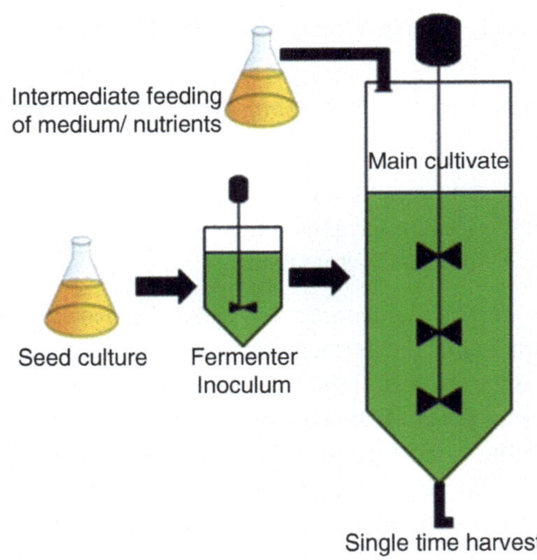

For commercial production of *Trichoderma*, batch fermentation is more convenient and most commonly used throughout the industries worldwide (Lejeune and Baron 1995). The selection of the type of fermentation depends upon the final products.

5.4.2 Solid-State Fermentation

SSF is a type of fermentation process mainly used for microbial biomass production at large scale manufacturing (Fig. 5.5). SSF has very minimum free water or

Fig. 5.4 Typical schematic of continuous fermentation

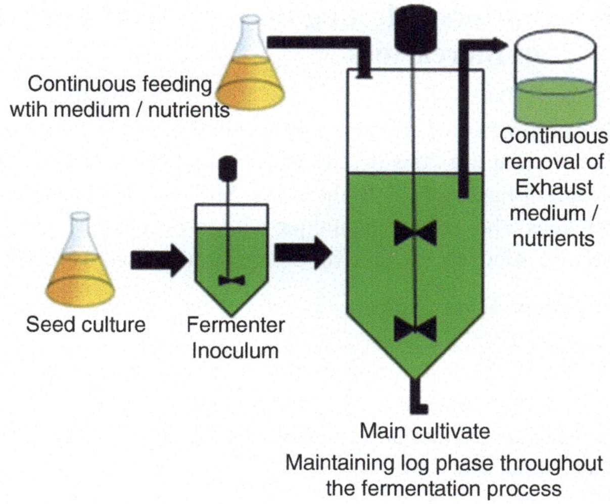

Fig. 5.5 Typical schematic of chamber type solid fermenter

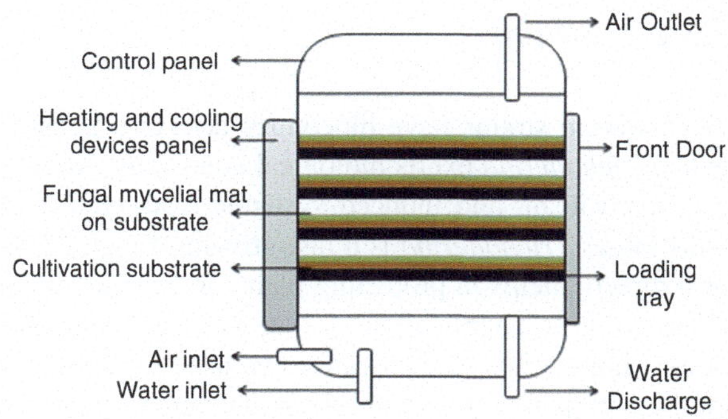

sometimes completely absent. The use of natural sources of carbon substrate is the source of energy for the microbes to grow. This is a major difference with SmF where free water is the medium. In SSF the substrate which is used contains sufficient moisture to support the metabolism a growth of the microorganism. In the case of *Trichoderma* spp. commercial production as a biocontrol agent (BCA), both SSF and SmF are used for large-scale production. In SSF the *Trichoderma* spp. mycelia grows on the surface and on the minor and major cavities of the substrate (Smits et al. 1996). Water activities are a major factor which influences the spore production. Higher moisture content will decrease the amount of aerial spores resulting in a low count of product per gram of harvest material. There are several other factors that eventually influence the productivity when grown as SSF culture. Culture condition, substrate type like particle size, moisture, pH, etc. plays a major role. *Trichoderma* is cultivated in Petri dish, flask, polypropylene bags, column reactors, tray reactors, and packed bed reactors.

5.5 Factors Affecting the Growth of *Trichoderma* spp. in Bioreactors

For production of effective bioformulation of *Trichoderma* spp. there are several key factors that are considered. The production is directly or indirectly influenced by different factors. In both the cases, SSF and SmF there are few common parameters whereas other factors are unique with respect to each process (Vrabl et al. 2019). Key factors affecting the cultivation are summarized as below:

1. Strain/microorganism
2. Nutritional factors
3. Physical factors—temperature, pH, agitation, oxygen requirement, moisture, etc.
4. Inoculum quantity, strength and age
5. Mechanical and design of bioreactor
6. Process scale-up

5.5.1 Strains

Trichoderma strains have biocontrol activities against several fungal plant pathogens. *Trichoderma* exerts biological control by different approaches like antibiosis, mycoparasitism, and induced systemic resistance to plants and by competing for nutrition as *Trichoderma* is a fast-growing fungus. Likewise, *Trichoderma* directly or indirectly helps in protecting plants from many plant diseases when applied as a foliar spray, soil application, or as a seed treatment. But the selection of strains is of utmost importance as the strain to be used for commercial activities needs to fulfill all the criteria of a biocontrol agent. Isolation source plays a crucial role in adapting the isolates into different cultivable condition (Kumari et al. 2014). Screening of strains is performed at the research phase through dual cultural techniques against severe plant pathogens like *Fusarium, Pythium, Pythophthora, Sclerotinia, Alternaria, Cylindrocladium,* etc. Selected strains are then considered for the developmental phase (Singh et al. 2013).

5.5.2 Nutritional Requirements

Trichoderma or any other microbial strains require any major and minor nutrients for growth and metabolism for SSF or SmF. Carbon and nitrogen are the major nutrient sources for growth along with P, K, Mg, Ca, Mn, Fe, etc. Different sources of carbon and nitrogen play a major role in the growth and sporulation of the fungus.

There are different forms of organic and inorganic sources of carbon and nitrogen. Organic sources of carbon are jaggery, molasses which are cheap and readily available, whereas glucose, maltose, sucrose are other types of carbon sources which are considered for production but are less used due to its expensive cost. Different sources of nitrogen are like yeast extract powder, meat extract, corn steep liquor, and inorganic sources like ammonium chloride, ammonium sulfate, etc. Other forms of trace elements are also added at a suitable combination to fasten upon the fungal growth. In SSF, there are other cheap sources of carbon and nitrogen sources which are commonly used like wheat bran, rice flour, and different pulses processed powder. SmF and SSF components for raw material varies as SmF requires more soluble particle for optimum growth whereas bigger size particle is required for SSF since *Trichoderma* utilizes the substrate and grows on the surfaces (Ming et al. 2019).

There are several forms of statistical approaches where the selection of ideal raw material for optimum growth is carried out. Factorial design and Plackett–Burman are the two most common tools for initial screening and optimization of raw materials. The value after initial screening is further statistically analyzed through response surface methodology (RSM). The statistical method of screening and optimization helps to understand the variables factors and its level and marginalizes the critical components essential for the growth of *Trichoderma*. The components of raw materials are sometimes selected on the basis of background and available research data which can save time which can otherwise be time-consuming studies.

5.5.3 Physical Factors: Temperature, pH, Moisture, Oxygen Concentration

As mentioned previously that raw material and nutrients play a crucial role in the production of *Trichoderma*, physical factors like temperature, pH, and oxygen concentration are the major influence for the optimum growth and ideal development of inoculants. Temperature plays a crucial role in growth where it influences the spore production along with other metabolic pathway determination (Mishra and Khan 2015; Singh et al. 2014a). The optimum pH range for the production of *Trichoderma* is 25–30 °C. pH on the other hand is equally important for the ideal manufacturing of *Trichoderma* formulations. It grows well at pH range between 4.5 and 7.0. The most important is oxygen concentration especially in the case of SmF where oxygen transfer and its uptake has a vital influence on the growth. Less or excessive oxygen concentration can also damage or delay the bioprocess. Each factor discussed above is the driving factors for the ideal production of *Trichoderma* formulation in SmF or SSF. Moisture is a limiting factor in SSF and moisture percentage in the range of 50–80% is ideally suited for proliferation and growth of *Trichoderma*. Optimum moisture of 75% is ideally suited for the spore production of *Trichoderma*. Moisture above 80% in solid-state fermentation is reported to decrease the sporulation. Oxygen is supplied to the fermenter vessel through specialized pipelines with a perforation at the bottom known as sparger in SmF. The dispersion of oxygen is carried out by impellers where air or O_2 bubbles are broken into small size for immersion into the suspension. Oxygen is an important factor for the growth of *Trichoderma*. In SSF oxygen supply was a constraint so far but with modern equipments, the supply has become efficient where oxygen is suplied externally.

5.5.4 Inoculum and Seed Culture

The seed culture, i.e., from slant to plate and to liquid broth for preparing a first-generation liquid culture sets is the first step to the production process (Singh and Nautiyal 2012). The inoculum plays a crucial role where cell concentration, stage of inoculation decides the take of the bioprocess. Percentage inoculum is optimized at the research phase prior to production through various studies in shake flask fermentation.

5.5.5 Mechanical Factors

Mechanical factors are mainly considered for the bioreactors used for the cultivation of microbes. The bioreactors used for solid-state fermentation or submerged fermentation have significant effects or the final yield and productivity (Flodman and Noureddini 2013; Olaniyi and Oyesiji 2015). The yield in submerged fermentation depends on the design of the reactors. Ideally, a manufacturing unit should have similar types of symmetrical bioreactors for commercial production. This reduces the variability and ensures standardization of bioprocess. Microbial fermenter for production has a specific dimension ratio of height and diameter. A typical microbial submerged bioreactor has 2.5:1 to 3:1. It consists of a middle vertical shaft with an impeller. The impellers are rotatory blades that rotates at a fixed axis which helps in oxygen supply to microbes form growth. Microbial bioreactors have Ruston turbines impellers of four to six blades. The supply of oxygen is exerted through the bottom ring sparger. Spargers are hollow pipes connected inside the fermenter vessel with a perforation at the lower surface of the ring. Vessel design influences the scaling-up strategies of the bioprocess. On the other hand in SSF, the structure is less complicated but the raw material plays a crucial role. Oxygen transfer, temperature, pH, and water content are limitations in the case of SSF. There are different types of SSF reactors like tray bioreactor, rotating drum bioreactor, packed bed bioreactor, and other forms varies industry-wise. For the production of *Trichoderma* trays, rotator drums and poly bags reactors are most common across industries. And stirred tank bioreactor is the most used submerged fermenter for *Trichoderma* commercial production.

5.5.6 Scale-Up of Bioprocess

Any microbial process which needs to be commercialized for the manufacturing of products undergoes a series of events, popularly known as the scale-up process. The laboratory and pilot study are operational and research activities emphasize on the standardization of research and development phases (Ortiz et al. 2015; Hardy et al. 2017). A laboratory shake flask study is initiated for initial screening and evaluation of input v/s output. Studies for optimization of raw material, temperature, pH, O_2 concentration, inoculum percentage are conducted. The initial investigating parameters set forth the process for the developmental phase, i.e., for piloting the bioprocess. Shake flask fermentation is a miniature form of bioreaction that significantly lays down the prototype for bioprocess.

At the development phase, laboratory-scale bioreactors of varying sizes are used for scaling up of *Trichoderma* formulation from 1 to 100 L. The configuration of bench-scale and pilot-scale bioreactors are exactly similar to commercial-scale bioreactor where cultivation of *Trichoderma* is controlled effectively. *Trichoderma* production using SSF trays and bag are used for commercial production which is less complicated in scale-up process as compared to the submerged one.

For scale-up process of *Trichoderma*, maintenance of an optimum oxygen concentration is of utmost importance in phases of the life cycle (Schmidt 2005). As oxygen is poorly soluble in water, the mechanical factors play a crucial role in the supply of O_2 from air to cell through water as a medium. Oxygen transfer rate (OTR) and oxygen uptake rate (OUR) have to be proportional at the exponential phase of *Trichoderma* cultivation. The exponential phase reaches at 30–48 h from the start of the bioprocess. The cell count per gram or milliliter is 2×10^7 to 5×10^8 with a spore count of 5–8 crores/ml. The wet cell biomass at this stage is ideally from 80 to 200 g/l. All the values obtained indicate a proper optimization of a *Trichoderma* production process. The process of scale-up is controlled through a constant factor known as the volumetric mass transfer coefficient, also popularly known as KL*a* (Tribe et al. 1995). This KL*a* is determined at various phases of scale-up through various calculations with respect to vessel volume or dissolve oxygen concentration.

$$\mathrm{KL}a = \mathrm{OTR} - \mathrm{OUR}$$

where KL is the overall transfer coefficient, "*a*" is the area of bubble per unit of liquid volume (mm^2/ml).

In terms of oxygen flux = KL*a* ($C^* - C$) per unit area of bubble.

C^* = dissolved oxygen concentration which would be at equilibrium with the bubble and C = dissolved oxygen in the liquid phase.

The *Trichoderma* production requires high oxygen supply during the early sporulation stages. The agitation and aeration rate is controlled in such a way that the growth is optimum. High agitation must be avoided for maintaining an optimum dissolved oxygen (DO) concentration as this may cause shearing of mycelia that reduces the spore production. The DO can be maintained through the supply of pure oxygen in certain cases where the air is not sufficient enough. Too much production of mycelia biomass of *Trichoderma* is also not recommended for commercial formulation. An ideal *Trichoderma* inoculant must contain plenty of spores/conidia as they are more stable and have longer shelf life during the storage period.

Modern bioreactors used for the production of *Trichoderma* are controlled with gas monitoring system (GMS) which is automatically regulated using PLC (programmable logic control). The data is managed and recorded through SCADA (a software and program for complete control and record of data for bioprocess). The airflow or oxygen flow is set as per the scale-up parameter as calculated in VVM (vessel volume per minute). Scale-up of *Trichoderma* production process required precision and zero deviation to optimize from pilot scale to commercial scale. The calculations throughout the scale-up study are important as it impacts the optimum output. The required agitation for oxygen supply is calculated throughout the geometric volume of bioreactors with respect to the tip speed of the impeller. This is dependent on the volume of the bioreactor and the diameter of the impeller. The value for tip speed can be calculated as:

$$3.14DN\ (D = \text{diameter of impeller}, N = \text{RPM}).$$

Also, power and volume *P*/*V* calculation is considered for minimizing the stress and shearing caused due to high agitation. This further ensures the *Trichoderma* produced is of high quality for commercial use. In case of SSF sclae up, quantity of substrate in tray or polybags need to be optimized, also water activity and oxygen supply are critical for ustilization of entire substrate and to produce high quantity of spores.

5.6 Factors Affecting Quality of *Trichoderma* Product

The quality of the bio inoculants is of utmost importance. Any potential strain is considered as ideal strain only if it is able to survive, multiply, and performs its desired function. In the process of manufacturing, many aspects are to be taken into consideration to achieve the above characters of ideal inoculants. Among the various factors which affect the quality of the *Trichoderma* when manufactured under large scale includes virulent potential of the strain, sporulation percentage, CFU count of the final product, moisture percentage, pH of the final formulated product, contaminations present in the final formulated product, kind of raw materials used for the production of *Trichoderma* in the fermenters, quality of carriers used for powder and granular product.

5.7 Conclusion

For better utilization of the potential of *Trichoderma*, research in future should focus on testing the suitability of commercially produced *Trichoderma* for management of both foliar and soilborne pathogens, development of better delivery system preferably liquid/oil formulations with long shelf life and product with good resilience potential against harsh conditions. In addition, percolation and better penetration to end users can be achieved by the support of policymakers who by fastening the registration process can help industries in scaling up the process.

References

Al-Taweil HI, Osman MB, Aidil AH, Yussof WM (2009) Optimizing of *Trichoderma viride* cultivation in submerged state fermentation. Am J Appl Sci 6(7):1284

Bhat KA, Anwar A, Lone GM, Hussain K, Nazir G (2009) Shelf life of liquid fermented product of *Trichoderma harzianum* in talc. J Mycol Plant Pathol 39:263

Flodman HR, Noureddini H (2013) Effects of intermittent mechanical mixing on solid-state fermentation of wet corn distillers grain with *Trichoderma reesei*. Biochem Eng J 81:24–28

Fraceto LF, Maruyama CR, Guilger M, Mishra S, Keswani C, Singh HB, de Lima R (2018) *Trichoderma harzianum*-based novel formulations: potential applications for management of next-gen agricultural challenges. J Chem Technol Biotechnol 93(8):2056–2063

Hardy N, Augier F, Nienow AW, Béal C, Chaabane FB (2017) Scale-up agitation criteria for *Trichoderma reesei* fermentation. Chem Eng Sci 172:158–168

Keswani C, Bisen K, Singh V, Sarma BK, Singh HB (2016) Formulation technology of biocontrol agents: present status and future prospect. In: Bioformulations: for sustainable agriculture. Springer, New Delhi, pp 35–52

Kumari TGV, Basu K, Nitya TG, Varma A, Kharkwal AC (2014) Isolation and screening of alkali tolerant trichoderma spp. as biocontrol agent for alkaline agriculture soil. Int J Pharm Pharm Sci 6(10,1):512–516

Kumar S, Thakur M, Rani A (2014) *Trichoderma*: mass production, formulation, quality control, delivery and its scope in commercialization in India for the management of plant diseases. Afr J Agric Res 9(53):3838–3852

Lejeune R, Baron GV (1995) Effect of agitation on growth and enzyme production of *Trichoderma reesei* in batch fermentation. Appl Microbiol Biotechnol 43(2):249–258

Ming S, Rong J, Zhang C, Li C, Zhang C, Zhang Y et al (2019) The solid fermentation state's optimization of *Trichoderma Harzianum* M1. IOP Conf Ser Mater Sci Eng 612(2):022111

Mishra PK, Khan FN (2015) Effect of different growth media and physical factors on biomass production of *Trichoderma viride*. People 8(2):11

Olaniyi OO, Oyesiji YV (2015) Stimulatory effect of physicochemical factors on the expression of cellulase by *Trichoderma viride* NSPRT23. Microbiol J 5(58):67

Ortiz GE, Guitart ME, Cavalitto SF, Albertó EO, Fernández-Lahore M, Blasco M (2015) Characterization, optimization, and scale-up of cellulases production by *Trichoderma reesei* cbs 836.91 in solid-state fermentation using agro-industrial products. Bioprocess Biosyst Eng 38 (11):2117–2128

Prasad RD, Rangeshwaran R (2000) A modified liquid medium for mass production of *Trichoderma harzianum* by fermentation process. Plant Dis Res 15(2):209–211

Ramanujam B, Prasad RD, Sriram S, Rangeswaran R (2010) Mass production, formulation, quality control and delivery of *Trichoderma* for plant disease management. J Plant Prot Sci 2(2):1–8

Schmidt FR (2005) Optimization and scale up of industrial fermentation processes. Appl Microbiol Biotechnol 68(4):425–435

Singh PC, Nautiyal CS (2012) A novel method to prepare concentrated conidial biomass formulation of *Trichoderma harzianum* for seed application. J Appl Microbiol 113(6):1442–1450

Singh HB, Singh BN, Singh SP, Sarma BK (2013) Exploring different avenues of *Trichoderma* as a potent bio-fungicidal and plant growth promoting candidate-an overview. Annu Review Plant Pathol 5:315

Singh A, Shahid M, Srivastava M, Pandey S, Sharma A, Kumar V (2014a) Optimal physical parameters for growth of *Trichoderma* species at varying pH, temperature and agitation. Virol Mycol 3(1):127–134

Singh A, Sarma BK, Singh HB, Upadhyay RS (2014b) *Trichoderma*: a silent worker of plant rhizosphere. In: Biotechnology and biology of *Trichoderma*. Elsevier, Amsterdam, pp 533–542

Smits JP, Rinzema A, Tramper J, Van Sonsbeek HM, Knol W (1996) Solid-state fermentation of wheat bran by *Trichoderma reesei* QM941. Microbiol Biotechnol 46(5–6):489–496

Tribe LA, Briens CL, Margaritis A (1995) Determination of the volumetric mass transfer coefficient (kLa) using the dynamic "gas out–gas in" method: analysis of errors caused by dissolved oxygen probes. Biotechnol Bioeng 46(4):388–392

Vrabl P, Schinagl CW, Artmann DJ, Heiss B, Burgstaller W (2019) Fungal growth in batch culture-what we could benefit if we start looking closer. Front Microbiol 10:2391

Chapter 6
Trichoderma Species: A Blessing for Crop Production

Ramji Singh, P. Anbazhagan, H. S. Viswanath, and Ajay Tomer

Abstract *Trichoderma* species are such a soil fungi which are present worldwide. A wide range of soil habitats ranging from cool temperate to tropical climates can be colonized by them. These include niches covered with field crops, orchards, forests, pasture, and also the soils of desert environment. The saprophytic nature of *Trichoderma* makes it capable of surviving in the soil's uppermost layer (F and H) where mycelium can be recovered in a huge quantity. Some *Trichoderma* species have also been recovered from the habitat representing very adverse ecosystems like mangrove swamps, salt marshes, and estuarine sediments. Survival in such an environment with adverse osmotic potential is a real challenge for *Trichoderma*. *T. viride* has been noticed to widely colonize such environments as it is probably the most widespread in nature. Species of *Trichoderma* have been found to be of immense benefit for the crop. These species have been found to promote the plant growth in addition to their capability of disease management, abiotic stress management, and also for enhancing the rate of seed germination. The role of *Trichoderma* spp. in managing the abiotic stress has now gained momentum. Several species of *Trichoderma* have been found to alleviate the drought and heat stress in crops like rice and wheat by interfering in the scavenging activities of free radicals and reactive oxygen species generated as a result of drought or heat exposure. *Trichoderma* treated plants have also been found to properly compensate for the water losses, thus saving the plants against excess evapotranspiration of water under a water deficit environment. In some of the recent studies, they have also been found to be of greater use for inducing plant defense responses against plant diseases. They also have been found to interfere in the regulation of gene expression mechanisms for disease management and abiotic stress management as well. Nano science is another recent domain where *Trichoderma* has a role for nano-particle synthesis. Several enzymes and other secondary metabolites produced by *Trichoderma* species have

R. Singh (✉) · P. Anbazhagan · H. S. Viswanath
Department of Plant Pathology, College of Agriculture, SVPUAT, Meerut, India

A. Tomer
Department of Plant Pathology, Lovely Professional University, Jalandhar, Punjab, India

C. Manoharachary et al. (eds.), Trichoderma: *Agricultural Applications and Beyond*,
Soil Biology 61, https://doi.org/10.1007/978-3-030-54758-5_6

been found to be of a greater role and importance in industrial use such as clothing and textiles along with food industries. *Trichoderma* spp. are also used as sources of transgenes for developing disease-resistant varieties through genetic engineering. These activities need to be exploited for increasing the high value and quality crop production and also for industrial applications. However, there is a strong need to concentrate on exploring the biodiversity of *Trichoderma* to develop some novel potential strains/isolates than the existing one or it should be developed using genetic engineering and molecular tools.

Keywords *Trichoderma* · Mechanism of action · Defense response · Disease management · Bioremediation · Drought tolerance · Salinity tolerance · Genetic engineering · Transgenes · Nanoparticles

6.1 Introduction

As it has been demonstrated in various literature and scientific communications so far, *Trichoderma* spp. are recognized as an opportunistic avirulent fungal microorganism. Management of several plant diseases and plant growth and yield enhancement are the major responses for which these are widely used as seed treatment, seedling root dip, and soil application (Harman et al. 2004a, b). Recent publications related to *Trichoderma* have been focused on their abilities to combat abiotic stresses through specific mechanisms, which include multiple stresses like osmotic, salinity, chilling, and heat stress. *Trichoderma* has also been found to combat physiological stress such as poor seed quality due to aging of seed. Accumulation of toxic reactive oxygen species (ROS) is very common and it can negatively affect plants under these abiotic stresses, and *Trichoderma* has the abilities to mitigate the damages caused due to accumulation of ROS in stressed plants. Accumulation of lipid peroxides in seedlings under osmotic stress has been found to be reduced due to seed treatment with *Trichoderma*. Among the beneficial effects of *Trichoderma*–plant interaction, enhancement of plant resistance to abiotic stresses has recently attracted more scientific attention; potentially useful tools for enhancing crop production can be achieved by properly understanding this phenomenon. Recently some of the *Trichoderma* strains have shown potential for their industrial use for the production of several enzymes, growth hormone, and some useful secondary metabolites. They are also now being used in genetic engineering for developing transgenics. Most recent advances and development in understanding the diverse functions of *Trichoderma* have been compiled along with metabolites-plants in interactions and how they manifest important modifications in favor of plants to protect them against different challenges.

6.2 Application for the Management of Plant Diseases

In its natural habitat, *Trichoderma* performs some specialized actions at very low dosages, i.e., @ micrograms per liter, and can also mediate chemical communication among soil-inhabiting microorganisms in various ecological habitats, facilitating a beneficial relationship with microflora, higher animals, insects, and plants. Recent studies revealed that *T. harzianum* induces the biosynthesis of chlorophyll, plant enzymes, and phytohormones under several biotic stress in plants (Rawat et al. 2011; Zhang et al. 2013; Hashem et al. 2014), and enable the plant to strongly face the challenges by providing them additional strength while they are under stress. *Trichoderma harzianum* is a biological control agent of fungal origin and being used against a vast range of economically important diseases caused by airborne and soilborne plant pathogens. The harmful plant pathogens are suppressed by using *Trichoderma* spp. possibly due to its abilities of competition, mycoparasitism, and antibiosis (lysis of fungal mycelium by producing enzymes responsible for cell wall degradation), induced systemic resistance, plant growth enhancement, siderophore production, endophytic activities, or a combination of all these antagonistic activities.

Since plant diseases adversely affect agriculture, and thereby the food crop production and its supply also get adversely affected, hence in order to reduce the use of chemical pesticides, there is an urgent need to formulate a sustainable strategy for plant disease management. Under this situation, the use of *Trichoderma* spp. as biological control agents seems to be useful and feasible alternative. Ejechi (1997) investigated certain isolates of *T. viride* with an ability to protect the decay of obeche (*Triplochiton sceleroxylon*) wood by the wood-decaying fungi *Gloeophyllum* sp. and *G. sepiarium* in dry and wet season both in a tropical environment for 11 months through mycoparasitism and competition for fungal nutrients. Considering the potential of genus *Trichoderma* to produce a diverse range of many secondary metabolites useful in application against phytopathogens, Keswani et al. (2014) reported that secondary metabolites of *Trichoderma* spp. are capable of inhibiting pathogen's growth and can be used in any geographic allocation. They further reported that such bioformulations can also be produced with a longer shelf life. However, frequent application of poisonous fungicides for plant disease management can adversely affect the efficiency of the bioagents. Therefore, several groups of scientists have tested the efficacy of *Trichoderma* against fungicides so that the fungicide-resistant *Trichoderma* strains can be effectively used for disease management in the area where fungicide application is quite intensive (Sawant and Mukhopadhyay 1990; Pandey and Upadhyay 1998; Sharma et al. 1999; Nallathambi et al. 2001; Tomer et al. 2018). Effect of integrated application of different fungicides coated with *Trichoderma* spp. for plant disease management has also been studied. Several *Trichoderma* spp. have been found to be quite effective in mitigating the adverse effect caused by fungicides with broad spectrum as compared to many other soil microbes because of its capacity to more rapidly colonize the soil contaminated with pesticides (Oros et al. 2011). *Trichoderma* with bacterial combination or their immobilized formulations can exhibit greater efficacy, because of the fact that several unwanted contaminants can be inhibited simultaneously and the same can be applied for a wider group of pathogens, hence this technology may prove to be more cost-effective.

6.3 How *Trichoderma* Performs Its Action in Plant Disease Management

6.3.1 Competition

It is an important aspect and basics of biological control and occurs when two or more microorganisms demand more of the same resources than it is available immediately. Competition between an introduced antagonist and the native microflora may take longer time for the establishment of introduce one (Papavizas et al. 1984; Howell 2003; Vinale et al. 2008). *Trichoderma* spp. bears persistent type of conidia and have an ability to utilize a broad spectrum of substrate.

6.3.2 Competition for Nutrient and Space

Trichoderma sp. has the ability to grow rapidly with persistent type of conidia and ability to utilize a broad spectrum of substrate. They also potentially compete for nutrition and space with other microorganisms responsible for causing plant diseases and result in inhibiting them (Harman et al. 2004a, b). Among competing microorganisms, of course, the one which is weaker gets starved and die. These fungi produce several siderophores capable of chelating iron and as consequence chelated iron becomes unavailable to the competing pathogenic microorganisms and their growth gets stopped. *Trichoderma* strains also compete for space and also for very important exudates from seeds and roots. These exudates have the capacity to stimulate the germination of propagules of plant pathogenic fungi in soil. *Trichoderma* also has a capacity of utilizing a broad spectrum of substrates including herbicides, fungicides, and phenolic compounds (Chet et al. 1997).

6.3.3 Root Colonization

Plants like mono- and dicotyledonous species usually exhibited a high degree of resistance against pathogen attack especially when they are pretreated with *Trichoderma* (Harman et al. 2004a, b). Root colonization with *Trichoderma* spp. decreases the disease-producing capacity of different pathogens at the site of inoculation (induced localized acquired resistance, LAR). *Trichoderma* has the capacity to colonize the roots of mono- and dicotyledonous plants, which may lead to significant changes in plant metabolic pathways which alter the level of hormones, soluble sugars, phenolic compounds, and amino acids, rate of photosynthesis and transpiration along with water content. It is most likely that there may be an extensive exchange of molecular messages during root colonization, along with the deposition of fungal elicitors in the root cell apoplast. This can have a profound impact on plant disease management and crop yield (Rey et al. 2001).

6.4 Nutrient Solubilization

Trichoderma harzianum enables the treated plants for comparatively better uptake of nutrients (macro- and microelements) by different crop plants. It ensures the availability of phosphorus and other micronutrients viz., Fe, Mn, Cu, and Zn to plants by solubilizing them. It has the capability to solubilize the minerals in four different mechanisms which include acidification through organic acids, chelation using siderophores production, redox using ferric reductase, and hydrolysis through phytase. It can ensure the solubilization of phytase, Fe_2O_3, CuO, and metallic Zn whereas $Ca_3(PO_4)_2$ or MnO_2 did not get solubilized. The presence of lactic acid, citric acid, tartaric acid, and succinic acid were detected using HPLC in *Trichoderma* cultures. *Trichoderma* increases dry matter in plant biomass (92%) and uptake of Cu (42%) in comparison to control plants where it was not applied. Some isolates have been found to solubilize the insoluble tricalcium phosphate to a greater extent in chickpea. *Trichoderma harzianum* also helps to increase phosphorus uptake in plants when treated with it. Gluconic and citric acids produced by *Trichoderma* help in increased solubilization of phosphates, micronutrients, and mineral matters such as Fe, Mg, and Mn by lowering the soil pH (Benitez et al. 2004; Harman et al. 2004b; Vinale et al. 2008).

6.5 Mycoparasitism

Weindling (1932) was the first to recognize *Trichoderma* spp. as a biocontrol agent besides he also noticed mycoparasitism in *Trichoderma*. Wells et al. (1972) also noticed and reported hyphal coiling in *T. lignorum* (*viride*) with *R. solani* and killing it. It is a complex process where *Trichoderma* species grows and attaches chemotropically toward the specific hosts and coils around its hyphae, followed by penetration at the contact points. Recent ultrastructural and/or histochemical approaches described that localized cell wall get lysed at points of contact of host hyphae where *Trichoderma* hyphae has been coiled. Electron microscopic observations revealed that during the parasitism of *Trichoderma* spp. with *Sclerotium rolfsii* and *Rhizoctonia solani*, the host cell walls are enzymatically digested by the antagonist. Several studies have revealed that *Trichoderma* spp. grown on cell walls of pathogenic fungi produce a vast variety of extracellular enzymes such as β-(1,3)-glucanases, chitinases, lipases, and proteases (Mukherjee et al. 2008). A number of secondary metabolites such as non-ribosomal peptides, terpenoids, pyrones, and indole-derived compounds are also produced by major species of *Trichoderma*. Exchange and recognition of signaling molecules by *Trichoderma* and plants may alter the physiological and biochemical aspects of plants and antagonists both. For example, several *Trichoderma* strains provide an increased level of fungal auxin-like compounds to the plant rhizosphere which result in induced root branching and increased shoot biomass. The presence of auxin-like compounds induces extra cell division, expansion, and differentiation. In addition, *Trichoderma* sp. colonizing plant roots can trigger systemic resistance and improve plant nutrient uptake along with suppression of disease.

6.6 Antibiosis

Trichoderma spp. produce and release some volatile compounds, toxic to surrounding pathogenic microorganisms, and under the toxic effect of volatile compounds, these pathogenic organisms get killed and either disease does not occur or get suppressed. *Trichoderma* produces secondary metabolites that help it to compete with other micro- and macroorganisms. Symbiotic activities, transport of metal, differentiation, etc. are other benefits of these volatile metabolites (Demain and Fang 2000). It is a major mechanism of biological control, in which the antagonist produces substances that could be an antibiotic, lytic enzymes (degrades plant cell wall), volatile substances, or toxin that effectively targets and destroys pathogen. *Trichoderma* spp. produces Trichodermin, Viridin, Viridiol, Gliotoxin, Gliovirin, Alkyl pyrones, Volatile compounds of lactones, alcohols, terpenes, etc.

6.7 Induction of Disease Resistance by *Trichoderma* spp.

It is well proven that *Trichoderma* spp. induce expression of some very important genes responsible for the production of chitinase, glucanase, and peroxidase which help the plants to counter against pathogenic microorganisms (Yedidia et al. 2003; Hanson and Howell 2004; Harman et al. 2004b). Seed biopriming with *Trichoderma* has been found to increase the level of resistance in plants against various diseases (Harman et al. 2004a, b). *Trichoderma* sp. is very fast growers, prolific spore bearers, and opportunistic invaders too. They produce such enzymes that are capable of cell wall degradation viz., cellulases, chitinases, and glucanases and also capable of producing antibiotics (Vinale et al. 2008). Moreover, pretreatment of plants with *Trichoderma* spp. results in the induction of hypersensitive response, systemic acquired resistance (SAR), and induced systemic resistance (ISR) in plants (Benitez et al. 2004; Vinale et al. 2008). It has been observed that when tomato plants were treated with *Trichoderma*, there was some active and induced systemic changes in the physiology of the plant, which ultimately resulted in an increased level of resistance to disease (Alfano et al. 2007). Cucumber plants pretreated with *Trichoderma* spp. exhibited induced systemic response of two defense genes responsible for encoding phenylalanine and hydroperoxide lyase and also for systemic accumulation of phytoalexins against *Pseudomonas syringae* pv. *lachrymans* (Yedidia et al. 2003). Several studies have also confirmed that *Trichoderma* spp. may have an indirect contribution by providing systemic acquired resistance in the treated plants (Ahmed et al. 2000; Lo et al. 2000). Harman et al. (2004a, b) mentioned that *Trichoderma* induces localized or systemic resistance against plant disease and provides a considerable level of decline in the level of disease. Thus, it can be stated that managing the disease by root-colonizing *Trichoderma* spp. involves a complex phenomenon where the host plant, the pathogen, the biocontrol agent, and several environmental factors interact together (Harman et al. 2004a, b; Hoitink et al. 2006; Alfano et al. 2007).

Successful exploitation of any specific habitat by an organism chiefly depends on its potential to defend its food source and its ability to compete with surrounding microflora for nutrients, space, and light. Several biocontrol agents and especially *Trichoderma* are masters of this activity (Herrera-Estrella and Chet 2004; Harman 2006; Vinale et al. 2008). Defense mechanisms of *Trichoderma* spp. has enabled it to become efficient competitors, antagonists, and biocontrol agents and these are the exploitable characteristics that make *Trichoderma* spp. very useful. Even metabolites of these fungi can be used to fight against important fungal plant diseases (Spiegel and Chet 1998; Vinale et al. 2006, 2009; Navazio et al. 2007).

6.8 Gene Regulatory Mechanisms Triggering the Defense by *Trichoderma*

The pathways responsible for expression of genes conferring the activity of disease suppression in plants and also for mycoparasitic activity have been worked out and confirmed in several *Trichoderma* isolates which effectively induce the chain of reaction leading to the formation of mitogen-activated protein kinase (MAPK) and also the cAMP pathway (Zeilinger and Omann 2007). The MAP-kinase TVK1 has been specially characterized in *T. virens* (Mendoza-Mendoza et al. 2003; Mukherjee et al. 2004) as well as the genes of similar ancestry in *T. asperellum* (Viterbo et al. 2005) and *T. atroviride* (Reithner et al. 2007), is of utmost importance for regulating signal transduction pathways and gene expression which is very important in conferring the biocontrol activity. Activity of the genes in *T. virens* and *T. asperellum* involved with the functions of plant disease suppression get activated and increased upon interaction with plant roots against *Rhizoctonia solani* (Viterbo et al. 2005). Even though withdrawal of these respective genes cause a reduction in the efficiency of mycoparasitism, the capabilities of biological control in such isolates get increased. Inoculation of *Botrytis cinerea* on plants, pretreated with *Trichoderma* resulted in increased activity of the genes related to jasmonate and as a consequence, systemic resistance to the pathogen get boosted. *Trichoderma* spp. have been exhaustively studied with respect to four genes responsible for governing their biocontrol capacity: viz., TGA1 of *T. atroviride* and TgaA of *T. virens* TGA3 of *T. atroviride* and GNA3 of *T. reesei*. The gene, TGA1, triggers a trait that helps *Trichoderma* to coil around the hyphae of targeted host and for the production of antifungal metabolites also. The absence of TGA1 resulted in increased inhibition of host growth (Rocha-Ramirez et al. 2002; Reithner et al. 2005). Mukherjee et al. (2004) reported MAP-kinases as triggered by the gene TgaA. The gene, TGA3 is also crucial for biocontrol potentiality in *Trichoderma* as deletion of this gene resulted in occurrence of avirulent strains of the antagonist (Zeilinger et al. 2005). The gene GNA3 reported in *T. reesei* has been found to positively govern the activity of mycoparasitism which results in successful plant disease suppression. Attempts were made to differentiate effective and noneffective strains, collected and isolated from the surrounding environment (Nagy et al. 2007; Scherm et al. 2009).

6.9 *Trichoderma* as a Protector of Plant Health

Most beneficial effect of *Trichoderma* spp. is to fight against pathogens directly apart from living as symbionts in association of plants by imparting systemic resistance of plants against several diseases (Yedidia et al. 1999; Shoresh et al. 2010). The response of systemic resistance is governed by proteins of ceratoplatanin type family (Djonović et al. 2006; Seidl et al. 2009). *Trichoderma* itself having genes which encode the activity of MAPK signaling which play a direct role in full induction of systemic response against diseases in plants (Viterbo et al. 2005). The protein, "swollenin" secreted by *Trichoderma* is directly responsible for a significant increase in colonization capacity while invading new niches in the soils (Brotman et al. 2008). This interaction between plant rhizosphere and *Trichoderma* results in enhanced and proliferated root, increased growth, and protection of plants against chemical toxicity. Keeping these positive effects of *Trichoderma* in mind, the same can be applied to overcome the problem of soil pollution and also water pollution by applying the antagonist's spores at an appropriate place (Harman et al. 2004a, b).

6.10 Plant Disease Suppression

In this regard, biological control using natural enemies like *Trichoderma* has been accepted as an ecologically safe, economic method with social acceptability for managing the pest (Routray et al. 2016). A very important role in the protection against plant pathogens are played by volatile compounds produced by these fungal antagonists. *A. thaliana* plants have been protected against the necrotroph *B. cinerea* using volatile organic compounds released by *T. virens*. *B. cinerea* is known to produce some complex symptoms like necrosis, chlorosis, and death of affected seedlings. *T. virens* treated plants were protected against the pathogen as only 15% plants get affected. Volatile compounds from *T. virens* could suppress the growth of *B. cinerea* by 12%. *Trichoderma* protects wheat from *Fusarium* head blight, reduces mycotoxin accumulation, and also the severity of root rot by 51.7%. Effect of a fungicide, i.e., Vitaflo-280 on wheat head blight and root rot (Registered fungicide for the effective control of root rot and head blight in the field trials) and those by volatile compounds from *Trichoderma* were at par to each other. Thus, instead of fungicide, these volatile compounds may be used for seed treatment (Baroncelli et al. 2016; Xue et al. 2017). Rust diseases are among the most important limiting factor of wheat production because of its widespread prevalence and airborne nature of pathogens. *Trichoderma* sprays significantly increased the spike weight, grains weight/spike, and 1000 kernel weight and it elevates the resistance of susceptible genotypes of wheat against leaf rust (*Puccinia triticina*).

Diseases affecting fruits and foliar parts have been effectively controlled through *T. harzianum* (T-22) using them as a spray application to affected plant parts. Cotton plants infected with *R. solani* have been protected using *T. virens* G-6 and disease was also reduced by 78%. Similarly, rice plants infected with blast and bacterial blight has also been protected using *T. harzianum*. They are commercially marketed as biopesticides, biofertilizers and also for the purpose of soil amendment. Once *Trichoderma lignorum* (later found to be *T. atroviride*) reported and published to act as a parasite on other disease-causing fungi in 1932 by Weindling, research progressed at a very rapid and faster rate on the line of antagonistic properties of *Trichoderma* spp. At present *T. atroviride*, *T. harzianum*, *T. virens*, and *Trichoderma asperellum* are the species which are extensively being for the purpose of biological control of plant diseases (Benitez et al. 2004) Availability of recombinant strains of *T. ressei* and established molecular techniques for this strain has facilitated the scientists to use this particular species as a model organism (Seidl et al. 2006). *Trichoderma* spp. are reported to control a broad range of plant pathogens representing all major classes of the fungal kingdom (Monte 2001; Benitez et al. 2004), along with major phytonematodes (Dababat et al. 2006; Kyalo et al. 2007; Goswami et al. 2008). *Trichoderma harzianum* has also been found effective in minimizing the losses due to bacterial leaf blight in rice (Gangwar 2013).

Some lytic enzymes are released by *Trichoderma* spp. make them capable of playing a defensive role against some important plant pathogens (Kubicek et al. 2001; Viterbo et al. 2002). They also produce protein degrading enzymes (Kredics et al. 2005; Suárez et al. 2007; Chen et al. 2009), ABC transporter membrane pumps (Ruocco et al. 2009), volatile or nonvolatile compounds (Calistru et al. 1997; Eziashi et al. 2006), and some more different metabolites of secondary origin (Reino et al. 2008). These compounds play an active role to combat the harmful effect of several plant pathogenic microorganisms (Benitez et al. 2004). Surrounding temperature directly affects the action and success of volatile, nonvolatile, and enzymatic compounds produced by *Trichoderma* spp. (Mukherjee and Raghu 1997), and this knowledge can be of utmost importance for the use of a particular organism as a biocontrol agent in different climates. *Trichoderma* spp. secrete some more important enzymes with potential effectivity under stressed conditions. These enzymes include superoxide dismutase (Grinyer et al. 2005) and amino acid oxidase (Tseng et al. 2008) as revealed through a series of studies related to gene expression. *Trichoderma* also provides protection to its host against several stress like nitrogen-deficient conditions, control of cross pathway, metabolism of lipids, and processes of signaling (Seidl et al. 2009).

6.11 *Trichoderma* as an endophyte for the disease suppression

Trichoderma spp. are known as endophytic to plants with the activity of symbionts also and they are extensively used as biocontrol agents against various kinds of diseases in crops (Harman 2011; Afzal et al. 2013). Endophyte or symbionts are often bacterial or fungal microorganisms, that continue their lives inside the plant over a considerable period of time and they are non-pathogenic too. It may positively affect the growth, uptake, and translocation of nutrients and minerals and also impart the capacity of tolerating abiotic stresses along with protection against biotic stresses. For example, *Trichoderma* inoculation resulted in a significant reduction of disease symptoms by *Botrytis cinerea* and *Cylindrocarpon destructans* and also induces ginsenoside biosynthesis in ginseng plants (Nicol et al. 2002).

6.12 Suppression of Disease Through Siderophore Production and Other Secondary Metabolites

Trichoderma is well known for siderophores production. Growing *Trichoderma* in an iron-deficient medium resulted in a culture filtrate which contains coprogen, coprogen B, and ferricrocin. The level of siderophores production may vary according to the strains, which may range from 270 mg/l to 2080 mg/l. Coprogen, ferricrocin, and a new coprogen derivative which carried a palmitoyl instead of an acetyl group are common siderophores produced by *Trichoderma* (Howell 2003).

6.13 Secondary Metabolites

Secondary metabolites like fungal enzymes produced by *Trichoderma* spp. help in the survival and also to compete with them in their ecological niche (Vinale et al. 2008). In addition to the production of potential antibiotics such as peptaibols, some fungal toxins and several other toxic compounds acting as antibiotics, amino acids, and polypeptide-based metabolites (Sivasithamparam and Ghisalberti 1998) have also been detected from several isolates of *Trichoderma* spp. Peptide-based antibiotic "paracelsin" is the first secondary metabolite produced by *Trichoderma* spp. which has been chemically characterized. (Bruckner and Graf 1983; Bruckner et al. 1984). Thereafter, peptaibols were identified in *Trichoderma* which is a different kind of secondary metabolite (Degenkolb et al. 2008; Stoppacher et al. 2010). Four species of *Trichoderma*, viz., *T. brevicompactum*, *T. arundinaceum*, *T. turrialbense*, and *T. protrudens* were found to produce a mycotoxin known as "trichothecene." These four species are rather distantly related to the *Trichoderma* species, which are being applied as biocontrol agents. This indicated that practices of biocontrol for crop diseases were not at all a risk and also that these mycotoxins are not important from the viewpoint of defense mechanisms against pathogens (Nielsen et al. 2005; Degenkolb et al. 2008). The recent attention of research scientists involved in the field of biocontrol using *Trichoderma* has now been focused on volatile organic compounds produced by *Trichoderma* spp. (Stoppacher et al. 2010). The regulation of peptaibol biosynthesis in *Trichoderma* spp. are greatly influenced by several important factors like light, pH of the substrate, nutrients availability, and mechanical injury. The efficient production of peptaibols is correlated with conidia formation (Kubicek et al. 2007; Tisch and Schmoll 2010). Growth regulators such as the molecules with a resemblance to cytokinin and gibberellin which are responsible for growth promotion, and enhances the root and shoot length or biomass leaf expansion have also been observed in *Trichoderma* treated plants (Howell 2003).

6.14 Use of *Trichoderma* spp. for Bacterial Disease Management

Trichoderma spp. are widely known for their inhibitory effects against a vast range of plant pathogens including different fungal pathogens and plant parasitic nematodes. Although literature available on the lines of using *Trichoderma* spp. against plant bacterial diseases are limited, there are enough reports supporting the efficiency of *Trichoderma* spp. against plant pathogenic bacteria not only in in vitro conditions but also reducing the severity of bacterial diseases even in glasshouse and field studies.

Trichoderma spp. are known to produce different types of secondary metabolites and volatile compounds that are found to inhibit the growth of several phytopathogenic bacteria by the mechanism of antibiosis. Apart from this, they are capable of inducing the acquired systemic resistance in plants by activating defense mechanisms in plants, thereby suppressing the infection caused by bacterial pathogens.

Different isolates of *Trichoderma asperellum* were found to delay the bacterial wilt symptoms development caused by *Ralstonia solanacearum* in tomato by effectively decreasing the disease incidence under field conditions. Application of

T. asperellum was found to increase the activities of peroxidase (POX), phenylalanine ammonia lyase (PAL), polyphenol oxidase (PPO), β-1,3-glucanase and total phenol after the inoculation with the pathogen under field conditions by induction of systemic acquired resistance in tomato plants (Konappa et al. 2018). Application of *T. asperellum* to the cucumber root resulted in the induction of systemic acquired resistance against angular leaf spot (*Pseudomonas syringae* pv. *lachrymans*) and severity of the disease was significantly decreased (Yedidia et al. 2003). *Trichoderma hamatum* was found to provide consistent protection against bacterial leaf spot of tomato caused by *Xanthomonas euvesicatoria* by alterations in the physiology and inducing disease resistance response through systemic modulation in the expression of genes related to stress and metabolism (Alfano et al. 2007).

Five *Trichoderma* species viz., *Trichoderma viride, T. hamatum, T. harzianum, T. lignorum, T. koningii* were tested against cotton bacterial blight caused by *Xanthomonas axonopodis* pv. *malvacearum* by dual culture technique in vitro. All the *Trichoderma* species except *T. viride* and *T. lignorum* were found to inhibit the population dynamics of *X. axonopodis* pv. *malvacearum* after 3 days of incubation (Jagtap et al. 2012). *T. harzianum* representing 10 different agro-ecological zones of Karnataka showed potential inhibition of *Xanthomonas* sp. In vitro by suppressing its growth (Tallapragada and Gudimi 2011). *T. harzianum* was found to inhibit the soft rot causing bacteria in vegetable and tuber crops caused by *Erwinia carotovora*, when tested by a well diffusion method as well as on postharvest stored products (Rashid et al. 2013).

6.15 Use of *Trichoderma* spp. for Nematode Disease Management

Phytonematodes are regarded as one of the most important damaging pathogens affecting various crops. These nematodes generally cause direct parasitism or they may be associated as a vector of other plant disease-causing agents like fungi, bacteria, viruses. Management of nematodes is a difficult task because of their feeding behavior on underground plant parts. Most of the nematicides have been withdrawn from the markets due to their high residual toxicity as they were causing harmful effects on public health and high level of environmental pollution. Furthermore, the application of nematicides will not be able to reach the required depths of soil in inhibiting the nematode population and helps them in developing a high degree of resistance. So, biological control may be one of the eco-friendly alternatives to manage plant-parasitic nematodes and now such approaches are gaining importance particularly in nematode management.

It is already an established fact that among different biocontrol agents widely used, *Trichoderma* spp. occupy a substantial position as antagonistic fungi against different plant pathogens like fungi, bacteria, etc. Apart from them, they also play a vital role in nematode management. *Trichoderma* shows direct parasitism activity against nematodes by hyphal coiling around the nematode body. *Trichoderma* also parasitizes nematode by conidia which are attached to nematodes. Gelatinous matrix of nematodes plays an important role in the attachment of conidia and parasitization process by forming carbohydrate–lectin-like interactions. *Trichoderma* also produces many secondary metabolites like fungal toxins and antibiotics like malformin, hadacidine, gliotoxin, viridian, and penicillin, which can contribute through intoxicating which results in inactivating and immobilizing them and ultimately control of nematodes are achieved.

Application of *T. harzianum* and *T. viride* in tomato crop significantly reduced the density of galls, total egg masses, egg number per egg mass, and reproductive factors of *M. incognita*, the incitant of root-knot nematode (Mukhtar 2018). Different concentrations of culture filtrates of *T. harzianum* were found to inhibit the eggs hatching of *M. javanica*. Inhibition of egg hatching was increased with the increasing concentration of culture filtrates. *M. javanica* eggs were directly parasitized by *T. harzianum* which decreased the pathogenicity of nematodes (Naserinasab et al. 2011). Culture filtrates of different *Trichoderma* spp. viz., *T. harzianum*, *T. viride*, *T. koningii*, *T. reesei*, and *T. hamatum* were found to inhibit the females and egg masses in the reniform nematode (*Rotylenchus reniformis*) and root-knot nematode under in vitro *conditions* due to the effect of toxic metabolites. They also suppressed the nematode activity and movements in both the genera under greenhouse conditions by inhibiting their penetration and development (Bokhari 2009). Eggs and second-stage juveniles (J2s) of *Heterodera avenae* in wheat crops were inhibited by *Trichoderma longibrachiatum* (TL6) by parasitizing their eggs by covering with dense mycelium, lysis of egg contents and penetration of a large number of hyphae through cuticle leading to deformation of J2s, respectively. In the greenhouse experiments, wheat seedlings treated with TL6 had reduced *H. avenae* infection and increased plant growth by inhibiting cysts and juveniles in soil (Zhang et al. 2017). *Trichoderma longibrachiatum* was found to inhibit *Scutellonema* sp. and *Helicotylenchus* by forming an appressorium-like structure on the nematode by penetrating the cuticle of the nematode without coiling but grew instead along the cuticle. The affected nematodes rapidly lost turgor and collapsed due to the penetration of cuticle disintegrating both cuticle and body contents. Whereas, *Trichoderma viride* and *Trichoderma harzianum* caused rapid and excessive coiling of mycelium with its networks of constricting rings/hyphal loops at the nematode body anterior and the head region, making constrictions that might be due to absorptive consumption of body contents suppressing the cuticle of *Scutellonema*s p. and *Helicotylenchus*. Endo- and exochitinases produced by *Trichoderma koningii* help in the penetration of hyphae through the body cuticle of the nematode. Strains of *T. virens*, *T. atroviride*, and *T. rossicum* observed highly efficient in inhibiting *Xiphenema index* population, and the application of *T. viride* had also reduced the potato cyst nematode (*Globoderarosto chinensis*) population in the soil (Daragó et al. 2013; Umamaheswari et al. 2012).

6.16 Application of *Trichoderma* for Mitigating Abiotic Stresses

Salinity and drought are the two most complex and major abiotic stress which not only adverse in nature but also severely affect plant growth and biomass production since long (Haggag et al. 2015). These stresses are the most prominent factors responsible for limiting agricultural/crop productivity. Crop plants need to be managed in such a way that they should be able to tolerate the adverse conditions due to the environment and poor soil health without losing the economic return from these crops. Some microorganisms capable of surviving in most diverse environments are strong enough to mitigate abiotic stresses through their different metabolic activities. In the living ecosystem, interactions of microorganisms with plants are an important part of rhizosphere activities; they are the partners capable of modulating the local and systemic mechanisms in plants which ultimately results in defense under harmful external conditions. *Trichoderma* leads to a variety of changes in the synthesis of secondary metabolites in plants such as plant growth regulators and osmolyte proline under drought. Tomato seeds bioprimed with *Trichoderma* resulted in increased root and shoot growth and chlorophyll contents than those observed in untreated control and also in those plant which undergone a drought exposure. An increase in proline and soluble protein was also noticed in *T. harzianum* treated plants under normal and drought exposure conditions as well. Application of *Trichoderma* also resulted in increasing phenol and flavonoid contents along with an enhanced amount of phytohormones like IAA, IBA, and GA under drought stress have also been reported. Improved level of secondary metabolites production due to *Trichoderma* has also been reported. These secondary metabolites play an integral role in providing tolerance to stress and can also protect the plant membranes from free radicals (ROS) and can help the plant to attain enhanced growth by acquiring more nutrients by robust root growth.

6.17 *Trichoderma harzianum* Can Interfere with Drought by Increasing Relative Water Content in Plants

Secondary metabolites of plants play a crucial role to overcome the harmful effect of various stresses. The availability of relative water content (RWC) indicates the presence of overall water content in the plant systems. The availability of relatively higher RWC in plant cells and tissues indicates the capability of a plant for comparatively more stress tolerance, whereas relatively less RWC indicates reduced stress tolerance. In some of the studies conducted at the Department of Plant Pathology, SVPUAT Meerut, from 2011 to 2018 in rice and wheat crop, it was noticed that seed treatment with diverse isolates of *T. harzianum*, there was comparatively less reduction of RWC as compared to check. These mechanisms are also found in drought-tolerant rice varieties. Relatively increased RWC has been reported in those wheat cultivars which are drought tolerant (Martin et al. 1997). Deka (2000) also reported the effect of drought on the physiological traits of upland rice cultivars at the vegetative stage. Leaf area and relative water content (RWC) decreased significantly under drought stress.

6.18 *Trichoderma* Can Interfere in Drought by Increasing Plant's Membrane Stability Index

Membrane stability index is another very important biochemical parameter which is the deciding factor for stress tolerance, especially drought and heat stress. Higher the membrane stability index, tolerance to water stress will be more as compared to those having lower membrane stability index. Lesser the level of reduction in membrane stability index, the greater will be the tolerance to water stresses. In the stress condition, there is always a greater reduction in membrane stability index; however, seed treatment with different strains of *T. harzianum* has been found to increase the level of membrane stability index in our studies conducted at the Department of Plant Pathology, SVPUAT, Meerut, India from 2011 to 2018 in rice and wheat crop (unpublished). T*richoderma* as seed biopriming resulted in rhizosphere colonization and finally an increase in membrane stability index (MSI) of rice (Shukla et al. 2014; Rawat et al. 2012).

6.19 *Trichoderma* Can Alleviate Drought Tolerance by Interfering the Proline Content

Every important biochemical indicator which is directly related to the capability of a plant to tolerate the drought is the amount of proline available in the plant. The presence of higher proline content in the plant tissues is directly related to stress in plants. When a plant comes under stress, the proline accumulation in the plant tissues gets increased and thus leaves very little space for water retention in the plant tissues and the plant faces water shortage and gets adversely affected due to drought. In some experimental trials on rice and wheat during 2011–2018 at SVPUAT Meerut, India, varying degrees of enhancement of proline content was noticed under stress and also in the plants exposed to a varying degree of drought. Seed biopriming with *T. harzianum* resulted in the accumulation of less proline content thus leaving spare space retention of water in plant tissues which help to provide drought tolerance in the treated crops. After 5, 8, 11, and 14 days of drought exposure also, seed biopriming with *T. harzianum* protected the wheat against drought by decreasing the level of proline and thus facilitating greater accumulation of water in the wheat plant. Greater accumulation of water is beneficial and in favor of plants under stress conditions. A decreased level of proline content in rice, chickpea, and *Arabidopsis* due to *Trichoderma* colonization in the rhizosphere was reported by many workers (Shukla et al. 2014; Cornejo et al. 2014; Chozin et al. 2014).

6.20 *Trichoderma* Can Alleviate Drought by Interfering with Reactive Oxygen Species

Reactive oxygen species (ROS) and other free radicals are increased and cross the limit of scavenging capacity by the host plant and accumulate to the levels that can disrupt the cell components especially those made up of lipids viz., plasma membrane, mitochondrial membrane, and nuclear membrane through the process of lipid peroxidation especially when plants are under severe stress. Probably the endophytic fungi colonizing roots can manipulate the deleterious levels of reactive oxygen species and in this way, the same can limit the symptoms expressed due to biotic and abiotic stress. Levels of antioxidant compounds and antioxidative enzymes get increased whereas levels of hydrogen peroxide get decreased in the roots and leaves of *Trichoderma* inoculated plants. Application of *Trichoderma* spp. also results in increased protection against the damages incurred due to ROS possibly by increasing scavenging abilities in the treated plants. In some experimental trials during 2016–2018 at SVPUAT, Meerut, India, varying degrees of increase in catalase and peroxidase activity was noticed in wheat treated with *Trichoderma* spp. under no stress and also in the plants exposed to varying degrees of drought. Enhanced peroxidase and catalase activity may act as a scavenger of free radicals, which are released when plants become stressed. Proteomics analysis of *Trichoderma* inoculated plant roots showed increased activity of superoxide dismutase along with increased levels of other important enzymes like peroxidase, glutathione reductase, glutathione-S-transferase, and other detoxifying enzymes. A peroxidase gene was also get activated in *Trichoderma* treated and pathogen-infected cucumber plants. Cucumber seed exposed to oxidative stress resulted in a comparatively much poor vigor, but vigor get restored after subsequent treatment with *Trichoderma*. Recently a study showed that seed treatment with *T. harzianum* has resulted in enhanced germination of tomato under osmotic stress (Mastouri 2010). Increased level of tolerance to drought was also noticed in rice due to *Trichoderma harzianum* T35 application by Gusain et al. (2014), and they also noticed that *T. harzianum* promoted the activity of antioxidant enzymes such as superoxide dismutase, catalase, ascorbate peroxidase and thus can prevent oxidative damage to rice through very fast elimination of reactive oxygen species.

6.21 Increased Salinity Tolerance by *Trichoderma*

Soil salinity is such a condition that has become a major problem for crop cultivation, as salt-affected area for crop production which is accounted for more than 20% of irrigated land worldwide. In saline soils, plant growth and physio-biochemical attributes in mustard seedlings get adversely hampered and finally manifest to decreased biomass. However, application of *Trichoderma harzianum* to seedlings of mustard treated with NaCl exhibited increased shoot and root length. The percentage of oil was drastically reduced by salt stress in mustard; however, *Trichoderma harzianum* bioprimed plants resulted in increased oil content from 19.4 to 23.4%. Thus, *T. harzianum* imparts tolerance to mustard plants against salinity through improved absorption of essential nutritional elements, modulation of osmolytes and compounds acting as antioxidants. A relatively higher concentration of sodium salt (NaCl) especially in the soil can not be tolerated by even salt-sensitive crops (Prasad et al. 2000). *Trichoderma* treatment upregulated expression of antioxidant enzymes, including two chloroplast superoxide dismutases and a chloroplast glutathione reductase (4–6 times higher expression in symbiotic plants than non-symbiotic plants). The importance of these enzymes in the water–water cycle and protection of chloroplasts from oxidative damage prompted scientists to examine the role of this symbiosis on the protection of photosynthetic machinery during abiotic stresses. While salinity and water deficit reduced shoot biomass, the biomass of symbiotic plants was similar to control plants (unstressed plants). Application of rhizosphere competent strain of *T.harzianum* has been resulted in increased root growth in field bean because of root colonization by *T. harzianum* (Amin et al. 2010; Shukla et al. 2014).

6.22 Altered Gene Expression Due to *Trichoderma harzianum*

Some strains of *Trichoderma* spp. are plant symbionts that enhance resistance to biotic and abiotic stresses. They colonize roots and change plant gene expression. *Trichoderma* treatment improved seedling growth and redox state of glutathione and ascorbate compared to untreated control plants. Water deficit, on the other hand, reduced concentration and the ratio of reduced and oxidized form of both molecules. The activity of glutathione–ascorbate cycle enzymes increased in response to T22. Similarly, the expression of several genes encoding isoforms of glutathione–ascorbate cycle enzymes and chloroplast Fe and Cu/Zn superoxide dismutase was increased. This data suggest that T22 enhancement of tolerance to water deficit tightly correlates with enhancement in the glutathione–ascorbate cycle, a mechanism that may also improve tolerance level against other stresses be it biotic or abiotic (Mastouri et al. 2010).

6.23 Application for Better Crop Productivity

During the recent past, *Trichoderma*-based products are increasingly being used in vegetable-based cropping systems. Common users of these products are commonly being used by crop growers, extension functionaries, and also by the scientists for sustainable crop production. A bio-stimulatory effect of *Trichoderma* has been noticed in the greenhouse, demonstrating that inoculations of *Trichoderma* to lettuce can significantly increase the yield in all fertilizer application levels. Application of strain GV41 had the most significant effect on yield under both low and optimal N application (Woo et al. 2014; Colla et al. 2015; Lorito and Woo 2015).

6.24 Plant Growth Promotion

These fungi possess the capacity to promote plant growth and development. The ability of *Trichoderma* spp. to induce the development of an exhaustive root system has been noticed and documented very well. Some *Trichoderma* strains can promote even more than a meter-long deep roots in rice below the soil surface. In one of the studies conducted on wheat, rice, tomato, and chilli at the Department of Plant Pathology, SVPUAT, Meerut, India, during 2008–2018 a considerable level of increase in the length of root and shoot, root and shoot biomass, chlorophyll content, area of flag leaf and leaf was recorded in all these crops (unpublished data). In addition, an increased number of buds in tomato and chilli, increased number and size of chilli and tomato fruits have also been noticed. Bioprimed seeds with *T. harzianum* showed improved germination and increased seedling vigor. *T. viride* treated seeds resulted in the enhancement in fresh and dried weight of shoots, roots, and number of nodules in broad bean (Yehia et al. 1985). Conway and Khan (1990) reported the enhancement in weight of broccoli stem and root after application of *T. harzianum* chlamydospores @ 5 kg per hectare. Harman et al. (2004a, b) also noticed frequent root enhancement after *Trichoderma* spp. application. Increased root and shoot lengths, dry weight, and plant height in rice have also been recorded in rice after application of talc formulations of *Trichoderma viride*, when applied as seed biopriming (Mathivanan et al. 2006). *Trichoderma virens* and *T. harzianum*both as seed biopriming have been found to be effective for increasing seed germination, length in root and shoot and of rice seedling fresh weight as well. In addition, dipping the root of seedling in the spore suspension of the biocontrol agents prior to transplanting was also effective for enhancing plant growth and vigor (Mishra and Sinha 2007; Biswas et al. 2008; Simon and Bhandari 2009; Agrawal and Kotasthane 2012). Improved grain yield in chickpea has also been reported due to seed biopriming with *T. viride* (Reddy et al. 2011). Secondary metabolites produced by *Trichoderma koningi* and *T.harzianum* are also capable of acting as plant growth hormones. Metabolites from both the fungi at a relatively higher concentration (10^{-3} M) were able to significantly inhibit the growth of etiolated wheat coleoptiles but not effective at lower dosages (range from 10^{-4} to 10^{-7}) (Ahluwalia et al. 2014; Mishra et al. 2016; Devi et al. 2012; Vahabi et al. 2011).

6.25 Application of *Trichoderma* for Bioremediation

Microorganisms constitute a most important and integral component of the soil. The living part of the soil is constituted by microbes which are also responsible for dynamics of transformation and development of soil structure and organic matter as well. Among the microorganisms, actinomycetes, algae, bacteria, fungi, and protozoa are chiefly available in the soils. A fertile soil is one that contains an adequate reserve of nutrients to be utilized by the plants; otherwise, it should have such a microbial population which is capable of releasing nutrients and facilitate in making them available for plants so that a good plant health can be maintained (Probioma 2006). Any area where agriculture is practiced, the development of healthy and fertile soil is of utmost importance. A wide range of factors affect the soils in an adverse way by a wide range of factors such as poor management practices, pollution by chemical pesticide application, mining or industrial activities, and desertification and climate change. *Trichoderma* spp. are capable to withstand and to adapt to adverse ecological conditions, persisting in the soil and increasing its population. *Trichoderma* spp. also improves the soil's properties contaminated with heavy metals and other industrial effluents and wastes. It also helps to improve the soil, adversely affected by other city and urban wastes. Microbial biotechnology based on the use of beneficial microorganisms like biodecomposers can contribute to the bioremediation of soil and prove to be a healthy alternative. Thus, *Trichoderma* spp. can be an important and easy-to-apply tool of applied biotechnology for the restoration of polluted soils (Uqab et al. 2016).

6.26 Application in Genetic Engineering

Studies on gene discovery and functional genomics have resulted in improved productivity along with tolerance/resistance to abiotic/ biotic stresses (Jewell et al. 2010). Transgenic tobacco and tomato plants with the endochitinase gene from *Trichoderma virens* have been developed using *Agrobacterium*-mediated genetic transformation (Sharad et al. 2015). The integration of endochitinase gene in the genome of transgenic plants was confirmed with the help of PCR and Southern-blot technique, whereas expression of this gene was confirmed through RTPCR. Concentration of endo-chitinase enzyme in transgenic tobacco and tomato plants were tenfold higher than control plants. Endo-chitinase activity was found to be higher in transgenic tomatoes. The level of resistance to fungal pathogens was also greater in transgenic plants (Punja 2006).

6.27 *Trichoderma* as a Source of Transgenes

Microbes capable of biocontrol agents contain a large number of genes which encode for such proteins which facilitate biocontrol. *Trichoderma* species are a good source of such proteins that help the plant to tolerate the stress and thereby facilitate the plant with an ability of survival in stress conditions either biotic or abiotic. Several genes from *Trichoderma* spp. can be utilized to produce transgenic crops with resistance to several plant diseases; these genes have been cloned also. These potentials of *Trichoderma* spp. or any other beneficial microorganisms can be harnessed for organic crop protection and production. A very useful gene from *Trichoderma*, namely hsp70 has been successfully transferred to develop heat and other abiotic stresses tolerant *Arabidopsis thaliana* plant through genetic engineering (Martinez et al. 2008). This gene (hsp70) encoded a protein that is responsible for enhancing the level of tolerance to heat osmotic pressure, high salt concentration, and also against oxidative reactions. The gene "Thkel1" isolated from *T. harzianum*, encoding putative kelch repeat protein, isolated from *T. harzianum* was found to regulate the activity of glucosidase along with mitigating salt and osmotic stresses tolerance in *Arabidopsis thaliana* plants (Hermosa et al. 2011). A number of other proteins related to stress viz., mitogen-activated protein kinase, Sm1 (Small Protein 1), 4-phosphor pantetheinyl transferase, and PKS/NRPS hybrid enzyme from *Trichoderma virens* have also been identified and utilized for developing resistance against several diseases of soil- born nature and also some of those which are foliar pathogens (Howell et al. 2000; Perazzoli et al. 2012; Viterbo et al. 2005).

6.28 *Trichoderma* spp. as Industrial Workhorses

In the near future, the effect of global warming on the environment and human health will have to be effectively tackled and the production of biofuel may be one of the most eco-friendly ways. This will greatly reduce the cost of expenditure on the energy sector and will also tackle the global warming effects. *T. reesei* is the most important genus to be used as a tool for cellulase production through biotechnology and also as a basic research model for the production of protein at industrial scale (Ahamed and Vermette 2009; Li et al. 2013). A novel way of metabolic engineering has been obtained through the molecular mechanism of the cellulose degradation and genome sequencing of *T. reesei* (Kubicek et al. 2009). *T. reesei* possesses quantitative genes that encode a protein (enzyme) capable of degrading the plant cell wall within Sordariomycetes (Martinez et al. 2008). In days to come, research and development may be focused on utilizing *T. reesei* or other strains/other microbes such as yeast for further fermentation as an alternative way of manufacturing biofuels using agricultural waste products with the help of cellulases and hemicellulases (Schuster and Schmoll 2010). Immediately after the discovery of *T. viride* QM6a isolate, during World War II by the US army (Reese 1976), industrial applications of cellulases enzymes have led to extensive research on the line of industrial applications of these enzymes. Since an Army man Elwyn *T. Reese* was behind the discovery of this species (*T. viride*), it was renamed *T. reesei* in his honor (Simmons 1977) and now most efficiently being used for cellulase production worldwide.

6.29 Cellulases and Other Plant Cell Wall-Degrading Enzymes

Production of biofuel has now-a-days became a top grade attention of society, policymakers and scientists as well because of rising costs of energy production and also the adverse climate change (Somerville et al. 2010; Rubin 2008). Some isolates of *Trichoderma* can be used for producing cellulose from cellulosic waste material, thus bioethanol can be produced at relatively cheaper cost (Kumar et al. 2008). Thus, they may also be used for the manufacturing of pulp in paper and textile industries. Induction of high-level gene expression for cellulase and hemicellulase may help to cultivate cellulose, xylan, or a mixture of plant polymers (Mach and Zeilinger 2003) and lactose as well (Seiboth et al. 2007), which are the major byproducts of agricultural and industrial activities. Metabolic engineering is another area where *Trichoderma* spp. are getting attention in recent years. Some *Trichoderma* isolates can be used to produce beneficial enzymes which are mainly required to degrade cell wall material in the plants (Foreman et al. 2003; Martinez et al. 2008). Thus in the very short time to come, economically reasonable products like second-generation biofuels can be produced from waste products (rural, urban, and agricultural wastes) with the help of *Trichoderma* spp.

6.30 Application of *Trichoderma* in Food and Other Industries

Trichoderma spp. efficiently produce extracellular enzymes. Several species of *Trichoderma* are commercially being utilized for the production of some important enzymes which are useful for the degradation of complex polysaccharides such as cellulases etc. Due to their enzyme production capacity *Trichoderma* spp. are frequently being used for product development in food and textile industries. Cellulases are the main enzyme produced by these fungi which are mainly used for "*biostoning*" of denim fabrics and as a result, treated denim become soft and whitened showing stone-washed appearance. Digestibility of hemicelluloses available with barley or other cereals/grains available in poultry feeds, also get increased after mixing these enzymes. Production of food additives and related products have also been benefitted with the use of some isolates of *Trichoderma* spp. (Nevalainen et al. 1994; Blumenthal 2004). Recently *Trichoderma* is also being used in brewing as β-glucanase enzyme produced by *Trichoderma* can improve this process. Macerating enzymes such as pectinases, cellulases, and hemicellulases are being utilized in fruit juice production and also for the production of xylanases to be used as a feed additive in livestock feed. Cellulase producing isolates have been extensively utilized for making of biscuits, bread, and other bakery products, malted products, and also for the production of alcohol from cereal grains (Galante et al. 1998). In addition to enzymes, secondary metabolites are also synthesized by *Trichoderma* spp. to be used as food additives. Some of the secondary metabolite-like materials produced by *T. viride* are having quite similar characteristics resembling aroma like coconut (a 6-pentyl-α-pyrone) exhibiting properties of antibiotics, are also of potential use in the food industry. Because of their antifungal effect, few isolates of *T. harzianum* can be used as food preservatives (Fuglsang et al. 1995). Accumulation of mutant in dental plaque can also be prevented by using some mutant isolates of *T. harzianum* in the toothpaste (Wiater et al. 2005).

6.31 Potential Applications in Modern Agriculture and Sustainable Environment

Genus *Trichoderma* is ubiquitous and this nature is achieved by its ability of diversified metabolic pathways which make it capable of producing several enzymes and secondary metabolites. Production of enzymes with commercial importance viz., amylases, cellulases, beta1–3 glucanases, and chitinases have been studied in a greater way and this technology is continuously being upgraded (Harman et al. 2004a, b; Ahamed and Vermette 2008; Sandhya et al. 2004). Recently, *Trichoderma* spp. have been found to be useful in nanotechnology for the biosynthesis of silver nanoparticles (Vahabi et al. 2011), thus opening another and alternate way of plant disease management through nanoparticles. Bioremediation of soil contaminated with certain pesticides using fungi has now become relatively older. Evidences of various *Trichoderma* spp. degrading polycyclic aromatic hydrocarbons (PAHs) are in plenty. Katayama and Matsumura (1993) have already reported the degradation potential of rhizosphere-competent *Trichoderma* strains against several synthetic dyes, pentachlorophenol, Endosulfan, and dichlorodiphenyltrichloroethane (DDT). These contaminants get degraded by the enzymes like hydrolyases, peroxidase, lactases and other lytic enzymes produced by *Trichoderma* spp. *Trichoderma* spp. not only help to improve the soil and plant health, but also a source for sustainable protection of crop yield. Inherent resistance ability of *Trichoderma* spp. to many toxic compounds such as fungicides, herbicides, insecticides, and phenolic compounds make it capable of growing rapidly in the soil where it is inoculated (Chet et al. 1997). *Trichoderma* strains efficiently degrade the pesticide-contaminated soil in addition to its ability to degrade a wide range of insecticides such as organochlorines, organophosphates, and carbonates. *Trichoderma* strains may express resistance mechanisms against tested noxious compounds because of their ability of ABC transporter protein systems (Harman et al. 2004a, b).

6.32 Biosynthesis of Nanoparticles from Trichoderma spp.

Nanotechnology is one of the dynamic and fast-growing branches of recent science having the potential to revolutionize many disciplines like science, technology, medicine, and agriculture as well. Conversion of macro-materials to nanosized particles having sizes ranging between 1 and 100 nm gives rise to change in their characteristics which are new and different from macro-materials. As sizes of nanoparticles are very small and sometimes even smaller than many viruses, their action toward target sites is very quick, deep, and accurate. Although chemical synthesis of nanoparticles is an efficient means, it involves a greater amount of risks like the involvement of toxic materials, hazardous chemical compounds, and high energy requirement which makes their production process very costly. So, biosynthesis of nanoparticles is a safer alternative to chemical methods of nanoparticle production because of its non-toxicity and involvement of less production costs. Nanoparticles had broad-spectrum applications in several fields of science and technology. In agriculture, they were found to act as potent fungicides, and antibacterial by inhibiting various plant pathogenic fungi and bacteria apart from their role in the detection and diagnosis of plant diseases as biosensors (Mishra 2017).

Trichoderma spp. are one of the most prevalent microorganisms present in all kinds of soils across the globe. Various species of *Trichoderma* are known to produce different types of extracellular metabolites and reductase enzymes which help in reducing the size of various metal ions to their elemental nano size with higher stability in a lesser time. Devi et al. (2013) isolated 75 *Trichoderma* isolates representing five different species viz., *Trichoderma virens*, *T. asperellum*, *T. harzianum*, *T. longibrachiatum*, and *T. pseudokonningi* from different locations, where they found that all the isolates produced silver nanoparticles with sizes ranging between 1 and 45 nm and they also reported that nitrate reductase enzyme is responsible to convert Ag2+ to Ag0. Different isolates of *Trichoderma virens* were found to be capable of synthesizing nanoparticles in silver (AgNP's) but the intensity of production varied among the isolates and was also found that the isolated extracellular culture filtrate of *T. virens* is responsible for the reduction of nanoparticles (Peeran et al. 2017). Culture filtrate of the fungal antagonist, *Trichoderma asperellum* was found to synthesize silver nanoparticles (Ag NPs) by using silver nitrate as the precursor chemical compound as confirmed by UV-Vis spectroscopy and TEM (Ahmed and Dutta 2019). The reaction of cell filtrate of agriculturally important bioagent *Trichoderma harzianum* with 1 mM silver nitrate solution, gave the formation of silver nanoparticles within 3 h which was observed visually by the change in the color of the solution. Transmission electron microscopy (TEM) showed polydisperse spherical and occasionally ellipsoid nanoparticles in the size ranging from 19 to 63 nm (Ahluwalia et al. 2014; Shelar and Chavan 2015). *Trichoderma reesei*, an industrially important fungus was found to synthesize AgNPs (silver nanoparticles) in much higher amounts than other fungi by reducing toxic silver $Ag+$ ions to such ametallicAgNPs which is nontoxic, through catalyzing the reduction process with the help of extracellular enzyme and metabolites of the fungus (Vahabi et al. 2011). Culture filtrates of *Trichoderma viride*, a potent biocontrol fungus and a filamentous fungus (*Trichoderma koningii*) were found to synthesize magnesium oxide nanoparticles (MgNPs), silver nanoparticles (AgNPs), and gold nanoparticles (AuNPS) involving various extracellular metabolites and enzymes. The sizes of particles were determined by SEM and UV spectroscopy. The time required for the biosynthesis of gold nanoparticles was found to be less than a minute with *Trichoderma viride* (Mishra 2017). *Trichoderma inhamatum* and *Trichoderma atroviride* were found to synthesize silver (AgNPs) and copper nanoparticles (CuNPs), respectively (Gnanamangai et al. 2017; Hussein 2016).

6.33 Conclusions

Several qualities of *Trichoderma* spp. make them a tool with great potential which can be efficiently used in agriculture for diverse beneficial effects, such as managing abiotic stresses, enhancing uptake of nutrients in plants, increased nitrogen-use efficiency, and increased photosynthetic efficiency in different crops. Use of *Trichoderma* spp. has been expanded worldwide as biocontrol agents for plant pathogens and plant growth promoters, besides its industrial application. Many useful genes have been located in the genome of *Trichoderma* spp. which makes it adaptive to the harsh environments (soil, water, dead tissues, inside the plants, etc.). The metabolic pathways of *Trichoderma* spp. are quite complex, especially for the production of secondary metabolites but with the help of more advanced and improved molecular approaches, exploring new pathways has become a little easy and bit possible. These secondary metabolites have broader potential use in agriculture. Mapping of gene-specific translation to protein in *Trichoderma* spp. and its interactions have been done successfully and the same information has been utilized to develop new synergistic products of the living fungus with its secreted metabolites. These new formulations are considered to be more effective than older products and effective on a broader range of plant pathogens.

Some of the *Trichoderma* strains are reported to be resistance inducers and growth promoters. An average yield increase of approximately 5% has been noticed in maize, treated with *Trichoderma*. However, some varietal differences were also observed, as some maize lines responded neutral or even negative growth responses to *Trichoderma* application. Moreover, knowing specific gene products associated with a favorable effect on plants will lead to rapid assays of the critical genes expression and decoded proteins on a field scale. This will help to create a management tool that will afford to acquire a reliable knowledge of the possible interaction. It is believed that the potential of these fungi viz., (a) ability of resistance induction against biotic stresses like disease and abiotic stresses like drought and salinity and (b) enhanced nutrient uptake will make them highly useful tools and plant productivity can be enhanced to a greater extent by using them. Thus, they can be used to ensure food security, and also to protect the ecosystem. *Trichoderma* spp. have a specific ability to produce nitrogenous compounds from unused fertilizer in the soil and it is this ability which can minimize the dosages of nitrogenous fertilizer in the crops, thus can help in reducing nitrate pollution in soil, water bodies, and the same can check air pollution also. It is an antagonistic fungus that has been used successfully for controlling onion diseases like white rot, pink rot, Fusarium basal rot, onion smudge, and damping off. Cucumber, bell pepper, and strawberry yields were increased significantly following the application of *Trichoderma harzianum* in the root zone of these crops. It has been also found to tolerate a vast range of pollutants like heavy metals, toxic agrochemicals, and polyaromatic compounds which are very difficult to be managed. Thus, *Trichoderma* can be considerable as a biological weapon for managing plant diseases and promoting agriculture and environmental sustainability.

References

Afzal S, Samrah T, Viqar S, Jehan A, Syed EH (2013) Managing the root diseases of okra with endo-root plant growth promoting *Pseudomonas* and *Trichoderma viride* associated with healthy okra roots. Pak J Bot 45:1455–1460

Agrawal T, Kotasthane AS (2012) Differential response of rice genotypes to bioinoculants. J Mycol Plant Pathol 42(3):310–313

Ahamed A, Vermette P (2008) Culture-based strategies to enhance cellulase enzyme production from *Trichoderma reesei* RUT-C30 in bioreactor culture conditions. Biochem Eng J 40:399–407

Ahamed A, Vermette P (2009) Effect of culture medium composition on *Trichoderma reesei's* morphology and cellulase production. Bioresour Technol 100:5979–5987

Ahluwalia V, Kumar J, Sisodia R, Shakil NA, Walia S (2014) Green synthesis of silver nanoparticles by *Trichoderma harzianum* and their bio-efficacy evaluation against Staphylococcus aureus and Klebsiella pneumonia. Ind Crop Prod 55:202–206

Ahmed AA, Dutta P (2019) *Trichoderma asperellum* mediated synthesis of silver nanoparticles: characterization and its physiological effects on tea [*Camellia sinensis*(L.) Kuntze var. *assamica* (J. Masters) Kitam.]. Int J Curr Microbiol App Sci 8(04):1215–1229

Ahmed AS, Sanchez CP, Candela ME (2000) Evaluation of induction of systemic resistance in pepper plants (*Capsicum annum*) to *Phytophthora capsici* using *Trichoderma harzianum* and its relation with capsidiol accumulation. Eur J Plant Pathol 106:817–829

Alfano G, Ivey ML, Cakir C, Bos JIB, Miller SA, Madden LV, Kamoun S, Hoitink HAJ (2007) Systemic modulation of gene expression in tomato by *Trichoderma hamatum* 382. Phytopathology 97(4):429–437

Amin F, Razdan VK, Mohid Din FA, Bhat KA, Banday S (2010) Potential of *Trichoderma* species as biocontrol agents of soil borne fungal propagules. J Phytol 2(10):38–41

Baroncelli R, Zapparata A, Piaggeschi G, Sarrocco S, Vannacci G (2016) Draft whole-genome sequence of *Trichoderma gamsii* T6085, a promising biocontrol agent of Fusarium head blight on wheat. Genome Announc 4(1):e01747–e01715. https://doi.org/10.1128/genomeA.01747-15

Benitez T, Rincon AM, Limon MC, Codon AC (2004) Biocontrol mechanism of *Trichoderma* strains. Int Microbiol 7:249–260

Biswas SK, Ratan VED, Srivastava SSL, Singh R (2008) Influence of seed treatment with biocides and foliar spray with fungicides for management of brown leaf spot and sheath blight of paddy. Indian Phytopathol 61(1):55–59

Blumenthal CZ (2004) Production of toxic metabolites in *Aspergillus niger*, *Aspergillus oryzae*, and *Trichoderma reesei*: justification of mycotoxin testing in food grade enzyme preparations derived from the three fungi. Regul Toxicol Pharmacol 39:214–228

Bokhari FM (2009) Efficacy of some *Trichoderma* species in the control of *Rotylenchulus reniformis* and *Meloidogyne javanica*. Arch Phytopathol Plant Protect 42(4):361–369

Brotman Y, Briff E, Viterbo A, Chet I (2008) Role of swollenin, an expansin-like protein from *Trichoderma*, in plant root colonization. Plant Physiol 147(2):779–789

Bruckner H, Graf H (1983) Paracelsin, a peptide antibiotic containing alpha-aminoisobutyric acid, isolated from *Trichoderma reesei* Simmons. Part A. Experientia 39:528–530

Bruckner H, Graf H, Bokel M (1984) Paracelsin; characterization by NMR spectroscopy and circular dichroism, and hemolytic properties of a peptaibol antibiotic from the cellulolytically active mold *Trichoderma reesei*. Part B. Experientia 40:1189–1197

Calistru C, McLean M, Berjak P (1997) *In vitro* studies on the potential for biological control of Aspergillus flavus and Fusarium moniliforme by *Trichoderma* species. A study of the production of extracellular metabolites by *Trichoderma* species. Mycopathologia 137(2):115–124

Chen LL, Liu LJ, Shi M, Song XY, Zheng CY, Chen XL, Zhang YZ (2009) Characterization and gene cloning of a novel serine protease with nematicidal activity from *Trichoderma pseudokoningii* SMF2. FEMS Microbiol Lett 299(2):135–142

Chet I, Inbar J, Hadar I (1997) Fungal antagonists and mycoparasites. In: Wicklow DT, Söderström B (eds) The mycota IV: environmental and microbial relationships. Springer, Berlin, pp 165–184

Chozin MAM, Lubis I, Junaedi A, Ehara H (2014) Some physiological character responses of rice under drought conditions in a paddy system. J Int Soc Southeast Asian Agric Sci 20(1):104–114

Colla G, Rouphael Y, Di Mattia E, El-Nakhel C, Cardarelli M (2015) Co-inoculation of Glomus intraradices and Trichoderma atroviride acts as a biostimulant to promote growth, yield and nutrient uptake of vegetable crops. J Sci Food Agric 95(8):1706–1715

Conway KE, Khan BA (1990) Enhanced growth of broccli transplants by the biocontrol fungi *Trichoderma harzianum* and *Laetisariaarvalis*. Phytopathology 80:434

Cornejo CHA, Rodriguez ML, Cuevas AR, Bucio LJ (2014) *Trichoderma* spp. improves growth of Arabidopsis seedlings under salt stress through enhanced root development, osmolite production, and na+ elimination through root exudates. Mol Plant-Microbe Interact 27(6):503–514

Dababat AA, Sikora RA, Hauschild R (2006) Commun use of *Trichoderma harzianum* and *Trichoderma viride* for the biological control of Meloidogyne incognita on tomato. Agric Appl Biol Sci 71(3 Pt B):953–961

Daragó Á, Szabó M, Hrács K, Takács A, Nagy PI (2013) In vitro investigations on the biological control of Xiphinema index with Trichoderma species. Helminthologia 50(2):132–137

Degenkolb T, von Dohren H, Nielsen KF, Samuels GJ, Bruckner H (2008) Recent advances and future prospects in peptaibiotics, hydrophobin, and mycotoxin research, and their importance for chemotaxonomy of *Trichoderma* and *Hypocrea*. Chem Biodivers 5:671–680

Deka M (2000) Effect of drought on physiological traits of upland ahu (rabi) rice (*Oryza sativa* L.) cultivars at vegetative stage. Crop Res 19(3):434–439

Demain AL, Fang A (2000) The natural functions of secondary metabolites. Adv Biochem Eng Biotechnol 69:1–39

Devi R, Kaur N, Gupta AK (2012) Potential of antioxidant enzymes in depicting drought tolerance of wheat (*Triticum aestivum* L.). Indian J Biochem Biophys 49(4):257–265

Devi TP, Kulanthaivel S, Kamil D, Borah JL, Prabhakaran N, Srinivasa N (2013) Biosynthesis of silver nanoparticles from *Trichoderma* species. Indian J Exp Biol 51:543–547

Djonović S, Pozo MJ, Dangott LJ, Howell CR, Kenerley CM (2006) Sm1, a proteinaceous elicitor secreted by the biocontrol fungus *Trichoderma virens* induces plant defense responses and systemic resistance. Mol Plant-Microbe Interact 19(8):838–853

Ejechi BO (1997) Biological control of wood decay in an open tropical environment with *Penicillium* sp. and *Trichoderma viride.* Int Biodeterior Biodegradation 39(4):295–299

Eziashi EI, Uma NU, Adekunle AA, Airede CE (2006) Effect of metabolites produced by *Trichoderma* species against *Ceratocystis paradoxa* in culture medium. Afr J Biotechnol 5:703–706

Foreman PK, Brown D, Dankmeyer L et al (2003) Transcriptional regulation of biomass-degrading enzymes in the filamentous fungus *Trichoderma reesei.* J Biol Chem 278:31988–31997

Fuglsang CC, Johansen C, Christgau S, Adler-Nissen J (1995) Antimicrobal enzymes: applications and future potential in the food industry. Trends Food Sci Technol 6:390–396

Galante YM, Conti A, Monteverdi R (1998) Application of *Trichoderma* enzymes in the food and feed industries. In: Harman GE, Kubicek CP (eds) *Trichoderma* and *Gliocladium.* Taylor and Francis, London, pp 327–342

Gangwar GP (2013) Growth promotion of rice seedlings by fungal and bacterial bioagents effective against bacterial leaf blight of rice. J Appl Nat Sci 5(2):430–434

Gnanamangai BM, Ponmurugan P, Jeeva SE, Manjukarunambika K, Elango V, Hemalatha K, Kakati JP, Mohanraj R, Prathap S (2017) Biosynthesised silver and copper nanoformulation as foliar spray to control bird's eye spot disease in tea plantations. IET Nanobiotechnol 11 (8):917–928

Goswami J, Pandey RK, Tewari JP, Goswami BK (2008) Management of root knot nematode on tomato through application of fungal antagonists, *Acremonium strictum* and *Trichoderma harzianum.* J Environ Sci Health 43(3):237–240

Grinyer J, Hunt S, McKay M, Herbert BR, Nevalainen H (2005) Proteomic response of the biological control fungus *Trichoderma atroviride* to growth on the cell walls of *Rhizoctonia solani.* Curr Genet 47(6):381–388

Gusain YS, Singh US, Sharma AK (2014) Enhance activity of stress related enzymes in rice (*Oryza sativa* L.) induced by plant growth promoting fungi under drought stress. Afr J Agric Res 9 (19):1430–1434

Haggag WM, Abouziena HF, Abd-El-Kreem F, El Habbasha S (2015) Agriculture biotechnology for management of multiple biotic and abiotic environmental stress in crops. J Chem Pharm Res 7(10):882–889

Hanson LE, Howell CR (2004) Elicitors of plant defence responses from biocontrol strains of *Trichoderma virens.* Phytopathology 94:171–176

Harman GE (2006) Overview of mechanisms and uses of *Trichoderma* spp. Phytopathology 96:190–194

Harman GE (2011) Multifunctional fungal plant symbionts: new tools to enhance plant growth and productivity. New Phytol 189:647–649

Harman GE, Petzoldt R, Comis A, Chen J (2004a) Interactions between Trichoderma harzianum strain T22 and maize inbred line M017 and effects of these interactions on diseases by *Pythium ultimum* and *Collectotrichum graminicola.* Phytopathology 94:147–153

Harman GE, Howell CR, Viterbo A, Chet I, Lorito M (2004b) *Trichoderma* species—opportunistic, avirulent plant symbionts. Nat Rev Microbiol 2(1):43–56

Hashem A, Abd-Allah EF, Alqarawi AA, Al Huqail AA, Egamberdieva D (2014) Alleviation of abiotic salt stress in *Ochradenusbaccatus* (Del.) by *Trichoderma hamatum* (Bonord.) Bainier. J Plant Interact 9:857–868. https://doi.org/10.1080/17429145.2014.983568

Hermosa R, Botella L, Keck E, Jiménez JÁ, Montero-Barrientos M, Arbona V, Gómez-Cadenas A, Monte E, Nicolás C (2011) The overexpression in *Arabidopsis thaliana* of a Trichoderma harzianum gene that modulates glucosidase activity, and enhances tolerance to salt and osmotic stresses. J Plant Physiol 168(11):1295–1302

Herrera-Estrella A, Chet I (2004) The biological control agent *Trichoderma*—from fundamentals to applications. In: Arora DK (ed) Fungal biotechnology in agricultural, food and environmental applications. Marcel Dekker, New York, pp 147–156
Hoitink HAJ, Madden LV, Dorrance AE (2006) Systemic resistance induced by Trichoderma spp.: interactions between the host, the pathogen, the biocontrol agent, and soil organic matter quality. Phytopathology 96:186–189
Howell CR (2003) Mechanisms employed by Trichoderma species in the biological control of plant diseases: the history and evolution of current concepts. Plant Dis 87:4–10
Howell CR, Hanson LE, Stipanovic RD, Puckhaber LS (2000) Induction of terpenoid synthesis in cotton roots and control of *Rhizoctonia solani* by seed treatment with *Trichoderma virens*. Phytopathology 90:248–252
Hussein M (2016) Silver tolerance and silver nanoparticle biosynthesis by Neoscytalidium novaehollandae and *Trichoderma inhamatum*. Eur J Biol Res 6(1):28–35
Jagtap GP, Jangam AM, Deya U (2012) Management of bacterial blight of cotton caused by *Xanthomonas axonopodis* pv. *malvacearum*. Sci J Microbiol *1*(1):10–18
Jewell MC, Campbell BC, Godwin ID (2010) Transgenic plants for abiotic stress resistance. In: Kole C, Michler CH, Abbott AG, Hall TC (eds) Transgenic crop plants. Springer, Berlin
Katayama A, Matsumura F (1993) Degradation of organochlorine pesticides, particularly endosulfan, by *Trichoderma harzianum*. Environ Toxicol Chem 12(6):1059–1065
Keswani C, Mishra S, Sarma BK, Singh SP, Singh HB (2014) Unraveling the efficient applications of secondary metabolites of various *Trichoderma* spp. Appl Microbiol Biotechnol 98 (2):533–544
Konappa N, Krishnamurthy S, Siddaiah CN, Ramachandrappa NS, Chowdappa S (2018) Evaluation of biological efficacy of *Trichoderma asperellum* against tomato bacterial wilt caused by Ralstonia solanacearum. Egypt J Biol Pest Control 28(1):63
Kredics L, Antal Z, Szekeres A, Hatvani L, Manczinger L, Vágvölgyi C, Nagy E (2005) Acta Microbiol Immunol Hung 52(2):169–184
Kubicek CP, Mach RL, Peterbauer CK, Lorito M (2001) *Trichoderma*: from genes to biocontrol. J Plant Pathol 83:11–23
Kubicek CP, Komon-Zelazowska M, Sandor E, Druzhinina IS (2007) Facts and challenges in the understanding of the biosynthesis of peptaibols by *Trichoderma*. Chem Biodivers 4:1068–1082
Kubicek CP, Mikus M, Schuster A, Schmoll M, Seiboth B (2009) Metabolic engineering strategies for the improvement of cellulase production by *Hypocrea jecorina*. Biotechnol Biofuels 2:19
Kumar R, Singh S, Singh OV (2008) Bioconversion of lignocellulosic biomass: biochemical and molecular perspectives. J Ind Microbiol Biotechnol 35:377–391
Kyalo G, Affokpon A, Coosemans J, Coynes DL (2007) Commun Biological control effects of Pochoniachlamysdosporia and *Trichoderma* isolates from Benin (West-Africa) on root-knot nematodes. Agric Appl Biol Sci 72(1):219–223
Li C, Yang Z, Zhang RHC, Zhang D, Chen S, Ma L (2013) Effect of pH on cellulase production and morphology of *Trichoderma reesei* and the application in cellulosic material hydrolysis. J Biotechnol 168(4):470–477
Lo CT, Liao TF, Deng TC (2000) Induction of systemic resistance of cucumber to cucumber green mosaic virus by the root-colonizing *Trichoderma* spp. Phytopathology 90:S47
Lorito M, Woo SL (2015) Discussion agronomic. In: Lugtenberg B (ed) Principles of plant-microbe interactions. Springer International, Berlin, pp 345–353. https://doi.org/10.1007/978-3-319-08575-3_36
Mach RL, Zeilinger S (2003) Regulation of gene expression in industrial fungi: *Trichoderma*. Appl Microbiol Biotechnol 60:515–522
Martin M, Morgan J, Zerbi G, Lecain D (1997) Water stress imposition rate affects osmotic adjustment and cell wall properties in winter wheat. Ital J Agron 1:11–20
Martinez D, Berka RM, Henrissat B, Saloheimo M, Arvas M, Baker SE, Chapman J, Chertkov O, Coutinho PM, Cullen D, Danchin EG (2008) Genome sequencing and analysis of the biomass-degrading fungus *Trichoderma reesei* (syn. *Hypocrea jecorina*). Nat Biotechnol 26(5):553

Mastouri F (2010) Use of *Trichoderma* spp. to improve plant performance under abiotic stresses. Ph.D. Thesis, Cornell Univeristy

Mastouri F, Björkman T, Harman GE (2010) Seed treatment with Trichoderma harzianum alleviates biotic, abiotic, and physiological stresses in germinating seeds and seedlings. Phytopathology 100(11):1213–1221

Mathivanan N, Prabavathy VR, Vijayanandraj VR (2006) Application of talc formulations of *Pseudomonas fluorescens* Migula and *Trichoderma viride* Pers. ex S.F. Gray decrease the sheath blight disease and enhance the plant growth and yield in rice. J Phytopathol 154 (11/12):697–701

Mendoza-Mendoza A, Pozo MJ, Grzegorski D, Martínez P, García JM, Olmedo-Monfil V, Cortés C, Kenerley C, Herrera-Estrella A (2003) Enhanced biocontrol activity of *Trichoderma* through inactivation of a mitogen-activated protein kinase. Proc Natl Acad Sci U S A 100 (26):15965–15970

Mishra A (2017) Journey of *Trichoderma:* plant stress amelioration to nanosynthesis. Genet Mol Biol Res 1(1):1

Mishra DS, Sinha AP (2007) Plant growth-promoting activity of some fungal and bacterial agents on rice seed germination and seedling growth. Trop Agric 77(3):188–191

Mishra G, Sapre S, Sharma A, Tiwari S (2016) Amelioration of drought tolerance in wheat by the interaction of plant growth-promoting rhizobacteria. Plant Biol 18(6):992–1000

Monte E (2001) Understanding *Trichoderma:* between biotechnology and microbial ecology. Int Microbiol 4(1):1–4

Mukherjee PK, Raghu K (1997) Effect of temperature on antagonistic and biocontrol potential of *Trichoderma* sp. on *Sclerotium rolfsii.* Mycopathologia 139(3):151–155

Mukherjee PK, Latha J, Hadar R, Horwitz BA (2004) Role of two G-protein alpha subunits, TgaA and TgaB, in the antagonism of plant pathogens by *Trichoderma virens.* Appl Environ Microbiol 70(1):542–549

Mukherjee KP, Nautiyal CS, Mukhopadhyay AN (2008) Molecular mechanisms of plant and microbe coexistence. Springer, Heidelberg

Mukhtar T (2018) Management of root-knot nematode, *Meloidogyne incognita*, in tomato with two *Trichoderma* species. Pak J Zool 50(4):1589–1592

Nagy V, Seidl V, Szakacs G, Komoń-Zelazowska M, Kubicek CP, DruzhininaI (2007) Application of DNA bar codes for screening of industrially important fungi: the haplotype of *Trichoderma harzianum* sensustricto indicates superior chitinase formation. Appl Environ Microbiol 73 (21):7048–7058

Nallathambi P, Padmanaban P, Mohanraj D (2001) Fungicide resistance in sugarcane associated *Trichoderma* isolates. J Mycol Plant Pathol 31:125

Naserinasab F, Sahebani N, Etebarian HR (2011) Biological control of *Meloidogyne javanica* by *Trichoderma harzianum* BI and salicylic acid on tomato. Afr J Food Sci 5(4):276–280

Navazio L, Baldan B, Moscatiello R, Zuppini A, Woo SL, Mariani P, Lorito M (2007) Calcium-mediated perception and defense responses activated in plant cells by metabolite mixtures secreted by the biocontrol fungus *Trichoderma atroviride*. BMC Plant Biol 7:41

Nevalainen H, Suominen P, Taimisto K (1994) On the safety of *Trichoderma reesei.* J Biotechnol 37:193–200

Nicol RW, Traquair JA, Bernards MA (2002) Ginsenosides as host resistance factors in American ginseng (*Panax quinquefolius*). Can J Bot 80(5):557–562

Nielsen KF, Grafenhan T, Zafari D, Thrane U (2005) Trichothecene production by *Trichoderma brevicompactum.* J Agric Food Chem 53:8190–8196

Oros G, Naar Z, Cserhati T (2011) Growth response of *Trichoderma* species to organic solvents. Mol Inf 30:276–285

Pandey KK, Upadhyay JP (1998) Sensitivity of different fungicides to *Fusarium udum*, *Trichoderma harzianum* and *Trichoderma viride* for integrated approach of disease management. Veg Sci 2:89–92

Papavizas GC, Dunn MT, Lewis JA, Beagle-Ristaino JE (1984) Liquid fermentation technology for experimental production of biocontrol fungi. Phytopathology 74:1171

Peeran MF, Kamil D, Prasad L (2017) Extracellular myco-synthesis of silver nanoparticles from *Trichoderma virens* and *Metarhizium nisopliae*. J Mycol Plant Pathol 47(4):424–429

Perazzoli M, Moretto M, Fontana P, Ferrarini A, Velasco R, Moser C, Delledonne M, Pertot I (2012) Downy mildew resistance induced by *Trichoderma harzianum* T39 in susceptible grapevines partially mimics transcriptional changes of resistant genotypes. BMC Genomics 13:660

Prasad S, Bagali P, Hittalmani S, Shashidhar H (2000) Molecular mapping of quantitative trait loci associated with seedling tolerance to salt stress in rice (*Oryza sativa* L.). Curr Sci 78(2):162–164

Probioma (2006) El SoyeroEcologico. Ed. 2 Noviembre. Santa Cruz

Punja ZK (2006) Recent developments toward achieving fungal disease resistance in transgenic plants. Can J Plant Pathol 28(S1):S298–S308

Rashid M, Chowdhury MSM, Sultana N (2013) *In-vitro* screening of some chemicals and biocontrol agents against *Erwinia carotovora* subsp. *carotovora*, the causal agent of soft rot of potato (*Solanum tuberosum*). Agriculturists 11(2):1–9

Rawat L, Singh Y, Shukla N, Kumar J (2011) Alleviation of the adverse effects of salinity stress in wheat (*Triticum aestivum* L.) by seed biopriming with salinity tolerant isolates of *Trichoderma harzianum*. Plant Soil 347:387–400. https://doi.org/10.1007/s11104-011-0858-z

Rawat L, Singh Y, Shukla N, Kumar J (2012) Seed biopriming with salinity tolerant isolates of *Trichoderma harzianum* alleviates salt stress in rice growth, physiological and biochemical characteristics. J Plant Pathol 94(2):353–365

Reddy ASR, Madhavi GB, Reddy KG, Yellareddygari SK, Reddy MS (2011) Effect of seed biopriming with *Trichoderma viride* and *Pseudomonas fluorescens* in chickpea (*Cicer arietinum*) in Andhra Pradesh, India. Plant growth-promoting rhizobacteria (PGPR) for sustainable agriculture. In: Proceedings of the 2nd Asian PGPR Conference, Beijing, 21–24, pp 324–429

Reese ET (1976) History of the cellulase program at the US Army Natick Development Center. Biotechnol Bioeng Symp 6:9–20

Reino JL, Guerrero RF, Hernandez-Galan R, Collado IG (2008) Secondary metabolites from species of the biocontrol agent *Trichoderma*. Phytochemistry 7:89–123

Reithner B, Brunner K, Schuhmacher R, Peissl I, Seidl V, Krska R, Zeilinger S (2005) The G protein alpha subunit Tga1 of *Trichoderma atroviride* is involved in chitinase formation and differential production of antifungal metabolites. Fungal Genet Biol 42(9):749–760

Reithner B, Schuhmacher R, Stoppacher N, Pucher M, Brunner K, Zeilinger S (2007) Signaling via the *Trichoderma atroviride* mitogen-activated protein kinase Tmk 1 differentially affects mycoparasitism and plant protection. Fungal Genet Biol 44(11):1123–1133

Rey M, Delgado-Jarana J, Benítez T (2001) Improved antifungal activity of a mutant of *Trichoderma harzianum* CECT 2413 which produces more extracellular proteins. Appl Microbiol Biotechnol 55:604–608

Rocha-Ramirez V, Omero C, Chet I, Horwitz BA, Herrera-Estrella (2002) Trichoderma atroviride G-protein alpha-subunit gene tga1 is involved in mycoparasitic coiling and conidiation. Eukaryot Cell 1(4):594–605

Routray S, Dey D, Baral S, Das AP, Mahantheshwara B (2016) Genetic improvement of natural enemies: a review. Agric Rev 37(4):325–332

Rubin EM (2008) Genomics of cellulosic biomass. Nature 454:841–845

Ruocco M, Lanzuise S, Vinale F, Marra R, Turrà D, Woo SL, Lorito M (2009) Identification of a new biocontrol gene in *Trichoderma atroviride*: the role of an ABC transporter membrane pump in the interaction with different plant-pathogenic fungi. Mol Plant-Microbe Interact 22 (3):291–301

Sandhya C, Adapa LKK, Nampoothri M, Binod P, Szakacs G, Pandey A (2004) Extracellular chitinase production by Trichoderma harzianum in submerged fermentation. J Basic Microbiol 44:49–58

Sawant IS, Mukhopadhyay AN (1990) Integration of metalaxyl MZ with *Trichoderma harzianum* for the control of Pythium damping-off in sugarbeet. Indian Phytopathol 43:535–541

Scherm B, Schmoll M, Balmas V, Kubicek CP, Migheli Q (2009) Identification of potential marker genes for *Trichoderma harzianum* strains with high antagonistic potential against *Rhizoctonia solani* by a rapid subtraction hybridization approach. Curr Genet 55(1):81–91

Schuster A, Schmoll M (2010) Biology and biotechnology of *Trichoderma.* Appl Microbiol Biotechnol 87(3):787–799

Seiboth B, Gamauf C, Pail M, Hartl L, Kubicek CP (2007) The D-xylose reductase of *Hypocrea jecorina* is the major aldose reductase in pentose and D-galactose catabolism and necessary for beta-galactosidase and cellulase induction by lactose. Mol Microbiol 66:890–900

Seidl V, Schmoll M, Scherm B, Balmas V, Seiboth B, Migheli Q, Kubicek CP (2006) Antagonism of Pythium blight of zucchini by *Hypocrea jecorina* does not require cellulase gene expression but is improved by carbon catabolite derepression. FEMS Microbiol Lett 257(1):145–151

Seidl V, Song L, Lindquist E, Gruber S, Koptchinskiy A, Zeilinger S, Schmoll M, Martínez P, Sun J, Grigoriev I, Herrera-Estrella A, Baker SE, Kubicek CP (2009) Transcriptomic response of oparasitic fungus *Trichoderma atroviride* to the presence of a fungal prey. BMC Genomics 10:567

Sharad U, Sharma G, Moger N, Bhat S, Krishnaraj PU (2015) Functional validation of plant transformation vector with stacked ech42 and bgn from *Trichoderma* in tomato for fungal disease resistance. Indian J Genet Plant Breed 75(1):86–92

Sharma DD, Gupta VP, Chandrashekhar DS (1999) Compatibility of certain biocontrol agents with chemical pesticides and fertilizers. Indian J Seric 38:79–82

Shelar GB, Chavan AM (2015) Myco-synthesis of silver nanoparticles from *Trichoderma harzianum* and its impact on germination status of oil seed. Biolife 3(1):109–113

Shoresh M, Harman GE, Mastouri F (2010) Induced systemic resistance and plant responses to fungal biocontrol agents. Annu Rev Phytopathol 48:21–43

Shukla N, Awasthi RP, Rawat L, Kumar J (2014) Seed biopriming with drought tolerant isolates of *Trichoderma harzianum* promote growth and drought tolerance in *Triticum aestivum.* Ann Appl Biol 166(2):171–182

Simmons EG (1977) Classification of some cellulase-producing *Trichoderma* species. In: Bigelow HE, Simmons EG (eds) 2nd international mycological congress. University of South Florida, Tampa, p 618

Simon LS, Bhandari G (2009) Efficacy of bio-agents for management of *Hirschmanniella mucronata* on rice. Ann Plant Protect Sci 17(2):522–523

Sivasithamparam K, Ghisalberti EL (1998) Secondary metabolism in *Trichoderma* and *Gliocladium.* In: Harman GE, Kubicek CP (eds) *Trichoderma* and *Gliocladium.* Taylor and Francis, London, pp 139–192

Somerville C, Youngs H, Taylor C, Davis SC, Long SP (2010) Feedstocks for lignocellulosic biofuels. Science 329:790–792

Spiegel Y, Chet I (1998) Evaluation of *Trichoderma* spp. as a biocontrol agent against soilborne fungi and plant–parasitic nematodes in Israel. Integr Pest Manag Rev 3:169–175

Stoppacher N, Kluger B, Zeilinger S, Krska R, Schuhmacher R (2010) Identification and profiling of volatile metabolites of the biocontrol fungus *Trichoderma atroviride* by HS-SPME-GC-MS. J Microbiol Methods 81(2):187–193

Suárez MB, Vizcaíno JA, Llobell A, Monte E (2007) Characterization of genes encoding novel peptidases in the biocontrol fungus *Trichoderma harzianum* CECT 2413 using the TrichoEST functional genomics approach. Curr Genet 51(5):331–342

Tallapragada P, Gudimi M (2011) Phosphate solubility and biocontrol activity of *Trichoderma harzianum.* Turk J Biol 35(5):593–600

Tisch D, Schmoll M (2010) Light regulation of metabolic pathways in fungi. Appl Microbiol Biotechnol 85:1259–1277

Tomer A, Singh R, Durga P (2018) Compatibility of *Trichoderma harzianum* with systemic and two non systemic fungicides of *in vitro.* Asian J Crop Sci 10(4):174–179

Tseng SC, Liu SY, Yang HH, Lo CT, Peng KC (2008) Proteomic study of biocontrol mechanisms of *Trichoderma harzianum* ETS 323 in response to *Rhizoctonia solani*. J Agric Food Chem 56 (16):6914–6922

Umamaheswari R, Somasekhar N, Manorama K, Joseph TA (2012) Eco-friendly management of potato cyst nematodes in the Nilgiris of Tamil Nadu. Potato J 39(2):185–190

Uqab B, Mudasir S, Nazir R (2016) Review on bioremediation of pesticides. J Bioremed Biodegr 7 (3):343

Vahabi K, Mansoori GA, Karimi S (2011) Biosynthesis of silver nanoparticles by fungus *Trichoderma reesei*: a route for large-scale production of AgNPs. Insci J 2011:65–79

Vinale F, Marra R, Scala F, Ghisalberti EL, Lorito M, Sivasithamparam K (2006) Major secondary metabolites produced by two commercial *Trichoderma* strains active against different phytopathogens. Lett Appl Microbiol 43:143–148

Vinale F, Sivasithamparam K, Ghisalberti EL, Marra R, Woo SL, Lorito M (2008) *Trichoderma*–plant–pathogen interactions. Soil Biol Biochem 40:1–10

Vinale F, Ghisalberti EL, Sivasithamparam K, Marra R, Ritieni A, Ferracane R, Woo S, Lorito M (2009) Factors affecting the production of *Trichoderma harzianum* secondary metabolites during the interaction with different plant pathogens. Lett Appl Microbiol 48:705–711

Viterbo A, Ramot O, Chemin L, Chet I (2002) Significance of lytic enzymes from *Trichoderma* spp. in the biocontrol of fungal plant pathogens. Antonie Van Leeuwenhoek 81(1–4):549–556

Viterbo A, Harel M, Horwitz BA, Chet I, Mukherjee PK (2005) *Trichoderma* mitogen-activated protein kinase signaling is involved in induction of plant systemic resistance. Appl Environ Microbiol 71(10):6241–6246

Weindling R (1932) *Trichoderma lignorum* as a parasite of other soil fungi. Phytopathology 22 (8):837–845

Wells, H.D., Bell, D.K. and Casimir, A.J. (1972). Efficacy of Trichoderma harzianum as a biocontrol for Sclerotiumro/fsii. Phytopathology, 62: 442–447.

Wiater A, Szczodrak J, Pleszczynska M (2005) Optimization of conditions for the efficient production of mutan in streptococcal cultures and post-culture liquids. Acta Biol Hung 56:137–150

Woo SL, Ruocco M, Vinale F, Nigro M, Marra R, Lombardi N et al (2014) *Trichoderma*-based products and their widespread use in agriculture. Open Mycol J 8:71–126. https://doi.org/10.2174/1874437001408010071

Xue AG, Guo W, Chen Y, Siddiqui I, Marchand G, Liu J, Ren C (2017) Effect of seed treatment with novel strains of *Trichoderma* spp. on establishment and yield of spring wheat. Crop Prot 96:97–102

Yedidia II, Benhamou N, Chet (1999) Induction of defense responses in cucumber plants (*Cucumis sativus* L.) By the biocontrol agent *Trichoderma harzianum*. Appl Environ Microbiol 65 (3):1061–1070

Yedidia I, Shoresh M, Kerem Z, Benhamou N, Kapulnik Y, Chet I (2003) Concomitant induction of systemic resistance to *Pseudomonas syringae* pv. *lachrymans* in cucumber by *Trichoderma asperellum* (T-203) and accumulation of phytoalexins. Appl Environ Microbiol 69:7343–7353

Yehia AH, El-Hassan SA, El-Bahadli AH (1985) Biological seed treatment to control *fusarium* root rot of broad bean. Egypt J Phytopathol 14:59–66

Zeilinger S, Omann M (2007) *Trichoderma* biocontrol: signal transduction pathways involved in host sensing and mycoparasitism. Gene Regul Syst Biol 1:227–234

Zeilinger S, Reithner B, Scala V, Peissl I, Lorito M, Mach RL (2005) Signal transduction by Tga3, a novel G protein alpha subunit of *Trichoderma atroviride*. Appl Environ Microbiol 71 (3):1591–1597

Zhang F, Yuan J, Yang X, Cui Y, Chen L, Ran W et al (2013) Putative *Trichoderma harzianum* mutant promotes cucumber growth by enhanced production of indole acetic acid and plant colonization. Plant Soil 368:433–444. https://doi.org/10.1007/s11104-012-1519-6

Zhang S, Gan Y, Ji W, Xu B, Hou B, Liu J (2017) Mechanisms and characterization of *Trichoderma longibrachiatum* T6 in suppressing nematodes (*Heterodera avenae*) in wheat. Front Plant Sci 8:1491. https://doi.org/10.3389/fpls.2017.01491

Chapter 7
Trichoderma: An Effective and Potential Biocontrol Agent for Sustainable Management of Pulses Pathogens

R. K. Mishra, Sonika Pandey, Monika Mishra, Utkarsh Singh Rathore, Kulbhusan Mani Tripathi, and Krishna Kumar

Abstract Pulses play a major role in meeting the projected targets relating to food and nutritional security worldwide. The complementation of cereal-based food with grain legumes keeps the vegetarian diet nutritionally balanced. However, the productivity of these crops is severely impacted by a number of biotic, mesobiotic, and abiotic stresses causing substantial economic losses globally. Among the biotic/mesobiotic constraints, diseases and insect pests remain the most crucial factors affecting all parts of the plant at different growth stages. For the management of these insect pests and diseases, chemical pesticides are extensively employed across the world. However, the global risk associated with the environmental pollution and health hazards posing toxicity to man, plants, domestic animals, and wildlife render these chemical-based interventions ecologically unacceptable. In the context, *Trichoderma* is one of the most effective and attractive biological control agents (BCAs) as well as an alternative to conventional fungicides. These *Trichoderma*-based BCAs are economically viable and environment-friendly and represent the most competent means to sustain the existing level of agricultural production system.

Keywords *Trichoderma* · Disease management · Pulse crops

7.1 Introduction

In Indian Agriculture, pulses play an important role in maintaining soil fertility and supplying protein to the large vegetarian population of the country. Thus, they provide nutritional security to the people and also to the soil. Their cultivation has remained an integral component of subsistence farming systems since time immemorial due to the ability to thrive under harsh fragile ecosystem of drylands in India.

R. K. Mishra (✉) · S. Pandey · M. Mishra · U. S. Rathore · K. M. Tripathi · K. Kumar
ICAR-Indian Institute of Pulses Research, Kanpur, India

C. Manoharachary et al. (eds.), Trichoderma: *Agricultural Applications and Beyond*, Soil Biology 61, https://doi.org/10.1007/978-3-030-54758-5_7

The importance of pulses in producing protein-rich food is more in India as the majority of the population is religiously dependent on pulse protein and follows the religious taboos in eating meat. Many pulse crops are grown in different parts of the country. Chickpea (*Cicer arietinum* L.), lentil (*Lens culinaris* Medik.), faba bean (*Vicia faba* L.), grass pea (*Lathyrus sativus* L.), and field pea (*Pisum sativum* L.) are grown in cool season (rabi), and pigeon pea (*Cajanus cajan* (L.) Millsp.), black gram (*Vigna mungo* (L.) Hepper), mungbean (*V. radiate* (L.) Wilczek), cowpea (*V. unguiculata* (L.) Walp) in the rainy season (*kharif*). Mungbean and urdbean are also grown in the spring/summer season in many parts of the country especially where irrigation is available. Several biotic stresses synergistically hamper the pulse production and productivity in the country (Table 7.1 and Fig. 7.1). *Trichoderma* spp. are common soil inhabitants that are well known for their biocontrol potential. They inhibit the growth of soilborne phytopathogens, through various mycoparasitic mechanisms such as antibiosis, secondary metabolite production, enzyme secretion, competition for food and space, etc. In rhizosphere, the exchange and recognition of signaling molecules between *Trichoderma* and plants alter the physiology and biochemistry of both. *Trichoderma* in association with plant roots can trigger systemic resistance and improve plant health. These are also colonizers of cellulosic substances and thus they can be easily found on decaying materials (Kubicek et al. 2008; Jaklitsch 2009). *Trichoderma* spp. are characterized by rapid growth, mostly bright green conidia, and a repetitively branched conidiophore structure (Gams and Bissett 1998). Despite the early suggested link between *Trichoderma* and Hypocrea (Tulasne and Tulasne 1865), this anamorph–teleomorph relationship was only confirmed more than 100 years later for *Trichoderma reesei* and *Hypocrea jecorina* (Kuhls et al. 1996). More than a decade was passed until the sexual cycle was discovered in *Trichoderma* species (Seidl et al. 2009). Because of the industrial importance of *T. reesei* sexual cycle discovery was a very important discovery for the genus. *Trichoderma* species colonize their habitats very efficiently because they utilize the substrate efficiently and secrete metabolites, antibiotics, and enzymes.

Trichoderma first came into light during 1794 (Persoon 1794) and description of their sexual stage *Hypocrea* came into the light during 1865 (Tulasne and Tulasane 1865). However, it is very difficult to distinguish different species of *Trichoderma/Hypocrea* morphologically. In 1969 concept for the identification of species was started (Rifai 1969; Samuels 2006). After that numerous *Trichoderma* species were investigated and to date around more than 100 species of *Trichoderma* are known. But morphological identification of *Trichoderma* sometimes cause confusion, with the name *Trichoderma harzianum* many *Trichoderma* species have been misidentified. However, it is very difficult to identify these isolates morphologically. In present time safe identification of new *Trichoderma* species is generally facilitated by the development of an oligonucleotide barcode (trchOKEY) and a customized similarity search tool (Tricho BLAST0) (Druzhinina et al. 2005). The aim of this chapter is to discuss the details scope and mechanisms of biocontrol agents and their exploitation for the effective management of pests and pathogens of pulse crops.

Table 7.1 List of major biotic stresses of pulses that can effectively be managed by biocontrol agents

S.N.	Name of crop	Name of pests	Causative agents
1	Pigeon pea	Phytophthora stem blight	*Phytophthora drechsleri* f. sp. *cajani.*
		Wilt	*Fusarium oxysporum* f. sp. *udum* (Butler) Snyd. and Hans.
		Leaf spot	*Cercospora* spp.
		Powdery mildew	*Erysiphe polygoni* DC
		Gram pod borer	*Helicoverpa armigera* Hubner
		Legume pod borer	*Maruca vitrata* Fabricius
		Pod fly	*Melanagromyza obtusa* Malloch
		Stem rot	*Sclerotinia sclerotiorum*
2	Chickpea/gram	Wilt	*F. oxysporum* f. sp. *ciceri* Matuo and Sato
		Dry root rot	*Rhizoctonia bataticola*
		Stem rot	*Sclerotinia sclerotiorum*
		Rust	*Uromyces ciceris-arietinii (Grogn.)* Jacz. & boy.
		Wet root rot	*Rhizoctinia solani*
		Botrytis gray mold	*Botrytis cinerea*
		Gram pod borer	*H. armigera* Hubner
3	Lentil	Wilt	*F. oxysporum* f. sp. *lentis*
		Root rot	*Macrophomina phaseolina*
		Rust	*Uromyces fabae*
		Ascochyta blight	*Ascochyta fabae* f. sp. *lentis*
		Pod borer	*Etiella zinckenella* Treitschke
4	Field pea	Rust	*U. fabae*
		Powdery mildew	*E. polygoni*
		Root rot	*R. solani*
		Pod borer	*H. armigera*
5	Mungbean/ urdbean	Root rot	*M. phaseolina*
		Leaf spot	*Cercospora* spp.
		Legume pod borer	*Maruca vitrata*
		Pod borer	*Helicoverpa armigera*

7.2 Status of Biocontrol Agents (BCAs) Research and Development

An increase in selection pressure consequent to the indiscriminate application of chemical pesticides leads to the emergence of pesticide résistance. In such conditions, alternate options of pest/disease management are much sought. During the past two decades, an urgent need was realized for management strategies that are safe vis-a-vis the environment and the land. Farmers are shifting toward eco-friendly technology for the management of pests, i.e., BCAs or BCAs-based formulations,

Fig. 7.1 Symptoms of major biotic stresses affecting pulses: (**a**) dry root rot infected chickpea field (**b**) Chickpea rust (**c**) Chickpea stem rot (**d**) Chickpea wilt (**e**) *Aschochyta* blight of chickpea (**f**) *Alternaria* blight on pigeon pea (**g**) Pigeon pea wilt affected field (**h**) *Phytophthora* blight of pigeon pea (**i**) Pigeon pea stem rot (**j**) Pea rust (**k**) Lentil rust (**l**) Anthracnose affected bean (**m**) Rajmash powdery mildew (**n**) *H. armigera* infestation on pigeon pea (**o**) chickpea (**p**) *Maruca* infestation on pigeon pea

referred to as biopesticides. Examples include *Trichoderma* spp., *Pseudomonas* spp., *Bacillus* spp., *Agrobacterium radiobacter*, nonpathogenic *Fusarium* spp., *Coniothyrium.* spp. and *Aspergillus niger, Bacillus thuringiensis* (Bt), *Metarhizium* spp., *Beauveria bassiana,* and nuclear polyhedrosis virus (NPVs), which are popularly used in plant protection. According to a recent report (NAAS 2013), nearly 1400 BCAs products were sold and 175 biopesticide active ingredients and 700 products were registered worldwide for their commercialization. A growing body of

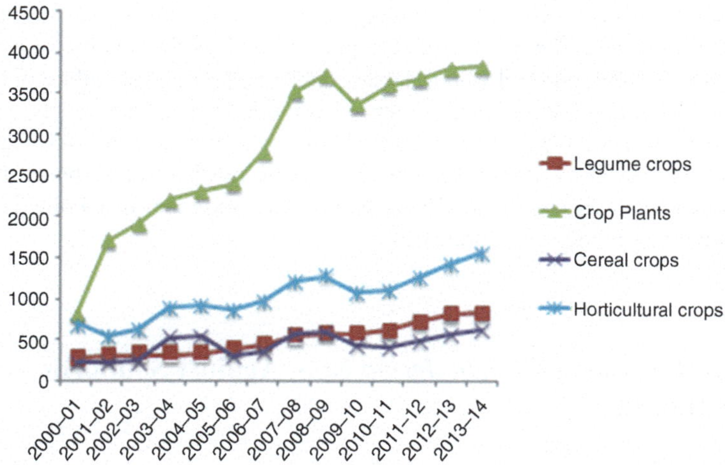

Fig. 7.2 Increasing the use of BCAs in different crop plants as reflected in the form of patents and publications (From year 2000–01 to 2014–15), a rising trend is seen in the number of patents and publications pertaining to BCAs in plants

research articles report on the identification and efficacy of different BCAs against a number of pests and pathogens; however, their slow embrace is evident from the fact that only 2% biopesticides are currently used for crop protection worldwide. On the positive side, the usage of BCAs has witnessed an increasing trend. The rise is notably high at the international level. Though literature on BCAs is growing worldwide (Fig. 7.2), the filing of patents on BCA technology is not in sync with the number of publications.

The development of a stable and economically viable bioformulation remains central to biological control. Of the various carrier-based formulations available worldwide, alginate pellet- and talc-based formulations of BCAs have emerged as the most important carrier for the application in the management of crop diseases. Still a majority of the reports on BCA document one target disease. This, however, has yielded inconsistent performance given that a single agent might not remain active in all soil conditions. Further, enabling mass production with a high level of microbial count and viability also assumes greater significance.

7.3 Biocontrol Mechanisms of *Trichoderma* Spp.

Trichoderma imparts the growth of pathogens through the following mechanisms.

Mycoparasitism—This process involves the direct attack of one fungus on the other fungi. This is the main mechanism which is used by *Trichoderma* to manage the phytopathogens. Process of mycoparasitism is a cascade of following events: chemotrophic growth of *Trichoderma*, recognition of the host, coiling and appressoria formation, cell wall degrading enzyme secretion, hyphal penetration, and lysis of host cell. The process of recognition involves the binding of carbohydrates in the *Trichoderma* cell wall to lectins on the target fungus, coiling of hyphae around pathogen involve a higher amount of osmotic solutes such as glycerol and attack the hyphae of the pathogen through the secretion of cell wall degrading enzymes (Harman et al. 2004a, b). This synergistic mechanism involves the host cell lysis and death. There are 20–30 genes, proteins, and other metabolites that are directly involved in the mycoparasitism.

7.4 Mechanisms of *Trichoderma* Spp. Against Soilborne Pathogens

7.4.1 Antibiosis

This process involves the production of low molecular weight diffusible secondary metabolites and antibiotics that are harmful to the pathogens. Variety of antibiotics such as gliovirin, gliotoxin, viridin, viridol, koninginins, pyrones, and peptaibols, against fungal phytopathogens are secreted by *Trichoderma* (Howell 2003; Harman et al. 2004a, b). Trichokonins via peptabiol is secreted by *Trichoderma pseudokoningii* SMF2, it exerts antibiotic action through extensive apoptotic programmed cell death in pathogen (Shi et al. 2016). *Trichoderma* secondary metabolites gliotoxin and gliovirin which belong to P and Q family are the main secondary metabolites which play role in phytopathogenic action. P group is secreted by *T. virens* and is active against *P. ultimum* but not against *R. solani* and Q group is active against *R. solani* but not against *P. ultimum* (Howell 2003). Koninginin D also inhibited the growth of soilborne plant pathogens, such as *R. solani*, *P. cinnamomi*, *Pythium middletonii*, *Fusarium oxysporum,* and *Bipolaris sorokiniana* (Dunlop et al. 1989). Viridins isolated from diverse *Trichoderma* spp. (*T. koningii, T. viride, T. virens*) prevent spore germination of *Botrytis alli*, *Colletotrichum lini*, *Fusarium caeruleum*, *Penicillium expansum*, *Aspergillus niger,* and *Stachybotrys atra* (Singh et al. 2005). Harzianic acid isolated from a *T. harzianum* strain showed in vitro antibiotic activity against *Pythium irregulare*, *Sclerotinia sclerotiorum,* and *R. solani* (Vinale et al. 2009). The isolation and overexpression of the tri5 (trichodiene synthase) gene in *T. brevibacterium Tb4ltri5, a* transformant has been found to increase the production of trichodermin and antifungal activity against *Aspergillus fumigates* and *Fusarium spp* (Tijerino et al. 2011). *T. asperellum* strain produces two asperelines (A and E) and five trichotoxins (T5D2, T5E, T5F, T5G, and 1717A) which can be associated with antibiosis (Brotman et al. 2013).

7.4.2 Starvation

It is the most common reason for the death of microorganisms, thus completion for nutrients results in the biological control of phytopathogens. Rhizospheres are rich sources of sugars, amino acids, iron, vitamins, organic acids, etc. Iron is essential for many filamentous fungi and for the iron uptake from the environment many fungi secrete low molecular weight ferric ion chelators termed as siderophore; siderophore helps in the mobilization of environmental iron. *Trichoderma* species secrete siderophore which binds to the insoluble iron and converts it to the soluble form for plant absorption and thus inhibits the growth of pathogens by depriving them (Leong 1986). Through this mechanism, *Trichoderma* decreases the growth of *Pythium. T. harzianum* T35 controls *Fusarium oxysporum* by competing for both rhizospheric colonization and nutrients. Competition has proved very efficient for the control of various phytopathogens such as *Botrytis, Fusarium, Phytium,* etc. *Trichoderma* is more efficient than any other fungi to mobilize and take soil nutrients. Promoter gene analysis related to antagonism revealed the presence of consensus sequences for transcription factors responsible for carbon (CreaA), nitrogen (AreA), Stress (Msn2/Msn4), pH (PacC), and mycoparasitism (MYC) regulation. Appropriate manipulation of these regulators will give an alternative way to isolate more competitive BCAs.

7.4.3 Induced Resistance

Induced resistance is the most indirect form of antagonism. Induced resistance can be local or systemic. Salicylic acid (SA) and non-expressor of pathogenesis-related genes (1) (NPR1) are key players in systemic acquired resistance. *T. harzianum* when inoculated on to roots or on to leaves of grapes provides control of diseases caused by *Botrytis cinerea* on leaves spatially separated from the site of application of the BCA (Deshmukh et al. 2006). Many classes of compounds are released by the *Trichoderma* sp. into the zone of interaction that induces resistance in plants. The first class is "proteins with enzymatic or other activity". Fungal proteins such as xylanase, cellulases, and swollenins are secreted by *Trichoderma* species (Martinez et al. 2001). *Trichoderma* endochitinase can also enhance defense, probably through induction of plant defense-related proteins. Active rhizosphere colonizers *T. harzianum* and *T. virens* produce some cell wall degrading enzymes, antibiotics such as gliotoxin and viridin (Tronsmo and Harman 1992) and also certain biologically active heat-stable metabolites such as ethyl acetate (Claydown et al. 1987) which inhibit various pathogens present in the soil. In another study, T. hamatum found to be an effective biocontrol agent for management of lentil wilt caused *F. oxysporum* f. sp. lentis. It inhibits growth of wilt pathogen by production of antifungal enzymes, competition for key nutrients and/or ecological niches, complex mechanisms of mycoparasitism, growth promotion, and a combination of these effects. During mycoparasitism, *Trichoderma* produces appressoria-like structures to penetrate the target fungus cell wall (Chet 1987).

7.4.4 Biocontrol Genes from Trichoderma

Whole-genome sequencing of *Trichoderma* species has provided many valuable data for the understanding of various genes. To date whole-genome sequence of seven *Trichoderma* species is available viz. *Trichoderma harzianum, Trichoderma asperellum, Trichoderma reesei, Trichoderma virens, Trichoderma atroviride, Trichoderma longibrachiatum, and Trichoderma citrinoviride* are available (Sharma et al. 2009). These studies have revealed that *Trichoderma* species contain many valuable genes that help in the control of growth of phytopathogens and plant growth promotional activities. Cell wall degrading genes-A gene named tvsp1 responsible for encoding serine protease from *Trichoderma virens* has been cloned. Serine protease helps in the degradation of the fungal cell wall. It has an important role in pathogenesis against *Rhizoctonia solani* (Pozo et al. 2004). tri5 gene (trichodiene synthase gene) isolated and characterized from *Trichoderma harzianum* is responsible for the secretion of enzyme trichothecene; this enzyme stops the DNA and protein synthesis in the cells of pathogens and hence stops their growth and ultimately kills them. This gene has shown phytotoxicity against *Fusarium* species. *Trichoderma asperellum* expresses the gene tag83 which is responsible for the secretion of exo-beta-glucanase; this gene with *R. solani* has shown inhibition action (Marcello et al. 2010). Beta-1,3 and beta-1,6 are the glucanase enzymes, which are coded by coding TvBgn2 and TvBgn3 genes. These enzymes help in cell wall degradation of phytopathogens. From *Trichoderma atroviridae* a gene named gluc 78 which encodes 1,3 beta-glucosidase was isolated cloned and sequenced; this gene is responsible for pathogen cell wall degradation (Donzelli et al. 2001). Crel is a glucose repressor gene that was isolated from *Trichoderma harzianum*. This gene causes the repression of cellulase and xylanase encoding genes; these two are the key cell wall degrading enzymes. β-tubulin genes are isolated and characterized from *T. harzianum*; β-tubulin genes are structural components of most cell walls. Three-dimensional model of β-tubulin gene was done by a Swiss model automated comparative protein modeling server (Li and Yang 2007). Sm1, a cysteine-rich protein, was isolated from *T. virens*; it is responsible for disease suppression in dicot and monocot plants (Buensanteai et al. 2010). From *T.harzianum* a gene named SL41 has been cloned and expressed; serine proteases play a key role in fungal biology and biocontrol activity. ThPG1 encodes endopolygalacturonase, isolated from *T. harzianum* is responsible for cell wall degradation of pathogens like *R. solani* and *P.ultimum*. G protein alpha subunit genes tGaA and TgaB were isolated from *T. virens* and cloned and characterized. These genes have antagonistic activity against *R. solani* and *Sclerotium rolfsi*.

7.5 Genes for Abiotic and Biotic Stress

Trichoderma has several genes which help the plant to survive against abiotic and biotic stress. A gene that is responsible for the resistance against heat, salt tolerance, osmotic and oxidative tolerance named hsp70 has been isolated from *T. harzianum* has been cloned and characterized (ManteroBarrientos et al. 2008). From *Trichoderma harzianum* a gene named thkel1 has been isolated and characterized. This gene codes putative kelch-repeat protein which helps in regulating the glucosidase activity and enhances tolerance to salt and osmotic stresses in *Arabidopsis thaliana* plants (Hermosa et al. 2011). TvGST (glutathione transferase gene) has been cloned from *T. virens*; this acts as a cadmium-tolerant gene (Dixit et al. 2011). From *T. virens* an adenalyte cyclase encoding gene, tac1 had been isolated and cloned, this gene has its role in mycoparasitism against *R. solani* and *P. ultimum* (Mukherjee et al. 2007). ThPTR2 gene obtained from *T. harzianum* has a significant role in mycoparasitism against *Botrytis cinerea*. Taabc2 obtained from *Trichoderma atroviride* has a significant role in ATP binding cassette transporter in cell membrane pump that helps in mycoparasitic activity. ech42 and prb1 genes have been isolated and characterized from *Trichoderma harzianum*; these genes are very effective against *R. solani* and *S. rolfsii*. egl1 which is isolated from *Trichoderma longibrachiatum* encodes beta 1,4-endoglucanase. This gene shows good biocontrol activity (Migheli et al. 1998). Mitogen-activated protein kinase encoding gene TmkA isolated from *T. virens* has been found to cause mycoparasitic action against *R. solani* and *S. rolfsi* (Mukherjee et al. 2003).

Genes for Antifungal activity—A transcription factor gene named Thctf1 was isolated from *T. harzianum*; this gene involved in the function production of 6-pentyl-2 h-pyran (6PP) has antifungal activity against *R. solani*, *B. cinerea*, and *S. rolfsii*. Trip5, a trichodiene synthase gene encodes by *T.brevicomactum*, is responsible for the production of trichodermin which has antifungal activity against *S. cerevisiae*, *Kluyveromyces marxianus*, *Candida albicans*, *C. glabrata*, *C. tropicalis* and Th-chit encoded by *T. harzianum* has antifungal activity against transgenic tobacco; this gene confers antifungal activity against *A.alternata* (Saiprasad et al. 2009). *T. harzianum* encodes erg1 gene which regulate the synthesis of enzyme squalene epoxidase which help in the synthesis of ergosterol and silencing of this gene provides resistance to antifungal compound terbinafine. Monooxygenase gene from *T. hamatum* has been isolated this gene has antifungal activity against *S. sclerotiorum*, *Sclerotinia minor*, and *Sclerotium cepivorum* (Carpenter et al. 2008). *T. reesei* encodes a gene named TrCCD1 which involves in carotenoid metabolism and helps in conidiospore and hyphal growth (Zhong et al. 2009).

7.6 Plant–*Trichoderma* Interaction

Plant growth promotional activity—plant growth promotional effect has been shown by various *Trichoderma* species. Koninginins, 6-pentyl-a-pyrone, trichocaranes A–D, harzianopyridone, cyclonerodiol, harzianolide, and harzianic acid are examples of isolated compounds that affect plant growth in a concentration-dependent manner (Vinale et al. 2014). Seed treatment with *Trichoderma* triggers the release of enzymes and phytohormones which enhance seed germination. Enhanced germination percent has been found in okra, maize, beans, mustard, chilli, soybean, chickpea, tomato, etc. (Mukhtar 2008; Okoth et al. 2011; Rehman et al. 2012; Lalitha et al. 2012; Kumar et al. 2014; Babychan and Simon 2017). Any pathogens like *Pythium* are unable to attack the host due to fast seed germination (Matsouri et al. 2010). *Trichoderma* produces a growth hormone called gliotoxin which has a function similar to gibberellic acid. *Trichoderma* enhances seed germination directly by releasing enzymes and phytohormones and indirectly by altering the soil microflora.

7.7 Effect on Plant Morphology

Trichoderma species are capable of enhancing the growth of plant components like plant height, leaf number, root length, and root fresh weight. In maize, it enhances root biomass and root hair formation (Bjorkman et al. 1998; Harman et al. 2004a, b). *Trichoderma harzianum* and *Trichoderma virens* have been found for growth promotional activity which was correlated with the prolific formation of lateral roots (Contreras-Cornejo et al. 2009; Pieterse et al. 2009). *Trichoderma* species produce phytohormones which are responsible for plant development (Chowdappa et al. 2013). *T. virens* produces indole-3-acetic acid (IAA), indole-3-acetaldehyde (IAAld), and indole-3-ethanol (IEt), which are responsible for plant growth and development (Contreras-Cornejo et al. 2009). In 2013 Cai and coworker reported that harzianolide produced by *Trichoderma* spp. helps in the development of root length in rice plant treated with *Trichoderma* spp. high leaf number and tiller number was found (Doni et al. 2014). *Trichoderma* enhances root length through environmental buffering. Phosphate solubilization, organic matter decomposition, and chelation and siderophore production are the mechanisms through which *Trichoderma* helps in the plant growth promotion. *Trichoderma* species are able to control several physiological processes such as photosynthetic rate, transpiration, nutrient uptake, etc. (Doni et al. 2014; Saba et al. 2012). *Trichoderma* significantly increases the ability of rice plant to tolerate drought stress and water holding capacity (Shukla et al. 2012). Approximately threefold increase in net photosynthetic rate and stomatal conductance and twofold increase in water use efficiency in *Trichoderma*-treated rice plants as compared to NPK-treated plants have been observed (Doni et al. 2014). High water efficiency is correlated with high photosynthetic rate and low transpiration rate (Thakur et al. 2010; Doni et al. 2014).The activity of *Trichoderma* spp. that contributes to the enhancement of root growth and distribution was also considered as a key factor in the prolonged photosynthetic activity and the delayed senescence in rice plants (Mishra and Salokhe 2011). *Trichoderma* releases cellulase which degrades cellulose and enhances the organic matter and nutrients in the rhizosphere. Phosphate solubilization and chelation of minerals can enhance nutrient availability which results in the enhancement of plant physiological activity (Harman et al. 2004a, b). *Trichoderma* has a profound effect on the vegetative and reproductive growth of plants. It enhances the number of branches, spikes, flowers, and fruits. High yield has been occurring in plants treated with *Trichoderma* such as mustard, wheat, corn, sugarcane, tomato, okra, etc. (Haqu et al. 2012; El-Katatny and Idres 2014; Naznin et al. 2015; Srivastava et al. 2006; Tucci et al. 2011; Idowu et al. 2016). Most of the *Trichoderma* species release MAMPs for molecular recognition and thus control signal cascade by signaling molecule such as salicylic acid (SA), jasmonic acid (JA), and ethylene (ET). JA, SA, and/or ET are the signaling molecules for *Trichoderma*-induced resistance. In Arabidopsis roots treated with *Trichoderma*-induced systemic resistance by SA, JA and ET signaling pathway has been found (Yoshioka et al. 2011). Elicitors released by *Trichoderma* spp. are involved in triggering expressions of defense protein within the plant (Thakur and Sohal 2013). *Trichoderma* acts locally and systematically that involves signaling cascade and activation and accumulation of defense-related antimicrobial compounds which include phenyl ammonium lysate, peroxidase, polyphenol oxidase, lipoxygenase, proteins as PR, terpenoid, phytoalexin antioxidants, etc. (Howell et al. 2000a, b).

7.8 Management of Pulse Pathogens Through *Trichoderma* Spp.

Several efforts have been made to manage different diseases of pulses by incorporating bioagents-colonized natural substrates into the soil. Among the antagonistic fungi, *Trichoderma* species are the most studied bioagents in the pulse ecosystem. They are successfully used to control wilts, root rots, collar rot, and stem rot diseases incited by different *Fusarium* spp., *Rhizoctonia solani*, *R. bataticola*, *Sclerotium rolfsii*, *Sclerotinia sclerotiorum*, *Ascochyta*, *Cercospora*, *Alternaria* spp. *Phytophthora* spp., and *Pythium* spp. in different pulses and other field crops (Table 7.2). Researchers at the Indian Institute of Pulses Research (IIPR), India has identified several native potential strains of *Trichoderma* spp. (*Trichoderma harzianum*, *T. asperellum*, *T. afroharzianum*, *T. longibrachiatum*, *T. asperelloides*, *T . brevicompactum*, *T. atrobrunneum*, and *T. aureoviride*) and plant growth-promoting rhizobacteria (PGPRs) from rhizospheres in major pulse growing areas in India and evaluated these for their antagonistic potential against a variety of pathogens (Fig. 7.3). Accordingly, mass production technology has been developed

Table 7.2 Biocontrol efficacy of *Trichoderma*/PGPRs against pulses' pathogens

Crops	Pests/pathogens	Effective *Trichoderma* spp./BCAs
Chickpea	*F. oxysporum* f. sp. *ciceri*, *Sclerotium rolfsii*, *R. bataticola*, *R. solani* *A. rabei* *H.armigera*	*Trichoderma harzianum*, *T. asperellum*, *T. koningii*, *T. virens*, *Gliocladium virens*, *Pseudomonas* sp., *Bacilus* sp. HaNPV, Bt, *B. bassiana*
Pigeonpea	*F. udum* *P. drechsleri* f. sp. *cajani* *H. armigera*, *M. vitrata*	*T. harzianum*, *T. asperellum*, *G. virens*, *Bacillus subtilis*, *T. koningii*, *T. hamatum*, *A. niger*, *Penicillium* sp., *Glomus mosseae*, *Verticillium chlamydosporum*, *Glomus mosseae*, *P*, *fluorescens*, *HaNPV*, *Bt*, *B. bassiana* *M. Anisopliae*
Lentil	*F. oxysporum* f. sp. *lentis* *M. phaseolina* *M. phaseolina* *A. fabae* f. sp. *lentis* *H.armigera*	*T. asperellum*, *T. harzianum*, *G. virens*, *Pseudomonas fluorescens* *Rhizobium leguminoserum* HaNPV, Bt, *B.bassiana*
Pea	*U. fabae* *E. polygoni* *R. solani*	*T. harzianum*, *T. viride*, *T. virens*, *T. hamatum* *Rhizobium leguminoserum*
Mungbean/ urd bean	*M. phaseolina* *Cercospora* spp. *H. armigera* *M. vitrata*	*T. asperellum*, *T. harzianum*, *T. konninghii*, *T. longibrachiatum*, *P. fluorescens*, *Bacillus subtilis*, *A. niger* and *Pencillium citrinum* HaNPV, *Bt*, *B. bassiana* *Bt*, *B. bassiana*

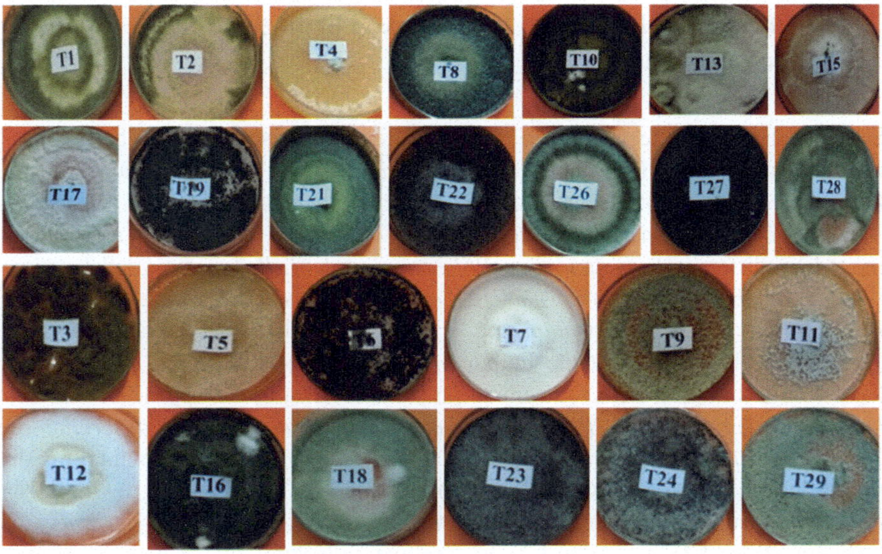

Fig. 7.3 Diversity among *Trichoderma* spp. associated with pulses rhizosphere

and popularized among the pulse-growing farmers in different agroecosystems (Chaudhary and Prajapati 2004; Mishra et al. 2015, 2016a, b; Mishra et al. 2018a, b).

7.9 Chickpea Diseases

Chickpea wilt caused by *Fusarium oxysporum* f. sp. *ciceri*, the major factor for limiting chickpea production worldwide is widely distributed in chickpea growing areas. Many *Trichoderma* (*Trichoderma viride, T. harzianum,* and *T. virens*) isolates from different locations were screened against races of *F. oxysporum* f. sp. *ciceri* commonly prevalent in India. Ranchi isolates of *T. viride* followed by T. *harzianum* inhibited maximum mycelial growth of the pathogen. The integrated application of *T. harzianum* (10 g /kg seed) and carboxin (2 g/kg seed) as a seed treatment reduced wilt incidence (44.1–60.3%) and increased seed germination by 12.0–14.0 percent and grain yields by 42.6–72.9% (Dubey et al. 2009). Application of tea leaves formulation of *T. harzianum* by seed treatment at 10 gm per kg seeds was found effective for the management of chickpea wilt followed by wheat bran saw dust formulation (Singh et al. 2007). *T.virens* based seed dressing formulation, viz., Pusa 5SD and soil application formulations, viz., Pusa Biopellet 16G (PBP 16G) and Pusa Biogranule 6 (PBG 6) were used for management of wet root rot caused by *R. solani* (Dubey et al. 2009). Other diseases like damping-off, root and/or stem rot caused by various organisms can be effectively managed by integrating *T. viride* (isolate TVM2).

7.10 Pigeonpea Diseases

Wilt disease of pigeon pea caused by *Fusarium udum* is an important disease-causing substantial yield reduction throughout the country. *Trichoderma* is predominantly used against pigeon pea wilt (Mishra et al. 2015, 2016a, b; Mishra et al. 2018a, b). Mandhare and Suryawanshi (2005) reported that 63.25% of wilt reduction was achieved by seed and soil application of *Trichoderma* spp. Roy and Sitansu (2005) utilized mutant isolates of *T. harzianum*, viz., 50Th3I1, 75Th4IV, and 125Th4l for the management of wilt disease. Shukla and Chaudhary (2007) studied the efficacy of 30 different *Trichoderma* isolates against *F. udum* and found variation in the antagonistic property of these isolates. Two isolates showed the highest colony growth reduction of 65.5 to 67.2%, 7 between 50 to 60%, and 21 isolates between 41 and 50%. They further observed that 11 isolates reduced conidia production by >90%. The culture filtrates of *Trichoderma* spp. also reduced >90% colony growth of *F. udum*. Phytophthora stem blight (PSB), caused by *Phytophthora drechsleri f. sp. cajani*, is another economically important disease of pigeon pea (*Cajanus cajan*), especially when excessive rains fall within a short span of time and hot and humid weather persists during the crop season. Several isolates of *Trichoderma* species were characterized and evaluated against the *Phytophthora* of pigeon pea (Mishra et al. 2018a, b).

7.11 Lentil Diseases

Fusarium oxysporum f. sp. *lentis* causes disease in wilt worldwide and is a major limiting factor in its production. In India, it causes yield loss of up to 50% (Bhat et al. 2006). The fungal bioagent, *T. hamatum* was reported to reduce up to 33% of mortality due to wilt in lentil (El-Hassan and Gowen 2006). Kumar et al. (2014) suggested that the chemical spray of Bayleton + *T. harzianum* showed significant response with respect to the disease severity, 1000-grain weight (g), and yield (kg/ha) followed by Mancozeb + *P. fluorescens* and Captaf + *T. harzianum*.

7.12 Mungbean/Urdbean Diseases

Root rots are caused by various fungal pathogens, viz., *Rhizoctonia bataticola, R. solani, S. rolfsii, M. phaseolina*, and *F. solani*. Soil application of *Trichoderma virens* based formulations, Pusa Biogranule 6 and Pusa Biopellet16G (PBP 16G) and seed dressing formulation, Pusa 5SD (PBG6) were found to be superior in reducing wet rot disease incidence and increasing seed germination and shoot and root lengths in mungbean. Combined application of PBP 16G (*T. virens*) and Pusa 5SD (*T. virens*) + Carboxin was superior compared to individual treatments (Dubey et al. 2009). However, *T. harzianum* based formulations, Pusa 5SD for seed dressing and Pusa Biopellet 10G effectively combined in integrated management of dry root rot of mungbean (Dubey et al. 2009). The bioagent, *T. viride* is reported to be effective against various soilborne pathogens of mungbean, viz., *R. solani, S. rolfsii, M. phaseolina*, and *F. solani* by confrontation assay (Mishra et al. 2011). Cercospora leaf spot of mungbean can be effectively managed by the integration of *T. viride* along with fungicides (Dubey and Singh 2006).

7.13 Pea and Bean Diseases

Root rot disease complex is considered as the most destructive in pea and bean as it affects its initial plant stand and more than 20 different pathogens have been reported to be associated with the disease from different parts of the world. Among these, *Fusarium oxysporum* f. sp. *pisi* and *Pythium* spp. are highly destructive. Different strains of *Trichoderma* effectively manage the root rot disease of field pea caused by *F. solani* f. sp. *pisi* (Di Pietro et al. 1992; Kapoor et al. 2006; Kumar and Dubey 2001).

7.14 Preparation of *Trichoderma* Formulations and Shelf Life

Trichoderma is the most commonly used biocide, which has been mass-produced by both solid (Singh et al. 2007) and liquid fermentation (Jin et al. 1996) by manipulating media composition for increasing desiccation tolerance. Normally, liquid fermentation-based formulations of *Trichoderma* spp. desiccate faster compared to solid-state fermentation-based formulations. Novel concentrated formulation of *T. harzianum* MTCC-3841 (NBRI-1055) was developed by a Simple scrapping method to maintain high colony forming units (CFU), long shelf life, and efficient in root colonization. In this method, conidiophores were harvested from 7–10 days grown *Trichoderma* on PDA plate with a scrapper and mixed with 10 g autoclaved talc powder. Eight percent moisture in the formulation maintained by adding sterile distilled water. Spores harvested from PDA plates after 10 days of incubation were highest in productivity giving CFU of 2.26×10^7 spores ml^{-1} (Singh et al. 2007). Singh et al. (2007) developed formulations based on agricultural wastes. Farm waste substrates were soaked in tap water overnight and excess water was drained out (moisture content was approximately 80%). The substrates were autoclaved twice at 15 psi for half an hour on two consecutive days and *T. harzianum* injected into the autoclavable bags with a syringe. The pore in the bag was sealed with sterile cellophane tape and incubated at room temperature (25–30 °C). After 15 days, *Trichoderma* colonized substrates were dried at 35 °C and ground to powder using a laboratory blender. Among these substrates, the population count after 30 days was maximum in tea leaves substrates. However, shelf life was found to be maximum in wheat bran-sawdust. The shelf life of formulation was enhanced by the addition of colloidal chitin at the rate of 0.2% (w/v) into talc-based formulation for additional 12 months (Sriram et al. 2010). The shelf life of *Trichoderma* formulation can also be increased up to 7 and 12 months by the addition of osmaticants like glycerol at 3 and 6%, respectively, compared to 4–5 months shelf life in formulations derived without the addition of glycerol (Sriram et al. 2011).

7.15 Delivery System of *Trichoderma* Formulations in Pulses

Biocontrol formulations are delivered through several means based on the survival nature and mode of infection of the pathogen. It is delivered through seed treatment, soil application, and foliar application or through a combination of several methods.

7.16 Seed Treatment

This is the most effective method for the management of phytopathogens. In seed treatment, seed priming is the process in which hydration of seed is controlled to a level that permits pre-germinative metabolic activity to take place without the emergence of the radical. Treating pigeon pea and chickpea seeds with a talc-based formulation of *T. harzianum, Trichoderma viride, T. hamatum, T. virens, Bacillus subtilis,* and *Pseudomonas fluorescens* facilitates management of Fusarium wilt in both crops (Chet and Baker 1981; Chand et al. 1991; Kaur and Mukhopadhayay 1992; Vidhyasekaran et al. 1997; El-Hassan and Gowen 2006; Khan et al. 2012). On-farm demonstrations have evidenced that the seed treatment with 2% talc-based formulation of *Trichoderma harzianum* (IPT-31) led up to 32.1 and 14.3% decrease in root rot incidence in chickpea and lentil, respectively, thus correspondingly improving crop yield by 16.6 and 12.6% (Purushottam et al. 2014).

7.17 Soil Application

A fully active growing population of bioagents is applied in the soil at the time of sowing. *Trichoderma* can be applied as powder form as well as a drench at the initial stage of the crops. According to Vidhyasekaran and Muthamilan (1995), soil application of peat-based formulation with *P. fluorescens* (Pf1) at 2.5 kg of formulation mixed with 25 kg of well-decomposed farmyard manure improved management of chickpea wilt. Combining *P. fluorescens* with safer fungicides reduced the wilt complex in pigeon pea (Siddiqui et al. 1998).

7.18 Foliar Spray

Several researchers have reported that the application of *Trichoderma* sp. in bean leaves reduces the incidence of bean rust (*Uromyces phaseoli*). In a similar manner, seed treatment and foliar application of *P. fluorescens* (Pf1) reduced the severity of *Puccinia arachidis* of groundnut and rust (*U. fabae*) of field pea under field conditions (Meena et al. 2002; Mishra and Pandey 2010).

7.19 Conclusions

Chemical fungicides that are normally used to control plant disease are very harmful to the environment. They pollute air, water, and soil causing harm to human health. The use of biocontrol agents makes the environment safe. *Trichoderma* is a well-known biocontrol agent and it is effective against a variety of pathogens. There are many biocontrol genes in *Trichoderma* which help the plant in developing resistance against plant pathogens. *Trichoderma*-based bioformulations are cheap, safe, and environmentally friendly. However, there is a need for more work to be done to develop stable, improved, and cost-effective *Trichoderma* formulations which can control pathogens more effectively and efficiently.

In the present agriculture scenario, the use of biocontrol agents (BCAs) is of utmost importance in pulses for biotic stresses management, but their potential is yet to be exploited fully mainly because the research in this area is still confined to the laboratory and very little attention has been paid to produce the commercial formulations of biocontrol agents. Moreover, whatever has been commercially produced has not been used efficiently by the farmers owing to the lack of information regarding its application and importance. So, the concept of biological control and their different formulations need to be popularized at the field level.

References

Babychan M, Simon S (2017) Efficacy of *Trichoderma* spp. against *Fusarium oxysporum f. sp. lycopersici*. (FOL) infecting pre-and post-seedling of tomato. J Pharmacognosy and Phytochemistry 6:616–619

Bhat NA, Beig MA, Maheshwari SK, Masoodi SD (2006) Screening and yield of lentil germplasm as influenced by Fusarium wilt. Ann PI Protec Sci 14:139–141

Bjorkman T, Blanchard LM, Harman GE (1998) Growth enhancement of shrunken-2 sweet corn when colonized with *Trichoderma harzianum* 1295-22 effect of environmental stress. J American Soc Horticultural Sci 123:35–40

Brotman Y, Landau U, Cuadros-Inostoza A, Takayuki T, Fernie A R et al. (2013) *Trichoderma*-Plant Root Colonization: Escaping early plant defense responses and activation of the antioxidant machinery for saline stress tolerance. https://doi.org/10.1371/journal.ppat.1003221

Buensanteai N, Mukherjee PM, Horwitz BA, Cheng C (2010) Expression and purification of biologically active *Trichoderma virens* proteinaceous elicitor Sm1 in *Pichia pastoris*. Protein Express 72:131–138

Carpenter MA, Ridgway HJ, Stringer AM, Hay AJ, Stewart A (2008) Characterization of a *Trichoderma hamatum* monooxygenase gene involved in antagonistic activity against fungal plant pathogens. Curr Genet 53:193–205

Chand H, Chhabra ML, Jalali BL (1991) Promising biocontrol agents for the control of chickpea wilt. Indian Phytopathol 46:36–39

Chet I (1987) *Trichoderma* application, mode of action, and potential as biocontrol agent of soil-borne pathogenic fungi. In: Innovative approaches to plant disease control. I. Chet, ed. John Wiley, New York, pp 137–160

Chet I, Baker R (1981) Isolation and biocontrol potential of *Trichoderma hamatum* from soil naturally suppressive to Rhizoctonia solani. Phytopathology 71:286–290

Chaudhary RG, Prajapati RK (2004) Comparative efficacy of fungal, biological agents *Fusarium udum*. Annals of Plant Protection Sciences 12(1):75–79

Chowdappa P, Kumar SPM, Lakshmi MJ, Upreti KK (2013) Growth stimulation and induction of systemic resistance in tomato against early and late blight by *Bacillus subtilis* OTPB1 or *Trichoderma harzianum* OTPB3. Biol Control 65:109–117

Claydown KL, Emerson OH, Sauthwell RJ (1987) The isolation of a toxic substance from the culture filtrate of *Trichoderma*. Phytopathology 36:1068

Contreras-Cornejo HA, Macías-Rodríguez L, Cortés-Penagos C (2009) *Trichoderma virens*, a plant beneficial fungus enhances biomass production and promotes lateral root growth through an auxin dependent mechanism in *Arabidopsis*. Plant Physiol 149:1579–1592

Deshmukh S, Hueckelhoven R, Schaefer P, Imani J, Sharma M (2006) The root endophytic fungus *Piriformospora indica* requires host cell death for proliferation during mutualistic symbiosis with barley. Proc Natl Acad Sci U S A 103:18450–18457

Di Pietro A, Gut-Rella M, Pachlatko JP, Schwinn FJ (1992) Role of antibiotics produced by *Chaetomium globosum* in biocontrol of *Pythium ultimum*, a causal agent of damping-off. Phytopathology 82:131–135

Dixit P, Mukherjee PK, Ramachandran V, Eapen S (2011) Glutathione transferase from *Trichoderma virens* enhances cadmium tolerance without enhancing its accumulation in transgenic *Nicotiana tabacum*. PLoS One 6(1):e16360. https://doi.org/10.1371/journal.pone.0016360

Doni F, Isahak A, Radziah C, Zain CM, Ariffin SM, Nurashiqin W, Wan M, Mohtar W, Wan Y (2014) Formulation of *Trichoderma* sp. SL2 inoculants using different carriers for soil treatment in rice seedling growth. Springer Plus 3:1–5

Donzelli BGG, Lorito M, Scala F, Harman GE (2001) Cloning, sequence and structure of a gene encoding an antifungal glucan 1,3- β-glucosidase from *Trichoderma atroviride* (*T. harzianum*). Gene 277:199–208

Druzhinina IS, Kopchinskiy AG, Komon M, Bissett J, Szakacs G, Kubicek CP (2005) An oligonucleotide barcode for species identification in *Trichoderm*a and *Hypocrea*. Fungal Genet Biol 42:813–828

Dubey SC, Singh B (2006) Integrated management of *cercospora* leaf spots and yellow mosaic of urdbean (*Vigna mungo*). Indian J Agric Sci 76:485–489

Dubey SC, Bhavani R, Singh B (2009) Development of Pusa 5SD for seed dressing and Pusa biopellet 10G for soil application formulation of *T. harzianum* and their evaluation for integrated management of dry root rot of mungbean (*Vigna radiata*). Biol Control 50:231–242

Dunlop RW, Simon A, Siwasithamparam D, Ghisalberti EL (1989) An antibiotic from *Trichoderma koningii* active against soil-borne plant pathogens. J Nat Prod 52:67–74

El-Hassan SA, Gowen SR (2006) Formulation and delivery of the bacterial antagonist *Bacillus subtilis* for management of lentil vascular wilt caused by *Fusarium oxysporum f. sp. Lentis*. J Phytopathol 154:148–155

El-Katatny MH, Idres MM (2014) Effects of single and combined inoculations with *Azospirillum brasilense* and *Trichoderma harzianum* on seedling growth or yield parameters of wheat (*Triticum vulgaris L.*, Giza 168) and corn (*Zea mays L.*, hybrid 310). J Plant Nutr 37:1913–1936

Gams W, Bissett J (1998) Morphology and identification of *Trichoderma*. In: Harmann GE, Kubicek CP (eds) *Trichoderma* and *Gliocladium*. Taylor and Francis, London, pp 3–34

Haqu MM, Ilias GNM, Molla AH (2012) Impact of *Trichoderma*-enriched biofertilizer on the growth and yield of mustard (*Brassica rapa* L.) and tomato (*Solanum lycopersicum* mill.). The Ariculturists 10:109–119

Harman GE, Howell CR, Viterbo A, Chet I, Lorito M (2004a) *Trichoderma* species opportunistic, avirulent plant symbionts. Nat Rev Microbiol 2:43–56

Harman GE, Lorito M, Lynch JM (2004b) Uses of *Trichoderma* spp. to alleviate or remediate soil and water pollution. Adv Appl Microbiol 56:313–330

Hermosa R, Botella L, Keck M, Jimenez JA, Montero Barrientos M, Arbona V, GomezCadenas A, Monte E, Nicolas C (2011) The overexpression in *Arabidopsis thaliana* of a *Trichoderma harzianum* gene that modulates glucosidase activity, and enhances tolerance to salt and osmotic stresses. J Plant Physiol 168:1295–1302

Howell CR, Hanson LE, Stipanovic RD, Puckhaber LS (2000a) Induction of terpenoid synthesis in cotton roots and control of *Rhizoctonia solani* by seed treatment with *Trichoderma virens*. Phytopathology 90:248–252

Howell CR (2003) Mechanisms employed by *Trichoderma* species in the biological control of plant diseases: the history and evolution of current concepts. Plant Dis 87:4–10

Howell CR, Hanson LE, Stipanovic RD, Puckhaber LS (2000b) Induction of terpenoid synthesis in cotton roots and control of *Rhizoctonia solani* by seed treatment with *Trichoderma* virens. Phytopathology 90:248–252

Idowu OO, Olawole OI, Idumu OO, Salami AO (2016) Bio-control effect of *Trichoderma asperellum* (Samuels) Lieckf. and *Glomus intraradices* Schenk on okra seedlings infected with *Pythium aphanidermatum* (Edson) Fitzp and *Erwinia carotovora* (Jones). American J Experimental Agriculture 10:1–12

Jaklitsch WM (2009) European species of Hypocrea. Part I. the green-spored species. Stud Mycol 63:1–91

Jin X, Harman GE, Taylor AG (1996) Conidial biomass and desiccation tolerance of Trichoderma harzianum produced at different medium water potentials. Biol Control 1:237–243

Kapoor AS, Paul YS, Singh A (2006) Integrated management of white rot and root rot-wilt disease complex of pea. Indian Phytopath 59:467–474

Kaur NP, Mukhopadhyay AN (1992) Integrated control of chickpea wilt complex by *T. harzianum* and chemical control methods in India. Int J Pest Manag 38:372–375

Khan RA, Bhat T, Kumar K (2012) Management of chickpea (*Cicer arietinum* L.) dry root rot caused *by Rhizoctonia bataticola* (Taub.) Butler. Int J Res Pharmaceut Biomed Sci 3(4):1539–1548

Kubicek CP, Komon-Zelazowska M, Druzhinina IS (2008) Fungal genus *Hypocrea/Trichoderma*: from barcodes to biodiversity. J Zhejiang Univ Sci B 9:753–763

Kuhls K, Lieckfeldt E, Samuels GJ, Kovacs W, Meyer W, Petrini O, Gams W, Borner T, Kubicek CP (1996) Molecular evidence that the asexual industrial fungus Trichoderma reesei is a clonalderivative of the ascomycete Hypocrea jecorina. Proc Natl Acad Sci U S A 93:7755–7760

Kumar D, Dubey SC (2001) Management of collar rot of pea by the integration of biological and chemical methods. Indian Phytopath 54:62–66

Kumar V, Shahid M, Singh A, Srivastava M, Mishra A, Srivastava YK, Pandey S, Shrarma A (2014) Effect of biopriming with biocontrol agents *Trichoderma harzianum* (Th.Azad) and *Trichoderma* viride on chickpea genotype (Radhey*)*. J Plant Pathology & Microbiology 5:1–4

Lalitha P, Srujana, Arunalakshmi K (2012) Effect of *Trichoderma viride* on germination of mustard and survival of mustard seedlings. Int J Life Sci Biotechnol Pharma Res 1:137–140

Leong J (1986) Siderophores: their biochemistry and possible role in biocontrol of plant pathogen. Annu Rev Phytopathol 24:187–209

Li M, Yang Q (2007) Isolation and characterization of a *β*-tubulin gene from *Trichoderma harzianum.* Biochem Genet 45:529–534

Mandhare VK, Suryawanshi AV (2005) Standarization of storage conditions to increase the shelf life of Trichoderma formulations. Agri Sci Digest 25:71–73

ManteroBarrientos M, Hermosa R, Nicolas C, Cardoza RE, Gutierrez S, Monte E (2008) Overexpression of a *Trichoderma* HSP70 gene increases fungal resistance to heat and other abiotic stresses. Fungal Genet Biol 45:1506–1513

Marcello CM, Steindorff AS, Silva SP, Silva RN, Bataus LAM (2010) Expression analysis of the exo-β-1,3-glucanase from the mycoparasitic fungus *Trichoderma asperellum.* Microbiol Res 165:75–81

Martinez C, Blanc F, Le CE, Besnard O, Nicole M, Baccou JC (2001) Salicylic acid and ethylene pathways are differentially activated in melon cotyledons byactive or heat-denatured cellulase from Trichoderma longibrachiatum. Plant Physiol 127:334–344

Matsouri F, Björkman T, Harman GE (2010) Seed treatment with *Trichoderma harzianum* alleviates biotic, abiotic and physiological stresses in germinating seeds and seedlings. Phytopathology 100:1213–1221

Meena B, Radhajeyalakshmi R, Marimuthu T, Vidhyasekaran P, Velazhahan R (2002) Biological control of groundnut late leaf spot and rust by seed and foliar applications of a powder formulation of *P. fluorescens*. Biocontrol Sci Tech 12:195–204

Migheli Q, GonzalezCandelas L, Dealessi L, Camponogara A, Ramon Vidal D (1998) Transformants of *Trichoderma longibrachiatum* overexpressing the β-1,4-endoglucanase gene *egl* 1 show enhanced biocontrol of *Pythium ultimum* on cucumber. Biol Control 88(7):673–677
Mishra RK, Pandey KK (2010) Effect of application of PGPR and neemazal on management of pea rust (Uromyces fabae). J Basic Appl Microbiol 1&2:115–119
Mishra A, Salokhe VM (2011) Rice growth and physiological responses to SRI water management and implications for crop productivity. Paddy Water Environment 9:41–52
Mishra BK, Mishra RK, Mishra RC, Tiwari AK, Yadav RS, Dikshit A (2011) Biocontrol efficacy of *Trichoderma viride* isolates against fungal plant pathogens causing disease in *Vigna radiata* L. Arch Appl Sci Res 3:361–369
Mishra RK, Bohra A, Naimuddin KK, Sujayanand GK, Saabale PR, Naik SJ Satheesh, Sarma BK, Kumar D, Mishra Monika, Srivastava DK and Singh NP. (2018a) Utilization of biopesticides as sustainable solutions for management of pests in legume crops: achievements and prospects. Egyptian Jr Biological Pest Control 28:3. https://doi.org/10.1186/s41938-017-0004-1
Mishra RK, Monika M, Naimuddin, Krishna K (2018b) *Trichoderma asperellum:* a potential biocontrol agents against wilt of pigeon pea caused by *Fusarium udum* Butler. J Food Legumes 31(1):50–53
Mishra RK, Naimuddin, Mishra M, Sharma S (2016a) *Trichoderma*: potential biocontrol agents for pulses. Dalhan Alok 14:80–86
Mishra RK, Naumuddin, Mohd A, Sachan DK, Monika M, Krishna K (2016b) Characterization of indigenous *Trichoderma* spp. through ITS based sequences. Pulses News Letter 27(2):5
Mishra RK, Saabale PR, Naimuddin, Jagadeeswaran R, Mishra O. (2015) Potential *Trichoderma* sp. from pulses Rhizosphere. Pulses News Letter. April–June pp: 3
Mukherjee M, Mukherjee PK, Kale PS (2007) cAMP signalling is involved in growth, germination, mycoparasitism and secondary metabolism in *Trichoderma virens*. Microbiology 153:1734–1742
Mukherjee PM, Latha J, Hardar R, Horwitz BA (2003) *Tmk* a, mitogen activated protein kinase of *Trichoderma virens* is involved in biocontrol properties and repression of conidiation in the dark. Eukaryot Cell 2:446–455
Mukhtar I (2008) Influence of *Trichoderma* species on seed germination in okra. Mycopathology 6:47–50
NAAS (2013) Biopesticides-quality assurance. Policy paper no. 62. National Academy of Agricultural Sciences, New Delhi, p 20
Naznin A, Hossain MM, Anju-Man Ara K, Hoque A, Islam M, Hasan T (2015) Influence of organic amendments and bio-control agent on yield and quality of tuberose. J Horticulture 2:4. https://doi.org/10.4172/2376-0354.1000156
Okoth SA, Otadoh JA, Ochanda OJ (2011) Improved seedling emergence and growth of maize and beans by *Trichoderma harziunum*. Tropical and Subtropical Agroecosystems 13:65–71
Persoon CH (1794) Disposita methodica fungorum. Römer's Neues Mag Bot 1:81–128
Pieterse CM, Leon-Reyes A, Vander Ent S, Van Wees SC (2009) Networking bysmall-molecule hormones in plant immunity. Nat Chem Biol 5:308–316
Pozo MJ, JongMin B, Garcia JM, Kenerley CM (2004) Functional analysis of *tvsp1*, a serine protease-encoding gene in the biocontrol agent *Trichoderma virens*. Fungal Genet Biol 41:336–348
Purushottam A, Swarnalakshmi K, Saabale PR, Ninawe AS (2014) On-farm demonstrations of *Trichoderma harzianum* in pulse crops under rainfed conditions of Bundelkhand—a case study. Int J Curr Microbiol App Sci 3(11):471–478
Rehman SU, Lawrence R, Kumar EJ, Badri ZA (2012) Comparative efficacy of *Trichoderma* viride, *T. harzianum* and carbendazim against damping-off disease of cauliflower caused by *Rhizoctonia solani* Kuehn. J Biopest 5:23–27
Rifai MA (1969) A revision of the genus Trichoderma. Mycol Pap 116:1–56
Saba H, Vibhash D, Manisha M, Prashant KS, Farhan H, Tauseef A (2012) *Trichoderma* – a promising plant growth stimulator and biocontrol agent. Mycosphere 3:524–531

Saiprasad GVS, Mythli JB, Anand L, Sunetha C, Rashmi HJ, Naveena C, Ganeshan G (2009) Development of *Trichoderma harzianum* gene constructs conferring antifungal activity in transgenic tobacco. Indian J Biotechnol 8:199–206

Samuels GJ (2006) *Trichoderma*: systematics, the sexual state, and ecology. Phytopathology 96:195–206

Seidl V, Seibel C, Kubicek CP, Schmoll M (2009) Sexual development in the industrial workhorse *Trichoderma reesei*. Proc Natl Acad Sci U S A 106:13909–13914

Sharma K, Mishra AK, Misra RJ (2009) Morphological, biochemical and molecular characterization of Trichoderma harzianum isolates for their efficacy as biocontrol agents. J Phytopathol 157:51–56. https://doi.org/10.1111/j.1439-0434.2008.01451.x

Shi WL, Chen XL, Wang LX et al (2016) Cellular and molecular insight into the inhibition of primary root growth of Arabidopsis induced by peptaibols, a class of linear peptide antibiotics mainly produced by *Trichoderma* spp. J Exp Bot 67(8):2191–2205. https://doi.org/10.1093/jxb/erw023

Shukla N, Chaudhary RG (2007) Reduction of inoculums potential of Fusarium udum by Trichoderma spp. Abstract of Paper presented during National Symposium on "Plant Pathogens: Exploitation and Management" at R. D. University, Jabalpur (M.P.), India during January 16–18, 2007. *Indian Phytopath* 60: 392

Shukla N, Awasthi RP, Rawat L, Kumar J (2012) Biochemical and physiological responses of rice (Oryza sativa L.) as influenced by *Trichoderma harzianum* under drought stress. Plant Physiol Biochem 54:78–88

Siddiqui ZA, Irshad Mahmood SH, Mahmood I, Hayat S (1998) Biocontrol of *Heterodera cajani* and *Fusarium udum* on pigeon pea using Glomus mosseae, Paecilomuces lilacinus and *Pseudomonas fluorescens*. Thai J Agric Sci 31:310–321

Singh A, Srivastava S, Singh HB (2007) Effect of substrates on growth and shelf life of *Trichoderma harzianum* and its use in biocontrol of diseases. Bioresour Technol 98:470–473

Singh S, Dureja P, Tanwar RS, Singh A (2005) Production and antifungal activity of secondary metabolites of *Trichoderma* virens. Pesticides Res J 17:26–29

Sriram S, Palanna KB, Ramanujam B (2010) Effect of chitin on the shelf-life of *Trichoderma harzianum* in talc formulation. Indian J Agri Sci 80:930–932

Sriram S, Roopa KP, Savitha MJ (2011) Extended shelf life of liquid formulations of *Trichoderma harzianum* with addition of glycerol. Crop Prot 30:1334–1339

Srivastava SN, Singh V, Awasthi SK (2006) *Trichoderma* induced improvement in growth, yield and quality of sugarcane. Sugar Tech 8:166. https://doi.org/10.1007/BF02943654

Thakur AK, Uphoff N, Antony E (2010) An assessment of physiological effects of system of rice intensification (sri) practices compared with recommended rice cultivation practices in India. Exp Agric 46:77–98

Thakur M, Sohal BS (2013) Role of elicitors in inducing resistance in plants against pathogen infection: a review. ISRN *Biochemistry*:1–10. https://doi.org/10.1155/2013/762412

Tijerino A, Cardoza RE, Moraga J, Malmierca MG, Vicente F, Aleu J, Collado IG, Gutierrez S, Monte E, Hermosa R (2011) Overexpression of the trichodiene synthase gene *tri5* increases trichodermin production and antimicrobial activity in *Trichoderma brevicompactum*. Fungal Genet Biol 48:285–296

Tronsmo A, Harman GE (1992) Coproduction of chitinases and biomass for biological control by *Trichoderma harzianum* on media containing chitin. Biol Control 2:272–277

Tucci M, Ruocco M, De Masi L, De Palma M, Lorito M (2011) The beneficial effect of *Trichoderma* spp. on tomato is modulated by the plant genotype. Mol Plant Pathol 12:341–354. https://doi.org/10.1111/j.1364-3703.2010.00674

Tulasne LR, Tulasne C (1865) Selecta fungorum carpologia, vol 3. Paris Museum, Paris

Vidhyasekaran P, Muthamilan M (1995) Development of formulations of *Pseudomonas fluorescens* for control of chickpea wilt. Plant Dis 79:782–786

Vidhyasekaran P, Sethuraman K, Rajappan K, Vasumathi K (1997) Powder formulation of *Pseudomonas fluorescens* to control pigeonpea wilt. Biol Control 8:166–171

Vinale F, Flematti G, Sivasithamparam K et al (2009) Harzianic acid, an antifungal and plant growth promoting metabolite from *Trichoderma harzianum*. J Nat Prod 72:2032–2035

Vinale F, Marra R, Ruocco M (2014) *Trichoderma* secondary metabolites active on plants and fungal pathogens. The Open Mycology Journal 8:127–139

Yoshioka Y, Ichikawa H, Naznin HA, Kogure A, Hyakumachi M (2011) Systemic resistance induced in Arabidopsis thaliana by *Trichoderma asperellum* SKT-1, a microbial pesticide of seedborne diseases of rice. Pest Manag Sci. https://doi.org/10.1002/ps.2220

Zhong YH, Wang TH, Wang XL, Zhang GT, Yu HN (2009) Identification and characterization of a novel gene, *TrCCD1*, and its possible function in hyphal growth and conidiospore development of *Trichoderma reesei*. Fungal Genet Biol 46:255–263

Chapter 8
Management of Diseases of Medicinal and Aromatic Plants Using High Shelf Life Formulation of *Trichoderma*

Akanksha Singh and Rakesh Pandey

Abstract Management of plant diseases holds a crucial cost component in crop productivity. Traditional approaches used for managing the diseases include the use of resistant varieties and various hazardous chemicals and pesticides. However, due to the increasing global concerns about the perilous effects of fungicides upon environment, residual toxicity, and nontarget organisms, strong desire to replace such chemicals gave way to the biological substitutes. One such organism is *Trichoderma* which has shown tremendous potential not only in terms of enhancing plant growth but has also proven its mettle in successfully managing the plant diseases. In this context, medicinal and aromatic plants (MAPs) are no exception since they too are attacked by various pathogens thereby hampering their total yield and quality of final product. The chapter has thus been written with the aim of unravelling the past and current day research done in managing the different diseases of MAPs with the usage of *Trichoderma* spp. and what future does it hold in the area of successful MAPs cultivation.

Keywords Medicinal and aromatic plants · Biocontrol agents · *Trichoderma* · Disease management · Secondary metabolites

8.1 Introduction

Plant and crop productivity, survival rate, and reproductive status are greatly influenced by various environmental stresses. The various forms of environmental stress are basically categorized into two areas namely biotic and abiotic stress. Being sessile in nature, the plants have been gifted with an inbuilt defense mechanism to combat the various stresses (Atkinson and Urwin 2012). But, many times such mechanisms fail to respond, and changing climatic conditions is one such situation.

A. Singh (✉) · R. Pandey
Department of Microbial Technology and Nematology, CSIR- Central Institute of Medicinal and Aromatic Plants, Lucknow, India

C. Manoharachary et al. (eds.), Trichoderma: *Agricultural Applications and Beyond*, Soil Biology 61, https://doi.org/10.1007/978-3-030-54758-5_8

It is being reported that as the climatic conditions are changing, many new diseases have popped up which are reported to cause heavy losses to the crop productivity (Montoya and Raffaelli 2010). Medicinal and aromatic plants (MAPs) are no exception to this as they too are attacked by various pathogens thereby resulting in substantial losses in the form of total yield and important bioactive molecules.

Many preventive and control methods have been extensively used in the past decade for plant disease management like the use of resistant varieties through breeding techniques, cultivation of genetically engineered plants, usage of chemicals, and various physical methods. However, the above-mentioned methods have their own limitations. The limitations like time taken up for developing resistant varieties, policy issues regarding the release of genetically engineered plants, and toxicity issues regarding usage of agrochemicals have led people to adopt some eco-friendly measures for disease management (Narayanasamy 2002). Biological control is thus being looked upon as a potential eco-friendly alternative for not only managing the plant diseases but also for reducing the use of hazardous chemical compounds in agriculture. Among the various microorganisms utilized globally for managing the plant pathogens, genus *Trichoderma* being blessed with versatile traits holds an important place in the list. *Trichoderma* has shown "power packed" performance against various plant pathogens like different viruses, phytoplasmas, bacteria, fungi, soil-inhabiting nematodes, and higher parasitic plants (Fig. 8.1). The fast attacking nature of the genus is mainly because of the presence of various secondary metabolites, extracellular enzymes, competition with other

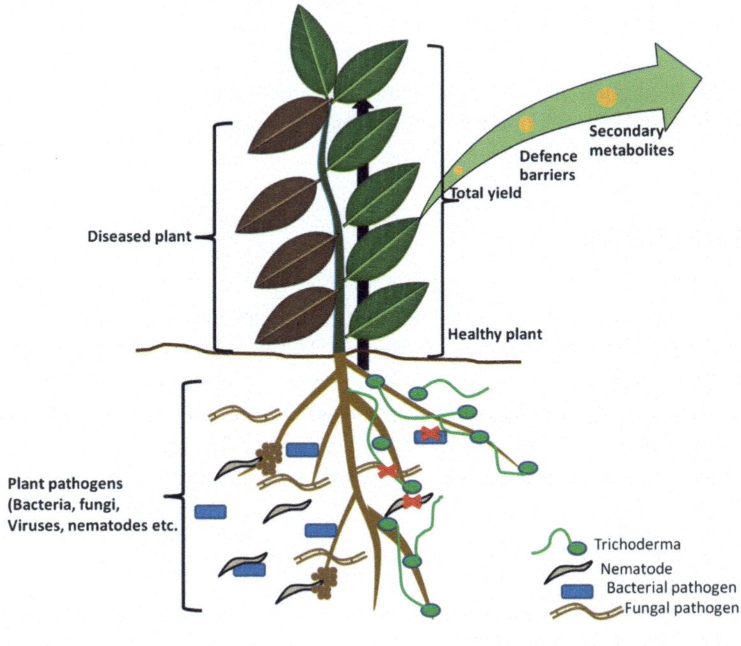

Fig. 8.1 Schematic illustration of the role of beneficial and pathogenic microbes on plant health

inhabitants along with being a strong systemic inducer (Al-Ani 2018). The application of *Trichoderma* holds great promise in MAPs cultivation too as most of the time medicinal plants are used in raw form and spraying of chemicals many times results in the accumulation of toxic compounds both in plant parts as well as surrounding (Amini et al. 2014). Thus, the present chapter has been written with the aim of highlighting the importance as well as creating awareness about the adoption of *Trichoderma* for MAPs cultivation as well as disease management.

8.2 Medicinal and Aromatic Plants: Usefulness and Cultivation

MAPs hold an important position across the various communities globally. MAPs in general can be described as plants whose active components having medicinal attributes to prevent many diseases and maintain the overall health of the person. Apart from being used extensively for the therapeutic purpose, MAPs also find usage in culinary, cosmetics, perfumery, and aromatherapy (Lubbe and Verpoorte 2011). MAPs are found growing in terrestrial and aquatic ecosystems but due to increasing demands survival of many wild species is threatened. According to the global forecast, the medicinal plant market is expected to reach a figure of USD 1,29,6893 million by 2023. The above-mentioned figure has a fair chance to reach its goal as approximately 80% of the world's total population is dependent on the plant-based products for health care needs (Organization 2002). The usefulness of the herbal medicine can also be further elucidated from the fact that the most important 150 proprietary drugs which find usage in the United States, 57% of them consist of at least one major important phytomolecule. In addition, approximately 50% of the drugs given nod for approval in the last 30 years have direct or indirect connections with the plants.

The MAPs cultivation sustainably produces industrial raw materials of high value thereby providing higher returns and gains to the farmers of developing nations like India. Even though such plants have been in usage since prehistoric times for curing diseases, recently, advancements in scientific technologies and validation through scientific studies have led to soaring up of the market value of such crops and crop products. The benefits of growing MAPs are numerous as they do not need exhaustive agricultural inputs and are also reported to grow exuberantly even in stress conditions (Lubbe and Verpoorte 2011). It is because of the above reasons that though India is growing MAPs in a small area the returns are pretty decent.

8.3 Diseases of Medicinal and Aromatic Plants

Like other agricultural crops, growers of MAP also face many constraints in cultivating them as many biotic and abiotic factors severely affect their production. The situation is pretty worse for the MAP growers as not much research has been done keeping in mind the diseases affecting the production of MAPs. In addition, the rising global temperature is further creating problems since many nonpathogenic microbes have become pathogenic giving rise to new epidemic diseases (Singh et al. 2016).

Among the diverse range of phytopathogens attacking the MAPs, fungal, bacterial, viral, and phytonematodes cause extensive damage to the plants. The phytopathogens not only result in yield loss but also affect the quality of raw materials and secondary metabolite status of the medicinal plant. The most common category of phytopathogen attacking the plants belongs to the fungal group. Fungi are eukaryotic, achlorophyllous filamentous microorganisms that usually cause local general symptoms or some host-specific ones. But, the most common observation is degeneration in the overall quality of the raw material as a result of metabolic and physiological disturbances happening in the plant organ (Abtahi and Nourani 2017). A range of fungal pathogens are reported to infect the aerial, foliar parts and underground plant parts of MAPs which majorly include the powdery mildews primarily reported to infect the leaves and fresh stems. Similarly, rust is another category of fungi infecting aerial parts which are reported to produce pustules on both upper and lower surface of leaves. In the group of fungal pathogens, other groups that are reported to cause heavy losses are the leaf spot and blight diseases (Shubhra et al. 2014).

The next very important category of phytopathogen affecting the MAPs cultivation is the plant-parasitic nematodes which attack the root system thereby causing significant yield and growth reductions. Majorly most of the MAPs are under the threat of parasitic nematodes namely *Withania*, *Bacopa*, *Coleus*, *Mentha*, and *Ocimum*, etc. (Fig. 8.2). Usually, the three species of nematodes that cause heavy

Fig. 8.2 Root galls formed in different medicinal plants due to nematode infection. (**a**) *Withania somnifera* (**b**) *Bacopa monnieri* seedlings (**c**) *Coleus forskholii* (**d**) *Ocimum sanctum* (**e**) *Mentha arvensis* sucker (**f**) *M. arvensis* runner

losses to the plants belong to the category of root-knot nematodes (*Meloidogyne incognita* and *M. javanica*), stunt nematode (*Tylenchorhynchus vulgaris*), and root-lesion nematode (*Pratylenchus thornei*) (Pandey 2017).

Apart from fungal, bacterial, and nematode diseases, plant viruses and phytoplasma too cause a range of diseases that severely affect the total yield, quality, and overall production of MAPs worldwide. Huge numbers of MAPs are reported to be infected by plant viruses like *Rauvolfia serpentina*, *Hyoscyamus albus*, *Catharanthus roseus*, *Mentha* spp. show varying symptoms from being topical to systemic in nature. The most common visible symptoms include chlorosis, malformations, mosaic and patchy leaves (Tzanetakis et al. 2010; Khan et al. 2015).

8.4 Diverse Methods Adopted for Management of Medicinal and Aromatic Plant Diseases

MAPs' importance is increasing with each passing day because of the significant role they play in public health care. Among the various countries fulfilling the global demand, India is one such country which holds an important position in the list because of being an 8% shareholder of global biodiversity. However, the situation is pretty different as it looks to be since there is a huge gap between demand and supply as most of the MAPs are collected from wild along with being susceptible to various biotic and abiotic factors that further create a problem (Marimuthu et al. 2018).

Pest and diseases like mites, viruses, fungi, nematodes, phytoplasmas, bacteria, etc. are believed to take a grave toll on these plants, but literature reports suggest that not much attention has been given so far for safeguarding such plants from them. Thus, the management aspect needs to be looked upon with extreme care and astuteness as the use of hazardous chemicals for controlling the pest and diseases are reported to not only alter the qualitative traits but also the quantitative composition of the active biomolecules which finally in turn diminishes their therapeutic values. The pesticides which are most frequently used in MAPs are chlorpyriphos, dichlorvos, ethion, malathion, dicofol, etc. and it has been shown that residual toxicity may be further promoted to the final product which subsequently reaches to the end consumers (Zuin and Vilegas 2000). Therefore, in this context Integrated Pest Management (IPM) is the best substitute for managing pest and disease in these crops since it entails extensive monitoring of populations of pest, disease epidemiology, identification through key traits, and formulating a perfect combination of methods to keep away the pest/disease populations. The methods adopted for developing a combination may include chemical, cultural, biological, or mechanical methods which may either be used in combination or alone.

8.5 Biological Control Agents (BCAs): An Eco-Friendly Approach for Disease Management in MAPs

Biological control for the management of plant pathogens as well as for plant growth promotion has been comprehensively studied over the past decades and is becoming a realistic alternative to chemical pesticides and fertilizers in sustainable agriculture (Weller 2007). Quite a good number of microbial inoculants for agricultural crops have been commercialized successfully but not much has been worked upon for MAPS which are in majority affected by various soilborne pathogens (Berg et al. 2013; Mendes et al. 2013). Among the various biological control agents (BCAs), spore-forming microbes especially the ones belonging to genus *Streptomyces, Bacillus,* and *Trichoderma* show promising results because of high tolerance level in field conditions along with being good antibiotic producers (Raaijmakers and Mazzola 2012; Köberl et al. 2013). Talking about the most prevalently present microbes in the rhizosphere, *Pseudomonas* sp. and *Bacillus* sp. hold the most important position which benefits the health of plants by acting as growth promoters along with being suppressors of plant pathogens in various medicinal and agricultural crops (Table 8.1) (Noori and Saud 2012; Shanmugam et al. 2011).

Medicinal plants are a rich source of various secondary metabolites which support a great diversity of microflora in their rhizosphere including plant growth-promoting microbes (PGPM) (Ahmed et al. 2014). Antagonistic fungi belonging to genus *Trichoderma* are the most regularly studied genus when we talk about plant protection via biological means (Hemashenpagam and Selvaraj 2011). Scientific reports suggest the involvement of mycoparasitism and the production of volatile and nonvolatile antibiotics as the main mode of action for biocontrol activity in *Trichoderma* sp. The use of *Trichoderma* has two major benefits: (i) they enhance the yield of crops along with managing the diseases and (ii) being environmentally friendly in nature they help in reducing the pesticide application thereby minimizing the residue problem in raw produce.

8.6 Case Study: High Shelf Life *Trichoderma* for Disease Management and Enhancement of Yield and Secondary Metabolites in MAPs

(i) Role of *Trichoderma* in management of fungal and bacterial diseases of MAPs
Trichoderma is one genus that has not only shown its efficacy in managing the diseases in the field but also in glasshouse conditions too (Fig. 8.3). A group of researchers showed that *Trichoderma harzianum* in in vitro conditions showed promising biocontrol activity against *Fusarium equiseti*, *Sclerotinia* sp., and *Rhizoctonia solani* which are reported to cause huge yield losses in some Chinese medicinal plants namely *Astragalus membranaceus*, *Glehnia littoralis*, and *Panax quinquefolium*, respectively (Ding et al. 2003). Application of

Table 8.1 List of some important plant pathogens managed by beneficial microbes

S. No.	Beneficial microbe used	Medicinal plant	Pathogen managed	Reference
1	*Bacillus subtilis, Pseudomonas chlororaphis, P. fluorescens*	*Phyllanthus amarus*	*Corynespora casiicola*	(Mathiyazhagan et al. 2004)
2	*Glomus fasciculatum* and *Pseudomonas*	*Coleus forskohlii*	*Fusarium chlamydosporum* and *Ralstonia solanacearum*	(Singh et al. 2013)
3	*Trichoderma viride* and *Glomus mosseae*	*C. Forskohlii*	*Fusarium chlamydosporum*	(Boby and Bagyaraj 2003)
4	*Pseudomonas aeruginosa* WS-1	*Withania somnifera*	*Alternaria dianthicola*	(Maiti et al. 2012)
5	*Trichoderma* spp.	*Glehnia littoralis*	*Sclerotinia sclerotiorum*	(Ding et al. 2003)
6	*Trichoderma* spp.	*W. somnifera*	*Meloidogyne incognita*	(Poornima and Rakesh 2009)
7	*Bacillus megaterium, Glomus intraradices, Pseudomonas fluorescens,* and *T. harzianum*	*Chlorophytum borivilianum*	*M. incognita*	(Pandey and Saikia 2014)
8	*T. harzianum, Pseudomonas monteilii, Bacillus megaterium* and *Azotobacter chroococcum*	*Pogostemon cablin*	*Rhizoctonia*	(Singh et al. 2013)
9	*T. harzianum* and *B. megaterium*	*Bacopa monnieri*	*M. incognita*	(Gupta et al. 2015)
10	*P. fluorescens* and *Paecilomyces lilacinus*	*Coleus forskohlii*	*M. incognita*	(Seenivasan and Deevrajan 2008)

T. harzianum in glasshouse conditions controlled sclerotium root rot in *Glehnia littoralis* by 83.6% while in field condition the control effect was up to 72.5%. Similarly, for *Astragalus membranaceus* and *Panax quinquefolium* the disease control was by 80% and 60%, respectively, in the field conditions. Similarly, in another report *Trichoderma* spp. showed strong antagonistic activity against damping-off pathogen *R. solani* reported to cause heavy losses in medicinal plant *P. quinquefolium* (Wanlong and Qiuyi 1994). In another study on *Coleus forskohlii, a* group of researchers found significant control of root rot/wilt upon application of *Trichoderma viride, G. fasciculatum, P. fluorescens,* and *G. mosesae* with respect to the control plants having chemical treatments (Boby and Bagyaraj 2003; Singh et al. 2009).

Fusarium wilt is reported to cause heavy losses in *Withania* and therefore people have worked on various management practices to curb the disease. In an important study by a group of researchers, the effect of *Trichoderma virens, T. harzianum,* and *Aspergillus niger* for controlling Fusarium wilt was

Fig. 8.3 Effect of *Trichoderma* on *Mentha arvensis* and *Pelargonium graveolens* infected fields (B and D), respectively, with respect to their nematode infected fields (A and C)

observed. Among the three bioagents, maximum reduction in wilt incidence was recorded by *T. virens* followed by *T. harzianum.* Similar results were replicated with seed treatment on the three different strains of *Trichoderma* for seed germination, plant height, thickness of root, and number of branches per plant (Ashraf and Zuhaib 2014). Combinational approach for managing wilt has shown promising leads as a group of researchers demonstrated that the application of *T. viride* with neem-based products resulted in a significant lowering of wilt in *C. forskohlii* (Kulkarni et al. 2007). *T. viride* and neem-based products (Neemto) were also found to significantly lower colony-forming units of both *Fusarium chlamydosporum* and *Rhizoctonia bataticola* along with nematode population in C. *forskohlii* (Ramaprasad 2007). Likewise, in another investigation, *Pseudomonas aeruginosa* strain WS-1 isolated from the rhizospheric region of *W. somnifera* plant showed both in vitro and in vivo antagonistic activity against the leaf blight pathogen. The antifungal activity of the bacterial isolate was found to be linked with the production of iron-chelating compounds and extracellular enzymes. The isolate not only showed its efficacy in lab conditions but excellent results were obtained in field conditions as well since the foliar application of the talc-based formulation reduced the disease harshness by 80% compared to the non-treated control (Maiti et al. 2012).

(ii) Role of *Trichoderma* in management of nematode diseases of MAPs

Trichoderma is being used extensively for managing notorious nematodes which are reported to survive in the soil for years and years (Fig. 8.3). Along with managing bacterial and fungal pathogens, *Trichoderma* has shown potential in managing root-knot nematode, *Meloidogyne incognita* when applied in combination with farmyard manure, neem oilseed cake, cow urine, and vermicompost individually and in different combinations in *W. somnifera* (Pandey et al. 2011). *W. somnifera* is one such crop that is very much susceptible to the attack of *M. incognita* resulting in gall formation, stunted appearance of the plant with a lowering in overall plant productivity (Pandey and Kalra 2003). In one investigation study conducted on *W. somnifera* and *M. incognita* negative interaction, potential effects of *T. harzianum, Paecilomyces lilacinus,* and *Arthrobotrys oligospora* were observed when applied in combination with neem compound mixture (Poornima and Rakesh 2009; Pandey and Kalra 2003).

At CSIR-CIMAP, a potential strain of *T. harzianum* (Accession number PTA-3701) was identified which showed promising nematicidal, fungicidal, and plant growth-promoting activities (Kalra et al. 2002). Similarly, in another report potentiality of rhizosphere inhabiting microbes namely *Bacillus megaterium*, *Glomus intraradices*, *P. fluorescens*, and *T. harzianum* CIMAP-RPN01, both singly as well as in different combinations were tested against *M. incognita* in *Chlorophytum borivilianum*. The study revealed maximum control of the pest in a dual combination of *T. harzianum* + *P. fluorescense* followed by *T. harzianum* + *B. megaterium* thereby enhancing the yield and quality of important phytochemicals of the crop (Pandey and Saikia 2014). Likewise, the authors in another experiment tested the efficacy of five important rhizospheric microbes for managing *M. incognita* in another important medicinal plant, *W. somnifera* out which four of them showed promising results against the nematode along with potential plant growth-promoting activities (Saikia et al. 2013).

The role of *Trichoderma* along with other microbes was also assessed for decomposing the distillation waste of patchouli. The group of workers in the above study found that *T. harzianum* with other compatible microbes remarkably boosted biodegradation of hard and tough degrading structures in plant systems namely lignin, cellulose, and hemicelluloses with final effect on the total yield of vermicompost which was increased by 15% as compared to the control set. In addition, vermicompost enriched with microbes not only significantly reduced *Rhizoctonia* root rot but also augmented the total oil yield with respect to the control set (Singh et al. 2013). Similarly, the application of bioagents like *T. viride* and *P. fluorescens* through soil significantly helped in lowering the nematode population along with increasing the yield and growth parameters of *C. forskohlii* crop (Senthamarai et al. 2008).

(i) Role of *Trichoderma* in MAPs growth promotion

Yield attributes hold an important place in MAPs cultivation as most of the active ingredients are either in foliage parts or in the roots. Being chemical in nature most of the compounds used for plant growth-promoting activities are being highly discouraged nowadays. Thus, biological substitutes are being looked upon as potential alternatives for enhancing growth and yield parameters in the field. Thus, working in the direction of fulfilling the current demand of the consumers as well environmentalist, a group of researchers investigated the effect of *B. megaterium* (ATCC No. 13525), *T. viride* (MTCC No. 167), and *P. fluorescens* (ATCC No. 14581) on the total yield parameters of *Ocimum tenuiflorum* L. cv. CIM-Ayu both in alone as well as in a combined set of treatments. The authors found a considerable increase in the oil percentage and nutrient uptake capacity of plants in combined treatments with respect to alone and control treatments (Saikia and Pandey 2014). Recently, a co-inoculation study was conducted by a group of researchers at CSIR CIMAP where *T. harzianum* in combination with plant growth-promoting rhizobacteria (PGPRs) enhanced the growth and essential oil content of *Mentha arvensis* both in glasshouse and field conditions (Singh et al. 2019). Likewise, the authors in another report found a significant enhancement in total plant yield, the status of secondary metabolites, total chlorophyll, and carotenoids content and total antioxidant status of *P. graveolens* treated with a synergistic combination of bio inoculants namely *T. harzianum* ThU, *G. intraradices,* and *Bacillus subtilis* CIM as compared with the control set (Gupta et al. 2016).

(ii) Role of *Trichoderma* in secondary metabolite status of MAPs

In a study conducted by Arpana and Bagyaraj (2007), two beneficial soil microbes namely *Glomus mosseae* and *T. harzianum* affected the health of *Andrographis paniculata* in a positive way by increasing the fresh growth parameters, phosphorus uptake, dry weight, and total andrographolide content compared to uninoculated plants (Poornima and Rakesh 2009). In another study to validate the role of microbes on plant health, a group of workers investigated the effect of secondary metabolites of *Streptomyces* sp. MTN14 and *T. harzianum* ThU on *W. somnifera*. The authors found a significant enhancement in biomass yield and Withanolide A content in *Trichoderma* and *Streptomyces* metabolites treated plants with respect to the control set (Singh et al. 2016). Likewise, the authors conducted a study on *P. graveolens* and found a significant upregulation in citronellol and geraniol content along with an increase in the number of glandular trichomes in *T. harzianum* ThU, *G. intraradices*, and *B. subtilis* CIM treated plants with respect to the control one (Gupta et al. 2016).

Artemisia annua L. is one crop that is primarily widely known for a key bioactive secondary metabolite, Artemisinin, having a role in malarial treatment without any case of resistance reported against it (De Ridder et al. 2008). It is because of the global demand of the bioactive molecule that researchers around the globe are working toward enhancing the percentage yield of the molecule in the plants. In an experiment, a group of workers used three microbes namely *B. megaterium*, *Streptomyces,* and *T. harzianum* and observed increment in growth, total phenolic content, and antioxidant potential of *A. annua* plants. The most significant observation was an increase in artemisinin content upon inoculation of all the three microbes in combination with respect to control plants

(Gupta et al. 2016). Similarly, in another report content of silychristin, silymarin, and isosilybin was significantly enhanced upon treatment with *Trichoderma* M7 strain with respect to non-inoculated *Silybum marianum* plants (Hasanloo et al. 2010). Likewise, forskolin content of the roots was reported to be enhanced in C. *forskohlii* upon inoculation of bio inoculants like *T. viride*, *G. mosesae*, *P. fluorescens*, and *G. fasciculatum* (Boby and Bagyaraj 2003; Singh et al. 2009).

8.7 Conclusion

Trichoderma is one genus that is reported to significantly manage plant diseases along with the added advantage of enhancing the yield and growth of the crops. However, a huge gap exists when we talk about their application in MAPs as scanty literature exists on them. In addition, with mounting public apprehension on the subject of environmental pollution and day-by-day increase in pesticide costs, efforts for the adoption of nonchemical alternatives are the need of the hour. Thus, popularization and adoption of *Trichoderma* based cultivation of MAPs for yield enhancement and disease control should be encouraged globally as it will not only help in fetching good returns for the raw material but will also be in harmony with the natural resources.

Acknowledgment The authors are thankful to the Director, CSIR-CIMAP, for his encouragement and support. AS is also grateful to the Department of Science and Technology, Govt. of India for providing fellowship under the DST Women Scientist Scheme (SR/WOS-B/2017/254-AAS).

References

Abtahi F, Nourani SL (2017) The most important fungal diseases associated with some useful medicinal plants. In: Medicinal plants and environmental challenges. Springer, Cham, pp 279–293

Ahmed EA, Hassan EA, El Tobgy K, Ramadan E (2014) Evaluation of rhizobacteria of some medicinal plants for plant growth promotion and biological control. Annals of Agricultural Sci 59(2):273–280

Al-Ani LKT (2018) *Trichoderma*: beneficial role in sustainable agriculture by plant disease management. In: Plant microbiome: stress response. Springer, Singapore, pp 105–126

Amini Y, Mohammadi A, Zafari D (2014) *Trichoderma* species associated with medicinal plants. Int J Adv Biol Biomed Res 2(9):2566–2568

Arpana J, Bagyaraj DJ (2007) Response of kalmegh to an arbuscular mycorrhizal fungus and a plant growth promoting rhizomicroorganism at two levels of phosphorus fertilizer. Am-Euras J Agriculture Environ Sci 2:33–38

Ashraf S, Zuhaib M (2014) Efficacy of rhizospheric microorganism against wilt of Ashwagandha (*Withania somnifera* DUNAL) and their influence on its growth. Trends in Biosciences 7 (16):2165–2167

Atkinson NJ, Urwin PE (2012) The interaction of plant biotic and abiotic stresses: from genes to the field. J Exp Bot 63(10):3523–3543

Berg G, Zachow C, Müller H, Philipps J, Tilcher R (2013) Next-generation bio-products sowing the seeds of success for sustainable agriculture. Multidisciplinary Digital Publishing Institute:648–656

Boby V, Bagyaraj D (2003) Biological control of root-rot of *Coleus forskohlii* Briq. Using microbial inoculants. World J Microbiol Biotechnol 19(2):175–180

De Ridder S, Van der Kooy F, Verpoorte R (2008) *Artemisia annua* as a self-reliant treatment for malaria in developing countries. J Ethnopharmacol 120(3):302–314

Ding WL, Cheng HZ, Chen J (2003) Presearch on preventing the medicinal plant diseases with *Trichoderma harzianum* preparation. China J Chinese Materia Medica 28(1):24–27

Gupta R, Singh A, Gupta M, Pandey R (2016) Cumulative role of bioinoculants on growth, antioxidant potential and artemisinin content in *Artemisia annua* L. under organic field conditions. World J Microbiol Biotechnol 32(10):167

Gupta R, Singh A, Pandey R (2016) Microbe-based technology ameliorates glandular trichomes, secondary metabolites and antioxidants in *Pelargonium graveolens* L'Hér. J Sci Food Agric 96 (12):4151–4159

Gupta R, Tiwari S, Saikia SK, Shukla V, Singh R, Singh S, Kumar PA, Pandey R (2015) Exploitation of microbes for enhancing bacoside content and reduction of *Meloidogyne incognita* infestation in *Bacopa monnieri* L. Protoplasma 252(1):53–61

Hasanloo T, Kowsari M, Naraghi SM, Bagheri O (2010) Study of different *Trichoderma* strains on growth characteristics and silymarin accumulation of milk thistle plant. J Plant Interact 5 (1):45–49

Hemashenpagam N, Selvaraj T (2011) Effect of arbuscular mycorrhizal (AM) fungus and plant growth promoting rhizomicroorganisms (PGPR's) on medicinal plant *Solanum viarum* seedlings. J Environ Biol 32(5):579

Kalra A, Singh HB, Pandey R, Patra NK, Katiyar N, Gupta ML, Dhawan OP, Kumar S (2002) Strain of *Trichoderma harzianum* useful as nematode inhibitor, fungicide and plant growth promoter and a process for the isolation thereof. US Patent No 6,475,772

Khan A, Saeed S, Samad A (2015) New record of Catharanthus yellow mosaic virus and a betasatellite associated with lethal leaf yellowing of Kalmegh (*Andrographis paniculata*) in northern India. Plant Dis 99(2):292–292

Köberl M, Schmidt R, Ramadan EM, Bauer R, Berg G (2013) The microbiome of medicinal plants: diversity and importance for plant growth, quality and health. Front Microbiol 4:400

Kulkarni M, Ramprasad S, Hedge Y, Laxminarayan H, Hedge N (2007) Management of collar rot complex disease of *Coleus forskohlii* (wild) Briq. Using bioagents, organic amendments and chemicals. Biomedicine 2:37–40

Lubbe A, Verpoorte R (2011) Cultivation of medicinal and aromatic plants for specialty industrial materials. Ind Crop Prod 34(1):785–801

Maiti CK, Sen S, Paul AK, Acharya K (2012) *Pseudomonas aeruginosa* WS-1 for biological control of leaf blight disease of *Withania somnifera*. Arch Phytopathol Plant Protect 45 (7):796–805

Marimuthu T, Suganthy M, Nakkeeran S (2018) Common pests and diseases of medicinal plants and strategies to manage them. In: New Age Herbals. Springer, Singapore, pp 289–312

Mathiyazhagan S, Kavitha K, Nakkeeran S, Chandrasekar G, Manian K, Renukadevi P, Krishnamoorthy A, Fernando W (2004) PGPR mediated management of stem blight of *Phyllanthus amarus* (Schum and Thonn) caused by *Corynespora cassiicola* (Berk and Curt) Wei. Arch Phytopathol Plant Protect 37(3):183–199

Mendes R, Garbeva P, Raaijmakers JM (2013) The rhizosphere microbiome: significance of plant beneficial, plant pathogenic, and human pathogenic microorganisms. FEMS Microbiol Rev 37 (5):634–663

Montoya JM, Raffaelli D (2010) Climate change, biotic interactions and ecosystem services. Royal Soc, 2013–2018

Narayanasamy P (2002) Microbial plant pathogens and crop disease management. In: Narayanasamy P (ed) Microbial plant pathogens and crop disease management. CRC Press

Noori MSS, Saud HM (2012) Potential plant growth-promoting activity of *Pseudomonas* sp. isolated from paddy soil in Malaysia as biocontrol agent. J Plant Pathol Microbiol 3(2):1–4

Organization WH (2002) WHO traditional medicine strategy 2002–2005. World Health Organization, Geneva

Pandey R (2017) Diseases of medicinal and aromatic plants: insights in nematode biomanagement. Chief Editor 70 (1):12–21

Pandey R, Kalra A (2003) Root-knot disease of ashwagandha *Withania somnifera* and its eco-friendly cost effective management. J Mycol Plant Pathol 33:240–245

Pandey R, Mishra A, Tiwari S, Kalra A (2011) Nematode inhibiting organic materials and a strain of *Trichoderma harzianum* effectively manages *Meloidogyne incognita* in *Withania somnifera* fields. Biocontrol Sci Tech 21(12):1495–1499

Pandey R, Saikia SK (2014) Rhizospheric bioweapons for tuber yield enhancement in *Chlorophytum borivilianum* against *Meloidogyne incognita* infestation. J Plant Biochemistry Physiol 2:1–5

Poornima S, Rakesh P (2009) Biological control of root-knot nematode; *Meloidogyne incognita* in the medicinal plant; *Withania somnifera* and the effect of biocontrol agents on plant growth. Afr J Agric Res 4(6):564–567

Raaijmakers JM, Mazzola M (2012) Diversity and natural functions of antibiotics produced by beneficial and plant pathogenic bacteria. Annu Rev Phytopathol 50:403–424

Ramaprasad SA (2007) Studies on Collar Rot Complex of *Coleus forskohlii* (Wild.) Briq. University of Agricultural Sciences GKVK, Bangalore

Saikia SK, Pandey R (2014) Rhizospheric microflora escalating aroma constituents and yield attributes in *Ocimum tenuiflorum* (L.) cv. CIM-Ayu. Advances in Agriculture

Saikia SK, Tiwari S, Pandey R (2013) Rhizospheric biological weapons for growth enhancement and *Meloidogyne incognita* management in *Withania somnifera* cv. Poshita Biological Control 65(2):225–234

Seenivasan N, Deevrajan K (2008) Integrated approach for the management of root-knot nematode, *Meloidogyne incognita* in medicinal coleus. Indian J Nematol 38(2):154–158

Senthamarai M, Poornima K, Subramanian S, Sudheer MJ (2008) Nematode-fungal disease complex involving *Meloidogyne incognita* and *Macrophomina phaseolina* on medicinal coleus, *Coleus forskohlii* Briq. Indian J Nematol 38(1):30–33

Shanmugam V, Kanoujia N, Singh M, Singh S, Prasad R (2011) Biocontrol of vascular wilt and corm rot of gladiolus caused by *Fusarium oxysporum* f. sp. *gladioli* using plant growth promoting rhizobacterial mixture. Crop Prot 30(7):807–813

Shubhra B, Harsh N, Sharma A, Mao LP, Shikha T (2014) A database of diseases of medicinal plants in Uttarakhand. Indian Forester 140(5):518–527

Singh A, Gupta R, Saikia SK, Pant A, Pandey R (2016) Diseases of medicinal and aromatic plants, their biological impact and management. Plant Genetic Resources 14(4):370–383

Singh A, Gupta R, Srivastava M, Gupta M, Pandey R (2016) Microbial secondary metabolites ameliorate growth, in planta contents and lignification in *Withania somnifera* (L.) Dunal. Physiol Mol Biol Plants 22(2):253–260

Singh R, Parameswaran T, Prakasa Rao E, Puttanna K, Kalra A, Srinivas K, Bagyaraj D, Divya S (2009) Effect of arbuscular mycorrhizal fungi and *Pseudomonas fluorescens* on root-rot and wilt, growth and yield of *Coleus forskohlii*. Biocontrol Sci Tech 19(8):835–841

Singh R, Singh R, Soni SK, Singh SP, Chauhan U, Kalra A (2013) Vermicompost from biodegraded distillation waste improves soil properties and essential oil yield of *Pogostemon cablin* (patchouli) Benth. Appl Soil Ecol 70:48–56

Singh R, Soni SK, Kalra A (2013) Synergy between *Glomus fasciculatum* and a beneficial *Pseudomonas* in reducing root diseases and improving yield and forskolin content in *Coleus forskohlii* Briq. Under organic field conditions. Mycorrhiza 23(1):35–44

Singh S, Tripathi A, Maji D, Awasthi A, Vajpayee P, Kalra A (2019) Evaluating the potential of combined inoculation of *Trichoderma harzianum* and *Brevibacterium halotolerans* for increased growth and oil yield in *Mentha arvensis* under greenhouse and field conditions. Ind Crop Prod 131:173–181

Tzanetakis IE, Postman JD, Samad A, Martin RR (2010) Mint viruses: beauty, stealth, and disease. Plant Dis 94(1):4–12

Wanlong D, Qiuyi G (1994) Studies on the preventive action of *Trichoderma* spp. against rhizoctonia damping-off of american ginseng (*Panax quinquefolium*). Chin Tradit Herb Drug 2:16

Weller DM (2007) *Pseudomonas* biocontrol agents of soilborne pathogens: looking back over 30 years. Phytopathology 97(2):250–256

Zuin VG, Vilegas JH (2000) Pesticide residues in medicinal plants and phytomedicines. Phytotherapy Research: An International Journal Devoted to Pharmacological and Toxicological Evaluation of Natural Product Derivatives 14(2):73–88

Chapter 9
Trichoderma: A Globally Dominant Commercial Biofungicide

Jhumishree Meher, Rahul Singh Rajput, Raina Bajpai, Basavaraj Teli, and Birinchi Kumar Sarma

Abstract Biopesticides are the derived products prepared from the living microbes such as plants, animals, fungi, bacteria, viruses, and minerals or their bioproducts that are being widely used against devastating phytopathogens. The use of biopesticides in agricultural fields is a sustainable, economical, and eco-friendly approach compared to the synthetic fungicides because of its target-specificity, biodegradability, and environmental safety property. Among the different biocontrol agents (BCAs) used worldwide, *Trichoderma* spp. based biofungicides are presently considered as relatively potential and most dominating commercial biofungicide in the global biopesticide market. About 60% of all fungal-based BCAs are contributed by *Trichoderma*-based biopesticides which are available in different formulations. It is also more popular due to its diverse mechanisms of biocontrol that include antibiosis, colonization, competition, direct mycoparasitism, etc. against a wide range of soil (*Sclerotium, Rhizoctonia, Fusarium, Macrophomina*, *Phytophthora* spp.), foliar (*Phyllactinia, Colletotrichum, Cladosporium* spp.), and post-harvest phytopathogens (*Penicillium, Aspergillus, Rhizopus, Botrytis* spp.), etc. In this chapter, we have discussed the reasons behind the wide acceptance of *Trichoderma* sp. as a BCA and its dominance in the global biofungicides market.

Keywords *Trichoderma* · Commercial biopesticide

9.1 Introduction

The use of agrochemicals in the management of pests and diseases act as first aid tools to the farmers in order to maintain crop health and increase in yield. However, globally it is observed that for the management of most fungal plant pathogens, synthetic fungicides are frequently used as it is an easily available option and could

J. Meher · R. S. Rajput · R. Bajpai · B. Teli · B. K. Sarma (✉)
Department of Mycology and Plant Pathology, Institute of Agricultural Sciences, Banaras Hindu University, Varanasi, India

C. Manoharachary et al. (eds.), Trichoderma: *Agricultural Applications and Beyond*, Soil Biology 61, https://doi.org/10.1007/978-3-030-54758-5_9

reduce the impact of plant diseases extensively. In spite of its high effectiveness, repeated use of synthetic fungicides has brought problems such as the development of resistance, environmental pollution, residual toxicity, objectionable effects on human health and other non-target microbes, i.e., beneficial endophytes (includes bacteria), AMF (Arbuscular Mycorrhizal Fungi), *Trichoderma*, cyanobacteria, etc. (Aktar et al. 2009; Hahn 2014; Rani et al. 2017; Staley et al. 2015). The global market share of synthetic fungicides, insecticides, herbicides, and others accounts for 26, 16, 48, and 10%, respectively (Yoon et al. 2013).

The application of biopesticides is one of the powerful tools for attaining sustainability in agriculture and most likely to be adopted against the synthetic chemical pesticides (Singh et al. 2016). The major advantage of biopesticides over synthetic pesticides is its special properties like target-specificity, environmental safety, easy biodegradability, and high reproducibility (Bisen et al. 2016; Keswani et al. 2013). As per the definition of the US Environmental Protection Agency (USEPA), biopesticides are products of natural materials such as plants, animals, fungi, bacteria, and minerals. Based on the type of active ingredients, the EPA categorized biopesticides mainly into three major categories namely, plant-incorporated protectants, biochemical, and microbial pesticides (USEPA 2008). However, the definition of biopesticide as mentioned by USEPA is not followed at the global level, rather another term, i.e., biological control agents (BCAs) is used instead of biopesticide, which is being promoted by the International Organization for Biological Control (IOBC 2008) and International Biocontrol Manufacturer's Association (IBMA).

9.2 Market Status of Biopesticides

Increasing demand for safe and organic food is playing a key role in enhancing the growth of the biopesticide market globally. It is expected that growth of the global biopesticides market will reach US$ 6.6 bn by 2020 and also expected to register a CAGR of 18.8% during the forecast period, 2015–2020 (http://www.marketsandmarkets.com/Market-Reports/biopesticides-267.html). Whereas, in 2016, the global fungicide (synthetic and biofungicide) market was comprised of USD 16 bn and during the forecast period, 2018–2023, it is expected to register a compound annual growth rate (CAGR) of 5.7%. South America is expected to emerge as the fastest-growing fungicides market during the forecast period, followed by Asia-Pacific; however, it is likely to be slow in North America (https://www.mordorintelligence.com/industry-reports/global-fungicides-market-industry). However, in India the growth of the biopesticide market is increasing both in terms of volume and value and predicted to show CAGR of 18.3% and 19% during the forecast period, 2015–2020 (http://www.businesswire.com/news/home/20160217005892/en/Indian-BiopesticidesMarket-Growth-Trends-Forecast).

9.3 Types of Biopesticides

Based on the active ingredients or the biocontrol agents used, biopesticides have been classified into four groups by IBMA: (1) microbials (41%), (2) macrobials (33%), (3) semiochemicals (insect behavior-modifying agents), and (4) natural products (Guillon, 2003). Microbial pesticides are well-known BCAs because of its higher selectivity toward the host and lower or zero phytotoxicity as compared to conventional synthetic chemical pesticides (MacGregor 2006). Typically the microorganisms are the key active ingredients of a microbial pesticide, which are pathogenic in nature for specific pests. Out of the total biopesticides available in the market for different crops, 74% are derived from bacteria, 10% from fungi, 5% virus, 8% predators, and other biopesticides 3% (Thakore 2006). In present practices, microbial biopesticides may be incorporated in the field in various forms, viz., live organisms, spores, or dead organisms. In comparison to other microbial pesticides, fungal biopesticides are more advantageous as it does need not to be consumed for its effectiveness. It often requires only optimum conditions including moist soils and cool temperatures to proliferate their generations (Gilligan, 2004).

9.4 *Trichoderma*: An Overview

Persoon (1794) was the first to propose *Trichoderma* as a genus 200 years ago in Germany and described it as a mealy powder microbe encapsulated by a hairy growth. Thakur and Norris (1928) were the first to isolate *Trichoderma* in India. In today's agriculture, *Trichoderma* spp. holds the share of >60% in the registered biofungicides worldwide (Singh et al. 2009). The agriculture fields are prevailed by fungal diseases which act as a major constraint due to severe yield losses to farmers (Khandelwal et al. 2012). To manage these diseases effectively, BCAs like *Trichoderma* spp. (teleomorph *Hypocrea*) are very cheap and eco-friendly alternatives to the ill effects caused by synthetic chemicals (Verma et al. 2007).

Trichoderma is an asexually reproducing soil-dwelling fungi, i.e., nearly about 10^1 to 10^3 propagules are found per gram soil in all temperate and tropical climatic regions which can be isolated and cultured aseptically (Waghunde et al. 2016b, b). It is the most widely studied microorganisms, presently marketed as active ingredients of biofertilizers, biopesticides, stimulants of natural resistance, and growth promoter. This is due to the property of *Trichoderma* spp. to protect plants by reducing the population of phytopathogens in soil and enhancing the growth of plants via improving nutrient availability, decomposition, and biodegradation (Alfano et al. 2007; Perazzolli et al. 2008; Korolev et al. 2008; Hermosa et al. 2012; Shoresh et al. 2005; Meher et al. 2018a, b). It has also been used as a preferred input for Integrated Disease Management (IDM) systems (Mukherjee et al. 1997; Jayaraj and Ramabadran 1999).

9.5 Mode of Action of *Trichoderma* Spp. against Wide Range of Phytopathogens

To antagonize phytopathogenic fungi, *Trichoderma* uses several mechanisms including antibiosis, colonization, competition, direct mycoparasitism, etc. (Fig. 9.1) (Howell 2003; Jash and Pan 2007; Rudresh et al. 2005; Swathi et al. 2015; Singh et al. 2018). Competition for essential elements such as carbon and iron is an effective mode used by several strains of *Trichoderma* to combat plant pathogens (Alabouvette et al. 2009; Sarrocco, et al. 2009). Small size ferric-iron specific chelators, i.e., Siderophores are produced by most BCAs under iron starving conditions to mobilize the available iron from surrounding areas. This highly efficient iron chelator is also produced by some *Trichoderma* isolates which inhibit growth and multiplication of several phytopathogenic fungi (Chet and Inbar 1994). *T. harzianum* is reported to adopt strategies such as competing for rhizosphere nutrients and root colonization in hosts for controlling *Fusarium oxysporum* (Tjamos et al. 1922). Further, to control the growth of *Pythium* spp. effectively, *Trichoderma* spp. competes for the available iron in the soil. Similarly, *Trichoderma* has the potential to conquer ATP from various types of sugars from the surrounding environment that enables it for proficient utilization of accessible nutrients: glucan, chitin and cellulose, and others, turning all into glucose (Chet et al. 1997).

Some volatile compounds (VCs) produced by certain isolates of *T. viride,* i.e., typically of coconut smell, show inhibitory activity against many pathogens. These metabolites include alamethicins, peptaibols, antibiotics, 6-penthyl-α-pyrone, harzianic acid, gliovirin, heptelidic acid, glisoprenins, massoilactone, etc. (Raaijmakers et al. 2009; Vey et al. 2001). Daguerre et al. (2014) recently summarized different pathways for the production of secondary metabolites such as polyketide biosynthesis pathway, pyrone biosynthesis pathway, flocculosin terpenoid/steroid biosynthesis pathway, peptaibol biosynthesis pathway, gliotoxin and gliovirin biosynthesis pathways.

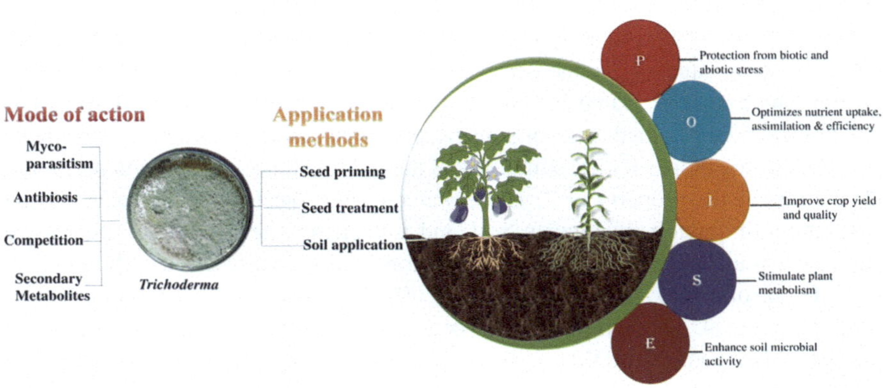

Fig. 9.1 Mode of action, application methods, and advantages of *Trichoderma*

Another important mechanism adopted by antagonistic *Trichoderma* is mycoparasitism. It includes sequential steps such as chemotropic growth of *Trichoderma* toward pathogen, recognition, extracellular enzyme secretion, hyphal penetration, and finally lysis of the host (Zeilinger et al. 1999). Growth of *Trichoderma* spp. toward fungal host is coordinated by its partial remote sensing activity due to the secretion of pathogenesis-related (PR) proteins, mostly chitinase and glucanase proteases (Harman et al. 2004). This process is affected by so many factors involving at least 20–30 proteins and other metabolites that are involved in the interaction. Gene-for-gene experiments are being used to study the involvement of chitinases and glucanases from *Trichoderma* spp. in the process of mycoparasitism (Daguerre et al. 2014).

In several different mono/dicots plants, activation of Induced System Resistance was observed in *Trichoderma* treated plants during the challenge by phytopathogenic fungi (*Rhizoctonia solani*, *Phytophthora* spp., *Magnaporthe grisea*, *Botrytis cinerea*, *Colletotrichum* spp., *Alternaria* spp., *Sclerotium* spp., etc.), bacteria (*Pseudomonas syringae*, *Xanthomonas* spp., etc.), and even some viruses like CMV. The first evidence of induced resistance in bean plants was observed with *T. harzianum* strain T-39, in which plants became resistant to fungal pathogens such as *Colletotrichum lindemuthianum* and *B. cinerea* in those plants where T-39 was incorporated in the soil, but not as any foliage treatment (Bigirimana et al. 1997). It was revealed by many researchers at the molecular level that the main reason behind resistance is due to upregulation in defense mechanisms which, in turn, enhanced production of secondary metabolites and enzymes, such as chalcone synthase (CHS), phenylalanine ammonia lyase (PAL), glucanase, chitinase and some proteins from cerato-platanin (CP) family and phytoalexins (HR response) synthesizing enzymes such as PKS/NRPS hybrid enzyme (Djonovic et al. 2006; Mukherjee et al. 2012; Seidi et al. 2006; Singh et al. 2017b, b; Yadav et al. 2019). These also comprise enzymes and pathogenesis-related proteins (PR) that participated in response to oxidative stress (Gajera et al. 2013). Production of commercial and industrially important enzymes such as amylases, cellulases, 1–3 beta glucanases, and chitinases produced from various *Trichoderma* spp. were extensively studied and their technology is being updated (Ahamed and Vermette 2009; Harman et al. 2004; Sandhya et al. 2004). Different kinds of enzymes are known to produce by different species of *Trichoderma*, which help in biocontrol activities such as antagonistic activity against plant pathogens, biotic and abiotic stress tolerance, hyphal growth, cell wall degradation, etc. Furthermore, genes responsible for enzyme production can be easily isolated, characterized, cloned, sequenced, and also their expression can be monitored by using advanced techniques of molecular biology such as RT-PCR (Sharma et al. 2014). Recently, *Trichoderma* spp. have been found useful in the production of silver nanoparticles as well (Maliszewska et al. 2009; Vahabi et al. 2011). For disease management under field conditions, the success of these bioagents has been greatly attributed by the development of strains with high rhizospheric competency, which can easily adapt to different agro-climatic zones of the country and establish in the root zones of different crops (Sharma et al. 1999).

9.6 *Trichoderma*: A Key Player of Biopesticide Industry

Trichoderma spp. based biofungicides are presently considered as potential type of BCAs. For efficient management of various soil, foliar and post-harvest plant pathogens such as *Sclerotium, Pythium, Ceratobasidium, Rhizoctonia, Fusarium, Macrophomina* and *Phytophthora* spp., *Trichoderma* spp. is being used worldwide as competent fungal biocontrol agents (Howell 2003; Meher et al. 2018a, b; Moni et al. 2019; Redda et al. 2018; Singh et al. 2016). Sixty percent of all fungal-based BCAs are contributed by *Trichoderma*-based products and an increasing number of *Trichoderma* spp. based BCAs products are registered regularly (Keswani et al. 2013). The inherent qualities of *Trichoderma*-based BCAs are constantly accumulating success (Singh et al. 2017b, b; Verma et al. 2007). Central Insecticide Board (CIB) is a regulatory body in India, under which 970 microbial-based biopesticides products from 34 microorganisms have been registered till date under sect. 9 (3B) and 9(3) of the Insecticides Act, 1968 Government of India (http://cibrc.nic.in/bpr.doc). A total of 354 *Trichoderma* spp. based products have been registered and available in the market for managing different kinds of soil and seed-borne diseases (Rajput et al. 2018).

T. harzianum is the most commonly used species of *Trichoderma* among the biological products that are being marketed worldwide. In Asia, this species is largely used as a BCA and particularly in India, 70% of the available products are covered by *T. harzianum*-based formulation (Lorito et al. 2010; Vinale et al. 2006; Woo et al. 2014). Biofungicides are also available in different combinations with other BCAs such as bacteria (*Pseudomonas fluorescens, Bacillus subtilis*, etc.), mycorrhizae (mainly *Glomus*), or other biological compounds. *Trichoderma* is also used with different combinations of BCAs of which *T. harzianum* (83%) is the most widely used species next to *T. koningii* (28%) and *T. viride* (55%).

In the developed nations, the liquid fermentation technique is more preferred for commercial production of *Trichoderma* spp. formulations, whereas, in developing countries, solid-state fermentation is used extensively due to its low initial cost. In India food grains including sorghum and bajra are used for commercial mass multiplication of *Trichoderma* spp. using solid-state fermentation technology at an industrial scale. Generally, *Trichoderma* spp. produces three types of spores, i.e., chlamydospores, hyphae, and conidia (Papavizas 1985). Among these three, conidia and chlamydospores have been preferred more as compared to hyphae, as hyphae cannot withstand drying and loses viability in a short span.

For agricultural application, the first step is to convince farmers with some substantial proof and the commercial formulization of BCA must possess several useful characters. These include easy preparation, high stability during storage and transportation, unfussy application, abundant viable propagules with high durability, sustained efficacy, adequate market potential, and reliable cost (Waghunde et al. 2016b, b). For preparation of commercial formulations of *Trichoderma*-based BCAs different carrier materials have been proved useful as it works as a food base. Among these, the most commonly used carrier material worldwide for commercial production of *Trichoderma* is talc (Jayaraj et al. 2006; Pandya et al. 2012). However, some natural and synthetic compounds were externally provided for improved crop protection or enhanced biological activity of formulations including chitosan, neem, and some amino acids.

9.7 Different Types of Formulations of *Trichoderma*

Generally, two kinds of formulations are available with biopesticides according to their physical state, i.e., dry or liquid formulations. Furthermore, dry formulations comprise six types of formulations such as dust (DP), wettable powder (WP), granules (GR), powders for seed dressing (DS), and water-dispersible granules (WG), whereas, suspension concentrates (SC), oil dispersions (OD), suspo-emulsions (SE), emulsions, capsule suspensions (CS), and ultralow volume formulations comes under liquid state formulation (Singh et al. 2014; Singh et al. 2016).

Worldwide, different formulations of *Trichoderma* spp. have been registered and also available in the market with different trade names (Table 9.1). Among the commercial formulations of *Trichoderma* spp., wettable powder (WP, 55.3%) is the most popular one, in which dried fungal conidia spores in the form of fine dust at a particular concentration have to be mixed with water for further use. Apart from this, other formulations commonly used are GR, liquid, and solids are 13.6, 10.3, and 6.2%, respectively, which constitutes organic substrates to support the growth and sporulation of *Trichoderma* culture, i.e., cereal grains such as rice, cocoa mat or peat moss or broken corn. Some other forms of BCA products include liquid suspensions, emulsions, pellets, dry flowable, powder, or talc, of which emulsions, granules, and suspensions can be used after addition of water into it to make a particularly suspension and can be applied as seed treatment, root drenching, dipping, hydroponics, irrigation; whereas dry flowables, pellets, and solid formulations are ready-to-use formulations which can be directly incorporated to the soil at the time of sowing or transplanting (Woo et al. 2014).

9.8 Conclusion

Globally, the production and utilization of biopesticides are increasing rapidly due to the growing interest of people in residue-free agricultural produce and organic farming. However, among the biopesticides, *Trichoderma*-based products contributed more than half of the total global biopesticide market, as it possesses many qualities which help to boost agricultural production such as enhancing nutrient-use efficiency in different crops, promoting plant growth, improving physiological response to both biotic and abiotic stresses, etc. Till date, 354 *Trichoderma*-based products have been registered in India under Central Insecticide Board (CIB) with various trade names and with different formulations, either in a solid or liquid state

Table 9.1 Different formulations of *Trichoderma* spp. with their trade name

Sl. No	Species name	Commercial product name	Formulation	Recommended against	References
1.	*T. harzianum* Rifai strain T-22	Rootshield	Wettable powder	Root pathogens like *Fusarium Pythium, Fusarium, Thielaviopsis, Rhizoctonia, Thielaviopsis,*	https://www.bioworksinc.com/products/shared/rootshield.pdf
2.	*T. harzianum* Rifai strain T-22	TRIANUM-P and TRIANUM-G	Water dispersible granule	Soil-borne disease *Pythium spp., Rhizoctonia sp., Fusarium sp.*, and *Sclerotinia*	https://www.koppert.com/trianum-p/
3.	*T. harzianum* Rifai strain T-39	Trichodex	Wettable powder	*Botrytis cinerea*	https://waset.org/publications/8370/biofungicide-trichodex-wp
4.	*T. harzianum (gamsii)* ATCC080	Bioten Remedier	Wettable powder	*Fusarium* spp., *Armillaria* sp., *Phytophthora* sp., *Pythium* sp., *Rosellinia* sp., *Sclerotinia* sp., *Rhizoctonia Verticillium* sp., *Sclerotium rolfsii, Thielaviopsis basicola*	https://www3.epa.gov/pesticides/chem_search/reg_actions/registration/decision_PC-119207_4-Mar-10.pdf
5.	*T. Harzianum*	Antagon WP	Wettable powder	Stem and collar rot, wilt and damping-off	https://www.indiamart.com/proddetail/antagon-trichoderma-4177313562.html

6.	*T. Harzianum*	Trichobiol	Wettable powder	*Pythium sp., Armelleria, Fusarium* sp., *Botrytis* sp.,	https://books.google.co.in/books?id=CciWDQAAQBAJ&pg=PA117&lpg=PA117&dq=Trichobiol+trichoderma&source=bl&ots=PifznzqcEC&sig=ACfU3U2pKjL1sYes0l0Hro4a84M0OW3T1w&hl=en&sa=X&ved=2ahUKEwiNg9DElrHiAhXn73MBHU3PBigQ6AEwB3oECAgQAQ#v=snippet&q=Trichobiol%20trichoderma&f=false
7.	*T. Harzianum*	Unite WP	Wettable powder	Damping-off and root rots	*Agrimm.Co.nz/wp/wp-content/uploads/Unite-15 kg.Pdf*
8.	*T. harzianum* DSM 14944	Foliguard	Suspension concentrate	*Cladosporium, Alternaria,Botrytis cinerea, Sphaeroteca pannosa, Oidium,*	Fraceto et al. 2018
9.	*T. harzianum* isolate DB 104	Romulus	Wettable powder	Nematicides	Abd-Elgawad and Askary 2018
10.	*T. harzianum* strain 21	Rootgard	Soluble powder	*Fusarium, Rhizoctonia, Pythium, and Sclerotium*	http://shambaza.com/listing/rootgard-sp.html
11.	*T. harzianum* strain kd	Eco-T	*Wettable powder*	Root pathogens like *Fusarium, Pythium, Rhizoctonia, Phytophthora*	http://plant-health.co.za/eco-t/
12.	*T. harzianum* strain SF	Bio-Tricho	Liquid or carrier-based	*Sclerotium, Pythium, Rhizoctonia Fusarium,* etc. causing root rot, damping-off, seedling rot, wilt, etc.	http://specialbiochem.com/page/bio-tricho-trichoderma-species-powder/
13.	*T. harzianum* strains ESALQ-1306, 1303	Trichodermil	Suspension concentrate	*Fusarium* sp., *Rhizoctonia solani*	Teixeira et al. 2012
14.	*T. Harzianum*	Supresivit	*Wettable powder*	Damping-off of ornamentals and forest species, fungal diseases of peas	Koch 1999

(continued)

Table 9.1 (continued)

Sl. No	Species name	Commercial product name	Formulation	Recommended against	References
15.	*T. harzianum* IIHR-Th-2	Ecosom-TH	*Wettable powder*	Fruit rot caused by *Botrytis* sp.	http://agrilife.in/biopesti_microrigin_ecosomth.htm
16.	*T. harzianum* strain B77	Eco-77	Wettable powder	*Botrytis* sp.	http://plant-health.co.za/plant-products/
17.	*T. harzianum* T-22	Tricho D	Wettable powder	Root pathogens	http://catalogo.procolombia.co/en/quimicos-y-farmaceuticos/otros-fertilizantes/tricho-d-wp
18.	*T. polysporum* IMI 206039	BINAB TF	Wettable powder	*B. cinerea*	https://horticulture.ahdb.org.uk/sites/default/files/research_papers/SF%20094_Report_Annual_2009.pdf
19.	*T. polysporum* Rifai ATTC 20475	Binab t	Wettable powder	Damping-off (*Pythium* and *Rhizoctonia* sp.)	https://www.eurekanetwork.org/project/id/1410
20.	*T. asperellum*	Ecohope	Dry wettable powder	Seed-born diseases of rice	https://www.researchgate.net/publication/250020692_Mode_of_action_of_Trichoderma_asperellum_SKT-1_a_biocontrolagent_against_Gibberella_fujikuroi
21.	*T. asperellum* ICC 012	Tenet	Wettable powder	Soil-borne diseases like *Phytophthora sp., Fusarium sp., Thielaviopsis basicola, Pythium sp., Rhizoctonia sp., Sclerotinia spp., S. rolfsii,, Verticillium sp., Armillaria sp.,* and *Rosellinia sp.*	http://www.blacksmithbiosciences.com/tenet-wp.html
22.	*T. asperellum*	Quality WG	Water dispersible granule	*Fusarium solani*	Teixeira et al. 2012

for managing various kinds of soil and seed-borne diseases. *T. harzianum* is the most widely used species of *Trichoderma* among biological products, being marketed worldwide and contributing about 70% of total *Trichoderma*-based products particularly in India. The major constraints in production and marketing of these products are unorganized stakeholders, initial cost and procedure of establishment and standardization of the industry, non-conformity to quality standards and likewise so on which limit the expansion of biopesticide industries but these constraints can be overcome with the help of massive research infrastructure, constructive public support and policies, creating awareness among farmers, and promoting the biopesticides through subsidies.

Acknowledgments We are grateful to DST-INSPIRE and UGC-RET scholarship for providing financial assistance.

References

Abd-Elgawad MM, Askary TH (2018) Fungal and bacterial nematicides in integrated nematode management strategies. Egyptian J Biol Pest Control 28(1):74

Ahamed A, Vermette P (2009) Effect of culture medium composition on *Trichoderma reesei's* morphology and cellulase production. Bioresour Technol 100:5979–5987

Aktar MW, Sengupta D, Chowdhury A (2009) Impact of pesticides use in agriculture: their benefits and hazards. Interdiscip Toxicol 2(1):1–12. https://doi.org/10.2478/v10102-009-0001-7

Alabouvette C, Olivain C, Migheli Q, Steinberg C (2009) Microbiological control of soil-borne phytopathogenic fungi with special emphasis on wilt-inducing *Fusarium oxysporum*. New Phytol 184:529–544

Alfano G, Ivey MLL, Cakir C et al (2007) Systemic modulation of gene expression in tomato by *Trichoderma hamatum* 382. Phytopathology 97(4):429–437

Bigirimana J, de Meyer G, Poppe J et al (1997) Induction of systemic resistance on bean (*Phaseolus vulgaris*) by *Trichoderma harzianum*. Med Fac Landbouww Univ Gent 62:1001–1007

Bisen K, Keswani C, Patel JS et al (2016) *Trichoderma* spp.: efficient inducers of systemic resistance in plants. In: Chaudhary DK, Verma A (eds) Microbial-mediated induced systemic resistance in plants. Springer, Singapore, pp 185–195

Chet I, Inbar J (1994) Biological control of fungal pathogens. Appl Biochem Biotechnol 48:37–43

Chet I, Inbar J, Hadar I (1997) Fungal antagonists and mycoparasites. In: Wicklow DT, Söderström B (eds) The Mycota IV: environmental and microbial relationships. Springer-Verlag, Berlin, pp 165–184

Daguerre Y, Siegel K, Edel-Hermann V et al (2014) Fungal proteins and genes associated with biocontrol mechanisms of soil-borne pathogens: a review. Fungal Biol Rev 28:97–125

Djonovic S, Pozo MJ, Dangott LJ et al (2006) Sm1, a proteinaceous elicitor secreted by the biocontrol fungus *Trichoderma virens* induces plant defense responses and systemic resistance. Mol Plant Microb Interact 19:838–853

Fraceto LF, Maruyama CR, Guilger M et al (2018) Trichoderma harzianum-based novel formulations: potential applications for management of next-gen agricultural challenges. J Chem Technol Biotechnol 93(8):2056–2063

Gajera H, Domadiya R, Patel S et al (2013) Molecular mechanism of *Trichoderma* as bio-control agents against phytopathogen system – a review. Curr Res Microbiol Biotechnol 1:133–142

Gilligan CA (2004) Modeling and analysis of disease induced host growth in the epidemiology of take all. Phytopathology 94:535–540

Guillon ML (2003) Regulation of biological control agents in Europe. In: Roettger U, Reinhold M (eds) International symposium on biopesticides for developing countries. CATIE, Turrialba, pp 143–147

Hahn M (2014) The rising threat of fungicide resistance in plant pathogenic fungi: botrytis as a case study. J Chem Biol 7(4):133–141. https://doi.org/10.1007/s12154-014-0113-1

Harman GE, Howell CR, Viterbo A (2004) *Trichoderma* species- opportunistic, avirulent plant symbionts. Nat Rev Microbiol 2:43–56

Hermosa R, Viterbo A, Chet I et al (2012) Plant-beneficial effects of *Trichoderma* and of its genes. Microbiology 158(1):17–25

Howell CR (2003) Mechanisms employed by *Trichoderma* species in the biological control of plant diseases: the history and evolution of current concepts. Plant Dis 87:4–10

IOBC (2008) International organization for biological control. IOBC Newsletter 84:5–7

Jash S, Pan S (2007) Variability in antagonistic activity and root colonizing behavior of *Trichoderma* isolates. J Trop Agric 95(2):29–35

Jayaraj J, Nv R, Velazhahan R (2006) Development of formulations of Trichoderma harzianum strain M1 for control of damping-off of tomato caused by *Pythium aphanidermatum*. Phytopathology and Plant Protection 39(1):1–8

Jayaraj J, Ramabadran R (1999) Rhizobium-Trichoderma interaction in vitro and in vivo. Indian Phytopath 52(2):190–192

Keswani C, Singh SP, Singh HB (2013) A superstar in biocontrol enterprise: *Trichoderma* spp. Biotech Today 3:27–30

Khandelwal M, Datta S, Mehta J et al (2012) Isolation, characterization and biomass production of *Trichoderma viride* using various agro products- a biocontrol agent. Adv Appl Sci Res 3:3950–3955

Koch E (1999) Evaluation of commercial products for microbial control of soil-borne plant diseases. Crop Prot 18(2):119–125

Korolev N, David DR, Elad Y (2008) The role of phytohormones in basal resistance and *Trichoderma*-induced systemic resistance to *Botrytis cinerea* in *Arabidopsis thaliana*. BioControl 53(4):667–683

Lorito M, Woo SL, Harman GE et al (2010) Translational research on *Trichoderma*: from 'Omics to the field'. Annu Rev Phytopathol 48(1):395–417

MacGregor JT (2006) Genetic toxicity assessment ofmicrobial pesticides: needs and recommended approaches. Intern Assoc environ mutagen Soc 1–17

Maliszewska I, Aniszkiewicz L, Sadowski Z (2009) Biological synthesis of gold nanostructures using the extract of *Trichoderma koningii*. Acta Physica Pol A 116:163–165

Meher J, Kashyap P, Sonkar SS et al (2018a) Studies on native isolates of fungal and bacterial bio-agents against collar rot of chickpea. Int J Curr Microbiol Appl Sci 7(1):226–238

Meher J, Sonkar SS, Singh SN (2018b) Growth promotion of chickpea plant on treatment with native isolates of *Trichoderma* spp. J Pharmacognosy and Phytochemistry 7(4):1631–1636

Moni RM, Rajput RS, Singh J et al (2019) Disease of aromatic grasses and their management. In: Pandey R, Mishra AK, Singh HB et al (eds) Diseases of medicinal and aromatic plants and their management. Today and Tomorrow Printers and Publisher, New Delhi, India, pp 47–65

Mukherjee PK, Buensanteai N, Moran-Diez ME et al (2012) Functional analysis of non-ribosomal peptide synthetases (NRPSs) in *Trichoderma virens* reveals a polyketide synthase (PKS)/NRPS hybrid enzyme involved in induced systemic resistance response in maize. Microbiology 158:155–165

Mukherjee PK, Haware MP, Raghu K (1997) Induction and evaluation of benomyl-tolerant mutants of *Trichoderma viride* for biological control of botrytis grey mould of chickpea. Indian Phytopath 50(4):485–489

Pandya JR, Sabalpara AN, Chawda SK et al (2012) Grain substrate evaluation for mass cultivation of *Trichoderma harzianum*. J Pure Appl Microbiol 6:2029–2032

Papavizas GC (1985) *Trichoderma* and *Gliocladium*: biology, ecology, and potential for biocontrol. Annu Rev Phytopathol 23(1):23–54

Perazzolli M, Dagostin S, Ferrari A et al (2008) Induction of systemic resistance against Plasmopara viticola in grapevine by *Trichoderma harzianum* T39 and benzothiadiazole. Biol Control 47 (2):228–234

Persoon CH (1794) Neuer Veersuch einer systematischen Eintheilung der Schwamme (Disposition methodica fungorum). Romer's news magazine 84:63–128

Raaijmakers JM, Paulitz TC, Steinberg C et al (2009) The rhizosphere: a playground and battle-field for soil-borne pathogens and beneficial microorganisms. Plant Soil 21:341–361

Rajput RS, Singh J, Moharana DP et al. (2018) Indian Biopesticide Industry: An Analytical Study of Current Scenario In: Singh J, Nigam R, Hasan W (eds.) Advances in Biodiversity Conservation for Sustainable Development. Parmar Publishers and Distributors, Jharkhand, India, pp 105–109

Rani A, Singh R, Kumar P et al (2017) Pros and cons of fungicides: an overview. Int J Eng Sci Res Tech 6(1)

Redda ET, Ma J, Mei J et al (2018) Antagonistic potential of different isolates of *Trichoderma* against *Fusarium oxysporum, Rhizoctonia solani* and *Botrytis cinerea.* Eur. Exp Biol 8(2):12

Rudresh DL, Shivaprakash MK, Prasad RD (2005) Potential of *Trichoderma* spp. as bio-control agents of pathogens involved in wilt complex of chickpea (*Cicer arietinum* L.). J Biol Control 19(2):157–166

Sandhya C, Adapa LKK, Nampoothri M et al (2004) Extracellular chitinase production by *Trichoderma harzianum* in submerged fermentation. J Basic Microbiol 44:49–58

Sarrocco S, Guidi L, Fambrini S et al (2009) Competition for cellulose exploitation between *Rhizoctonia solani* and two *Trichoderma* isolates in the decomposition of wheat straw. J Plant Pathol 91:331–338

Seidi V, Marchetti M, Schandl R et al (2006) EPL1, the major secreted protein of *Hypocrea atroviridis* on glucose, is a member of a strongly conserved protein family comprising plant defense response elicitors. FEBS J 273:4346–4359

Sharma DD, Gupta VP, Chandrashekhar DS (1999) Compatibility of certain biocontrol agents with chemical pesticides and fertilizers. Indian J Sericulture 38:79–82

Sharma P, Sharma M, Raja M et al (2014) Status of *Trichoderma* research in India: a review. Indian Phytopathol 14(67):1–19

Shoresh M, Yedidia I, Chet I (2005) Involvement of jasmonic acid/ethylene signaling pathway in the systemic resistance induced in cucumber by *Trichoderma asperellum* T203. Phytopathology 95(1):76–84

Singh V, Maharshi A, Singh DP et al (2018) Role of microbial seed priming and microbial Phytohormone in modulating growth promotion and defense responses in plants. In: Rakshit A, Singh HB (eds) Advances in seed priming. Springer, Singapore, pp 115–126

Singh J, Rajput RS, Bisen K et al (2017a) Role of *Trichoderma* secondary metabolites in plant growth promotion and biological control. In: Singh HB, Sarma BK, Keswani C (eds) Advances in PGPR research. CABI, New York, USA, pp 411–426

Singh V, Ray S, Bisen K et al (2017b) Unravelling the dual applications of *Trichoderma* spp. as biopesticide and biofertilizer. In: Singh HB, Sarma BK, Keswani C (eds) Advances in PGPR research. CABI, UK, pp 364–374

Singh A, Shahid M, Srivastava M et al (2014) Optimal physical parameters for growth of *Trichoderma* species at varying pH, temperature and agitation. Virol Mycol 3:1–7

Singh HB, Singh BN, Singh SP et al (2009) Biological control of plant diseases: current status and future prospects. In: Johri JK (ed) Recent advances in biopesticides: biotechnological applications. New India Pub, New Delhi, p 322

Singh V, Upadhyay RS, Sarma BK et al (2016) *Trichoderma asperellum* spore dose depended modulation of plant growth in vegetable crops. Microbiol Res 193:74–86

Staley ZR, Harwood VJ, Rohr JR (2015) A synthesis of the effects of pesticides on microbial persistence in aquatic ecosystems. Crit Rev Toxicol 45(10):813–836. https://doi.org/10.3109/10408444.2015.1065471

Swathi B, Patibanda AK, Prasuna RP (2015) Antagonistic efficacy of *Trichoderma* spp. on *Sclerotium Rolfsii in vitro*. IOSR Journal of Agriculture and Veterinary Science 8(7):19–22
Teixeira H, Júnior P, Vieira RF et al (2012) Trichoderma spp. decrease Fusarium root rot in common bean. Summa Phytopathol 38(4):334–336
Thakore Y (2006) The biopesticide market for global agricultural use. Ind Biotechnol 2:192–208
Thakur AK, Norris RV (1928) A biochemical study of some soil fungi with special reference to ammonia production. Journal of Indian Institute of Science 18:141–160
Tjamos EC, Papavizas GC, Cook RJ (1922) In: biological control of plant diseases. Progress and challenges for the future. Plenum press, New York: 222
USEPA (2008) What are biopesticides? http://www.epa.gov/pesticides/biopesticides/whatarebiopesticides.htm
Vahabi K, Mansoori GA, Karimi S (2011) Biosynthesis of silver nanoparticles by fungus *Trichoderma reesei*: a route for large scale production of AgNPs. Insciences J 1:65–79
Verma M, Brar SK, Tyagi RD et al (2007) Antagonistic fungi, Trichoderma spp.: panoply of biological control. Biochem Eng J 37:1–20
Vey A, Hoagland RE, Butt TM (2001) Toxic metabolites of fungal biocontrol agents. In: Butt TM, Jackson C, Magan N (eds) *Fungi as* biocontrol agents: Progress, problems and potential. CABI, Bristol, UK, pp 311–346
Vinale F, Marra R, Scala F et al (2006) Major secondary metabolites produced by two commercial *Trichoderma* strains active against different phytopathogens. Lett Appl Microbiol 43 (2):143–148
Waghunde R, Shelake R, Ambalal SN (2016b) Trichoderma: a significant fungus for agriculture and environment. Afr J Agric Res 11:1952–1960
Waghunde RR, Shelake RM, Sabalpara AN (2016a) Trichoderma: a significant fungus for agriculture and environment. Afr J Agric Res 11(22):1952–1965. https://doi.org/10.5897/AJAR2015.10584
Woo SL, Ruocco M, Vinale F et al (2014) *Trichoderma*-based products and their widespread use in agriculture. The Open Mycology Journal 8(1):71–126
Yadav RN, Rashid MM, Zaidi NW et al (2019) Secondary metabolites of *Metarhizium* spp. and *Verticillium* spp. and their agricultural applications. In: Secondary metabolites of plant growth promoting Rhizomicroorganisms. Springer, Singapore pp, pp 27–58
Yoon MY, Cha B, Kim JC (2013) Recent trends in studies on botanical fungicides in agriculture. Plant Pathol J 29(1):1–9
Zeilinger S, Galhaup C, Payer K et al (1999) Chitinase gene expression during mycoparasitic interaction of *Trichoderma harzianum* with its host. Fungal Genet Biol 26:131–140

Chapter 10
Modulation of Microbiome Through Seed Bio-priming

Deepranjan Sarkar, Arghya Chattopadhyay, Sonam Singh, O. Shiva Devika, Subhadip Pal, Manoj Parihar, Sumita Pal, Harikesh Bahadur Singh, and Amitava Rakshit

Abstract In natural ecosystems, plants harbour diverse microbial communities in different compartments (above- and below-ground) of their system. The microbes colonizing the plant parts form complex interactions leading to the formation of microbiomes in inner tissues (endosphere) and outer surfaces (ectosphere) of the host plant. As microbiome represents a key factor in ecological functions, e.g. nutrient cycling, plant growth and productivity, and development of stress or protection against them, the topic is gaining substantial interest among researchers. A general attempt is triggered in the process provoking thought of bringing some modifications in the microenvironments. Small interventions in agroecosystems are of prime importance since their implementation in field levels becomes easy. Bio-priming is one such technology emerging as rhizosphere engineering, which is capable of tackling several challenges in agriculture arising right from seed germination to field stand. Promising results have been obtained with this technique because it has various mechanisms to stimulate the physiological and metabolic processes in the plant system and other environmental processes associated with the host niche. In this chapter, we also aim to briefly discuss the basic interactions that take place between plants and microorganisms with particular attention to plant growth and health and soil health.

D. Sarkar · A. Chattopadhyay · S. Singh · O. S. Devika · S. Pal · A. Rakshit (✉)
Department of Soil Science and Agricultural Chemistry, Institute of Agricultural Sciences, Banaras Hindu University, Varanasi, Uttar Pradesh, India
e-mail: amitavar@bhu.ac.in

M. Parihar
Crop production division, ICAR-VPKAS, Almora, Uttarakhand, India

S. Pal
Department of Mycology and Plant Pathology, Institute of Agricultural Sciences, Banaras Hindu University, Varanasi, India

H. B. Singh
Department of Biotechnology, GLA University, Mathura, Uttar Pradesh, India

C. Manoharachary et al. (eds.), Trichoderma: *Agricultural Applications and Beyond*, Soil Biology 61, https://doi.org/10.1007/978-3-030-54758-5_10

Keywords Host–microbe interactions · Microbiome · Plant colonization · Seed bio-priming · Plant growth promotion

10.1 Introduction

A wide diversity of microorganisms is nurtured on Earth which makes possible the existence of life on this planet. However, with the advancement of technology, we have acquired little information about their habitat in the ecosystem. Faulty agricultural practices damage ecological functions along with depletion of diverse flora and fauna. Soon the importance of microbes and their positive applications in agro-ecosystems was realized. They have been marked as green solutions to maintain environmental sustainability as against the costly chemical inputs required in agriculture to meet the increasing food demand. In order to understand the potentials of microbial communities in harnessing sustainable and productive ecosystems, microbiome research got momentum and started providing us new insights about plant-associated microbes and their behaviour (Sergaki et al. 2018).

The structural and functional diversity of the microbes were studied so as to identify the beneficial and pathogenic microbes. Initially, they were studied using culture-based methods but nowadays, molecular-based approaches are common (Turner et al. 2013). This enabled us to explore the microbial interactions which are not only confined to outer plant surfaces (epiphytic) but also linked with the inner tissues (endophytic) of the host. The symbiotic relationships of plant and microbes are used in nutrient management, stress resistance, plant growth promotion, seed production, bioremediation, etc. (Kaul et al. 2017; Velmourougane et al. 2017).

In conventional agricultural practices, the application of synthetic chemicals leads to pollution of the environment, heavy tillage deteriorates soil structure, creating conditions unfavourable for microbiome development (Arora et al. 2018). Whereas, sustainable agricultural management with the help of bio-priming interventions favour microbiome-driven agriculture system (Fig. 10.1).

In the present chapter, we try to briefly discuss the basic microbiomes that develop in the agro-ecosystems and how we can shape those microenvironments with small techniques like seed bio-priming.

10.2 Plant–Microbe Interactions

In general, plant–microbe interactions may be viewed as beneficial, pathogenic, or neutral in nature. Good agricultural practices promote favourable environmental conditions to form positive interactions between the host (plants) and microbes (Pagano et al. 2017). This might be due to increased organic matter content, better structural stability, enhanced microbial biomass, and diversity of soil (Wang et al. 2017; Sergaki et al. 2018). However, the microbes colonizing the host is also a prime

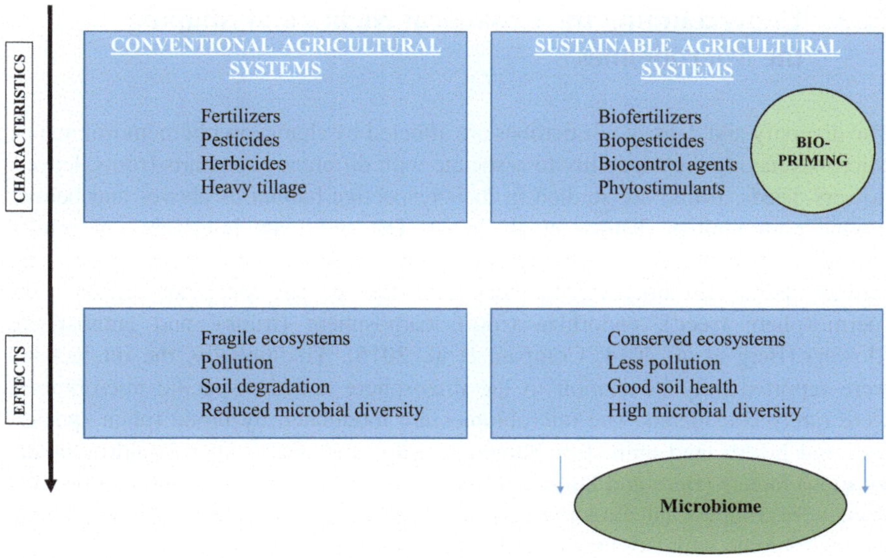

Fig. 10.1 Comparison of conventional agricultural systems with bio-priming interventions as sustainable practice in harnessing better microbiome

factor to determine the type of interaction (Akram et al. 2017). These studies started in the nineteenth century. Earlier, the symbiotic relationship was focused mainly on arbuscular mycorrhizae and nitrogen fixer organisms. With growing interests, other plant-associated non-pathogenic microbes were explored to find their novel traits useful to overcome new challenges in agriculture.

The complex web of interactions carried by beneficial microbes are mostly recognized of bacterial or fungal origin, belonging to genera *Rhizobium*, *Azospirillum*, *Azotobacter*, *Bacillus*, *Pseudomonas*, *Trichoderma*, etc. (Verma et al. 2017). A number of cross-talk occurs between host plants and microbial communities during the colonization process such as the release of signaling molecules, organic acids, phytohormones, secondary metabolites, cellulolytic enzymes, and quorum sensing (Lareen et al. 2016; Kandel et al. 2017). The multipartite interactions are not only important in terms of growth promotion, nutrition, and stress alleviation of plants but also required for maintaining soil quality, i.e., physical, chemical, and biological properties of soil. The negative interactions between host and microbes, also known as pathogenesis, are detrimental for plant health as they involve the proliferation of pathogenic microbes and the development of diseases in plants (Schirawski and Perlin 2018).

10.3 Understanding the Ecological Niches and Shaping the Microbiomes

The diversity and density of microbes are affected by change in microenvironments. Microbes having the capability to associate with different plant parts (roots, leaves, flowers, seeds, fruits) are studied in their respective habitat as above- and below-ground communities (Kumar et al. 2017). The soil–plant interface is generally known as rhizosphere, while the air–plant interface is termed as phyllosphere, and further microenvironments have been classified as per the plant organs, viz., spermosphere (seed), endorhiza (root), carposphere (fruits), and anthosphere (flower) (Berg et al. 2014; Compant et al. 2016). Traditionally, the interactions were reported only in relation to the rhizosphere and the specific microbiomes were rarely considered. The microbiomes are modulated by biotic (plant species, age, and health, and animal or human activity) and abiotic (soil health, climate, location) factors (Berg and Smalla 2009). Therefore, to reap the maximum benefits from these complex and attractive ecological niches, we have to shape or manipulate the microbiome with sustainable agricultural practices.

Farmers often face poor productivity in crops due to less availability of high-quality seeds. In order to overcome such problems, numerous seed enhancement technologies such as hydropriming, osmopriming, solid matrix priming, bio-priming, nutrient priming, etc. are adopted by them (Chatterjee et al. 2018). Bio-priming being a simple, eco-friendly, and low-cost technique, it is gaining popularity among the farming community as a pre-sowing technique (Sarkar et al. 2018b). The efficacy of beneficial microbes on crops is dependent upon their application methods. Bio-priming as a biological seed treatment process incorporates seed hydration before microbial coating and incubation, all in a controlled manner (Sarkar et al. 2017). The seed activates its metabolism during imbibition and offers several advantages after sowing like uniform and enhanced germination, good crop stand, stress tolerance, etc. (Mahmood and Kataoka 2018; Sarkar et al. 2018a). With the preloading of microbes in seed, the technique is able to regulate plant–microbe interactions because the colonized plants release organic compounds in the rhizosphere which enhance the microbial activity (Singh et al. 2018). Beneficial microbes as better competitors among diverse microbial groups further play a key role in nutrient cycling and improving plant health by producing phytohormones, developing systemic resistance, and inhibiting the growth of pathogens (Compant et al. 2010; Sarkar et al. 2017). Thus, bio-priming is rated as the most evolved process in achieving higher resource use efficiency and better resource conservation by promoting mutualistic interactions between plant and microbial flora and minimizing the use of external inputs (Rakshit et al. 2015). The mechanisms of bio-priming responsible for shaping different microbiomes are presented in Fig. 10.2.

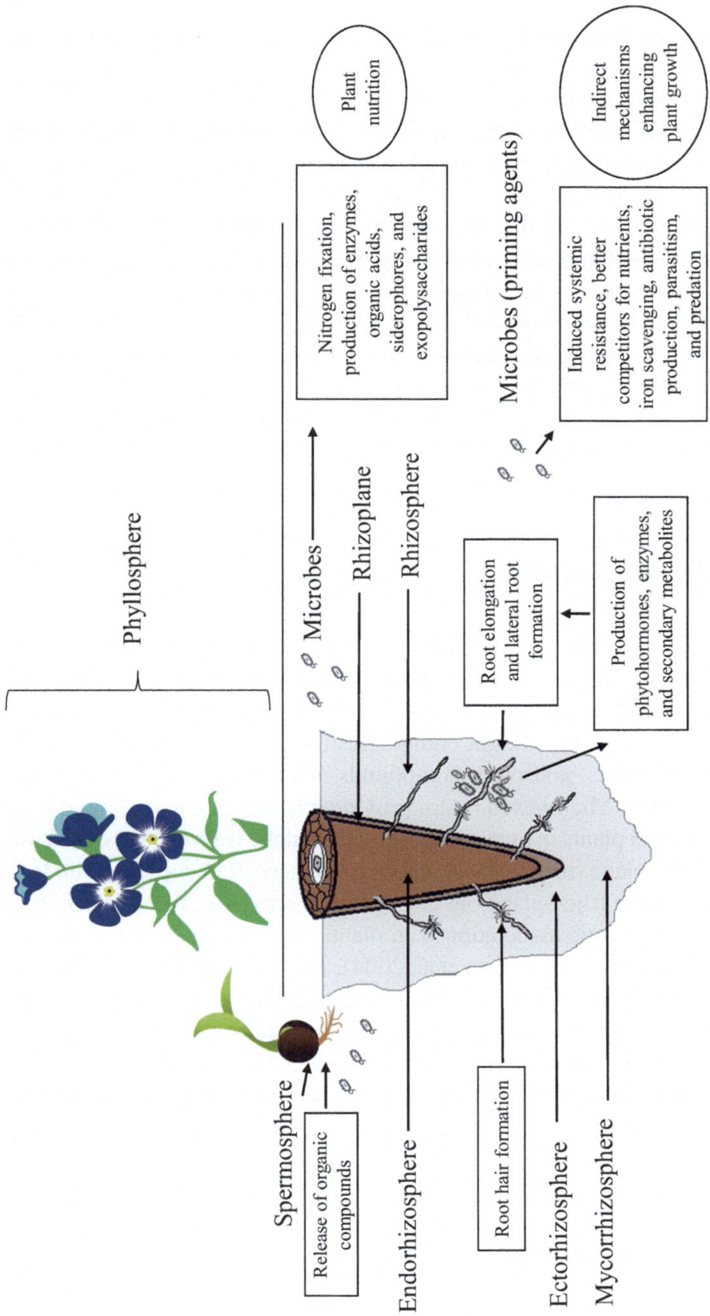

Fig. 10.2 The possible mechanisms of bio-priming involved in shaping better microbiomes

10.3.1 The Seed Microbiome

Seed is a basic input to continue the crop cycles in agricultural systems. It is home to diverse microbial groups and also an important means to disperse them in the environment, facilitating early colonization of plant species (Nelson 2018). However, the knowledge of the seed microbiome (spermosphere) is still in its inception stage. Microbes inhabiting seeds may be epiphytic or endophytic in nature (Nelson et al. 2018). They flourish during seed germination and move to soil, seedlings, and other parts of the plants during the growth season (Barret et al. 2015; Nelson et al. 2018). The composition of seed microbiomes is enlightening new prospects for research in plant–microbe interactions. Recently, Rybakova et al. (2017) concluded that studying the structure of such microenvironment can reveal plants capacity to develop colonization resistance against the pathogenic microbes, and thus, provide us new biomarkers in breeding programmes (Rybakova et al. 2017). A clear picture of seed-borne and soil-borne microorganisms and their interactions is also necessary to enhance agricultural productivity.

10.3.2 The Rhizosphere

The narrow zone of soil surrounded by the root system is known as the rhizosphere. It is considered as the largest ecosystem due to its high energy flux. The system is a centre of microbial colonization because of the availability of plant root exudates comprising low molecular weight compounds (amino acids, organic acids, sugars), secondary metabolites, and other compounds or rhizodeposits (Bertin et al. 2003; Akram et al. 2017). Microbes in return provide mineralization of organic matter and supply nutrients to plants in available forms. This niche is further classified into three zones, viz., rhizoplane (root surface), endorhizosphere (root cortex and endodermis), and ectorhizosphere (rhizoplane out into the bulk soil) (Fig. 10.2). When the fungus is involved in symbiotic association with plants, the term mycorrhizosphere is used in place of rhizosphere (Johansson et al. 2004). The characteristics of rhizosphere are different from the bulk soil and it is marked responsible for plant fitness (Barea et al. 2002). Soil properties such as pH, structure, texture organic matter, and nutrient status determine the selection of microbes by plants by forming favourable environments for root exudation and growth of particular microbes (Igiehon and Babalola 2018).

10.3.3 The Endosphere

Microbes living in the internal and peripheral tissues of host for at least some part of their life cycle without causing any apparent disease symptoms or harm on the plant are defined as endophytes (Sudheep et al. 2017; Varma et al. 2017). They colonize different plant parts and form an internal environment known as endosphere. A diverse group of endophytic organisms are reported in the literature. Some examples of bacterial genera are *Rhizobium, Bacillus, Burkholderia, Klebsiella, Pseudomonas, Rahnella, Pantoea, Gluconobacter, Azoarcus*, and *Herbaspirillum* (Kandel et al. 2017) and fungal genera include *Trichoderma, Penicillium, Alternaria, Fusarium, Aspergillus, Cladosporium, Curvularia*, etc. (Varma et al. 2017).

10.3.4 The Phyllosphere

The above-ground portion or aerial surface of the plant colonized by microbes is known as phyllosphere. This is a very dynamic microbiome as it faces huge fluctuations in radiation, moisture, and temperature during day and night time (Turner et al. 2013; Remus-Emsermann and Schlechter 2018). Leaf surfaces also harbour a dense microbial assemblages and the colonization pattern is affected by stomata, lea veins, and hairs. Such microenvironments are better studied with the help of high-throughput and fluorescence technologies.

10.4 Future Perspectives

In order to exploit the microbiome to the fullest, traditional approaches are inadequate to provide deeper insights. While those methods can be integrated with new approaches, viz., molecular (next-generation sequencing), bioinformatics, and modelling tools to have a better understanding of the plant–microbe interactions. Scientists are searching for extreme habitats and new beneficial microbes which can enrich our current knowledge on microbial associations. This will further increase the efficacy of the present microbial products and help in the synthesis of new customized products. Biotechnological tools are very crucial in unravelling new metabolites from the uncultured microorganisms and discovering the trans-kingdom cross-talks between plant and microbial species. Seed bio-priming as a microbial application technique should be taken in long-term studies to investigate the microbial community dynamics in soil and crop environment. The microbial formulation (concentration) and duration of bio-priming must be standardized before advancing towards the field levels.

10.5 Conclusions

The life cycle of plants and microorganisms are intricately and intrinsically connected with each other. Therefore, the study of this fascinating topic (e.g. microbiome) is very crucial to solve numerous challenges in agriculture, especially related to plant nutrition, protection, and production, and ecosystem restoration. Each microbiome has specific features and interconnected with each other; hence understanding them individually is very crucial to develop sound ecological strategies for reshaping their structure. Seed bio-priming has huge potential in engineering or modulating the microbiomes. Accelerated pollution and climate change direct the urgent use of green technology for food production to save our environment from its destruction. Bio-priming acting as a good delivery system for biological seed treatment is reliable not only in yielding quality seeds but also continuously depended as a protection tool to cope with several biotic and abiotic stresses in agroecosystems.

References

Akram MS, Shahid M, Tahir M, Mehmood F, Ijaz M (2017) Plant-microbe interactions: current perspectives of mechanisms behind symbiotic and pathogenic associations. In: Plant-microbe interactions in agro-ecological perspectives, Fundamental mechanisms, methods and functions, vol 1. Springer, Singapore, pp 97–126

Arora NK, Fatima T, Mishra I, Verma M, Mishra J, Mishra V (2018) Environmental sustainability: challenges and viable solutions. Environ Sustainability 1:309–340. https://doi.org/10.1007/s42398-018-00038-w

Barea JM, Azcón R, Azcón-Aguilar C (2002) Mycorrhizosphere interactions to improve plant fitness and soil quality. Antonie Van Leeuwenhoek 81(1–4):343–351. https://doi.org/10.1023/A:1020588701325

Barret M, Briand M, Bonneau S, Préveaux A, Valière S, Bouchez O, Hunault G, Simoneau P, Jacques MA (2015) Emergence shapes the structure of the seed microbiota. Appl Environ Microbiol 81(4):1257–1266

Berg G, Grube M, Schloter M, Smalla K (2014) Unravelling the plant microbiome: looking back and future perspectives. Front Microbiol 5:148. https://doi.org/10.3389/fmicb.2014.00148

Berg G, Smalla K (2009) Plant species and soil type cooperatively shape the structure and function of microbial communities in the rhizosphere. FEMS Microbiol Ecol 68:1–13. https://doi.org/10.1111/j.1574-6941.2009.00654.x

Bertin C, Yang X, Weston LA (2003) The role of root exudates and allelochemicals in the rhizosphere. Plant Soil 256(1):67–83. https://doi.org/10.1023/A:102629050

Chatterjee N, Sarkar D, Sankar A, Pal S, Singh HB, Singh RK, Bohra JS, Rakshit A (2018) On-farm seed priming interventions in agronomic crops. Acta agriculturae Slovenica 111(3):715–735. https://doi.org/10.14720/aas.2018.111.3.19

Compant S, Clément C, Sessitsch A (2010) Plant growth-promoting bacteria in the rhizo-and endosphere of plants: their role, colonization, mechanisms involved and prospects for utilization. Soil Biol Biochem 42(5):669–678. https://doi.org/10.1016/j.soilbio.2009.11.024

Compant S, Saikkonen K, Mitter B, Campisano A, Mercado-Blanco J (2016) Editorial special issue: soil, plants and endophytes. Plant Soil 405(1–2):1–11. https://doi.org/10.1007/s11104-016-2927-9

Igiehon N, Babalola O (2018) Rhizosphere microbiome modulators: contributions of nitrogen fixing bacteria towards sustainable agriculture. Int J Environ Res Public Health 15(4):574. https://doi.org/10.3390/ijerph15040574

Johansson JF, Paul LR, Finlay RD (2004) Microbial interactions in the mycorrhizosphere and their significance for sustainable agriculture. FEMS Microbiol Ecol 48(1):1–13. https://doi.org/10.1016/j.femsec.2003.11.012

Kandel SL, Joubert PM, Doty SL (2017) Bacterial endophyte colonization and distribution within plants. Microorganisms 5:77. https://doi.org/10.3390/microorganisms5040077

Kaul S, Choudhary M, Sharma T, Dhar MK (2017) Harnessing the plant microbiome: a key towards sustainable agriculture. In: Singh DP, Singh HB, Prabha R (eds) Plant-microbe interactions in agro-ecological perspectives, Microbial interactions and agro-ecological impacts, vol 2. Springer, Singapore, pp 307–322

Kumar J, Singh D, Ghosh P, Kumar A (2017) Endophytic and epiphytic modes of microbial interactions and benefits. In: Singh DP, Singh HB, Prabha R (eds) Plant-microbe interactions in agro-ecological perspectives, Fundamental mechanisms, methods and functions, vol 1. Springer, Singapore, pp 227–253

Lareen A, Burton F, Schäfer P (2016) Plant root-microbe communication in shaping root microbiomes. Plant Mol Biol 90(6):575–587. https://doi.org/10.1007/s11103-015-0417-8

Mahmood A, Kataoka R (2018) Potential of biopriming in enhancing crop productivity and stress tolerance. In: Rakshit A, Singh HB (eds) Advances in seed priming. Springer, Singapore, pp 127–145

Nelson EB (2018) The seed microbiome: origins, interactions, and impacts. Plant Soil 422 (1–2):7–34. https://doi.org/10.1007/s11104-017-3289-7

Nelson EB, Simoneau P, Barret M, Mitter B, Compant S (2018) Editorial special issue: the soil, the seed, the microbes and the plant. Plant Soil 422(1–2):1–5. https://doi.org/10.1007/s11104-018-3576-y

Pagano MC, Correa EJA, Duarte NF, Yelikbayev B, O'Donovan A, Gupta VK (2017) Advances in eco-efficient agriculture: the plant-soil mycobiome. Agriculture 7:14. https://doi.org/10.3390/agriculture7020014

Rakshit A, Sunita K, Pal S, Singh A, Singh HB (2015) Bio-priming mediated nutrient use efficiency of crop species. In: Rakshit A, Singh HB, Sen A (eds) Nutrient use efficiency: from basics to advances. Springer, Singapore, pp 181–191

Remus-Emsermann MNP, Schlechter RO (2018) Phyllosphere microbiology: at the interface between microbial individuals and the plant host. New Phytol 218(4):1327–1333. https://doi.org/10.1111/nph.15054

Rybakova D, Mancinelli R, Wikström M, Birch-Jensen AS, Postma J, Ehlers RU, Goertz S, Berg G (2017) The structure of the Brassica napus seed microbiome is cultivar-dependent and affects the interactions of symbionts and pathogens. Microbiome 5:104. https://doi.org/10.1186/s40168-017-0310-6

Sarkar D, Pal S, Mehjabeen M, Singh V, Singh S, Pul S, Garg J, Rakshit A, Singh HB (2018a) Addressing stresses in agriculture through bio-priming intervention. In: Rakshit A, Singh HB (eds) Advances in seed priming. Springer, Singapore, pp 107–113

Sarkar D, Pal S, Singh HB, Yadav RS, Rakshit A (2017) Harnessing bio-priming for integrated resource management under changing climate. In: Singh HB, Sarma BK, Keswani C (eds) Advances in PGPR research. CAB International, UK, pp 349–363

Sarkar D, Ray S, Singh NK, Rakshit A, Singh HB (2018b) Seed priming with bio-inoculants triggers nutritional enrichment in vegetables: a review. International Journal of Agriculture, Environment and Biotechnology, Special Issue, 727–735

Schirawski J, Perlin M (2018) Plant–microbe interaction 2017—the good, the bad and the diverse. Int J Mol Sci 19:1374. https://doi.org/10.3390/ijms19051374

Sergaki C, Lagunas B, Lidbury I, Gifford ML, Schäfer P (2018) Challenges and approaches in microbiome research: from fundamental to applied. Front Plant Sci 9:1205. https://doi.org/10.3389/fpls.2018.01205

Singh V, Maharshi A, Singh DP, Upadhyay RS, Sarma BK, Singh HB (2018) Role of microbial seed priming and microbial phytohormone in modulating growth promotion and defense responses in plants. In: Rakshit A, Singh HB (eds) Advances in seed priming. Springer, Singapore, pp 115–126

Sudheep NM, Marwal A, Lakra N, Anwar K, Mahmood S (2017) Fascinating fungal endophytes role and possible beneficial applications: an overview. In: Singh DP, Singh HB, Prabha R (eds) *Plant-Microbe Interactions in Agro-Ecological Perspectives*, volume *1*: fundamental mechanisms, methods and functions. Springer, Singapore, pp 255–273

Turner TR, James EK, Poole PS (2013) The plant microbiome. Genome Biol 14:209. https://doi.org/10.1186/gb-2013-14-6-209

Varma PK, Uppala S, Pavuluri K, Chandra KJ, Chapala MM, Kumar KVK (2017) Endophytes: role and functions in crop health. In: Singh DP, Singh HB, Prabha R (eds) Plant-microbe interactions in agro-ecological perspectives, Fundamental mechanisms, methods and functions, vol 1. Springer, Singapore, pp 291–310

Velmourougane K, Saxena G, Prasanna R (2017) Plant-microbe interactions in the rhizosphere: mechanisms and their ecological benefits. In: Singh DP, Singh HB, Prabha R (eds) Plant-microbe interactions in agro-ecological perspectives, Microbial interactions and agro-ecological impacts, vol 2. Springer, Singapore, pp 193–219

Verma P, Yadav AN, Kumar V, Singh DP, Saxena AK (2017) Beneficial plant-microbes interactions: biodiversity of microbes from diverse extreme environments and its impact for crop improvement. In: Singh DP, Singh HB, Prabha R (eds) Plant-microbe interactions in agro-ecological perspectives, Microbial interactions and agro-ecological impacts, vol 2. Springer, Singapore, pp 543–580

Wang Y, Li C, Tu C, Hoyt GD, DeForest JL, Hu S (2017) Long-term no-tillage and organic input management enhanced the diversity and stability of soil microbial community. Sci Total Environ 609:341–347. https://doi.org/10.1016/j.scitotenv.2017.07.053

Chapter 11
Opportunistic Avirulent Plant Symbionts *Trichoderma*: Exploring Its Potential Against Soilborne Phytopathogens

Narasimhamurthy Konappa, Soumya Krishnamurthy, Nirmaladevi Dhamodaran, Udayashankar C. Arakere, Srinivas Chowdappa, and Niranjana Siddapura Ramachandrappa

Abstract A major threat to agriculture is soilborne diseases which extensively decline the crop yield. Control of soilborne phytopathogens is challenging because these pathogens persist for numerous years as sclerotia in soil or in organic matter under numerous environmental conditions. Pathogen management with the application of chemical pesticides imposes environmental threats and is potentially dangerous to humans and other living forms. The employment of biological control agents for disease reduction and improved yield provides an alternative for the chemical pesticides and this is a key aspect of disease management of plant pathogens. Various control agents such as fungi and bacteria are involved in biocontrol activity among control agents, the fungal genus *Trichoderma* shows a major role in the control of phytopathogens. *Trichoderma* spp. are extensively applied as biocontrol agents for the management of soilborne phytopathogens in agriculture. The control effects of *Trichoderma* on soilborne pathogens are higher in comparison to synthetic fertilizers and they exhibit prolonged persistence in soil post application. The mechanisms of biocontrol exerted by *Trichoderma* are generally antibiosis, mycoparasitism, and competition for nutrients, induced defense responses, or systemic resistance responses in the plants. *Trichoderma* spp. are well known for the secretion of cell wall degrading enzymes (CWDEs) and these enzymes play key roles in the degradation of the cell wall of the pathogens and the biocontrol

N. Konappa (✉) · U. C. Arakere · N. S. Ramachandrappa
Department of Studies in Biotechnology, University of Mysore, Mysore, Karnataka, India

S. Krishnamurthy
Department of Microbiology, Field Marshal K. M. Cariappa College, A Constituent College of Mangalore University, Madikeri, Karnataka, India

N. Dhamodaran
Department of Microbiology, Ramaiah College of Arts, Science and Commerce, Bengaluru, Karnataka, India

S. Chowdappa
Department of Microbiology and Biotechnology, Jnanabharathi Campus, Bangalore University, Bangalore, Karnataka, India

C. Manoharachary et al. (eds.), Trichoderma: *Agricultural Applications and Beyond*, Soil Biology 61, https://doi.org/10.1007/978-3-030-54758-5_11

mechanism. Several genes and their products govern the biocontrol activity and are called biocontrol genes which are crucial for *Trichoderma* as a potent biological control agent.

Keywords *Trichoderma* · Soilborne pathogens · Secondary metabolites · Cell wall degrading enzymes · Biocontrol mechanisms · Biocontrol genes

11.1 Introduction

Since the beginning of agriculture, plant diseases have been anxious with mankind and found to show a serious part in the damage of natural means, contributing 20–30% losses in crop production worldwide (O'Rourke et al. 2009). Agronomy has remained facing the damaging actions of several pests and pathogens (biotic stress) from an initial time, biotic stresses indications not only to the decrease of the crop but also the esthetic value and food supply and agronomic production (Singh et al. 2014; Kashyap et al. 2017; Sharma et al. 2017). Soilborne disease pathogens are widely found in soil. Soilborne pathogens include bacteria, nematodes, fungi, viruses, and parasites that cause important crop losses. As a group, they can cause a wide array of diseases of the plant kingdom, containing fruits and ornamental plants, shrubs, vegetables, and trees. They invade the host through belowground structures but may also spread the aboveground parts of the host. Distinctive soilborne phytopathogens have a wide host variety and keep on for extensive periods in soil by strong latent structures. These types of pathogen infections are evident with different symptoms, containing wilt, lesions, rots, and finally lead to plant death. The soil environment including both natural and agricultural is the main factor, but the influence of decisive weather conditions decides the occurrence of an infection and its severity.

Globally, farmers exploit synthetic chemical pesticides to manage the phytopathogens in order to keep the quality and compensation of agronomic products (Junaid et al. 2013). Current agricultural practices to protect crops against the detrimental losses affected by plant diseases (Leadbeater 2015) follow increased usage of chemical pesticides leading to the accumulation of poisonous compounds possibly harmful to humans and environmental pollution (Haggag et al. 2015). Furthermore, the extensive usage of chemical pesticides has been noted to cause the appearance of microbial pathogens resistant to chemical pesticides (Naher et al. 2014) and distorted the biological equilibrium in soil by killing the beneficial microorganisms. The use of such chemical pesticides entails a convincing cost to the developing nations. Presently, the world has attention to find sustainable safe and eco-friendly alternative methods to control the plant diseases (Abdel-Ghany et al. 2017; Elshahawy et al. 2018; Abdel-Ghany and Alawlaqi 2018; Carmona-Hernandez et al. 2019). In this current situation, there has been an increasing attention in the application of innovative technique based on biological control agents (BCAs) or their genes and/or compounds (metabolites) to decrease or inhibit the harmful effects of phytopathogens (Viscardi et al. 2016; Iqbal and Ashraf 2017; Colla et al. 2017). Biological control encompasses the application of certain living organisms to defeat the development of soilborne plant pathogens.

Currently, several BCAs have been documented and are accessible as bacterial biocontrol agents such as *Pseudomonas* spp., *Agrobacterinum* spp., *Bacillus* spp., and as fungal BCAs, for example, *Aspergillus* spp., *Trichoderma* spp., *Candida* spp., *Gliocladium* spp., *Ampelomyces* spp., *Actinomycetes* spp., and *Coniothyrium* spp. (Naher et al. 2014). Among the fungal BCAs, *Trichoderma* spp. have gained much interest due to their great reproductive ability, persistence under harsh environments, productive producers of secondary metabolites, and capacity to resist phytopathogens (Contreras-Cornejo et al. 2016; Devi et al. 2017). *Trichoderma* spp. are significant filamentous fungal biocontrol agents having biocontrol abilities against economically significant soilborne pathogens in plants as active constituents in biopesticides and biofertilizers (Srivastava et al. 2014; Devi et al. 2017; Oladipo et al. 2018). They are free-living, filamentous, ascomycetes fungi usually widely distributed and ubiquitous in almost all types of soils (Etschmann et al. 2015; Moran-Diez et al. 2015) and grow saprophytically on various substrates such as decaying wood materials, bark or leaf, interact with plants and taking a progressive influence in hosts (Manganiello et al. 2018; Macías-Rodríguez et al. 2018). *Trichoderma* spp. grows quickly in soil application, meanwhile *Trichoderma* spp. are obviously resistant to various lethal composites (DDT, dieldrin, endosulfan, pentachloronitro benzene, and pentachlorophenol), herbicides, fungicides, and pesticides (Seethapathy et al. 2017), bio-remediation agents for heavy metal and xenobiotic contamination (Zhang et al. 2018).

The *Trichoderma* spp. used as biocontrol agents that successfully control soilborne phytopathogens have been well documented and about 90% of *Trichoderma* spp. (*T. koningii*, *T. viride*, *T. harzianum*, *T. hamatum*, and *T. virens*) (Saravanakumar et al. 2015). Recently, four novel *Trichoderma* spp. including *T. henanense*, *T. asterineum*, *T. odoratum*, and *T. pseudobritdaniae* were revealed and defined (Qin and Zhuang 2016). The *Trichoderma* has an extensive history and it was initially described and reported in 1794 and further recommended to ensure a connection with the sexual form of a *Hypocrea* spp. The significance of *Trichoderma* as a biocontrol agent was first defined by Weindling in 1932 (Pandya et al. 2011). Later, many new *Trichoderma* spp. were discovered by 2013 and the *Trichoderma* comprises above 200 phylogenetically identified based on rpb2 gene sequences (Atanasova et al. 2013). *Trichoderma* spp. is the most effective bio fungicide in present agronomy as above 60% of the recorded biofungicides worldwide arrived from *Trichoderma* formulations (Verma et al. 2007). In India only, more than 250 products *from Trichoderma* are accessible applications in fields but the percentage of bio fungicides' share is a small part of the fungicides found in the international market and dominated by synthetic chemicals (Woo et al. 2014).

The antimicrobial capacity of *Trichoderma* spp. has been broadly studied against different soilborne phytopathogens and projected antimicrobial mechanisms, both indirectly and directly, have been studied (Abbas et al. 2017; Zaidi and Singh 2018; Iqbal and Ashraf 2019). The indirect mechanism includes competition for space and

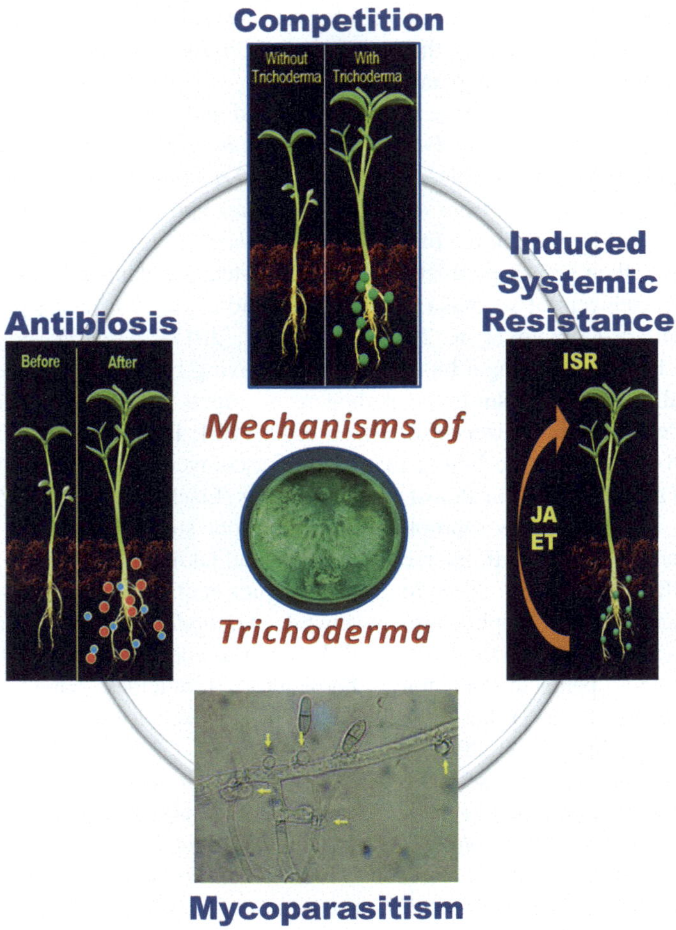

Fig. 11.1 Different mechanisms by *Trichoderma* species against soilborne pathogens

nutrients, tolerance to abiotic stress, antibiosis, and disease resistance toward pathogens (Shukla et al. 2015; Pandey et al. 2016; de Medeiros et al. 2017; Mendoza-Mendoza et al. 2018). Direct mechanisms include the facilitation or increase of nutrients uptake such as phosphate solubilization, iron sequestration, mycoparasitism, synthesis of secondary metabolites and lytic enzymes, synthesis of phytohormones and non-volatile or volatile complexes (Bisen et al. 2016; Garnica-Vergara et al. 2016; Guzmán-Guzmán et al. 2019) (Fig. 11.1). *Trichoderma*-based formulations have been useful as a soil application, seedling dip, wound dressing, seed biopriming, and foliar spray because of their distinctive plant defensive capacities (Kumar et al. 2014; Sharma et al. 2015; Oros and Naár 2017).

Colonization of *Trichoderma* spp. in roots of the plant can decrease diseases and abiotic stresses like drought and salinity, stimulate biomass gaining, plant growth, greater germination of seeds, augmented height of plant, growth of root, shoot dry mass and leaves number, improved crop yield and enhanced plant vigor (Waghunde et al. 2016; Chagas et al. 2017; Martìnez-Medina et al. 2017; Jogaiah et al. 2018). Several *Trichoderma* spp. have been found to have a mutualistic endophytic association with numerous plant species and have the ability to act as biological agents in the management of plant pathogens (Bae et al. 2011). Plant colonization by *Trichoderma* involves its entry into the outer coats of cells of the tissue of the root or it inhabits within the intracellular places and develops among the plasma membrane and the host cell wall (Nogueira-Lopez et al. 2018). *Trichoderma* spp. can prevent the colonization and development of phytopathogens in rhizospheric soil (Vinodkumar et al. 2017). *Trichoderma* species are distinguished by their competence to elicit induced systemic resistance (ISR) based on jasmonic acid and ethylene-dependent signaling pathways (Hermosa et al. 2012) and also some *Trichoderma* spp. elicit host resistances arbitrated by salicylic acid (SA) (Contreras-Cornejo et al. 2016).

Various reports have showed the potential activity of *Trichoderma* spp. as biocontrol agents antagonistic to an enormous number of soilborne phytopathogens, for example, *P. aphanidermatium*, *F. oxysporum*, *F. culmorum*, *G. graminis*, *S. rolfsii*, *S. sclerotiorum*, *Colletotrichum* spp., *P. cactorum*, *Monilochaetes infuscans*, *Alternaria alternate*, *Rhizoctonia solani*, *Thielaviopsis paradoxa*, *Botrytis cinerea*, *Alternaria* species, *Ralstonia solanacearum*, *Xanthomonas* spp., *Agrobacterium* spp., *Pseudomonas* spp., *Meloidogyne* spp., *Clavibacter* spp., and *Erwinia* spp., (Abbas et al. 2017; Koka et al. 2017; Majeed et al. 2017) (Table 11.1). This chapter designates the different mechanisms of *Trichoderma* spp. as biocontrol agents to different soilborne phytopathogens from agricultural crops.

11.2 Mechanisms of *Trichoderma* in the Control of Soilborne Pathogens

11.2.1 Mycoparasitism

The direct interaction among pathogens and *Trichoderma* is termed as mycoparasitism. Mycoparasitism or hyperparasitism (the ability to attack other fungi and utilize their nutrients) is the foremost antagonistic mechanism involved in *Trichoderma* as a biocontrol agent (Fig. 11.1). Mycoparasitism is a complex manner that involves a direct attack of one species fungus on another species (Fand et al. 2013). This type of hyphal interaction and parasitism of *Trichoderma* spp. with numerous soilborne phytopathogenic fungi has been recognized (Druzhinina et al. 2011). This mechanism is mediated by the physical intrusion of the mycoparasite into the pathogenic hyphae through the growth of atypical structures such as haustoria and production of several cell wall degrading enzymes (CWDEs) and secondary metabolites main to the breaks the pathogen followed by the uptake of

Table 11.1 *Trichoderma* spp. against soilborne plant pathogens

Trichoderma spp.	Pathogen/disease	Plant/host	References
T. hamatum	*B. cinerea*	*A. thaliana*	Mathys et al. (2012)
T. asperellum	*Pseudomonas syringae*	*A. thaliana*	Yoshioka et al. (2011)
T. harzianum	*B. cinerea, P. viticola*	*A. thaliana* tomato	Perazzoli et al. (2012)
T. asperellum T 34	*P. syringae*	Arabidopsis	Segarra et al. (2009)
T. harzianum *T. viride*	*Ralstonia solanacearum*	Banana	Ceballos et al. (2014)
T. viride *T. harzianum*	*Macrophomina phaseolina, Alternaria alternate*	Black gram	Dubey and Patel (2012)
T. harzianum *T. viride*	*Fusarium solani* *F. oxysporium*	Brinjal	Balaji and Ahir (2011)
T. harzianum	*R. solanacearum*	Brinjal	Barua and Bora (2008)
T. harzianum *T. longibrachiaum* *T. atroviride* *T. longibrachiaum*	*F. solani* *F. oxysporum*	Brinjal and tomato	Enespa and Dwivedi (2014)
T. harzianum *T. viride*	*S. sclerotiorum* *Rhizoctonia solani*	Cabbage	Sharma et al. (2003)
T. harzianum *T. viride*	*Alterneria alternate*	Capsicum	Kapoor (2008)
T. harzianum *T. viride*	*Pythium vexans* *R. solani*	*Cardamomum elettaria*	
T. viride *T. harzianum*	*Rhictonia solani,* *P. aphanidermatum*	Cauliflower	Ahuja et al. (2012)
T. harzianum *T. viride*	*R. solani*	Cauliflower	Sharma et al. (2003)
T. harzianum *T. viride* *T. ressei*	*F. oxysporium* *R. solani* *S. rolfsii* *M. phaseolina* *B. cinerea*	Chickpea	Pandey et al. (2003)
T. harzianum *T. viride*	*F. oxysporum, R. solani,* *Chaetomium* spp., *S. rolfsii*	Chickpea	Poddar et al. (2004)
T. hazianum	*F. oxysporum*	Chickpea	Verma et al. (2014)
T. harzianum *T. viride*	*F. oxysporium* *Pythium* spp. *R. solani*	Chilli	Kapoor (2008)
T. viride *T. harzianum*	*Sclerotium rolfsii* *F. oxysporum* *Pythium* spp., *R. solani*	Chilli	Vasanthakumari and Shivanna (2013)
T. viride, *T. harzianum* *T. pseudokoningii*	*S. rolfsii, F. oxysporum,* *Pythium* spp., *R. solani*	Chilli	Kapoor (2008)

(continued)

Table 11.1 (continued)

Trichoderma spp.	Pathogen/disease	Plant/host	References
T. harzianum *T. atroviride*	*S. delphinii*	Cotton	Mukherjee et al. (2013)
T. harzianum	*X. campestris* pv. *malvacearum*	Cotton	Raghavendra et al. (2013)
T. harzianum	*R. solani*	Cowpea	Pan and Das (2011)
T. harzianum	*P. aphanidermatum*	Cucumber	Devi and Shivprakash (2013)
T. harzianum	*Phytophthora melonis*	Cucumber	Sabbagh et al. (2017)
T. harzianum	*F. oxysporum*, *Botrytis cinerea*	Cucumber, *A. thaliana*	Alizadeh et al. (2013)
T. harzianum	*Botrytis cinerea and Podosphaera xanthii*	Cucumber, bean, tomato, strawberry	Levy et al. (2015)
T. harzianum *T. viride*	*P. aphanidermatum* *S .sclerotiorum*	Fenugreek	Sharma and Trivedi (2010)
T. harzianum	*P. aphanidermatum*	Ginger	Gupta et al. (2010)
T. harzianum, *T. viride*, *T. koningii* *T. longibrachiatum*	*R. solanacearum*	Ginger	Roop and Jagtap (2017)
T. harzianum *T. viride*	*Fusarium moniliforme*	Grapes	Senthil et al. (2011)
T. hazianum	*F. solani*, *F. oxysporum* *R. solani*, *Pythium* spp. *Sclerotium rolfsii*	Green bean	El-Mohamedy and Alla (2013)
T .longibrachiatum *T. harzianum* *T. viride*	*R. solani*	Groundnut	Sreedevi et al. (2012)
T. harzianum *T. viride*	*F. oxysporum* spp. *Phytophthora*	Guava	Misra (2007)
T. harzianum *T. viride*	*F. oxysporum* *F. proliferatum* *F. proliferatum*	In vitro	Ghanbarzadeh et al. (2014)
T. harzianum	*Sclerotinia sclerotiarum*	In vitro	Alkooranee et al. (2017)
T. virens	*R. solani*, *P. ultimum*	In vitro	Abbas et al. (2017)
T. harzianum *T. viride*	*Alterneria alternate*	Jute	Bhandari and Vishunavat (2013)
T. harzianum *T. viride*	*R. solani* *Alterneria alternate*	Maize	Bhandari and Vishunavat (2013)
T. harzianum	*F. verticillioides*	Maize	Ferrigo et al. (2014)
T. harzianum	*F. oxysporium*, *R. solani*, *H. tetramera*	Maize	Bhandari and Vishunavat (2013)
T. viride	*R. stolonifer*, *C. kimurae*, *F. oxysporum*	Mushroom	Rawal et al. (2013)
T. viride	*Botrytis ricini*	Neem	Rawal et al. (2013)
T. harzianum	*Sclerotinia sclerotiarum*	Oilseed rape	Alkooranee et al. (2017)

(continued)

Table 11.1 (continued)

Trichoderma spp.	Pathogen/disease	Plant/host	References
T. viride *T. reesei* *T. harzianum*	*A. alternate*, *S. vesicarium* *Cladosporium alliicepae*	Onion	Mishra and Gupta (2012)
T. viride, *T. reesei*, *T. harzianum*	*C. alliicepae*, *A. alternata*, *A. tenuissima*, *C. circinans*	Onion	Mishra and Gupta (2012)
T. viridae	*F. oxysporum* *Alternaria alternate*	Pigeon pea, moong bean	Rao et al. (2015)
T. viride *T. harzianum*	*R. solani*, *Streptomyces scabies*, *Ralstonia solanacearum*, *P. infestans*	Potato	Pandey and Pundhir (2013)
T. harzianum *T. viride*	*Erwinia carotovora*	Potato	Sandipan et al. (2015)
T. harzianum	*R. solanacearum*	Potato, tobacco tomato	Maketon et al. (2008)
T. viride *T. harzianum*	*R. solani*, *Fusarium* spp.	Rice	Chakravarthy et al. (2011)
Trichoderma spp	*Rhizoctonia solani*	Rice	Biswas and Datta (2013)
Trichoderma spp.	*Rhizoctonia solani*	Rice	Chakravarthy and Nagamani (2007)
T. viride *T. harzianum*	*B. theobromae* *R. solani* *F. solani* *F. oxysporum*	Saffron	Suhanna et al. (2013)
T. viride *T. harzianum*	*Pythium notatum*, *Pythium chrysogenum*, *F. moniliforme*, *F. oxysporium*	Sesame	Jeyalakshmi et al. (2013)
T. viride	*F. oxysporum*	Soybean	John et al. (2010)
T. citrinoviride	*S. sclerotiorum*	Soybean	Thakkar and Saraf (2014)
T. viride *T. harzianum*	*R. solani* *M. phaseolina* *Curvularia lunata* *P. arrhenomanes* *S. rolfsii*	Soybean	John et al. (2010), Anitha (2011), Jat and Agalave (2013)
T. hazianum	*R. solani*	Sugar beet	Kakvan et al. (2013)
T. viride *T. longibrachiatum* *T. reesei*, *T. koningii* *T. harzianum*	*S. rolfsii*	Sugarbeet	Paramasivan et al. (2013)
T. harzianum *T. viride*	*S. rostrata* *F. moniliformae* *S. scitamineum*	Sugarcane	Mahalingam et al. (2011)

(continued)

Table 11.1 (continued)

Trichoderma spp.	Pathogen/disease	Plant/host	References
T. harzianum *T. viride*	*F. moniliforme* *F. oxysporium*	Sunflower and safflower	Jat and Agalve (2013)
T. harzianum	*A. alternate*	Tobacco	Gveroska and Ziberoski (2012)
T. harzianum *T. viride*	*P. aphanidermatum* *R. solani* *F. oxysporiu*	Tobacco	Sumana and Devaki (2012)
T. harzianum,	*R. solanacearum*	Tobacco	Yuan et al. (2016)
T. harzianum *T. longibrachiatum* *T. virens* *T. viride*	*F. oxysporum,* *P. aphanidermatum,* *R. solani, S. rolfsii*	Tomato	Jayaraj et al. (2006)
T. harzianum *T. asperellum*	*X. campestris*	Tomato	Saksirirat et al. (2009)
Trichoderma spp.	*F. oxysporum,* *Sclerotinia* spp., *Pythium* spp. *R. solani*	Tomato	Komy et al. (2015) and Marzano et al. (2013)
T. harzianum *T. virens*	*Rhizoctonia solani*	Tomato	Kumar (2013)
T. harzianum *T. viride*	*Fusarium solani*	Tomato	Haggag and El-Gamal (2012)
T. hazianum	*F. oxysporum*	Tomato	Alwathnani and Perveen (2012)
T. harzianum	*F. oxysporum*	Tomato	Marzano et al. (2013)
T. harzianum	*F. oxysporum*	Tomato	Sriram et al. (2010)
T. viride	*R. solani, F. oxysporum* *F. verticilloid,* *A. alternate* *Mucorracemosus*	Tomato	Hafez et al. (2013)
T. arundinaceum	*B. cinerea; R. solani*	Tomato	Malmierca et al. (2012)
T. virens, *T. atroviride*	*A. solani, B. cinerea,* and *P. syringae*	Tomato	Salas-Marina et al. (2015)
T. harzianum *T. viride* *T. longibrachiatum*	*R. solani* *F. oxysporum* *A. solani*	Tomato	Enespa and Dwivedi (2013) and Jayaraj et al. (2006)
T. harzianum	*Ralstonia Solanacearum*	Tomato	Liza and Bora et al. (2009)
T. asperellum	*R. solanacearum*	Tomato	Narasimha Murthy and Srinivas (2013) and Narasimha Murthy et al. (2013, 2018)
T. asperelloides	*Pseudomonas syringae*	Tomato, cucumber	Brotman et al. (2012)

nutrient from hosts (Daguerre et al. 2014). In this process, *Trichoderma* spp. sense the pathogen and come into contact with host accompanied by morphological modifications like coiling and appressorium formation which develop holes on the surface of host or target pathogen (Omann and Zeilinger 2010). Subsequently, the mycoparasitic fungal hyphae discharge antibiotics that infuse the affected hyphae and prevent the cell wall resynthesis (Toghueoa et al. 2016). At the place of appressoria, creation of holes in the pathogen indicates the direct entry of *Trichoderma* hyphae into the host lumen and killing of the host pathogens (Kubicek and Druzhinina 2013).

In *Trichoderma*, the ATP binding cassette (ABC) transporter proteins are involved in mycoparasitism and uptake of nutrients (Sarma et al. 2014). The major CWDEs of *Trichoderma* involved in mycoparasitism comprise β-1, 3-glucanases and chitinolytic enzymes. Mycoparasites produce CWDEs which allow them to degrade the cell wall polysaccharides into small oligomers and in this way facilitate the hyperparasite to enter into the pathogenic fungal cytoplasm (Kubicek et al. 2011). Numerous chitinolytic enzymes have been described in *Trichoderma* spp. containing 1, 4-β-N-acetyl glucosaminidases, exochitinases, and endochitinases; these are induced through development in medium with chitin (Vos et al. 2015). The levels of enzymes such as N-acetylglucosaminidase, β-1, 3-glucanase, chitinase, endoglucanase, protease, amylase, cellulase, and glucosidase were shown to be improved in the occurrence of substrates. Overall, around 20–30 genes, proteins, and metabolites find direct implication in mycoparasitism. The role of numerous chitinases and glucanases in the progression of parasitism have been well deliberated from *Trichoderma* spp. by gene-for-gene investigations will help in a better understanding of this multipart progression (Daguerre et al. 2014).

Reports on the mycoparasitic capacity of *Trichoderma* spp. are well documented against phytopathogenic fungi such as *Pythium* spp., *Fusarium* spp., *A. alternate*, *B. cinerea*, *R. solani*, *Phytophthora* spp., and *S. sclerotiorum* (Bae et al. 2016) (Table 11.1). More than 1100 strains of *Trichoderma* have been found to be mycoparasite from molecularly defined 75 species (Druzhinina et al. 2011). Weindling (1934) first identified *T. lignorum* (*T. viride*) parasitizing *R. solani* hyphae and also recommended that application to soil with spores of *Trichoderma* to manage damping-off of citrus plant (Lo 1997). The growth of mycelium in *S. sclerotiorum* was inhibited by metabolites produced by *T. viride*. The chitinolytic capability of *T. harzianum* is connected with varied chitinase genes including *chi33*, *ech42*, *chi18-13nag1*, which governed the secretion of an array of enzymes, helping the mycoparasitic activity to different phytopathogens (Seidl et al. 2005).

T. atroviride strain P1 and *T. virens* strain 41 secrete the chitobiosidase, -acetyl-b-D- glucosaminidase and endochitinase which have a spore germination inhibitory activity and elongation of hyphae of numerous fungal pathogens such as, *B. cinerea*, *U. necator*, *U. avenae*, *Fusarium* spp., *Alternaria* spp. and almost all fungi having chitin in the cell wall (Schirmböck et al. 1994). The second most plentifully existing polymer of the fungal cell wall is β-1, 3-glucan (Latgé 2007) which is hydrolyzed by β-1,3-glucanases enzyme. β-1, 6-glucanase homolog *Bgn16.3* in *T. harzianum* CECT 2413 changed it to a good biocontrol agent with improved biocontrol effectiveness to *P. citrophthora*, *B. cinerea*, and *R. solani*. The β-1, 6-glucanases overexpressing strains of *T. harzianum* and *T. virens* also exhibited improved biocontrol against *R. solani*, *B. cinerea*, and *P. ultimum* (Ihrmark et al. 2010). Cellulolytic enzymes belonging to different classes are also produced by *Trichoderma* spp. that performs synergistic activity in the degradation of lignocellulose and cell walls of the phytopathogens such as *Phytophthora* spp. and *Pythium* spp. (Gajera et al. 2013).

Trichoderma spp. was able to deform and produce septation in *F. oxysporum* conidia, as well as to lyse and destroy conidiophores and spores. These effects may be related to enzymes secreted such as chitinase, α-glucanase, cellulase, proteases, and others (Harman 2006). Similar processes were reported against *Pythium oligandrum* Dreschler, *Rhizoctonia solani* Kühn, *F. oxysporum*, *Phytophthora megasperma Dreschler*, and *Pythium ultimum* Trow (Benhamou et al. 1999). A copy of several genes involved in cell wall deprivation synergistically was also described for *T. atroviride* in contact with *B. cinerea* and *P. capsici* (Reithner et al. 2011).

Recently relative study of the genome, secretome, and transcriptome of the three species, *T. virens*, *T. atroviride*, and *T. reesei* designated mycoparasitism as the familial existence of *Trichoderma* (Atanasova et al. 2013). In *T. atroviride*, *Tga3*, and *Tga1* two G-protein α-subunits from the cAMP signaling pathway, regulate coiling (Rocha-Ramírez et al. 2002). In addition, *Tga1* regulates the production of lytic enzymes and biosynthesis of antifungal metabolites that impact mycoparasitism while *Tga3* regulates secretion of CWDEs but not their biosynthesis (Zeilinger et al. 2016). The *T. harzianum*-T20 was extremely effective in displaying the activity of mycoparasitic to *C. falcatum*. The endophytic nature of *Trichoderma* spp. accomplished the *C. falcatum* mycelial development inhibition and decreasing red rot in sugarcane (Elamathi et al. 2017). Volatile secondary metabolites too have been confirmed to be critical in mycoparasitism by *Trichoderma* spp. (Stoppacher et al. 2010). *T. harzianum*'s useful effects on plants have been credited generally to the ability to inhibit pathogens through a blend of different mechanisms (Vos et al. 2015).

Extracellular enzymes containing chitinases, proteases β-1, 3 glucanase, and pectinases play important role in mycoparasitic capacity to antagonize the phytopathogen (Mukherjee et al. 2013). Mycoparasitic mode of action has been well known for inhibiting the growth of some various phytopathogens including *R. solani*, *S. sclerotiorum*, *S. rolfsii*, *F. oxysporium* (Harman et al. 2000) along with *C. capsici* (Saxena et al. 2015). Potential biocontrol properties of various native *Trichoderma* spp. such as *T. longibrachiatum*, *T. koningiopsis*, *T. harzianum*, *T. aureoviride*, and *T. asperellum* were recognized in the agricultural soils. Their biological control potential to three main soilborne phytopathogens, for example, *C. capsici*, *S. rolfsii*, and *S. sclerotiorum* was established through the dual culture plate method. All isolates showed effective mycoparasitic capacity in relation to hydrolytic enzyme production viz., chitinase, pectinase, lipase, β-1, 3 glucanases, amylase, and cellulase (Saxena et al. 2015). Recently, cytochrome P450s (477 numbers) were recognized from 7 *Trichoderma* spp. (Chadha et al. 2018). The cytochrome p450 action is required for the production of secondary metabolites and connected to the parasitic ability and/or its connotation with hosts. In *T. hamatum*, the enzyme coded by the G3 gene triggered on reaction to *Sclerotinia* spp. and *Sclerotium* spp. (Carpenter et al. 2008). Greater glucanase and chitinase activity was exhibited when *T. harzianum* was cultivated on media stimulated with *R. solani*, *S. rolfsii*, F. *oxysporum*, and *B. cinerea* cell walls. The mechanism of *T. atroviride* spp. control pathogen is probably by mycoparasitism of hyphae or sclerotia. Some isolates of *T. harzianum* have the capability to parasitize on nematodes and their egg masses, coiling around the second-stage juveniles of *Meloidogyne javanica* and entered them by forming haustoria like structures (Sahebani and Hadavi 2008). Smitha et al. (2017) defined that hydrolytic enzyme assays of *T. viride* co-cultured with pathogens exhibited an augmented activity of chitinase, cellulase, and pectinase over a monoculture, which confirmed the positive induction of enzyme discharge by *Pythium* spp., *Alternaria* spp., and *Fusarium* spp.

11.2.2 Competition

Competition is a phenomenon in which *Trichoderma* spp. and pathogens compete for limited nutrient and space availability. Competition for nutrients and space, for example, carbon and nitrogen is a significant antagonistic feature of *Trichoderma*. *Trichoderma* is generally considered as an aggressive competitor against soilborne fungal pathogens that grow very fast toward the pathogen and rapidly colonize it (Cuervo-Parra et al. 2014). During the competition process, *Trichoderma* may suppress the development of the pathogens in the rhizosphere and thus decrease the development of the disease. Starvation is the most general reason for destruction of soilborne pathogens, so it is for inadequate nutrient consequences in biological control of pathogens (Benitez et al. 2004) (Fig. 11.1). This mechanism for nutrients has been deliberated as a crucial mechanism of biocontrol by *Trichoderma* spp. (Harman 2000). Soils and plant exteriors establish nutrient-limited situation, thereby placing stress on a microbe to compete for the accessible nutrients. Competition for space and nutrients also depend on *Trichoderma* spp. and pathogen (Infante et al. 2013). *Trichoderma* spp. have the ability to synthesize and/or are resistant to metabolites that either hinder spore germination (fungistasis), cell destruction (antibiosis), or alter the rhizosphere, for example, by acidification of the soil, hindering pathogen cannot development (Benitez et al. 2004). The soil is a rich compendium of diverse organisms that have multifaceted roles in the ecological dynamics (Siyar et al. 2019).

Competition for nitrogen, carbon, and parallel with the competition for infection sites or space is an indirect mechanism by which *Trichoderma* controls phytopathogens (Vinale et al. 2008). Iron and carbon are the two vital essentials in most of the fungi, essential for their sustainability. This mechanism for space as well as nutrients is of major importance in the rhizosphere region. *Trichoderma* obtains ATP through the metabolism of different sugars by the production of different enzymes hydrolyzing polymers such as cellulose, glucan, chitin from the environment. The glucose thus formed is used for their carbon and energy requirements which make them strong competitors. Iron acts as a cofactor of several enzymes and a vital nutrient for the growth of hosts and other microorganisms. Iron acquirement is a significant constituent of microbial competition, particularly in the rhizosphere, where microbial actions are intense. *Trichoderma* involve in scavenging iron from the surroundings thus making it unavailable for competing microbes.

Trichoderma spp. produces extremely effective siderophores that chelate iron and impede the development of other pathogens (Benitez et al. 2004). The intracellular siderophore ferricrocin is accountable for iron storage and is intricate in the defense of cells from oxidative stress. *Trichoderma* spp. is known to produce extracellular siderophores of the fusigen and coprogen family (Jalal et al. 1987) along with a great variety of extracellular siderophores (Lehner et al. 2013). Isotope aided showing of *T. atroviride*, *T. polysporum*, *T. gamsii*, *T. hamatum*, *T. asperellum*, *T. harzianum*, *T. virens*, and *T. reesei* resulted in the identification of an average 12–14 siderophores such as dimerum acid, fusarinine A, fusigen, coprogren, and the intracellular siderophore ferricrocin being formed by all studied species (Lehner et al. 2013).

Nutrients competition linked to soil rhizosphere and competition for infection sites appear inside or on the roots of the plant. Root colonization is generally limited to diffusion into layers of cells (Brotman et al. 2008). Among all the mechanisms, nutrient competition is the most important (Verma et al. 2007) preventing pathogen infection. Root exudates and rhizosphere are rich sources of nutrients that include sugars, amino acids, iron, vitamins, organic acids, etc. Soil treatments with T. *harzianum* spores have been reported to inhibit infestations of F. *oxysporum* f. sp. *melonis* and F. *oxysporum* f. sp. *vasinfectum* (Sivan and Chet 1989). Competition for carbon is an active mode not only in *Trichoderma* but also in certain other fungi like strains of *R. solani*, *F. oxysporum* (Alabouvette et al. 2009). Colonies of *T. harzianum* inhibited the development of *F. culmorum* in altered environmental situations and the macroscopic study showed that *T. harzianum* competed with *F. culmorum* for competition (Saravanakumar et al. 2008). *T. harzianum* (T-22) strain revealed rhizosphere competent and capable to manage the numerous fungi containing *R. solani* and it decreased the early infection harshness by as much as 71% on various plants (Lo et al. 1996). The *T. harzianum* T35 strain inhibits *F. oxysporum* by competing for rhizosphere colonization and nutrients leading to biocontrol (Tjamos and Fravel 1995).

Competition has shown to be crucial in the biocontrol of phytopathogens such as *B. cinerea*, the leading fungal pathogenic agent in numerous countries (Latorre et al. 2001). Recently the antifungal activities of *Trichoderma* filtrates were used in the management of *C. paradoxa* infection in pineapple and sugarcane plants (Rahman et al. 2009). Production of proteins has been established to be vital in root colonization by *Trichoderma* as well as in competing with other root colonizing pathogens (Saloheimo et al. 2002; Brotman et al. 2008) and several of them aid in symbiotic association with host plants (Samolski et al. 2012). Vargas et al. (2009) reported the intracellular invertase from *T. virens* (TvInv) hydrolyzing sucrose was significant in deriving sucrose, nutritional supply to *Trichoderma*.

There are many reports showing siderophore involvement in the inhibition of soilborne pathogenic fungi (Vinale et al. 2013). Numerous *Trichoderma* spp. like *T. lignorum*, *T. viride*, *T. harzianum* are potent siderophore synthesizers than pathogens such as *F. solani* and *F. oxysporum*. So *Trichoderma* spp. accesses the little amounts of obtainable iron with high competence (Dutta et al. 2006). *Trichoderma* secrete siderophore, an iron-chelating compound, which binds with insoluble iron (Fe^{3+}) and converting to soluble form (Fe^{2+}) for plant absorption, thus inhibiting the development of phytopathogens by divesting them of iron sources. In the aerobic condition neutral pH and with oxygen, iron occurs mostly as Fe^{3+} and inclines to form insoluble ferric oxide, making it inaccessible for root absorption and microbial development (Miethke 2013). The iron competition has been defined as one of the main aspects of the antagonistic activity of *T. asperellum* to *F. oxysporum* and siderophores may be useful for plants because of their iron solubilizing action (Segarra et al. 2010). The novel siderophore harzianic acid produced by *T. harzianum*, improved the development of seedlings and antagonistic activity against certain plant pathogens, for example, *P. irregulare*, *R. solani*, *S. sclerotiorum*, even in iron lacking conditions (Vinale et al. 2013). The enhancement of plant growth was credited to the Fe^{3+} binding attraction of harzianic acid, aiding its easy uptake by numerous plants (Vinale et al. 2013).

By deceiving the Fe^{3+} from the communal niche, *Trichoderma* spp. can prevent the development and action of post-harvest and soilborne phytopathogens like *B. cinerea* (Harman et al. 2000). Tsahouridou and Thanassoulopoulos (2002) reported *T. koningii* as rhizosphere competent, when tomato seeds application with a conidial suspension of *T. koningii* was sown, resulted in decreased damping-off by the pathogen. The *Trichoderma* strains such as *T. virens* and *T. reesei* harbor an extra recognized gene cluster governing siderophore production (Mukherjee et al. 2012b) and they have two putative gene clusters containing an NRPS as the core member, whose orthologues (*SidD* and *NPS6*) are known to be intricate in siderophore production (Kubicek et al. 2011). Verónica Herrera-Téllez et al. (2019) reported that challenge inoculation with two fungal phytopathogens such as *B. cinerea* and *F. oxysporum* and application with *T. asperellum* ensued in less wilting and restricting symptoms than nontreated hosts. Application with *T. asperellum* formulation inhibited ROS production in reaction to the pathogens in comparison to plants that were only challenge inoculation with both pathogens.

11.2.3 Antibiosis

Antibiosis may be intricate and play a momentous role in the control of disease in plants by some bacteria and fungi. The mechanism has been defined as the communications connecting low molecular weight diffusible compounds or antibiotics or secondary metabolites formed by microorganisms that prevent and/or kill the other microbes (Mukherjee et al. 2012a; Bae et al. 2016; Contreras-Cornejo et al. 2016) (Fig. 11.1). *Trichoderma* produces antibiotics or secondary metabolites composed of different groups of chemical compounds and the compounds help the organism to compete with other macroorganisms, metal transport, symbiosis, differentiation, etc. (Demain and Fang 2000). In addition, they are involved in a progressive role in communicating with the plant, by inducing resistance systemically and stimulating plant development (Vinale et al. 2012). *Trichoderma* spp. produces numerous metabolites that decrease the colonization activity of phytopathogens (Antal et al. 2000). Secondary metabolites, containing antibiotics, that are not directly involved in natural development, reproduction, and are chemically dissimilar from natural compounds may play significant roles in the symbiosis, metal transport, competition against other microorganisms, defense response, differentiation, and stimulating or preventing spore formation and germination, etc. (Ajitha and Lakshmedevi 2010; Vinale et al. 2014). Based upon analytical reports from the genus *Trichoderma* over 180 SMs (natural products) have been characterized to date, representing an array of classes of compounds and with the structures of in excess of 100 compounds are reported (Reino et al. 2008).

Trichoderma produces many SMs with antibiotic activities and their production is strain-dependent (Mukherjee et al. 2012a; Zeilinger et al. 2016). A group of 43 compounds secreted by *Trichoderma* possessing antagonistic activity included isonitriles, peptaibols, alkyl pyrones, diketopiperazines, sesquiterpenes, polyketides, and steroids (Sivasithamparam and Ghisalberti 1998). Majority of the *Trichoderma* spp. secreted volatile and nonvolatile metabolites, for example, glucanase, cellulase, xylanase, lipase, pectinase, amylase, arabinase, and protease, 6-n-pentyl-2H-pyran-2-one (6PP/6-PAP), 6-penthyl-α-pyroneglisoprenins, epipolythiodioxopi perazines (ETPs) and antibiotics encompassing trichodermin, viridin, trichodermol, massoilactone, gliovirin, gliotoxin, pyrones, harzianic acid, formic aldehyde, alamethicins, heptelidic acid, herzianolide, peptaibols, and ethylene (Gajera et al. 2013; Hermosa et al. 2014; Goud et al. 2015; López-Bucio et al. 2015; Ghorbanpour et al. 2018; Tingting Li et al. 2020). Antagonistic effects of *Trichoderma* spp. toward *Pythium* spp. and *R. solani* were widely reported (Kotasthane et al. 2015; Waghunde et al. 2016; Naik et al. 2017; Rajendraprasad et al. 2017).

Different secondary metabolites engaged in the management of numerous soil-borne phytopathogens have been designated. In 1983, Howell defined and isolated a novel antibiotic, gliovirin, from *T. virens* with potent antagonistic activity to *P. ultimum* and a *Phytophthora* spp. The key antibiotic produced by *T. virens* is peptaibol compounds (Mukherjee et al. 2013). This type of antibiotic, class 14, 11, 18 mer from *Trichoderma* spp. are intricate in the inhibition of several phytopathogens like *P. capsici*, *S. rolfsii*, *A. solani*, *R. solani*, and *S. cepivorum* (Vinale et al. 2012). Lorito et al. (1996) examined the action of peptaibols and CWDEs secreted by *T. harzianum* in the antagonistic activity of *B. cinerea* pathogen. The gliotoxin is produced by *T. virens* Q strains while gliovirin is synthesized by the P strains; both have tough antimicrobial action (Mukherjee et al. 2012b; Scharf et al. 2016). Epipolythiodioxopiperazines (ETPs), described by a diketopiperazine ring, are certain extremely noxious SMs formed by *Trichoderma* spp. (Błaszczyk et al. 2014).

The extracellular compounds from *Trichoderma* spp. culture filtrates have been assessed for their antibiotic activity toward *P. capsici* phytotoxicity against pepper plants (Bae et al. 2011). Toghueoa et al. (2016) confirmed that from ethyl acetate extract solvent with *T. atroviridae* strain can prevent the germination *F. solani* spore.

Trichoderma spp. produce and discharge small molecules and gaseous volatile organic compounds that are able to diffuse through atmosphere and soil (Morath et al. 2012) and active against numerous phytopathogens, for example, *Alternaria* spp. and *Fusarium* spp. (Meena et al. 2017). *Fusarium* is described to induce the making of certain VOCs in *T. harzianum* (Zhang et al. 2014). A *T. harzianum* strain raised the points of enzymes and of α-pyrone augmented defense against *R. solani* in examinations of grape defense against *B. cinerea* pathogen in numerous measured ecological situations (Rey et al. 2001). The combination of antibiotics and hydrolytic enzymes achieve the maximum level of antagonism rather than individual mechanism (Karlsson et al. 2017; Nygren et al. 2018).

Reports have made known that *T. hamatum* and *T. harzianum* could be active as biocontrol agents for control of wilt in lentil affected by *F. oxysporum.* Along with antifungal enzymes secretion, it harmfully affects wilt causing pathogens by posing competition for important nutrients and/or biological places (El-Hassan et al. 2013). Koninginin D produced by *Trichoderma* spp. suppress the development of soilborne phytopathogens, for example, *P. middletonii*, *P. cinnamomi*, *R. solani*, *F. oxysporum*, and *B. sorokiniana* (Dunlop et al. 1989). Viridins are compounds secreted from varied *Trichoderma* spp. like *T. viride*, *T. koningii*, and *T. virens* stopping spore germination of different pathogens like *C. lini*, *P. expansum*, *F. caeruleum*, *B. allii*, and *S. atra* (Singh et al. 2005). Harzianic acid isolated from a *T. harzianum* strain showed in vitro antagonistic activity toward *S. sclerotiorum*, *P. irregulare*, and *R. solani* (Vinale et al. 2009). The isolation and expression of *tri5* gene from *T. brevicompactum Tb41 tri5* transformant improved the trichodermin synthesis and antifungal activity to *Fusarium* spp. (Tijerino et al. 2011). *T. asperellum* strain yields two asperelines (A and E) and five trichotoxins (*T5G*, *T5F*, *T5D2*, *1717A*, and *T5E*) which can be connected with antibiosis (Brito João et al. 2014).

The treatment of culture filtrates containing SMs synthesized by four *Trichoderma* spp. in agar plates reduced the growth of *Fusarium* mycelia. These results coincide with many other works and depend on the species tested (Reino et al. 2008; Moosa et al. 2016). Pyrones were reported from numerous *T. harzianum* spp. and have many antifungal features to *G. graminis* and *F. moniliforme*. 6-PP produced by *Trichoderma harzianum* destroys fusaric acid and mycotoxins, prevents the *F. moniliforme* mycelial growth (El-Hasan et al. 2008). Various *Trichoderma* strains like *T. harzianum*, *T. viride*, *T. atroviride*, and *T. koningii* are capable to yield 6-PP that goes to the group of volatile antibiotics playing an important role in biological control activity to *B. cinerea*, *F. oxysporum*, and *R. solani* (Reino et al. 2008). *T. atroviride* synthesized 6-PP compound which stimulated the growth of the plant and controlled root growth with the inhibition of primary root growth and prompting lateral root development (Garnica-Vergara et al. 2015).

Trichodermin was described to have antifungal action against the fungal plant pathogens *A. solani* and *R. solani* (Chen et al. 2007) in addition to other fungal spp. (Tijerino et al. 2011). The fungus *T. atroviride* uses an approach of parasitic communication mainly connecting antibiosis along with degrading enzymes, gliotoxin generating *T. virens* isolates is recommended to immediately target to kill the plant at a distance by harming *R. solani* with gliotoxin compound (Atanasova et al. 2013). Trichokonins VI, a kind of peptaibols in *T. pseudokoningii* SMF2, showed antibiotic actions by stimulating wide apoptotic automatic cell decease in pathogens (Shi et al. 2012).

Trichoderma atroviride SG3403 was improved for the synthesis of antifungal compounds; the culture filtrate inhibited *F. graminearum* by 54.22%. The improved making of definite antifungal and plant growth-promoting constituents improve the biocontrol influence of *Trichoderma*, offering an active biocontrol agent (Tingting Li et al. 2020). In vitro antagonistic effects of *Trichoderma* spp. toward plant pathogens, for example, *S. rolfsii*, *Fusarium oxysporum*, and *R. solani* have been reported. *Trichoderma* isolate Tvb1 was found to be the most efficient in tolerating high temperature up to 45 °C for 4 days, 1750 mM salt (NaCl) concentration, and pH up to 11 (Anwer et al. 2020). Napitupulu et al. (2019) described that the *T. harzianum* (ten isolates) strains isolated from several areas in Java were assessed through two in vitro antagonistic methods. All strains showed antifungal action to *F. oxysporum*, *T. harzianum* formed toxic volatile compounds that had important properties on the development and growth of the *F. oxysporum*. Dwivedi and Enespa (2013) reported the possibility of controlling *F. solani* and *F. oxysporum* using eight biocontrol agents viz., four species of *Aspergillus*, two types of *Trichoderma* (*T. viride* and *T. koningii*), and two species of *Penicillium*. Al-Mekhlafi et al. (2019) reported that a full of 200 soil samples from the rhizosphere were screened for the presence of *Trichoderma* spp. and they displayed antifungal action toward four phytopathogenic fungi like *A. solani*, *F. oxysporum*, *P. ultimum*, and *R. solani*. Among these isolates of *Trichoderma*, only 16 isolates showed strong antagonistic activity and inhibited four pathogenic fungi by more than 50%. Rai et al. (2016) exhibited those 20 strains of *Trichoderma* suppressed mycelial development of fungal phytopathogens (*F. oxysporum*, *A. alternate*, *C. gleosporoides*, and *R. solani*) by more than 50%. *T. asperellum* NVTA2 successfully prevented mycelial progression of *S. sclerotiorum* as well as stopped sclerotial formation (Vinodkumar et al. 2017). Markidahun Biam et al. (2019) reported the effects of *Trichoderma* spp. against *Pythium* spp. and *R. solani* affecting tomato damping-off disease. Rapid screening to *Pythium* spp. and *R. solani* revealed that 20 *Trichoderma* spp. showed maximum antagonism; TR 55 isolated from tomato rhizosphere was found to be the most effective isolate against both *Pythium* spp. and *R. solani* Kuhn. Ommati and Zaker (2012) described the biocontrol activity of *T. longibrachiatum* and *T. brevicompactum* to *F. solani* in pot culture conditions. Timila and Manandhar (2016) also described native *T. harzianum* strain in the control *Phytophthora* blight of pepper under field conditions.

Rajeswari (2019) reported the combining of different *Trichoderma* spp. and *P. fluorescens* like *T. viride* + *P. fluorescens*, *P. fluorescens* + *T. harzianum*, and

T. harzianum + *T. viride* to manage the *Fusarium* wilt. Amongst the three combinations, *P. fluorescens* + *T. viride* treated on leaves was found effective in controlling *F. oxysporum*. Jahangir Alam Liton et al. (2019) described that the *T. harzianum*, Chan-6 isolate was found to be the maximum active in preventing the radial development of *F. oxysporum*, *R. solani*, and *S. rolfsii*. *T. harzianum* fortified compost inhibited the soilborne pathogens and also enhanced soil fertility by the addition of organic matter. Abdel Ghany and Bakri (2019) described that the *T. harzianum* culture filtrates were suppressed the radial development of *A. solani* culture, but growth was not completely inhibited until a high filtrate percentage of 75% was reached. The microscopic study revealed that the *T. harzianum* culture filtrate and its spores changed the size, number, and shape of the *A. solani* conidiospores. Priyadharcini et al. (2018) reported that *Trichoderma* spp., for example, *T. harzianum* (TspT), *Trichoderma* sp. (TspK), and *T. viride* 1 performed well and suppressed the *S. rolfsii* Sacc mycelial growth efficiently.

11.3 Induction of Defense Response

Another mechanism generally associated with defense response of hosts by biological agents is the stimulation of the plant defense pathways (Fig. 11.1). The plant defense response induction mediated by *Trichoderma* has been well recognized (Harman et al. 2012). Induced systemic resistance (ISR) mechanism is one of the greatest mechanisms of biological control actions of *Trichoderma* spp. (Harman 2006). Strains of *Trichoderma* on application to the rhizosphere protect plants from various types of phytopathogens which points to the stimulus of defense mechanisms (Harman et al. 2012; Hermosa et al. 2013; Rubio et al. 2014; Sharma 2018). Any external interaction or dissemination in plant roots stimulates their immune system, but *Trichoderma* spp. alter the plant's immune system and recognized as nonpathogenic (Contreras-Cornejo et al. 2011; Sharma et al. 2017). The elicitors or inducers are several types of compounds, which may act as in the communications of many recognized *Trichoderma* strains with plants. These bioactive metabolites comprise proteins with enzymatic activities, phenyl ammonia lyase (PAL), xylanases, glucanases, chitinases, peroxidase (POX), polyphenol oxidase (PPO), and lipoxygenases (LOX), proteins as PR proteins or low molecular weight peptides (small proteins 1) (Salas-Marina et al. 2015), gene products like proteins coded by avirulent genes, compounds of indole, lipids, fatty acids, poly- or oligosaccharides containing chitin or chitin resembling composites, low molecular compounds which are produced from fungi or host cell walls by the action of enzymes of *Trichoderma* spp., phytoalexin accumulation in host plants (Tuão Gava and Pinto 2016), glycosphingolipids (Mukherjee et al. 2012a), Terpenoids, phytoallexin as rishitin, lubimin, phytotuberol, coumarin, solevetivone, resveratol and antioxidant as ascorbic acid, glutathione, etc. (Contreras-Cornejo et al. 2016; Tuão Gava and Pinto 2016; Birkenbihl et al. 2017).

Trichoderma spp. produces plant growth stimulating metabolites with the capability to improve biomass production, photosynthesis, and stimulate through regulation of gene expression (Martınez-Medina et al. 2010; Rubio et al. 2014). The pathways of defense comprise the development of definite outline recognition receptors for recognizing microorganism-based indications called as pathogen or microbe-associated molecular patterns (PAMPs/ MAMPs) (Saravanakumar et al. 2015). Certain *Trichoderma* sp. proteins intricate in root colonization can also act as MAMPs. *Trichoderma* spp. produce a variety of MAMPs for molecular recognition and may contribute to signal cascade with the help of signaling molecule within the plant-like JA, SA, and ET (Fig. 11.1) (Lorito et al. 2010). In the JA and SA pathways, the synthesis of pathogenesis-related proteins (PR) are triggered by the attack of pathogens and the wounding or necrosis inducing phytopathogens.

Two broad resistance classes, induced in plants by *Trichoderma* called ISR and SAR (systemic acquired resistance), can be distinguished by the biochemical pathways complex (Birkenbihl et al. 2017) and these are regularly phenotypically analogous (Contreras-Cornejo et al. 2011). The SAR is elicited on the prior disease by nonpathogens, while ISR is activated by prior *Trichoderma* spp. colonization from the rhizosphere. SAR generally could be activated by exposure of the plant to pathogens, in addition to avirulent microbes (Salas-Marina et al. 2015) and SAR is an SA-dependent pathway and might be identified by the necrotic wounds, hypersensitive reaction, and phytotoxicity in host plants (Birkenbihl et al. 2017). The SAR is generally elicited by local infection, associating with the connection of SA, the initiation of pathogenesis-related proteins (PR) genes (You et al. 2016). The ISR might be triggered by avirulent root inhabiting plant growth-promoting microorganisms and the pathways controlled by ET and JA play a key role in host defense stimulation (Contreras-Cornejo et al. 2011; Birkenbihl et al. 2017).

The stimulation of a systemic response to *Pseudomonas syringae* pv. *Lachryman*, the causative organism of leaf spot in cucumber, subsequent root treatment of *T. asperellum* isolate, linked with the accumulation of SMs of a phenolic characteristic, these prevented the pathogen development (Yedidia et al. 2003). *Trichoderma harzianum* (T-22) mediated ISR against phytopathogens in maize plants have also been described (Harman et al. 2012; Yoshioka et al. 2011). The endophytic *Trichoderma* described to trigger resistance to *P. capsici* in pepper and postponed the disease onset (Bae et al. 2011). The proteins in the form of swollenins with a cellulose-binding accomplished of exciting confined resistance reactions in roots and leaves of cucumber toward *B. cinerea* and *Pseudomonas syringae* were accessible (Brotman et al. 2008). Brotman et al. (2012) deliberated metabolic and transcriptional profile, the defense reaction of *A. thaliana* plants to the leaf pathogen *P. syringae* in tomato plants induced by *T. asperelloides* (T203). The induced resistance in tomato plant to bacterial leaf spot (*X. campestris pv. vesicatoria*) with 69.32% reduction in disease after 14 days post inoculation of *T. harzianum* (T9) and 14 days after inoculation, augmented the chitinase and glucanase actions (Saksirirat et al. 2009). *Trichoderma* spp. on colonization with roots can elicit resistance and increase nutrient uptake in the host (Contreras-Cornejo et al. 2016). The induction of systemic defense gets stimulated by *Trichoderma* spp. (*T. asperellum* and

T. harzianum) in *A. thaliana* plant against *B. cinerea* (Segarra et al. 2010). The genus *Trichoderma* elicits biochemical and molecular modifications representative of SAR, generally related to the appearance of PR proteins (PR2, PR5, PR1) (Hermosa et al. 2012).

Chitinolytic enzymes may be involved in resistance stimulation and impact on resistance induction in apple plant and cotton plant to *R. solani* (Kumar et al. 2009), *V. inaequalis* (Faize et al. 2003). Shoresh et al. (2006) described that in cucumber, a mitogen-activated protein kinase (MAPK) is stimulated by *T. asperellum* and this activity is essential for fighting the attack of *P. syringae*. *T. harzianum* increased JA and SA contents in melon plants and changed the host responses to *F. oxysporum* (Martınez-Medina et al. 2010). *Arabidopsis* mutant plants decreased in the biosynthesis of JA, exhibited an analogous level of root colonization to wild plants (Martìnez-Medina et al. 2017). Root colonization by *T. atroviride* improved the phytoalexin camalexin in *A. thaliana* plant (Contreras-Cornejo et al. 2011). *T. asperellum* T203 modified the appearance of the LOX1 genes (Lipoxygenase 1), a constituent of JA synthesis, PAL1 a component in the biosynthesis of SA and ETR1 and CTR1, both involved in ethylene (ET) signaling (Shoresh et al. 2005). The action of *T. longibrachiatum* cellulase stimulated the SA and ET pathways important to the noticeable induction of POX and chitinase actions in the melon plant (Martinez et al. 2001).

T. harzianum on application to roots or foliage of grapes affords management of diseases affected by *B. cineria* on leaves spatially divided (Deshmukh et al. 2006). The SA and nonexpressor of PR genes1 (NPR1) are important players in SAR. The different enzymes like aspartyl proteases and proteases were characterized in *T. asperellum* and *T. harzianum* strains intricated in both *Trichoderma* host symbiosis and mycoparasitism, increasing defense mechanisms (Viterbo et al. 2004). *T. harzianum* was described to limit *F. verticillioides* in maize through the stimulation of systemic resistance by initiating ethylene and JA signaling pathways (Ferrigo et al. 2014). The antagonistic activity and ISR by two *T. asperellum* isolates against *Ralstonia solanacearum* in tomato plants have been investigated (Narasimha Murthy and Srinivas 2013; Narasimha Murthy et al. 2013, 2018). The 18mer peptaibols from *T. virens* stimulate systemic induced defense responses in cucumber to *P. syringae* (Luo et al. 2011). *Trichoderma* induces SAR resistance even on the trigger of ISR defense and they can increase the plant resistance against plant pathogens, for example, *S. sclerotiorum*, where the mechanism of *Trichoderma* spp. associates with the secretion of CWDEs by the plant (Lopes et al. 2012).

In chickpea wilt, the seed inoculation with *T. harzianum* induced maximum soluble protein and glucanase activity in contrast to the untreated seedlings (Moradi et al. 2012). Jayalaksmi et al. (2009) reported increased PPO activity in chickpea by the treatment with *T. harzianum* strain L1, implicating it in induced resistance against root rot in chickpea. Saravanakumar et al. (2016) described the *Trichoderma* cellulase elicits the ISR in maize against leaf spot, increasing the gene expression related to JA or ET signaling pathways. Gallou et al. (2009) observed that the resistance response of *T. harzianum* challenge inoculation in potato against *R. solani* was reliant on JA/ET and SA pathways. Christopher et al. (2007) observed induction of defense enzymes containing POX, PAL, and PPO with seed application plus soil treatment of talc-based formulation of *T. viride* against *Fusarium* in tomato.

In vitro analysis, the activities of glucanase, chitinase revealed the higher production by *T. harzianum* and *T. hamatum* than untreated control. The inoculation with *T. atroviride* in maize and *T. viride* in black gram challenging *C. heterostrophus*, *F. oxysporum*, and *A. alternate* significantly improved the synthesis of defense enzymes (POX, SOD, and CAT) (Wang et al. 2015).

11.4 *Trichoderma* Genes

Genomic studies revealed that *Trichoderma* spp. contains hundreds of separate genes and gene products which can be used to deliver resistance to the abiotic and biotic conditions, excessive range of expression patterns, which allows the *Trichoderma* spp. uses of effective biocontrol organism and in plant growth advancement actions (Reithner et al. 2014). The current genome sequencing projects for *Trichoderma* spp. have targeted seven strains of *Trichoderma* like *T. atroviride*, *T. reesei*, *T. virens*, *T. harzianum*, *T. asperellum*, *T. longibrachiatum*, and *T. citrinoviride* (Srivastava et al. 2014; Baroncelli et al. 2016; Rai et al. 2016). The genes of *T. harzianum* encoding chitinases such as *chit42* and *chit33* are pivotal to the mycoparasitic action against phytopathogens particularly *F. oxysporum* (Mondejar et al. 2011). In certain genomes, they are established as multiple copies: *endoPG* gene families were first revealed in *S. sclerotiorum* and *B. cinerea*. The antagonistic activity was established to *R. solani*, *B. cinerea* fungal phytopathogens through the wild form and mutant isolates. The *Chit36* inhibits the *B. cinerea* spore germination and inhibits the progress of *S. rolfsii* and *F. oxysporum* pathogens (Viterbo et al. 2001).

The *T. harzianum* chitinase *Chit42* expression in potato and tobacco seedlings resulted in improved resistance to pathogens such as *A. alternata*, *B. cinerea*, *A. solani*, and also to soilborne *R. solani* phytopathogen (Howell 2003). The β-1, 6-glucanase *Bgn 16.2* transformants suppressed *R. solani* and *B. cinerea* development. *T. atroviride* endochitinase *Ech42* expression showed improved resistance to *Fusarium* spp. (McIntyre et al. 2004). The expression of *chit42* in lemon improved resistance to *P. tracheiphila* and *B. cinerea*, an important association between resistance and transgene expression being detected, with an upregulation of ROS and JA/ET responsive genes (Distefano et al. 2008). The homologous *chit42* gene from *T. virens* was capable to improve resistance to *R. solani* when it was expressed in rice plants (Shah et al. 2009). From *T. harzianum*, *T. virens* isolates overexpression of β-1, 6-glucanases resulted in more effective control of *R. solani*, *B. cinerea* pathogens (Ihrmark et al. 2010), *P. ultimum* (Djonovic et al. 2006). The glucanase *agn13.2* from *T. asperellum* and glucanase *bgn16.2* from *T. harzianum* has antagonistic activity to *B. cinerea* (Sanz et al. 2005). The *tag83* gene expression of encoding exoglucanase enzyme was identified in *T. asperellum* strain and the expression of *tag83* gene exhibited antagonistic activity to several pathogens (Marcello et al. 2010).

The *Gluc78* gene in *T. atroviride* P1 revealed great antimicrobial activity to a varied variety of fungal pathogens; it represented synergistically with new enzymes and *Tv-bgn1* and *Tv-bgn2*, these glucanases have been identified and cloned (Donzelli et al. 2001). The *qid74*gene in *T. harzianum* CECT has an important character in cell defense and affords observance to hydrophobic exteriors, supporting the fungus in mycoparasitic action to *R. solani* pathogen (Rosado et al. 2007). The *Bgn16.3* in *T. harzianum* CECT 2413 encoding β-1, 6-glucanase displayed more effective biocontrol in *B. cinerea* growth inhibition, destruction of *R. solani* and *P. citrophthora* (Montero et al. 2007). The genes coding for secondary metabolite biosynthesis in *Trichoderma* are arranged in clusters that can span more than 10 kb, while there are a few exceptions. Atanasova et al. (2013) described that the transcriptomic reactions of *T. reesei*, *T. atroviride*, and *T. virens* to the existence of *R. solani*. The *TmkA* MAPK in *T. Virens* is identified to function in mycoparasitic action to *S. rolsfii* and *R. solani* (Mukherjee et al. 2003).

Alamethicin is a 20mer peptaibol from *T. viride*, elicits SA and JA synthesis in lima bean seedlings (Engelberth et al. 2001), while 18mer peptaibols from *T. virens* stimulate systemic resistances in cucumber to *P. syringae* (Viterbo et al. 2007). A richness of genes encoding subtilisin-like serine proteases were also detected in a study of expressed sequence tags (ESTs) accumulated through the beginning of interaction among *T. atroviridis* with *R. solani* and *S. sclerotiorum* (Seidl et al. 2009). *T. asperellum* induced a systemic reaction of two resistance genes coding phenylalanine and hydroperoxide lyase and systemic buildup of phytoalexins to *P. syringae* in cucumber plants (Yedidia et al. 2003). Contreras-Cornejo et al. (2011) recommended the defense reactions stimulated in *Arabidopsis* plant by *Trichoderma* to *B. cinerea* pathogen intricate the JA as a significant aspect of increasing host protection. The soil treatment with *T. viride* in tomato seedlings with pathogens like *F. oxysporum*/*R. solani* improved the JA related genes (*PDF1* and *PDF2*) expression (Hafez et al. 2013). *Trichoderma* isolates have a tri cluster; these contain seven genes that encode transport and pointing enzymes essential for trichodermin production. The gene *Tri3* was responsible for the synthesis of the trichodermin compound and this gene coded enzyme catalyzed the acetylation response of the hydroxy group at C-4 of the trichodermin skeleton.

Signaling pathways/genes involved in mycoparasitism include the kinase *T. virens* (*Tvk1*/*TmkA*) and from *T. atroviride* (*Tmk1*), these are negative regulators of hydrolytic enzymes and antibiotics. The corresponding gene deletion mutants were more active in the management of plant pathogens caused by *R. solani* than the commercial chemical fungicides in beans (Mukherjee et al. 2013).

11.5 Conclusion

Soilborne diseases are critical in deciding the growth and yield of the majority of the agricultural crops. Managing these diseases is challenging due to various reasons such as their heterogeneous incidence, resistance exhibited by the pathogen, failure of management practices in the field, and so on. Chemical-based control strategies, although effective, pose potential hazards to the health of humans, non-target living forms, soil, and environment. Continuous efforts for the development and application of eco-friendly methods for pathogen control are highly recommended. The employment of microbial bio inoculants for the management of soilborne diseases is an alternative, safe, and natural approach. Among the various biocontrol agents, the application of *Trichoderma* for plant protection is one of the best ways to replace synthetic chemicals. An array of biocontrol and plant growth-promoting attributes of *Trichoderma* encompassing its potent antagonistic activity against pathogens, induction of disease resistance in the host, enhancement of plant nutrient uptake, and abiotic stress resistance and so on, make it an efficient and dependable biocontrol agent. The secretion of cell wall degrading enzymes and secondary metabolites are noteworthy in the mechanism of antibiosis by *Trichoderma*. The *Trichoderma* genome harbors several genes controlling the expression of important biocontrol traits. These decide the various direct or indirect mechanisms, which work synergistically in biocontrol as well as the adaptation of the fungus in response to the host and different environments. The genetic improvement of these biocontrol traits with enhanced efficacy needs a better understanding, which is crucial for the development of improved strains with better market and field potentials.

References

Abbas A, Jiang D, Fu Y (2017) *Trichoderma* spp. as antagonist of *Rhizoctonia solani*. Journal of. Plant Pathol 8:402

Abdel-Ghany TM, Alawlaqi MM (2018) Molecular identification of rhizospheric thermo-halotolerant *Aspergillus terreus* and its correlation to sustainable agriculture. Biol Resour 13 (4):8012–8023

Abdel Ghany T, Bakri M (2019) Effectiveness of a biological agent (*Trichoderma harzianum* and its culture filtrate) and a fungicide (methyl benzimacold-2-ylcarbamate) on the tomato rotting activity (growth, celluloytic, and pectinolytic activities) of *Alternaria solani*. BioRes 14 (1):1591–1602

Abdel-Ghany TM, El-Naggar MA, Ganash MA, Al Abboud MA (2017) PCR identification of *Aspergillus niger* with using natural additives for controlling and detection of malformins and maltoryzine production by HPLC. Biol Nano Sci 7(4):588–596

Ahuja DB, Ahuja RU, Srinivas P, Singh RV, Malik M, Sharma P, Bambawale OM (2012) Development of farmer–led integrated management of major pests of cauliflower cultivated in rainy season in India. J Agric Sci 4(2):79–90

Ajitha PS, Lakshmedevi N (2010) Effect of volatile and von-volatile compounds from *Trichoderma* spp. against *Colletotrichum capsici* incitant of anthracnose on bell peppers. Nat Sci 8:265–296

Alabouvette C, Olivain C, Migheli Q, Steinberg C (2009) Microbiological control of soil-borne phytopathogenic fungi with special emphasis on wilt-inducing *Fusarium oxysporum*. New Phytol 184:529–544

Alizadeh H, Behboudi K, Ahmadzadeh M, Javan-Nikkhah M, Zamioudis C, CMJ P, PAHM B (2013) Induced systemic resistance in cucumber and *Arabidopsis thaliana* by the combination of *Trichoderma harzianum* Tr6 and *Pseudomonas* sp. Ps14. Biol Control 65:14–23

Alkooranee JT, Aledan TR, Ali AK, Lu G, Zhang X, Wu J, Fu C, Li1 M (2017) Detecting the hormonal pathways in oilseed rape behind induced systemic resistance by *Trichoderma harzianum* TH12 to *Sclerotinia sclerotiorum*. PLoS One. https://doi.org/10.1371/journal.pone.0168850

Al-Mekhlafi NA, Abdullah QY, Al-Helali MF, Alghalibi SM (2019) Antagonistic potential of native *Trichoderma* species against tomato fungal pathogens in Yemen. Int J Mol Microbiol 2 (1):1–10

Alwathnani HA, Perveen K (2012) Biological control of Fusarium wilts of tomato by antagonist fungi and cyanobacteria. Afr J Biotechnol 11:1100–1105

Antal Z, Manczinger L, Szakacs G, Tengerdy RP, Ferency L (2000) Colony growth, *in vitro* antagonism and secretion of extracellular enzymes in cold-tolerant strains of *Trichoderma* species. Mycol Res 104:545–549

Anitha KN (2011) Physiological and biochemical basis of resistance to purple seed stain of soybean *Glycine max* (L.) Merrill. Karnataka J Agric Sci 25(4):557–608

Anwer MA, Singh K, Prasad BD, Yadav AK, Kumari P (2020) Abiotic stress tolerant *Trichoderma asperellum* Tvb1 from hot spring and its antagonistic potential against soil borne Phytopathogens. Int Arch Appl Sci Technol 11(3):53–59

Atanasova L, Le Crom S, Gruber S, Coulpier F, Seidl-Seiboth V, Kubicek CP et al (2013) Comparative transcriptomics reveals different strategies of *Trichoderma* mycoparasitism. BMC Genomics 14:121

Bae H, Roberts DP, Lim HS, Strem MD, Park SC, Ryu CM, Melnick RL, Bailey B (2011) Endophytic *Trichoderma* isolates from tropical environments delay disease onset and induce resistance against *Phytophthora capsici* in hot pepper using multiple mechanisms. Mol Plant-Microbe Interact 24(3):36–51

Bae SJ, Mohanta TK, Chung JY, Ryua M, Park G, Shim S, Hong SB, Seo H, Bae DW, Bae I, Kima JJ, Bae H (2016) *Trichoderma* metabolites as biological control agents against *Phytophthora* pathogens. Biol Control 92:128–138

Balaji LP, Ahir RR (2011) Evaluation of plant extracts and biocontrol agents against leaf spot disease of brinjal. Indian Phytopathol 64(4):378–380

Baroncelli R, Zapparata A, Piaggeschi G, Sarrocco S, Vannacci G (2016) Draft whole genome sequence of *Trichoderma gamsii* T6085, a promising biocontrol agent of Fusarium head blight on wheat. Genome Announc 4:e01747–e01715

Barua L, Bora BC (2008) Comparative efficacy of *Trichoderma harzianum* and *Pseudomonas fluorescens* against *Meloidogyne incognita* and *Ralstonia solanacearum* complex in Brinjal. Indian J Nematol 38(1):86–89

Benhamou N, Rey K, Picard Y (1999) Ultraestructural and cytochemical aspects of the interaction between the mycoparasite *Pythium oligandrum* and soil borne plant pathogens. Phytopathology 89:506–517

Benitez T, Rincon AM, Limon MC, Codon AC (2004) Biocontrol mechanisms of *Trichoderma* strains. Int Microbiol 7:249–260

Bhandari PC, Vishunavat K (2013) Screening of different isolates of *Trichoderma harzianum* and *Pseudomonas fluorescens* against *Fusarium moniliforme*. Pantnagar J Res 11(2):243–247

Biam M, Majumder D, Papang H (2019) *In vitro* efficacy of native *Trichoderma* isolates against Pythium spp. and *Rhizoctonia solani* (Kuhn.) causing damping-off disease in tomato (*Solanum lycopersicum* miller). Int J Curr Microbiol App Sci 8(2):566–579

Birkenbihl RP, Liu S, Somssich IE (2017) Transcriptional events defining plant immune responses. Curr Opin Plant Biol 38:1–9

Bisen K, Keswani C, Patel JS, Sarma BK, Singh HB (2016) *Trichoderma* spp.: efficient inducers of systemic resistance in plants. In: Chaudhary DK, Verma A (eds) Microbial-mediated induced systemic resistance in plants. Springer, Singapore, pp 185–195

Biswas S, Datta M (2013) Evaluation of biological control agents against sheath blight of rice in Tripura. Indian Phytopathol 66(1):77–80

Błaszczyk L, Siwulski M, Sobieralski K, Lisiecka J, Jędryczka M (2014) *Trichoderma* spp. application and prospects for use in organic farming and industry. J Plant Protect Res 54:309–317

Brito João PC, Ramada Marcelo HS, de Magalhães Mariana TQ, Silva LP, Ulhoa CJ (2014) Peptaibols from *Trichoderma asperellum* TR356 strain isolated from Brazilian soil. Springer Plus 3:600

Brotman Y, Briff E, Viterbo A, Chet I (2008) Role of swollenin, an expansin-like protein from *Trichoderma*, in plant root colonization. Plant Physiol 147:779–789

Brotman Y, Landau U, Pninic S, Lisec J, Balazadeh S et al (2012) The LysM receptor-like kinase LysMRLK1 is required to activate defense and abiotic-stress responses induced by overexpression of fungal chitinases in *Arabidopsis* plants. Mol Plant 5:1113–1124

Carmona-Hernandez S et al (2019) Biocontrol of postharvest fruit fungal diseases by bacterial antagonists: a review. Agronomy 9(3):121

Carpenter MA, Ridgway HJ, Stringer AM, Hay AJ, Stewart A (2008) Characterization of a *Trichoderma hamatum* monooxygenase gene involved in antagonistic activity against fungal plant pathogens. Curr Genet 53:193–205

Ceballos G, Álvarez E, Bolaños MM (2014) Reducción de poblaciones de Ralstonia solanacearum raza 2 (Smith) en plátano (Musa AAB Simmonds) con aplicación de extractos de *Trichoderma* sp. (Alexopoulus y Mims) bacterias antagonistas. Acta Agron 63(1):1–11

Chadha S, Mehetre ST, Bansal R, Kuo A, Aerts A, Grigoriev IV et al (2018) Genome-wide analysis of the cytochrome p450 of *Trichoderma* spp.: annotation and evolutionary relationships. Fungal Biol Biotechnol 4:12

Chagas LF, Orozco BS, Rodrigues G (2017) Rice growth influence by *Trichoderma* spp. with natural phosphate fertilization under greenhouse conditions. IJDR 7:13147–13152

Chen WQ, Xu SC, Liu TG, Lin RM, Wu LR, Jin SL, Co SQ (2007) Wheat stripe rust and its prospects for ecological control in China. In: Proceedings of the XVI International Plant Protection Congress, 15–18 October 2007, Glasgow, pp 812–813

Chakravarthy SK, Nagamani A (2007) Efficacy of nonvolatile and volatile compounds of *Trichoderma* species on *Rhizoctonia solani*. J Mycol Plant Pathol 37:82–86

Chakravarthy S, Nagamani K, Ratnakumari AR, Bramarambha YS (2011) Antagonistic ability against Rhizoctonia solani and pesticide tolerance of Trichoderma strains. Adv Environ Biol 5 (9):2631–2638

Christopher D, John Suthinraj T, Udhayakumar R (2007) Induction of defense enzymes in Trichoderma viride treated blackgram plants in response to *Macrophomina phaseolina* infection. Indian J Plant Prot 35(2):299–303

Colla G, Hoagland L, Ruzzi M, Cardarelli M, Bonini P, Canaguier R et al (2017) Biostimulant action of protein hydrolysates: unraveling their effects on plant physiology and microbiome. Front Plant Sci 8:2202

Contreras-Cornejo HA, Macías-Rodríguez L, Beltrán-Peña E, Herrera-Estrella A, López-Bucio J (2011) *Trichoderma*-induced plant immunity likely involves both hormonal and camalexin-dependent mechanisms in *Arabidopsis thaliana* and confers resistance against necrotrophic fungus *Botrytis cinerea*. Plant Signal Behav 6:1554–1563

Contreras-Cornejo HA, Macías-Rodríguez L, del-Val E, Larsen J (2016) Ecological functions of *Trichoderma* spp. and their secondary metabolites in the rhizosphere: interactions with plants. FEMS Microbiol Ecol 92:fiw03

Cuervo-Parra JA, Snchez-Lpez V, Romero-Cortes T, Ramrez-Lepe M (2014) Hypocrea/*Trichoderma viridescens* ITV43 with potential for biocontrol of *Moniliophthora roreri* Cif Par, *Phytophthora megasperma* and *Phytophthora capsici*. Afr J Microbiol Res 8:1704–1712

Daguerre Y, Siegel K, Edel-Hermann V, Steinberg C (2014) Fungal proteins and genes associated with biocontrol mechanisms of soil borne pathogens: a review. Fungal Biol Rev 28:97–125

de Medeiros HA, Filho JVDA, de Freitas LG, Castillo P, Rubio MB, Hermosa R, Monte E (2017) Tomato progeny inherit resistance to the nematode Meloidogyne javanica linked to plant growth induced by the biocontrol fungus *Trichoderma atroviride*. Sci Rep 7:40216

Demain AL, Fang A (2000) The natural functions of secondary metabolites. Adv Biochem Eng Biotechnol 69:1–39

Deshmukh S, Hueckelhoven R, Schaefer P, Imani J, Sharma M (2006) The root endophytic fungus *Piriformospora indica* requires host cell death for proliferation during mutualistic symbiosis with barley. Proc Natl Acad Sci USA 103:18450–18457

Devi TS, Shivaprakash MK (2013) Biopriming of cucumber seeds with *Trichoderma harzianum* and Pseudomonas fluorescens for enhanced plant defense against *Pythium aphanidermatum*. Green Farming 4:489–492

Devi S, Gupta C, Jat SL, Parmar MS (2017) Crop residue recycling for economic and environmental sustainability: the case of India. Open Agric 2:486–494

Distefano G, La Malfa S, Vitale A, Lorito M, Deng Z (2008) Defence related gene expression in transgenic lemon plants producing an antimicrobial *Trichoderma harzianum* endochitinase during fungal infection. Transgenic Res 17:873–879

Djonovic S, Pozo MJ, Dangott LJ, Howell CR, Kenerley CM (2006) Sm1, a proteinaceous elicitor secreted by the biocontrol fungus *Trichoderma virens* induces plant defense responses and systemic resistance. Mol Plant-Microbe Interact 19:838–853

Donzelli BG, Lorito M, Scala F, Harman GE (2001) Cloning, sequence and structure of a gene encoding an antifungal glucan 1, 3-beta-glucosidase from *Trichoderma atroviride* (T. *harzianum*). Gene 277:199–208

Druzhinina IS, Seidl-Seiboth V, Herrera-Estrella A, Horwitz BA, Kenerley CM, Monte E, Mukherjee PK, Zeilinger S, Grigoriev IV, Kubicek CP (2011) *Trichoderma*: the genomics of opportunistic success. Nat Rev Microbiol 9:749–759

Dubey SC, Patel B (2012) Evaluation of fungal antagonists against *Thanatephorus cucumeris* causing web blight of urd and mung bean. Indian Phytopathol 54(2):206–209

Dunlop RW, Simon A, Siwasithamparam D, Ghisalberti EL (1989) An antibiotic from *Trichoderma koningii* active against soil borne plant pathogens. J Nat Prod 52:67–74

Dutta S, Kundu A, Chakraborty M, Ojha S, Chakrabarti J, Chatterejee N (2006) Production and optimization of Fe(III) specific ligand, the siderophore of soil inhabiting and wood rotting fungi as deterrent to plant pathogens. Acta Phytopathol Entomol Hung 41:237–248

Dwivedi SK, Enespa S (2013) *In vitro* efficacy of some fungal antagonists against *Fusarium solani* and *Fusarium oxysporum* f. sp. *lycopersici* causing brinjal and tomato wilt. Int J Biol Pharm Res 4(1):46–52

Elamathi E, Malathi P, Viswanathan R, Ramesh Sundar A (2017) Molecular analysis of the suppression of *Colletrotrichum falcatum* by *Trichoderma harzianum* in sugarcane. J Plant Pathol 99(1):211–218

El–Hasan A, Walker F, Buchenauer H (2008) *Trichoderma harzianum* and its metabolite 6–pentyl–alpha–pyrone suppress fusaric acid produced by *Fusarium moniliforme*. J Phytopathol 156:79–87

El-Hassan SA, Gowen SR, Pembrok B (2013) Use of *Trichoderma hamatum* for bio-control of lentil wilts disease. J Plant Protect Res 53(1):12–17

El-Mohamedy RSR, Alla MA (2013) Bio–priming seed treatment for biological control of soil borne fungi causing root rot of green bean (*Phaseolus vulgaris* L.). J Agric Technol 9:589–599

Elshahawy IE, Saied NM, Abd-El-Kareem F, Morsy AA (2018) Field application of selected bacterial strains and their combinations for controlling onion and garlic white rot disease caused by *Stromatinia cepivora*. J Plant Pathol 100:493–503

Enespa, Dwivedi SK (2013) *In vitro* efficacy of some fungal antagonists against *Fusarium solani* and *Fusarium oxysporum* f. sp. *lycopersici* causing brinjal and tomato wilt. Int J Biol Pharm Res 4(1):46–52

Enespa, Dwivedi SK (2014) Effectiveness of some antagonistic fungi and botanicals against *Fusarium solani* and *Fusarium oxysporum* f.sp. *lycopersici* infecting brinjal and tomato plants. Asian J Plant Pathol 8:18–25

Engelberth J, Koch T, Schuler G, Bachmann N, Rechtenbach J, Boland W (2001) Ion channel-forming alamethicin is a potent elicitor of volatile biosynthesis and tendril coiling. Cross talk between jasmonate and salicylate signaling in lima bean. Plant Physiol 125:369–377

Etschmann MM, Huth I, Walisko R et al (2015) Improving 2- phenylethanol and 6-pentyl-α-pyrone production with fungi by microparticle-enhanced cultivation (MPEC). Yeast 32:145–157

Faize M, Malnoy M, Dupuis F, Chevalier M, Parisi L, Chevreau E (2003) Chitinases of *Trichoderma atroviride* induce scab resistance and some metabolic changes in two cultivars of apple. Phytopathology 93:1496–1504

Fand BB, Suroshes S, Gautam RD (2013) Fortuitous biological control of insect pests and weeds: a critical review. Bioscan 8(1):01–10

Ferrigo D, Raiola A, Piccolo E, Scopel C, Causin R (2014) *Trichoderma harzianum* T22 induces in maize systemic resistance against *Fusarium verticillioides*. J Plant Pathol 96:133–142

Gajera H, Domadiya R, Patel S, Kapopara M, Golakiya B (2013) Molecular mechanism of *Trichoderma* as bio-control agents against phytopathogen system—a review. Curr Res Microbiol Biotechnol 1:133–142

Gallou A, Cranenbrouck S, Declerck S (2009) *Trichoderma harzianum* elicits defense response genes in roots of potato plantlets challenged by *Rhizoctonia solani*. Eur J Plant Pathol 124:219–230

Garnica-Vergara A, Barrera-Ortiz S, Munoz-Parra E (2015) The volatile 6-pentyl-2H-pyran-2-one from *Trichoderma atroviride* regulates *Arabidopsis thaliana* root morphogenesis via auxin signaling and ETHYLENE INSENSITIVE 2 functioning. New Phytol 209:1496–1512

Garnica-Vergara A, Barrera-Ortiz S, Munoz-Parra E, Raya-Gonzalez J, Mendez-Bravo A et al (2016) The volatile 6-pentyl-2H-pyran-2-one from *Trichoderma atroviride* regulates Arabidopsis thaliana root morphogenesis via auxin signaling and ethylene insensitive 2 functioning. New Phytol 209:1496–1512

Ghanbarzadeh B, Safaie N, Goltapeh EM (2014) Antagonistic activity and hyphal interactions of *Trichoderma* spp. against *Fusarium proliferatum* and *F. oxysporum in vitro*. Arch Phytopathol Plant Protect 47:1979–1987

Ghorbanpour M, Omidvari M, Abbaszadeh-Dahaji P, Omidvar R, Kariman K (2018) Mechanisms underlying the protective effects of beneficial fungi against plant diseases. Biol Control 117:147–157

Goud TS, Raju RA, Karri S, Kumar YS (2015) Production and antagonistic effect of Trichoderma spp. on pathogenic microorganisms (*Botrytis cinerea, Fusarium oxysporium, Macrophomina phasealina* and *Rhizoctonia solani*). Afr J Biotechnol 14:668–675

Gupta M, Dohroo NP, Gangta V, Shanmugam V (2010) Effect of microbial inoculants on rhizome disease and growth parameters of ginger. Indian Phytopathol 63(4):438–441

Guzmán-Guzmán P, Porras-Troncoso MD, Olmedo-Monfil V, Herrera-Estrella A (2019) *Trichoderma* species: versatile plant symbionts. Phytopathology 109:6–16

Gveroska B, Ziberoski J (2012) *Trichoderma harzianum* as a biocontrol agent against *Alternaria alternata* on tobacco. Appl Technol Innov 7:67–76

Hafez EE, Meghad A, Elsalam HAA, Ahmed SA (2013) *Trichoderma viride*–plant pathogenic fungi interactions. World Appl Sci J 21:1821–1828

Haggag KHE, El–Gamal NG (2012) In vitro study on *Fusarium solani* and *Rhizoctonia solani* isolates causing the damping off and root rot diseases in tomatoes. Nat Sci 10:16–25

Haggag WM, Abouziena HF, Abd-El-Kreem F, Habbasha S (2015) Agriculture biotechnology for management of multiple biotic and abiotic environmental stress in crops. J Chem Pharm 7 (10):882–889

Harman GE (2000) Myths and dogmas of biocontrol: changes in perceptions derived from research on *Trichoderma harzianum* T–22. Plant Dis 84:377–393

Harman GE (2006) Overview of mechanisms and uses of *Trichoderma* spp. Phytopathology 96:190–194

Harman GE, Herrera-Estrella AH, Horwitz BA, Lorito M (2012) Special issue: *Trichoderma*—from basic biology to biotechnology. Microbiology 158:1–2

Hermosa R, Viterbo A, Chet I, Monte E (2012) Plant-beneficial effects of *Trichoderma* and of its genes. Microbiology 158:17–25

Hermosa R, Rubio ME, Cardoza MB, Nicolás E, Monte E, Gutiérrez S (2013) The contribution of *Trichoderma* to balancing the costs of plant growth and defense. Int Microbiol 16:69–80

Hermosa R, Cardoza MB, Rubio ME, Gutiérrez S, Monte E (2014) Secondary metabolism and antimicrobial metabolites of *Trichoderma*. In: Gupta VK, Schmoll M, Herrera-Estrella A, Upadhyay RS, Druzhinina I, Tuohy M (eds) Biotechnology and biology of *Trichoderma*. Elsevier, Netherlands, pp 125–137

Herrera-Téllez VI, Cruz-Olmedo AK, Plasencia J, Gavilanes-Ruíz M, Arce-Cervantes O, Hernández-León S, Saucedo-García M (2019) The protective effect of *Trichoderma asperellum* on tomato plants against *Fusarium oxysporum* and *Botrytis cinerea* diseases involves inhibition of reactive oxygen species production. Int J Mol Sci 20:2007

Howell CR (2003) Mechanisms employed by *Trichoderma* species in the biological control of plant diseases: the history and evolution of current concepts. Plant Dis 87:4–10

Ihrmark K, Asmail N, Ubhayasekera W, Melin P, Stenlid J, Karlsson M (2010) Comparative molecular evolution of *Trichoderma* chitinases in response to mycoparasitic interactions. Evol Bioinforma 6:1–26

Infante D, Martinez B, Peteira B, Reyes Y, Herrera A (2013) Molecular identification of thirteen isolates of *Trichoderma* spp. and evaluation of their pathogenicity towards *Rhizoctonia solani* Kühn. Biotecnol Apl 30:23–28

Iqbal MN, Ashraf AJPM (2017) Antagonism in Rhizobacteria: application for biocontrol of soil borne plant pathogens. PSM Microbiol 2(3):78–79

Iqbal MN, Ashraf A (2019) *Trichoderma*: a potential biocontrol agent for soil borne fungal pathogens. Int J Mol Microbiol 2(1):22–24

Jalal MAF, Love SK, Van der Helm D (1987) Siderophore mediated iron (III) uptake in *Gliocladium virens* (*Trichoderma* virens). 2. Role of ferric mono and dihydroxamates as iron transport agents. J. Inorgan Biochem 29:259–267

Jat JG, Agalave HR (2013) Antagonistic properties of *Trichoderma* species against oilseed–borne fungi. Sci Res Rep 3(2):171–174

Jayalaksmi SK, Raju S, Usha Rani S, Bengai VI, Sreeramulu K (2009) *Trichoderma harzianum* L1 as a potential source for lytic enzymes and elicitor of defense responses in chickpea against wilt disease caused by *Fusarium oxysporum* f. sp. ciceri. Aust J Crop Sci 3:44–52

Jayaraj J, Radhakrishnan NV, Velazhahan R (2006) Development of formulations of *Trichoderma harzianum* strain M1 for control of damping–off of tomato caused by *Pythium aphanidermatum*. Arch Phytopath Plant Prot 39(1):1–8

Jeyalakshmi C, Rettinassababady C, Nema S (2013) Integrated management of sesame diseases. J Biopesticides 6(1):68–70

Jogaiah S, Abdelrahman M, Tran LSP, Ito SI (2018) Different mechanisms of *Trichoderma* virens mediated resistance in tomato against Fusarium wilt involves the jasmonic and salicylic acid pathways. Mol Plant Pathol 19:870–882

John RP, Tyagi RD, Prévost D, Brar SK, Pouleur S, Surampalli RY (2010) Mycoparasitic *Trichoderma viride* as a biocontrol agent against *Fusarium oxysporum* f. sp. *adzuki* and *Pythium arrhenomanes* and as a growth promoter of soybean. Crop Prot 29:1452–1459

Junaid JM, Dar NA, Bhat TA, Bhat AH, Bhat MA (2013) Commercial biocontrol agents and their mechanism of action in the management of plant pathogens. Int J Mod Plant Anim Sci 1:39–57

Kapoor AS (2008) Biocontrol potential of *Trichoderma* spp. against important soil borne diseases of vegetable crops. Indian Phytopathol 61(4):492–498

Karlsson M, Atanasova L, Jensen DF, Zeilinger S (2017) Necrotrophic mycoparasites and their genomes. Microbiol Spectr 5. https://doi.org/10.1128/microbiolspec.FUNK-0016-2016

Kakvan N, Heydari A, Zamanizadeh HR, Rezaee S, Naraghi L (2013) Development of new bioformulations using *Trichoderma* and *Talaromyces* fungal antagonists for biological control of sugar beet damping–off disease. Crop Prot 53:80–84

Kashyap PL, Kumar S, Srivastava AK (2017) Nanodiagnostics for plant pathogens. Environ Chem Lett 15:7–13

Koka JA, Wani AH, Bhat MY, Parveen S (2017) *In vitro* efficacy of *Trichoderma* isolates against some fungi causing fungal rot disease of tomato. Int J Adv Res 5(3):2050–2053

Komy MHE, Saleh AA, Eranthodi A, Molan YY (2015) Characterization of novel Trichoderma asperellum isolates to select effective biocontrol agents against tomato Fusarium Wilt. Plant Pathol J 31(1):50–60

Kotasthane A, Agrawal T, Kushwah R, Rahatkar OV (2015) In vitro antagonism of *Trichoderma* spp. against *Sclerotium rolfsii* and *Rhizoctonia solani* and their response towards growth of cucumber, bottle gourd and bitter gourd. Eur J Plant Pathol 141:523–543

Kubicek CP, Druzhinina IS (2013) *Trichoderma*: genomic aspects of mycoparasitism and biomass degradation. Springer, Berlin

Kubicek CP, Herrera-Estrella A, Seidl-Seiboth V, Martinez DA, Druzhinina IS, Thon M et al (2011) Genome Biol. Comparative genome sequence analysis underscores mycoparasitism as the ancestral life style of *Trichoderma*. Genome Biol 12:R40

Kumar S (2013) Trichoderma: a biological control weapon for managing plant diseases and promoting sustainability. Int J Agric Sci Veter Med 1:106–121

Kumar V, Parkhi V, Kenerley CM, Rathore KS (2009) Defense-related gene expression and enzyme activities in transgenic cotton plants expressing an endochitinase gene from *Trichoderma virens* in response to interaction with *Rhizoctonia solani*. Planta 230:277–291

Kumar S, Roy PD, Lal M, Chand G, Singh V (2014) Mass multiplication and shelf life of *Trichoderma* species using various agro products. Int Quar J Life Sci 9:1143–1145

Latgé JP (2007) The cell wall: a carbohydrate Armour for the fungal cell. Mol Microbiol 66:279–290

Latorre BA, Lillo C, Rioja ME (2001) Eficacia de los tratamientos fungicidas para el control de *Botrytis cinerea* de lavid en función de la época de aplicación. Ciencia Investig Agrar 28:61–66

Leadbeater A (2015) Recent developments and challenges in chemical disease control. Plant Protect Sci 51:163–169

Lehner SM, Atanasova L, Neumann NK, Krska R, Lemmens M et al (2013) Isotope-assisted screening for iron-containing metabolites reveals a high degree of diversity among known and unknown siderophores produced by *Trichoderma* spp. Appl Environ Microbiol 79:18–31

Levy NO, Harel YM, Haile ZM, Elad Y, Rav–David E, Jurkevitch E, Katan J (2015) Induced resistance to foliar diseases by soil solarization and *Trichoderma harzianum*. Plant Pathol 64 (2):365–374

Li T, Tang J, Karuppiah V, Li Y, Xu N, Chen J (2020) Co-culture of *Trichoderma* atroviride SG3403 and *Bacillus subtilis* 22 improves the production of antifungal secondary metabolites. Biol Control 140:104122

Liton MJA, MKA B, Jannat R, Ahmed JU, Rahman MT, Tanbir Rubayet M (2019) Efficacy of *Trichoderma*-fortified compost in controlling soil-borne diseases of bush bean (*Phaseolus vulgaris* L.) and sustainable crop production. Adv Agric Sci 7(02):123–136

Liza B, Bora BC (2009) Comparative efficacy of *Thicoderma harzianum* and *Pseudomonas fluorescens* against *Meliodogyne incognita* and *Ralstonia solanacearum* complex on Bijnjal. Indian J Nematol 39(1):29–34

Lo CT (1997) Biological control of turfgrass diseases using *Trichoderma harzianum*. Plant Protect Bull 39:207–225

Lo CT, Nelson EB, Harman GE (1996) Biological control of turfgrass diseases with a rhizosphere competent strain of *Trichoderma harzianum*. Plant Dis 80:736–741

Lopes FAC, Steindorff AS, Geraldine AM, Brandao RS, Monteiro VN, Lobo M, Coelho ASG, Ulhoa CJ, Silva RN (2012) Biochemical and metabolic profiles of *Trichoderma* strains isolated from common bean crops in the Brazilian Cerrado, and potential antagonism against *Sclerotinia sclerotiorum*. Fungal Biol 116:815–824

López-Bucio J, Pelagio-Flores R, Herrera-Estrella A (2015) *Trichoderma* as biostimulant: exploiting the multilevel properties of a plant beneficial fungus. Sci Hortic 196:109–123

Lorito M, Farkas V, Rebuffat S, Bodo B, Kubicek CP (1996) Cell wall synthesis is a major target of mycoparasitic antagonism by *Trichoderma harzianum*. J Bacteriol 178:6382–6385

Lorito M, Woo SL, Harman GE, Monte E (2010) Translational research on *Trichoderma*: from 'omics to the field. Annu Rev Phytopathol 48:395–417

Luo Y, Ruan LF, Zhao CM, Wang CX, Peng DH, Sun M (2011) Validation of the intact zwittermicin A biosynthetic gene cluster and discovery of a complementary resistance mechanism in *Bacillus thuringiensis*. Antimicrob Agent Chemother 55:4161–4169

Macías-Rodríguez L, Guzmán-Gómez A, García-Juárez P, Contreras-Cornejo HA (2018) *Trichoderma atroviride* promotes tomato development and alters the root exudation of carbohydrates, which stimulates fungal growth and the biocontrol of the phytopathogen *Phytophtora cinnamomic* in a tripartite interaction system. FEMS Microbiol Ecol 94. https://doi.org/10.1093/femsec/fiy137

Mahalingam R, Prince L, Ambikapathy V, Panneerselvam A (2011) *In vitro* studies on biocontrol measures of seedling blight disease in sugarcane. World J Sci Tech 1(9):18–21

Majeed M, Hassan MG, Hassan M, Mohuiddin FA, Paswal S, Farooq S (2017) Damping off in chilli and Its biological management–a review. J Curr Microbiol Appl Sci 7(4):2175–2185

Maketon M, Apisitsantikul J, Siriraweekul C (2008) Greenhouse evaluation of *Bacillus subtilis* AP–01 and *Trichoderma harzianum* AP–001 in controlling tobacco diseases. Braz J Microbiol 39:296–300

Malmierca MG, Cardoza RE, Alexander NJ, McCormick SP, Hermosa R, Monte E et al (2012) Involvement of *Trichoderma trichothecenes* in the biocontrol activity and induction of plant defense–related genes. Appl Environ Microbiol 78:4856–4868

Manganiello G, Sacco A, Ercolano MR, Vinale F, Lanzuise S, Pascale A et al (2018) Modulation of tomato response to *Rhizoctonia solani* by *Trichoderma harzianum* and its secondary metabolite harzianic acid. Front Microbiol 9:1966

Marcello CM, Steindorff AS, Silva SP, Silva RN, Bataus LAM (2010) Expression analysis of the exo-β-1, 3-glucanase from the mycoparasitic fungus *Trichoderma asperellum*. Microbiol Res 165:75–81

Mathys J, De Cremer K, Timmermans P, Van Kerckhove S, Lievens B, Vanhaecke M, Cammue BPA, De Coninck B (2012) Genome-wide characterization of ISR induced in *Arabidopsis thaliana* by *Trichoderma hamatum* T382 against *Botrytis cinerea* infection. Front Plant–Microbe Interact 3:108

Martinez C, Blanc F, Le Claire E, Besnard O, Nicole M, Baccou JC (2001) Salicylic acid and ethylene pathways are differentially activated in melon cotyledons by active or heat-denatured cellulase from *Trichoderma longibrachiatum*. Plant Physiol 127:334–344

Martınez-Medina A, Pascual JA, Perez-Alfocea F et al (2010) *Trichoderma harzianum* and Glomus intraradices modify the hormone disruption induced by *Fusarium oxysporum* infection in melon plants. Phytopathology 100:682–688

Martìnez-Medina A, Fernandez I, Lok GB, Pozo MJ, Pieterse CMJ, Van Wees SCM (2017) Shifting from priming of salicylic acid- to jasmonic acid-regulated defences by *Trichoderma* protects tomato against the root knot nematode *Meloidogyne incognita*. New Phytol 213:1363–1377

Marzano M, Gallo A, Altomare C (2013) Improvment of biocontrol efficacy of *Trichoderma harzianum vs. Fusarium oxysporum f. sp. lycopersici* through UV induced tolerance to fusaric acid. Biol Control 67:397–408

Mcintyre M, Nielsen J, Arnau J, Brink H, Hansen K, Madrid S (2004) Proceedings of the 7th European conference on fungal genetics. Copenhagen, Denmark, pp 125–130

Meena M, Swapnil P, Zehra A, Dubey MK, Upadhyay RS (2017) Antagonistic assessment of *Trichoderma* spp. by producing volatile and non-volatile compounds against different fungal pathogens. Arch Phytopathol Plant Protect 50:629–648

Mendoza-Mendoza A, Zaid R, Lawry R, Hermosa R, Monte E, Horwitz A et al (2018) Molecular dialogues between *Trichoderma* and roots: role of the fungal secretome. Fungal Biol Rev 32:62–85

Miethke M (2013) Molecular strategies of microbial iron assimilation: from high affinity complexes to cofactor assembly systems. Metallomics 5:15–28

Misra AK (2007) Present status of important diseases of guava in India with special reference to wilt. ISHS 735:507–523

Mishra RK, Gupta RP (2012) *In vitro* evaluation of plant extracts bioagents and fungicides against purple blotch and Stemphylium blight of onion. J Med Plant Res 6(48):5840–5843

Mondejar R, Ros M, Pascual JA (2011) Mycoparasitism-related genes expression of *Trichoderma harzianum* isolates to evaluate their efficacy as biological control agent. Biol Control 56:59–66

Montero M, Sanz L, Rey M, Llobell A, Monte E (2007) Cloning and characterization of bgn16·3, coding for a β-1, 6-glucanase expressed during *Trichoderma harzianum* mycoparasitism. J Appl Microbiol 103:1291–1300

Moosa A, Sahi ST, Haq IU, Farzand A, Khan SA, Javaid K (2016) Antagonistic potential of *Trichoderma* isolates and manures against Fusarium wilt of tomato. Int J Veg Sci 23:207–218

Moradi H, Bahramnejad B, Amini J, Siosemardeh A, Haji-Allahverdipoor K (2012) Suppression of chickpea Fusarium wilt by *Bacillus subtillis* and *Trichoderma harzianum*. Plant Omics J 5:68–74

Moran-Diez ME, Trushina N, Lamdan NL, Rosenfelder L, Mukherjee PK et al (2015) Host-specific transcriptomic pattern of *Trichoderma virens* during interaction with maize or tomato roots. BMC Genomics 16:8

Morath SU, Hung R, Bennett JW (2012) Fungal volatile organic compounds: a review with emphasis on their biotechnological potential. Fungal Biol Rev 26:73–83

Mukherjee PM, Latha J, Hardar R, Horwitz BA (2003) TmkA, mitogen activated protein kinase of *Trichoderma virens* is involved in biocontrol properties and repression of conidiation in the dark. Eukaryot Cell 2:446–455

Mukherjee PK, Buensanteai N, Moran-Diez ME, Druzhinina IS, Kenerley CM (2012a) Functional analysis of non-ribosomal peptide synthetases (NRPSs) in *Trichoderma* virens reveals a polyketide synthase (PKS)/NRPS hybrid enzyme involved in the induced systemic resistance response in maize. Microbiology 158:155–165

Mukherjee PK, Horwitz BA, Kenerley CM (2012b) Secondary metabolism in *Trichoderma*—a genomic perspective. Microbiology 158:35–45

Mukherjee PK, Horwitz BA, Herrera-Estrella A, Schmoll M, Kenerley CM (2013) *Trichoderma* research in the genome era. Annu Rev Phytopathol 51:105–129

Naher L, Yusuf U, Ismail A, Hossain K (2014) *Trichoderma* Spp: a biocontrol agent for sustainable management of plant diseases. Pak J Bot 46(4):1489–1493

Naik MK, Amaresh YS, Ravikiran DS, Ashwathanarayan MG, Patil MD, Nair SP (2017) Morphological, molecular characterization of *Trichoderma* species isolated from different rhizosphere soils and its anti-pathogenic properties. Imp J Interdiscip Res 3:4

Napitupulu TP, Ilyas M, Kanti A, Sudiana IM (2019) *In vitro* evaluation of *Trichoderma harzianum* strains for the control of *Fusarium oxysporum* f. sp. *cubense*. Plant Pathol Quar 9(1):152–159

Narasimha Murthy K, Srinivas C (2013) Efficacy of *Trichoderma asperellum* against *Ralstonia solanacearum* under greenhouse conditions. Ann Plant Sci 2:342–350

Narasimha Murthy K, Uzma F, Srinivas C (2013) Induction of systemic resistance by *Trichoderma asperellum* against bacterial wilt of tomato caused by *Ralstonia solanacearum*. Int J Adv Res 1:181–194

Narasimha Murthy K, Soumya K, Chandranayak S, Niranjana SR, Srinivas C (2018) Evaluation of biological efficacy of *Trichoderma asperellum* against tomato bacterial wilt caused by *Ralstonia solanacearum*. Egypt J Biol Pest Control 28:63

Nogueira-Lopez G, Greenwood DR, Middleditch M, Winefield C, Eaton C, Steyaert JM et al (2018) The apoplastic secretome of *Trichoderma virens* during interaction with maize roots shows an inhibition of plant defense and scavenging oxidative stress secreted proteins. Front Plant Sci 5:409

Nygren K, Dubey M, Zapparata A, Iqbal M, Tzelepis GD, Durling MB et al (2018) The mycoparasitic fungus *Clonostachys rosea* responds with both common and specific gene expression during interspecific interactions with fungal prey. Evol Appl 11:931–949

O'Rourke TA, Scanlon TT, Ryan MH, Wade LJ, McKay AC, Riley IT, Li H, Sivasithamparam K, Barbetti MJ (2009) Severity of root rot in mature subterranean clover and associated fungal pathogens in the wheatbelt of Western Australia. Crop Pasture Sci 60:43–50

Oladipo OG, Awotoye A, Olayinka OO, Bezuidenhout CC, Maboeta MS (2018) Heavymetal tolerance traits of filamentous fungi isolated from gold and gemstone mining sites. Braz J Microbiol 49(1):29–37

Omann M, Zeilinger S (2010) How a mycoparasite employs G-protein signaling: using the example of *Trichoderma*. J Signal Transduction 2010:1. https://doi.org/10.1155/2010/123126

Ommati F, Zaker M (2012) *In vitro* and greenhouse evaluations of *Trichoderma* isolates for biological control of potato wilt disease (*Fusarium solani*). Arch Phytopathol Plant Protect 45 (14):1715–1723

Oros G, Naár Z (2017) Mycofungicide: *Trichoderma* based preparation for foliar applications. Am J Plant Sci 8:113–125

Pan S, Das A (2011) Control of cowpea (*Vigna sinensis*) root and collar rot (Rhizoctonia solani) with someorganic formulations of *Trichoderma harzianum* under field condition. J Plant Prot Sci 3(2):20–25

Pandey S, Pundhir VS (2013) Mycoparasitism of potato black scurf pathogen (*Rhizoctonia solani* Kuhn) by biological control agents to sustain production. Indian J Hort 70(1):71–75

Pandey G, Pandey RK, Pant H (2003) Efficacy of different levels of *Trichoderma viride* against root–knot nematode in chickpea (*Cicer arietinum* L.). Ann Plant Prot Sci 11:101–103

Pandey V, Ansari MW, Tula S, Yadav S, Sahoo RK et al (2016) Dose-dependent response of *Trichoderma harzianum* in improving drought tolerance in rice genotypes. Planta 243:1251

Pandya JR, Sabalpara AN, Chawda SK (2011) *Trichoderma*: a particular weapon for biological control of phytopathogens. J Agric Technol 7(5):1187–1191

Paramasivan M, Chandrasekaran A, Mohan S, Muthukrishnan N (2013) Ecological management of tropical sugar beet (TSB) root rot (Sclerotium rolfsii (Sacc.)) by rhizosphere *Trichoderma* species. Arch Phytopathol Plant Protect 47:1629–1644

Perazzoli M, Moretto M, Fontana P, Ferrarini A, Velasco R, Moser C, Delledonne M, Pertot I (2012) Downy mildew resistance induced by Trichoderma harzianum T39 in susceptible grapevines partially mimics transcriptional changes of resistant genotypes. BMC Genomics 13:660

Poddar RK, Singh DV, Dubey SC (2004) Integrated application of *Trichoderma harzianum* mutants and carbendazim to manage chickpea wilt (*Fusarium oxysporum f.sp. ciceris*). Indian J Agric Sci 74(6):346–348

Priyadharcini M, Akila R, Mini ML, Rajinimala N, Kannan R (2018) Exploration of *Trichoderma* spp. as an effective bio control agents against the Sclerotial wilt caused by *Sclerotium rolfsii* Sacc. Int J Curr Microbiol App Sci 7(8):1672–1682

Qin WT, Zhuang WY (2016) Seven wood-inhabiting new species of the genus *Trichoderma* (Fungi, Ascomycota) in Viride clade. Sci Rep 6:27074

Raghavendra VB, Siddalingaiah L, Sugunachar NK, Nayak C, Ramachandrappa NS (2013) Induction of systemic resistance by biocontrol agents against bacterial blight of cotton caused by *Xanthomonas campestris pv. malvacearum*. ESci J Plant Pathol 2:59–69

Rahman MA, Begum MF, Alam MF (2009) Screening of *Trichoderma* isolates as a biological control agent against *Ceratocystis paradoxa* causing pineapple disease of sugarcane. Mycobiology 37:277–285

Rai S, Kashyap PL, Kumar S et al (2016) Identification, characterization and phylogenetic analysis of antifungal *Trichoderma* from tomato rhizosphere. Springer Plus 5:1939

Rajendraprasad M, Vidyasagar B, Devi GU, Rao SRK (2017) Biological control of tomato damping off caused by *Pythium debaryanum*. Int J Chem Stud 5(5):447–452

Rajeswari P (2019) Combination of *Trichoderma viride* and *Pseudomonas fluorescens* for the enhanced control of Fusarium wilt disease caused by *Fusarium oxysporum* infecting *Arachis hypogaea* L. J Appl Nat Sci 11(1):138–143

Rao GS, Reddy NNR, Surekha C (2015) Induction of plant systemic resistance in legumes *Cajanus cajan, Vigna radiata, Vigna mungo* against plant pathogens *Fusarium oxysporum* and *Alternaria alternata* – a *Trichoderma viride* mediated reprogramming of plant defense mechanism. Int J Rec Sci Res Res 6:4270–4280

Rawal P, Sharma P, Singh ND, Joshi A (2013) Evaluation of fungicides, neem bioformulations and biocontrol agent for the management of root rot of safed musli caused by *Rhizoctonia solani*. J Mycol Plant Pathol 43(30):297

Reino JL, Guerrero RF, Hernandez-Galan R, Collado IG (2008) Secondary metabolites from species of the biocontrol agent *Trichoderma*. Phytochemistry 7:89–123

Reithner B, Ibarra-Laclette E, Mach RL, Herrera-Estrella A (2011) Identification of mycoparasitism-related genes in *Trichoderma atroviride*. Appl Environ Microbiol 77:4361–4370

Reithner B, Mach-Aigner AR, Herrera-Estrella A, Mach RL (2014) The transcriptional regulator *Xyr1* of *Trichoderma atroviride* supports the induction of systemic resistance in plants. Appl Environ Microbiol 80:5274–5281

Rey M, Delgado-Jarana J, Benitez T (2001) Improved antifungal activity of a mutant of *Trichoderma harzianum* CECT 2413 which produces more extracellular proteins. Appl Microbiol Biotechnol 55:604–608

Rocha-Ramírez V, Omero C, Chet I, Horwitz BA, Herrera-Estrella A (2002) *Trichoderma atroviride* G-protein α-subunit gene tag1 is involved in mycoparasitic coiling and conidiation. Eukaryot Cell 1:594–605

Roop S, Jagtap GP (2017) *In Vitro* evaluation of antibacterial chemicals and bioagents against *Ralstonia solanacearum* infecting bacterial wilt in Ginger. Int J Curr Microbiol App Sci 6:2034–2045

Rosado IV, Rey M, Codon AC, Govantes J, MorenoMateos MA, Benitez T (2007) QID74 cell wall protein of *Trichoderma harzianum* is involved in cell protection and adherence to hydrophobic surfaces. Fungal Genet Biol 44:950–964

Rubio MB, Quijada NM, Pérez E, Domínguez S, Monte E, Hermosa R (2014) Identifying beneficial qualities of *Trichoderma parareesei* for plants. Appl Environ Microbiol 80:1864–1873

Sabbagh SK, Roudini M, Panjehkeh N (2017) Systemic resistance induced by *Trichoderma harzianum* and *Glomus mossea* on cucumber damping–off disease caused by *Phytophthora melonis*. Arch Phytopathol Plant Protect 50:375–388

Sahebani N, Hadavi N (2008) Biological control of the root-knot nematode *Meloidogyne javanica* by *Trichoderma harzianum*. Soil Biol Biochem 40:2016–2020

Saksirirat W, Chareerak P, Bunyatrachata W (2009) Induced systemic resistance of biocontrol fungus, *Trichoderma* spp. against bacterial and gray leaf spot in tomatoes. Asian J Food Agro Ind 2:99–104

Salas-Marina MA, Isordia-Jasso M, Islas-Osuna MA, Delgado-Sánchez P, Jiménez-Bremont JF et al (2015) The Epl1 and Sm1 proteins from *Trichoderma atroviride* and *Trichoderma virens* differentially modulate systemic disease resistance against different life style pathogens in *Solanum lycopersicum*. Front Plant Sci 23:77

Saloheimo M, Paloheimo M, Hakola S, Pere J, Swanson B, Nyyssönen E, Bhatia A, Ward M, Swollenin MP (2002) A *Trichoderma reesei* protein with sequence similarity to the plant expansins, exhibits disruption activity on cellulosic materials. Eur J Biochem 269:4202–4211

Samolski I, Rincon AM, Pinzón LM, Viterbo A, Monte E (2012) The *qid74* gene from *Trichoderma harzianum* has a role in root architecture and plant biofertilization. Microbiology 158:129–138

Sandipan Prashant B, Chaudhary RF, Shanadre CM, Rathod NK (2015) Appraisal of diverse bioagents against soft rot bacteria of potato caused by *Erwinia carotovora* sub sp. *carotovora* under in vitro test. Eur J Pharm Med Res 2:495–500

Sanz L, Montero M, Redondo J, Llobell A, Monte E (2005) Expression of an alpha-1,3-glucanase during mycoparasitic interaction of *Trichoderma asperellum*. FEBS J 272:493–499

Saravanakumar D, Ciavorella A, Spadaro D, Garibaldi A, Gullino ML (2008) *Metschnikowia pulcherrima* strain MACH1 outcompetes *Botrytis cinerea*, *Alternaria alternata* and *Penicillium expansum* in apples through iron depletion. Postharvest Biol Technol 49:121–128

Saravanakumar K, Yu C, Dou K, Wang M, Li Y, Chen J (2015) Synergistic effect of *Trichoderma*-derived antifungal metabolites and cell wall degrading enzymes on enhanced biocontrol of *Fusarium oxysporum* f. sp. *cucumerinum*. Biol Control 94:37–46

Saravanakumar K, Fan L, Fu K, Yu C, Wang M, Xia H et al (2016) Cellulase from *Trichoderma harzianum* interacts with roots and triggers induced systemic resistence to foliar disease in maize. Sci Rep 10:35543

Sarma BK, Yadav SK, Patel JS, Singh HB (2014) Molecular mechanisms of interactions of *Trichoderma* with other fungal species. Open Mycol J 8:140–147

Saxena A, Raghuwanshi R, Singh HB (2015) *Trichoderma* species mediated differential tolerance against biotic stress of phytopathogens in *Cicer arietinum* L. J Basic Microbiol 55:195–206

Scharf DH, Brakhage AA, Mukherjee PK (2016) Gliotoxin e bane or boon? Environ Microbiol 18:1096–1109

Schirmböck M, Lorito M, Wang YL et al (1994) Parallel formation and synergism of hydrolytic enzymes and peptaibol antibiotics, molecular mechanisms involved in the antagonistic action of *Trichoderma harzianum* against phytopathogenic fungi. Appl Environ Microbiol 60:4364–4370

Seethapathy P, Kurusamy R, Kuppusamy P (2017) Soil borne diseases of major pulses and their biological management. An Int J Agric 2(1):1–11

Segarra G, Van der Ent S, Trillas I, Pieterse CM (2009) MYB72, a node of convergence in induced systemic resistance triggered by a fungal and a bacterial beneficial microbe. Plant Biol 11:90–96

Segarra G, Casanova E, Aviles M, Trillas I (2010) *Trichoderma asperellum* strain T34 controls Fusarium wilt disease in tomato plants in soilless culture through competition for iron. Microb Ecol 59:141–149

Seidl V, Huemer B, Seiboth B, Kubicek CP (2005) A complete survey of *Trichoderma* chitinases reveals three distinct subgroups of family 18 chitinases. FEBS J 272:5923–5939

Seidl V, Song L, Lindquist E et al (2009) Transcriptomic response of the mycoparasitic fungus *Trichoderma atroviride* to the presence of a fungal prey. BMC Genomics 10:567

Senthil R, Kuppusamy P, Lingan R, Gandhi K (2011) Efficacy of different biological control agents against major postharvest pathogens of grapes under room temperature storage conditions. Phytopathol Mediterr 50:55–65

Shah JM, Raghupathy V, Veluthambi K (2009) Enhanced sheath blight resistance in transgenic rice expressing an endochitinase gene from *Trichoderma virens*. Biotechnol Lett 31:239–244

Sharma KK (2018) *Trichoderma* in agriculture: an overview of global scenario on research and its application. Int J Curr Microbiol App Sci 7(08):1922–1933

Sharma P, Trivedi PC (2010) Evaluation of different fungal antagonists against *Fusarium oxysporum* infecting *Withania somnifera* (L.) Dunal. Biol Env Sci 6:37–41

Sharma P, Sain SK, James S (2003) Compatibility study of *Trichoderma* isolates with fungicides against damping-off of cauliflower and tomato caused by *Pythium aphanidermatum*. Pestic Res J 15(2):133–138

Sharma KK, Singh US, Sharma P, Kumar A, Sharma L (2015) Seed treatments for sustainable agriculture—a review. J Appl Nat Sci 7(1):521–539

Sharma S, Rai P, Rai S, Srivastava M et al (2017) Genomic revolution in crop disease diagnosis: a review. In: Singh SS (ed) Plants and microbes in an ever changing environment. Nova Science Publishers, Hauppauge, pp 257–293

Shi M, Chen L, Wang XW, Zhang T, Zhao PB, Song XY, Sun CY, Chen XL, Zhou BC, Zhang YZ (2012) Antimicrobial peptaibols from *Trichoderma pseudokoningii* induce programmed cell death in plant fungal pathogens. Microbiology 158:166–175

Shoresh M, Yedidia I, Chet I (2005) Involvement of jasmonic acid/ethylene signaling pathway in the systemic resistance induced in cucumber by *Trichoderma asperellum* T203. Phytopathology 95:76–84

Shoresh M, Gal-On A, Leibman D, Chet I (2006) Characterization of a mitogen-activated protein kinase gene from Cucumber required for *Trichoderma*—conferred plant resistance. Plant Physiol 142:1169–1179

Shukla N, Awasthi RP, Rawat L, Kumar J (2015) Seed biopriming with drought tolerant isolates of *Trichoderma harzianum* promote growth and drought tolerance in *Triticum aestivum*. Ann Appl Biol 166:171–182

Singh US, Mishra DS, Zaidi NW, Varshney S, Sharma R, Singh N (2005) Potential and effectiveness of fungi and bacteria as biocontrol agents for plant disease management. In: Integrated pest management: principles and application, vol 1: principles. CBS, New Delhi

Singh HB, Singh A, Sarma BK (2014) *Trichoderma viride* 2% WP formulation suppresses tomato wilt caused by *Fusarium oxysporum* f. sp. lycopersici and chilli damping-off caused by *Pythium aphanidermatum* effectively under different agroclimatic conditions. Int J Agric Environ Biotechnol 7:313–320

Sivan A, Chet I (1989) The possible role of competition between *Trichoderma harzianum* and *Fusarium oxysporum* on rhizosphere colonization. Phytopathology 79:198–203

Sivasithamparam K, Ghisalberti EL (1998) In: Kubicek CP, Harman GE (eds) *Trichoderma* and Gliocladium, vol 1, pp 139–191

Siyar S, Inayat N, Hussain F (2019) Plant growth promoting Rhizobacteria and plants' improvement—a mini-review. PSM Biol Res 4(1):1–5

Smitha C, Finosh GT, Rajesh R, Abraham PK (2017) Induction of hydrolytic enzymes of phytopathogenic fungi in response to *Trichoderma viride* influence biocontrol activity. Int J Curr Microbiol App Sci 3(9):1207–1217

Sreedevi B, Charitha Devi M, Saigopal DVR (2012) Production and optimization of chitinase by Trichoderma harzianum for control of the phytopathogenic fungus M. Phaseolina. Agri Sci Digest 32(3):224–228

Sriram S, Savitha MJ, Ramanujam B (2010) Trichoderma–enriched coco–peat for the management of Phytophthora and Fusarium diseases of chilli and tomato in nurseries. J Biol Control 24:311–316

Srivastava M, Shahid M, Pandey S, Singh A, Kumar V, Gupta S, Maurya M (2014) *Trichoderma* genome to genomics: a review. J Data Min Genomic Proteomic 5:1000172

Stoppacher N, Kluger B, Zeilinger S, Krska R, Schuhmacher R (2010) Identification and profiling of volatile metabolites of the biocontrol fungus *Trichoderma atroviride* by HS-SPME-GC-MS. J Microbiol Methods 81:187–193

Suhanna NY, Hartinee A (2013) Application of *Trichoderma* spp. to control stem end rot disease of mango var. J Trop Agric Fd Sc 41(1):159–168

Sumana K, Devaki NS (2012) In vitro evaluations of some bioagents against tobacco wilt pathogen. J Biopest 5(1):18–22

Thakkar A, Saraf M (2014) Development of microbial consortia as a biocontrol agent for effective management of fungal diseases in Glycine max L. Arch Phytopathol Plant Protect 48:459–474

Tijerino A, Cardoza RE, Moraga J, Malmierca MG, Vicente F et al (2011) Overexpression of the trichodiene synthase gene tri5 increases trichodermin production and antimicrobial activity in *T. brevicompactum*. Fungal Genet Biol 48:285–296

Timila RD, Manandhar S (2016) Biocontrol efficacy of *Trichoderma* spp. against *Phytophthora* blight of Pepper (Abstract). In: Proceedings of first international horticulture conference, 8–11 Apr, NARC, Kathmandu, pp 277

Tjamos EC, Fravel D (1995) Detrimental effects of sub-lethal heating and Talaromyces flavus on microsclerotia of *Verticillium dahliae*. Phytopathology 85:388–392

Toghueoa RMK, Ekea P, González IZ, de Aldana BRV, Nana LW, Boyom FF (2016) Biocontrol and growth enhancement potential of two endophytic *Trichoderma* spp. from *Terminalia catappa* against the causative agent of Common Bean Root Rot (*Fusarium solani*). Biol Control 96:8–20

Tsahouridou PC, Thanassoulopoulos CC (2002) Proliferation of *Trichoderma koningii* in the tomato rhizosphere and the suppression of damping-off by *Sclerotium rolfsii*. Soil Biol Biochem 34(6):767–776

Tuão Gava CA, Pinto JM (2016) Biocontrol of melon wilt caused by *Fusarium oxysporum* Schlect f. sp. *melonis* using seed treatment with *Trichoderma* spp. and liquid compost. Biol Control 97:13–20

Vargas WA, Mandawe JC, Kenerley CM (2009) Plant-derived sucrose is a key element in the symbiotic association between *Trichoderma virens* and maize plants. Plant Physiol 151:792–808

Vasanthakumari MM, Shivanna MB (2013) Biological control of anthracnose of chilli with rhizosphere and rhizoplane fungal isolates from grasses. Arch Phytopathol Plant Protect 46 (14):1641–1666

Verma M, Brar SK, Tyagi RD, Surampalli RY, Valero JR (2007) Antagonistic fungi, *Trichoderma* spp.: panoply of biological control. Biochem Eng J 37:1–20

Verma JP, Yadav J, Tiwari NK, Jaiswal KD (2014) Evaluation of plant growth promoting activities of microbial strains and their effect on growth and yield of chickpea (*Cicer arietinum* L.) in India. Soil Biol Biochem 70:33–37

Vinale F, Sivasithamparam K, Ghisalberti E, Marra R, Woo S, Lorito M (2008) *Trichoderma*–plant–pathogen interactions. Soil Biol Biochem 40:1–10

Vinale F, Ghisalberti EL, Sivasithamparam K, Marra R, Ritieni A, Ferracane R, Woo S, Lorito M (2009) Factors affecting the production of *Trichoderma harzianum* secondary metabolites during the interaction with different plant pathogens. Lett Appl Microbiol 48:705–711

Vinale F, Sivasithamparam K, Ghisalberti EL, Ruocco M, Woo S, Lorito M (2012) *Trichoderma* secondary metabolites that affect plant metabolism. Nat Prod Commun 7:1545–1550

Vinale F, Nigro M, Sivasithamparam K, Flematti G, Ghisalberti EL, Ruocco M, Varlese R, Marra R, Lanzuise S, Eid A, Woo SL, Lorito M (2013) Harzianic acid: a novel siderophore from *Trichoderma harzianum*. FEMS Microbiol Lett 347:123–129

Vinale F, Sivasithamparam K, Ghisalberti EL, Woo SL, Nigro M, Marra R, Lombardi N et al (2014) *Trichoderma* secondary metabolites active on plants and fungal pathogens. Open Mycol J 8:127–139

Vinodkumar S, Indumathi T, Nakkeeran S (2017) *Trichoderma* asperellum (NVTA2) as a potential antagonist for the management of stem rot in carnation under protected cultivation. Biol Control 113:58–64

Viscardi S, Ventorino V, Duran P, Maggio A, De Pascale S, Mora ML et al (2016) Assessment of plant growth promoting activities and abiotic stress tolerance of *Azotobacter chroococcum* strains for a potential use in sustainable agriculture. J Soil Sci Plant Nutr 16:848–863

Viterbo A, Haran S, Friesem D, Ramot O, Chet I (2001) Antifungal activity of a novel endochitinase gene (chit36) from *Trichoderma harzianum* Rifai TM. Microbiol Lett 200:169–174

Viterbo A, Harel M, Chet I (2004) Isolation of two aspartyl proteases from *Trichoderma asperellum* expressed during colonization of cucumber roots. FEMS Microbiol Lett 238:151–158

Viterbo A, Wiest A, Brotman Y, Chet I, Kenerley CM (2007) The 18mer peptaibols from *Trichoderma virens* elicit plant defence responses. Mol Plant Pathol 8:737–746

Vos CM, De Cremer K, Cammue BPA, De Coninck B (2015) The toolbox of *Trichoderma* spp. in the biocontrol of *Botrytis cinerea* disease. Mol Plant Pathol 16:400–412

Waghunde RR, Shelake MR, Sabalpara NA (2016) *Trichoderma*: a significant fungus for agriculture and environment. Afr J Agric Res 11:1952–1965

Wang M, Ma J, Fan L, Kehe F, Yu C, Gao J, Li Y, Chen J (2015) Biological control of southern corn leaf blight by *Trichoderma atroviride* SG3403. Biocontrol Sci Technol 25:1133–1146

Weindling R (1934) Studies on lethal principle effective in the parasitic action of *Trichoderma lignorum* on *Rhizoctinia solani* and other soil fungi. Phytopathology 24:1153–1179

Woo SL, Ruocco M, Vinale F, Nigro M, Marra R, Lombardi N, Pascale A, Lanzuise S, Manganiello G, Lorito M (2014) *Trichoderma*-based products and their widespread use in agriculture. Open Mycol J 8:71–126

Yedidia I, Shoresh M, Kerem Z et al (2003) Concomitant induction of systemic resistance to *Pseudomonas syringae* pv. *lachrymans* in cucumber by *Trichoderma asperellum* (T-203) and accumulation of phytoalexins. Appl Environ Microbiol 69:7343–7353

Yoshioka Y, Ichikawa H, Naznin HA, Kogure A, Hyakumachi M (2011) Systemic resistance induced in Arabidopsis thaliana by *Trichoderma asperellum* SKT–1, a microbial pesticide of seed borne diseases of rice. Pest Manag Sci. https://doi.org/10.1002/ps.2220

You J, Zhang J, Wu M, Yang L, Chen W, Li G (2016) Multiple criteria-based screening of *Trichoderma* isolates for biological control of *Botrytis cinerea* on tomato. Biol Control 101:31–38

Yuan S, Meiyun L, Fang Z, Yan L, Wen S et al (2016) Biological control of tobacco bacterial wilt using *Trichoderma harzianum* amended bioorganic fertilizer and the arbuscular mycorrhizal fungi *Glomus mosseae*. Biol Control 92:164–171

Zaidi NW, Singh US (2018) Trichoderma an impeccable plant health booster. In: Anwer MA (ed) Biopesticides and Bioagents: novel tools for pest management. Apple Academic Press, USA, pp 17–42

Zeilinger S, Gruber S, Bansal R, Mukherjee PK (2016) Secondary metabolism in *Trichoderma*—chemistry meets genomics. Fungal Biol Rev 30:74–90

Zhang F, Yang X, Ran W, Shen Q (2014) *Fusarium oxysporum* induces the production of proteins and volatile organic compounds by *Trichoderma harzianum* T-E5. FEMS Microbiol Lett 359 (1):116–123

Zhang X, Li X, Yang H, Cui Z (2018) Biochemical mechanism of phytoremediation process of lead and cadmium pollution with *Mucor circinelloides* and *Trichoderma asperellum*. Ecotoxicol Environ Saf 157:21–28

Chapter 12
Biotechnological Application of *Trichoderma*: A Powerful Fungal Isolate with Diverse Potentials for the Attainment of Food Safety, Management of Pest and Diseases, Healthy Planet, and Sustainable Agriculture

Charles Oluwaseun Adetunji and Ajit Varma

Abstract Numerous *Trichoderma* species have been highlighted as sustainable biotechnological tools exploited for the mass fabrication of biofertilizers and biofungicides as well as for the biological management of diseases and pests which constitute a major factor that mitigate against an increase in agricultural productivities. *Trichoderma* species possessed the capability to suppress the diseases that normally affect plant growth, and they could enhance crop improvement, thereby leading to an increase in agricultural production. Some of the various mechanisms of actions which enable them to perform such significant function includes myco-parasitism, induced systemic resistance, improved nutrient efficiency, and antibiosis. The application of various biotechnological techniques like proteomics, genomics, bioinformatics, and especially metabolomics has given a better insight into beneficial metabolites with uncountable active metabolites having antibacterial, antifungal, and bioremediation properties. This constitutes a major reason for their utilization for the biological control of diseases and pests affecting which has been a major constrain to increase agricultural productivities and even post-harvest management of agricultural commodities. The secretion of volatile metabolites from *Trichoderma* spp. has been shown to portend the capability to prompt resistance to plant pathogens, thereby enhancing plant health and their potential role for the bioremediation of heavily polluted agricultural soil. Therefore, this chapter intends to provide recent trends on the utilization of *Trichoderma* spp. as a biotechnological tool for the attainment of food safety, management of pest and

---

C. O. Adetunji (✉)
Applied Microbiology, Biotechnology and Nanotechnology Laboratory, Department of Microbiology, Edo University Iyamho, Auchi, Edo State, Nigeria
e-mail: adetunji.charles@edouniversity.edu.ng

A. Varma
Amity Institute of Microbial Technology, Amity University, Uttar Pradesh, Noida, India

C. Manoharachary et al. (eds.), Trichoderma: *Agricultural Applications and Beyond*,
Soil Biology 61, https://doi.org/10.1007/978-3-030-54758-5_12

diseases, healthy planet, bioremediation of polluted agricultural soil, and sustainable agriculture. Moreover, the mechanism of action utilized by these *Trichoderma* spp. against the pathogens and agricultural pests has been discussed in detail.

Keywords Biotechnological · *Trichoderma* · Food safety · Pest and diseases · Healthy planet · Sustainable · Agriculture · Environment

12.1 Introduction

Global agriculture has experienced a lot of tremendous achievements which might be linked to the recent trends and biotechnological application of sustainable agricultural technologies (Adetunji et al. 2017, 2018a, b, 2019a, b, 2020; Adetunji and Adejumo 2017, 2018, 2019). The unwarranted and unsuitable utilization of agrochemicals has irrefutably given rise to lots of hazards on human health and the ecosystem. Moreover, in an endeavor to increase yields of numerous agricultural crops, the farmer might have applied an overdose of agro-pesticides which might lead to a high rate of disease-resistant, high rate of soil degradation, pollution of groundwater, and contamination of ecosystem. The issue of pesticide residues has generated lots of concern about food safety among numerous consumers and has been a major barrier affecting trade for export crops (Ricardo et al. 2018; Adetunji et al. 2018a, b). Therefore, in view of the aforementioned, there is a need to search for a sustainable solution that could mitigate all the highlighted barriers that could ensure food security, food safety, and an increase in agricultural production while maintaining a healthy planet.

The world population has been forecasted to increase drastically to nine billion thereabout in the year 2050 (FAO 2017). Therefore, the application of biological techniques using beneficial microorganisms could be a sustainable and permanent replacement to all the adverse effects experienced whenever different agrochemicals are applied. The application of beneficial microorganisms in agriculture for pest management will be a fruitful strategy and a first step toward the realization of integrating biological control agents for adequate control of diseases and agricultural pests. Some of the factors that distinguished them as a potential candidate include growth promotion, aggressive trait against pathogens available in the microbial community around the rhizosphere region of a plant, improved host confrontation to abiotic stresses and biotic stress, ability to overpower invading pathogens, improved nutrient accessibility and uptake, and the overall enhancement of plant health (Harman 2000; Harman et al. 2004; Vinale et al. 2008). Some other significant features of these beneficial microorganisms include initiation of plant resistance mechanisms and growth, killing or parasitism of pathogens with the aid of antibiotics which work in a synergetic effect with cell wall-degrading enzymes which are produced extracellular, pugnacious for nutrients at various colonization sites (Howell 2003; Whipps 2001).

Trichoderma spp. has been reported as a biotechnological tool and natural sustainable resource that could be utilized for the prevention of several diseases and pests affecting increment in agricultural productivity. In order to quench these biotic stressors mitigating against an increase in agriculture productivity, the attention of several researchers has been diverted to implement research, innovatory discovery, and development in numerous fields that could be a sustainable solution to these monumental challenges. The application of *Trichoderma* spp. might be utilized for the prevention of parasitic nematodes, pest and pathogens, insects, and agricultural weeds (Gardener and Fravel 2002). Also, they have been shown to possess the capability to enhanced development and adequate root growth of plants, increased crop productivity, adequate absorption of nutrients, and management of biotic and abiotic stress (Harman et al. 2004). Numerous *Trichoderma* spp. have been documented to influence plant positively by enhancing plant growth, prevention of bacterial and fungal attack. They are utilized in the biological plant protection majorly for the production of biofungicides and for the bioremediation of polluted soil (Blaszczyk et al. 2014). *Trichoderma* spp. could prevent pests and pathogens by preventing detrimental pathogens available in the rhizospheric region by their effective mycoparasitic and antagonistic activities (Hermosa et al. 2012).

Moreover, several biotechnological potentials of *Trichoderma* spp. have been documented by several scientists from Nigeria. Fajola et al. (1975) tested the effectiveness of the cultural filtrates produced from *Trichoderma harzianum* Rifai for the management of *Pythium aphanidermatum* responsible for the damping-off disease of tobacco in most areas of Nigeria. The result obtained showed that it prevented the mycelia growth, zoospore germination, and inhibited the germ tube elongation. The authors also reported the biological control activities of *T. harzianum* against damping-off disease when carried out in unsterilized and sterilized soil.

Jonathan et al. (2017) assessed the tolerance and the reaction of *Trichoderma harzianum* isolated from the rhizosphere of grasses found in an environment that has been contaminated with crude oil. Their experiment indicated that strain Asemo J was the best isolated that could tolerate pollution among all the 50 isolates and was further subjected to molecular and morphological characteristics. The molecular characteristics were performed using amplified 18 s (1609–1627) and 28 s (287–266) rRNA regions by ITS1/ITS4. Also, manganese peroxidase (mnp) genes and lignin peroxidase (lig1–6) were detected from strain using RT-PCR methodology. The strain asemoJ showed that a maximum occurrence of 80% was observed when bioremediation of contaminated soil was later identified as *Trichoderma harzianum*. The active strain asemoJ was later deposited at the gene bank with accession number KY488466. The peroxidase gene was detected in the strain with molecular weight which varies between 900 and 1000 bp for lig2, lig4, and mnp genes, respectively, when compared to lig6 which demonstrated smaller sizes. The enzymatic activity by this strain was expressed as aliquots (U/mL). Moreover, the highest enzymatic activity of lignin peroxidase was 90 ± 0.87 U/mL while that of manganese peroxidase was 120 ± 1.23 U/mL, respectively. Their result implies that strain KY488466 isolated from Nigeria could be utilized for ecorestoration of oil-spilled polluted soil in addition to their other biotechnological utilizations in diverse sectors.

Arotupin and Ogunmolu (2012) evaluated the influence of nitrogen and carbon sources on the simultaneous synthesis of polygalacturonase and amylase, in order to select the best maximum nutrient that will support the highest production of enzymes. The strain of *Trichoderma viride* utilized for this experiment was strain BITRS-1001 of *Trichoderma viride* that has been previously reported for high production of polygalacturonase and amylase. The result obtained demonstrated that the various nitrogen and carbon sources significantly influence the production of polygalacturonase and amylase when performed in submerged fermentation. Moreover, it was detected that the amendment of the basal medial with the casein and maltose serving as the nitrogen and carbon sources influence the rate of enzyme production.

Gwa and Nwankiti (2017) tested the biocontrol effectiveness of *Trichoderma harzianum* against *Fusarium moniliforme* responsible for the rotting of *Dioscorea rotundata* tubers. Some of the rot-inducing fungus isolated from this area include *Pestalotia* sp., *Botryodiploidia theobromae*, *Penicillium purpurogenum*, *Aspergillus flavus*, *Fusarium oxysporum Penicillium purpurogenum*, *Aspergillus ochraceus*, and *Fusarium moniliforme.* The strain of *Fusarium moniliforme* was selected out of the several rots inducing microorganisms because the pathogenicity test performed showed that it had the highest occurrence and it is responsible for the high level of rottenness in yam, especially in some locations in Makurdi town in Nigeria. The result obtained shows that *Trichoderma harzianum* exhibited the highest percentage growth inhibition value of 58.70% against *F. moniliforme* after 48 h of incubation. Their study showed that *Trichoderma harzianum* could be used for the post-harvest management of yam tuber rot.

Terna et al. (2013) evaluated the biocontrol efficacy of *T. viride* isolated from the soil in Nigeria. The authors tested the effectiveness of *T. viride* against three common post-harvest rot pathogens in an in vitro assay. These rot-inducing microorganisms include *F. oxysporum*, *A. flavus*, and *A. niger*, respectively. The authors also screened the presence of three extracellular enzymes which include chitinase, cellulose, and protease from their enzyme-induced culture filtrates. The numerous lytic enzyme-induced culture filtrates present in the isolate of *T. viride* were tested against all the post-harvest rot fungi using Potato Dextrose Agar. The result obtained showed that the combined effect of three enzymes present in the cultural filtrates of *T. viride* influenced the maximum radial growth inhibitions against *A. flavus* with 67.89% and *A. niger* with 77.69%, respectively. Their study showed that *T. viride* strain exploited in their study could be utilized as a biological control agent against all the post-harvest rot pathogens used during their study.

Ekundayo et al. (2016) isolated *Trichoderma viride* from the maize cob which was utilized for chitinase production. The influence of some metal, pH, and temperature on the level of chitinase production was performed. The purification of chitinase was carried out using Diethylaminoethyl cellulose-cellulose ion-exchange chromatography, ammonium sulfate precipitation, and sephadex G-100 gel filtration, respectively. The optimum condition for the maximum production of chitinase was pH of 5 and temperature of 50 °C while there was a decrease in the level of chitinase production when exposed to $MnCl_2$ and EDTA as metal ions. The temperature of 40 and 50 °C was found as the most stable temperature that enhanced the stability of the chitinase enzyme.

Ekefan et al. (2000) evaluated the effect of four isolates of *Trichoderma harzianum* (Th-F, Th-G, Th-I, and Th-N) as a BCA for the management of *Colletotrichum capsici* responsible the anthracnose of pepper. The result obtained from an in vitro assay revealed that *Trichoderma harzianum* decreases the colony radius of *C. capsici* in comparison to treatment without any BCA. Moreover, it was observed that all the *Trichoderma harzianum* strains used in the seeds coating of *C. capsici* significantly enhance all the growth parameters tested as well as decrease the occurrence of the pathogen on seeds and soil. Also, all the strains of *Trichoderma harzianum* (Th-F, Th-G, Th-I, and Th-N) exhibited a more enhanced biological control activity when compared to the chemical control using benomyl. The author later suggested that there is a need to evaluate further the effectiveness of *Trichoderma harzianum* for the post-harvest management and field trial for the control of anthracnose of pepper.

Adedeji et al. (2008) evaluated the effect of five *Trichoderma* strains isolated from cocoa pod. The in vitro experiment was performed against *Phytophthora megakarya* responsible for the *Cacao* pod-rot using potato dextrose agar. The result obtained showed that the *Trichoderma* strains decrease the diseases from 95.0 to 25.0% when compared to the control. Moreover, it was observed that the Early Infection Index was minimal especially for all the treatment containing *Trichoderma* strains when compared to the control. Also, the maximum Early Infection Index of 16.03% was observed from strain NIG-T293 in comparison with the control that had 100%. Their study showed that the screened *Trichoderma* strains isolated from cocoa could be utilized as a BCA for the prevention of the *Cacao* pod-rot especially on the field.

This chapter describes a detailed and current insight on the application of *Trichoderma* spp. as a biotechnological tool for the attainment of food safety, management of pests and diseases, healthy planet, bioremediation of polluted agricultural soil, and sustainable agriculture. Moreover, the mechanism of action utilized by *Trichoderma* spp. used in the prevention of pests and pathogens has been discussed in detail.

12.2 Application of *Trichoderma* spp. as a Dependable Biotechnology Tool for the Post-harvest Prevention of Diseases and Attainment of Food Safety

Post-harvest diseases have been highlighted as a significant challenge mitigating against the achievement of an increase in the production of food. The biotic factor has been emphasized as a factor responsible for the various problems encountered during different pre- and post-harvest stages (Nunes 2012). The application of post-harvest treatment will go a long way in achieving an enhanced extension of post-harvest commodities, especially fruits and vegetables. The problem of the high cost of modified and control storage system which poor farmers could not afford in most developing countries, high level of illiteracy, limited transportation, and inadequate storage system encouraging the introduction of diverse pathogens constitute some of the major reasons behind the problem encountered in post-harvest management of commodities (Sharma et al. 2009). The utilization of synthetic pesticides has been known as a common means toward the management of pathogens mitigating against the development of agricultural commodities. Their utilization has been recognized to involve several challenges like a high level of pathogen resistance and environmental and health issues (Gutiérrez-Martínez et al. 2018). Therefore, there is a need to search for an alternative from a natural environment with biological control attributes toward various pests and pathogens affecting the development of agricultural commodities during various post-harvest, pre-harvest, and during-harvesting stages. Also, the isolation of a biocontrol agent from the extreme environment will also go a long way (Medina-Cordova et al. 2018; Sharma et al. 2009).

Trichoderma spp. has been identified as a sustainable biological control agent that could be utilized for the management of several post-harvest pathogens like guava (*Rhizopus* spp.), strawberry (*Botrytis cinerea*), banana (*Colletotrichum musae*), citrus (*Penicillium italicum*), and kiwifruit (*Botrytis cinerea*). The application of *Trichoderma* spp. for the pre- and post-harvest might be linked to some modes of action such as the release of the lytic enzyme used in parasitism, competition, antibiotics, and the stimulation of plant defenses. Moreover, the application of *Trichoderma* strain as antagonist, especially when combined with other control systems such as GRAS substances like encapsulation using some polymeric matrices like chitosan, is generally regarded as safe (Medina-Cordova et al. 2018; Howell 2003; Valenzuela et al. 2015; Singh et al. 2018).

Papaya (*Carica papaya*) has been identified as one of the most grown fruits globally but its mass production is hampered by post-harvest and pre-harvest diseases. In view of this, *Phytophthora palmivora* has been identified as one of the major casual organisms responsible for fruit rot affecting *Carica papaya.* Most of the treatment documented for the post-harvest management of this disease is fungicide applications but with numerous disadvantages like resistance to the diseases by the pathogen and exposure to hazard, especially the environment and human being. The application of *Trichoderma* as a natural biological control agent will go a long way as a permanent replacement to synthetic fungicides used in the management of *Phytophthora palmivora* and all the highlighted hazards. Santos de Oliveira et al. (2018) utilized four biological control agent containing *T. harzianum* strain THP, *T. asperellum* strain SF04, *T. longibrachiatum* strain 4088, and *T. virens* strain 255C1. The result obtained shows that all the tested *Trichoderma* strains significantly decrease the severity of diseases and the rate of incidence of diseases while the most active strain among all the *Trichoderma* was strain 4088 (*T. longibrachiatum*) with the highest biocontrol agent of *P. palmivora* and for the post-harvest management and shelf life extension of *Carica papaya* fruit. *Trichoderma gamsii* strain

6085 has been shown to possess lots of biological control attributes which has highlighted them as a great competitor, mycoparasite, and antagonist on natural substrates containing the mycotoxigenic strains of *Fusarium culmorum and Fusarium graminearum.*

Sarrocco et al. (2013) highlighted the significance of strain 6085 of *Trichoderma gamsii* as biological control agents have the capacity to reduce the level of deoxynivalenol production secreted by some pathogens. The authors highlighted that *T. gamsii* 6085 has the capacity to utilize deoxynivalenol as a substrate that could enhance their growth. The analysis performed using high-pressure chromatograph exhibited no significant changes in the level of growth medium when the inoculation was performed 72 h. They also investigated the function of PDR-ABC transporters toward the resistance of deoxynivalenol by strain 6085 of *T. gamsii* was examined. The field trial carried out in the years 2011 and 2012 showed that strain 6085 *T. gamsii* drastically decreased the incidence of the disease from these two pathogenic fungi and prevented the synthesis of mycotoxin from these fungal pathogens. Also, it was detected that the usage of the *T. gamsii* 6085 in the soil-applied before the sowing showed that the biologicalcontrol agent decreased the level of diseases severity by 57%. Their study showed that *T. gamsii* 6085 could be used as a biocontrol agent and pre-harvest biotechnological tool against the incidence of diseases caused by these *Fusarium* spp. used during this study.

Senthil et al. (2011) evaluated the application of some beneficial microorganisms for post-harvest management of some pathogenic diseases that affect grapes. Some of these strains was tested against some of these post-harvest pathogens like *Trichoderma*, *B. subtilis* strains EPCO-16 and EPC-8, yeast, and *Pseudomonas* while some of the post-harvest pathogen affecting grapes screened were *Fusarium moniliforme*, *Aspergillus carbonarius*, and *Penicillum expansum*. The result obtained during the in vitro assay showed that *B. subtilis* strains EPCO-16 and EPC-8, exhibited the highest inhibitory effect value of 88.8% against the mycelial growth development from *A. carbonarius* and *P. expansum*, respectively. The biocontrol agent used during their study was tested against the post-harvest pathogens of grapes using pre-, post-, and the synergetic inoculation of the biological control agents against the tested pathogens. The result obtained during the pre-inoculation reveals that strain EPC-8 of *B. subtilis* lead to a decrease in the occurrence of rots from *A. carbonarius*, *T. harzianum* showed effectiveness against *P. expansum* while *T. viride* exhibited effectiveness against *F. moniliforme*. Moreover, the same high level of biological control activity was demonstrated during the combined inoculation and the post-inoculation test.

Terna et al. (2013) assessed the effectiveness of the cultural filtrates obtained from *T. viride* against some post-harvest rot pathogens including *F. oxysporum*, *A. flavus*, and *A. niger* in an in vitro assay. The screening of the production of the enzymes from the *T. viride* strain isolated from the soil revealed that some of the enzymes secreted wereprotease, cellulose, and chitinase, respectively. The enzyme-induced culture assay showed that the synergetic effect of the three enzymes produced by the strain of *T. viride* induced and enhanced the high antagonistic activity with radial growth inhibitions values of 67.89% and 77.69% against

A. flavus and *A. niger*, respectively, while chitinase enzyme produced by *T. viride* enhanced the maximum inhibitory of 46.95% when tested against *F. oxysporum.* Their study showed that strain of *T. viride* could be used as a pre- and post-harvest prevention of the *F. oxysporum, A. flavus, and A. niger* pathogens.

12.3 Application of *Trichoderma* spp. as a Dependable Biotechnology Tool for the Management of Pests and Diseases and Their Role in Crop Protection

Beneficial microorganism has been confirmed to produce biologically active compounds that could influence the communications of plants with their pathogens. Some of these metabolites possess the potential to inhibit pest and pathogen affecting agricultural improvement. These beneficial microorganisms may also enhance disease resistance by stimulating systemic plant defense activity and promote the growth and yield of agricultural crops. The *Trichoderma* genus has been highlighted as a producer of secondary metabolites that possess an inhibitory effect against pathogens and pests. The usage of metabolomics has shield more light on the utilization of these metabolites for the prevention of post-harvest diseases and the promotion of plant crops. They are now recognized as a sustainable biotechnological tool that needs to be introduced into an integrated pest management system.

Bogumił et al. (2013) isolated and assessed the biological control effectiveness from 52 isolates of *Trichoderma* spp. against *Botrytis cinerea* in an in vitro assay using the dual culture method. The result showed that exhibited great potential to prevent the mycelia development of gray mold. Moreover, it was observed that the *Trichoderma* isolates had a decrease in the value of 45–78%, when the plant pathogen, *Botrytis cinerea* after 6 days of incubation at 25 °C. The maximum antifungal effectiveness recorded were 76% for *Trichoderma* strain Tr43 and 78% for *Trichoderma* strain Tr52. The result of the molecular and the biochemical characterization indicates the two isolates with the highest biological control activity were confirmed to be *T. atroviride.* The major differences obtained between Tr43 and Tr52 was found in their rate of exploiting of 11–96 various carbon sources, production of chitinases, siderophores, and indole-3-acetic acid. Furthermore, it was detected that none of these *Trichoderma* strains showed any potential to solubilized phosphate when cultures on Pikovskaya's medium.

Anand and Jayarama (2009) isolated 42 strains of *Trichoderma* sp. from cultivated lands around Bangalore and they were assessed for their antifungal activity against *Fusarium ciceri* and *Sclerotium rolfsii.* An in vivo assay was performed to establish their biocontrol effectiveness by utilizing chickpea (*Cicer argentums* cv. Annigeri) as an experimental plan. This was carried out using the roll paper towel technique. The result revealed that strains T25, T30, T35, and T40 exhibited the highest antifungal effectiveness and improved the rapid growth of the plant. The modes of action of these *Trichoderma* spp. were performed by evaluating their capability to produce chitinases that have the potential to disintegrate the cell walls of targeted plant pathogens. This was carried out using a media containing *Sclerotium rolfsii* cell wall extract and colloidal chitin and the best 10 *Trichoderma* spp. with the best antifungal activity was screened against the pathogen. It was established that all the *Trichoderma* sp. exhibited chitinolytic activity on the third day and on the fifth day, respectively. Furthermore, exochitinase and endochitinase present in the most active strain of *Trichoderma* sp. were tested using submerged fermentation with the aids of colloidal chitin amended broth. It was observed that strain T6 displayed the highest amount of exochitinase while T35 displayed the highest amount of endochitinase. It was also noted that strains T6 and T35 produced the maximum amount of cellulose enzyme, but almost all exhibited a potential to produce cellulose enzyme. The study shows that there was a greater correlation between the biological control efficiency of the tested *Trichoderma* sp. and their rate of enzyme production.

Faruq et al. (2014) assessed the effect of *Trichoderma harzianum* T22 combined with different soil amendments like cow dung, vermicompost, solarized sand, coco-dust, saw-dust, khudepana (*Azolla pinnata*), poultry waste, ash against *Fusarium* wilt disease of eggplant. The various formulated treatments were added to the soil at 15–30 days ahead of transplanting. The occurrence of wilt diseases was carried for a period of 55–95 days. It was reported that all this led to a decrease in the incidence of the wilt after performing transplanting on the plant, especially on the treated plots when compared to the control that had the highest rate of incidence because no treatment was applied. It was discovered that the treatment of *Trichoderma harzianum* followed by the poultry waste amended with *Trichoderma harzianum* exhibited the highest biological control effectiveness by suppressing the incidence of the wilt, increment in the growth rate of the plant, especially the yield of the fruit. Moreover, it was discovered that the plot treated with the poultry waste shows effectiveness when linked to the vermicompost produced by *Trichoderma harzianum* and followed by the treatment containing a coco-dust treated plot in terms of biological control effectiveness against the disease, enhanced growth, and lead to increase in yield. Their study shows that there was a positive relationship between wilt incidence and the day that transplanting was carried out.

Peccatti et al. (2019) evaluated *Trichoderma* spp. as a biotechnology tool for the enhancement of *Maytenus ilicifolia* seedlings as well as it on its germination rate and initial seedling growth. The experiment was performed in a greenhouse and in the laboratory. This was carried out with three different strains; T1 and T2 strains were identified as *Trichoderma* spp. while strain T10 was identified as *Trichoderma virens*. The laboratory experiment carried out showed that *Maytenus ilicifolia* seeds deprived of aryl were treated in the solution having fungal spores and dispersed in substrate paper and this was performed in four replicates with 25 seeds per treatment. The following parameters were carried out at 7, 14, and 21 days after incubation including percentages of accumulated dead seeds, first germination count, accumulated germination, and firm seeds. Moreover, the same experiment was performed in the greenhouse with all the *Trichoderma* spp. isolates but the experiment was conducted with 40 replicates after having a seedling as each replicate. Furthermore, some other growth parameters such as diameter at root collar, total height, and the number of leaves were assessed at various different days from 90 to 180 days after planting the seed. The result obtained showed that isolate coded T2 exhibited a positive correlation that the inoculation of the fungal isolates enhances the rate of germination and vigor when performed in the laboratory but it was observed that isolated coded T2 does not have any plant growth effect on *M. ilicifolia* seedlings in the greenhouse. The authors later suggested that more experiments need to be performed so as to establish the biological modes of interaction between the strain of *Trichoderma* isolates.

Jonglaekha and Vichittragoonthavorn (2009) emphasized the importance of using beneficial fungi as part of plant protection program. The biological control agent was given to all the farmers that participated in all the three production systems, i.e, organic, EUREPGAP, and GAP. It was discovered that most farmers that practiced farming on the highland were very poor and they did not have sufficient capital to perform intensive farming. The farmers were exposed to some biotechnology techniques through the application of biological control agents, mixing of the biocontrol with different composts from rice bran as a sustainable integrated pest management technique. The two selected strains utilized for this biological control activity were *Paecilomyces lilacinus* for the prevention of egg parasite from root-knot nematode while *Trichoderma harzianum* was utilized for the prevention of root rot diseases. The major reason attributed to the build-up of these pests might be linked to continuous planting, limited area, and monocropping system. It was also stated that the application of biological control agents without the application of integrated pest management might not lead to any observable success. It was also noted that the training given to the farmers might not lure the farmers to change their attitude from the utilization of chemical pesticides. It was discovered that practical demonstration of the biotechnological techniques through which the fungal isolates were grown on a solid substrate through solid-state fermentation using steamed sorghum seeds gave the farmers a better understanding of how it works and it enhanced the acceptability of the technology. This also makes the farmers continue with this technology. It was later discovered that Plant protection program revealed that the fungal isolate from *T. harzianum* starter given to the farmers on the highlands enhanced their increase from 563 to 1467 kg in the year 2005 while the application of *P. lilacinus* led to the increment of 17–63 kg in the year 2007. Moreover, it was discovered that the same trends were recorded for the two isolates during the year 2008 project.

Begum et al. (2008) isolated and screened ten fungal and one bacteria out of which the best strains were selected which were strain UPM40 of *Trichoderma harzianum*, strain UPM23 of *Trichoderma virens*, and strain UPM13B8 of *Pseudomonas aeruginosa* isolates. They were selected because they exhibited the biological control activities against *Colletotrichum truncatum*. The preliminary screening was performed using culture filtrate tests and dual culture assay. The modes of action were validated using light microscopy which shows that *T. harzianum and T. virens* possessed that capacity to prevent the development of *Colletotrichum truncatum*. It was observed that these biocontrol agents could swap around the fungal pathogen, penetrate in their hyphae, which later became malformed and swollen. Moreover, the bacterial isolates also showed distorted vacuole, swollen in, and the tips of hyphae. Also, it was established that *Pseudomonas aeruginosa* exhibited the highest biological control activity against the tested pathogen. These biological control strains did not show any inhibitory effect against the soybean seeds and seedlings but improved the rate of seedling establishment and seed germination but *P. aeruginosa* improved the dry weight and the wet weight of the soybean seeds/seedlings.

Babu and Pallavi (2013) assessed the potential of *Trichoderma candidum* for the prevention of plant pathogen and pest. This will go a long way as a permanent replacement to the synthetic pesticides as a typical example of integrated pest management and finally preventing the issues boarding around environmental pollution. They also highlight that some of the challenges boarding around the mass production and application of biological control agents from effective strain include effective methods for storage, development of cost-effective methods, transportation and its formulation, mass production, and selection of effective strains. The authors utilized various vegetable wastes obtained from different households for the multiplication of *Trichoderma candidum* so as to find an effective production technology that can easily be adopted.

El-Mougy et al. (2012) evaluated the biocontrol potential of some *Trichoderma* (*T. hamatum*, *T. viride*, *T. harzianum*) together with some other microorganisms including *Saccharomyces cerevisiae*, *Pseudomonas fluorescens*, and *Bacillus subtilis*. These biocontrol agents were screened against some pathogenic microorganisms that are responsible for the damping-off disease or root rot from several plants like pepper, cucumber, tomato, and cantaloupe that grew in a plastic house under the greenhouse. The effectiveness of the thyme oil and calcium chloride applied as seed coating materials were tested against the incidence of the diseases was carried out in a potted experiment after inoculating these root inducing pathogenic microorganisms under greenhouse condition. It was observed that the treatment applied led to a decrease in the incidence of rot during the pre- and post-emergence development of all the crops used during this experiment (pepper, cucumber, tomato, and cantaloupe) when compared to the control. Moreover, it was observed that the synergetic effect between the combination of thyme oil and calcium chloride induced drastically the incidence of rot from all the tested vegetable plant used during this study. Their study showed that there is a relationship between essential oil and calcium chloride when applied together with biological agents for the production of cheaper and safe biocontrol agents.

Barakat et al. (2014) isolated natural biological control agents inhabiting on the surface of the leaf of faba bean. The identified biocontrol agent consists of 26 isolates of *Trichoderma* species (*T. viride*, *T. album*, *T. harzianum*, *T. aureoviride*, *T. hamatum*) and 4 isolates from different genera containing *Epicoccum*, *Paecilomyces*, *Cladosporium*, and *Gliocladium*. The biological control effectiveness of *B. fabae* was performed using mycelia growth inhibition on potato dextrose agar plate. The result revealed that the mycelium inhibitory effect exhibited by *Trichoderma* spp. varied from 51.11 to 77.78%. Moreover, *T. album* (Isolate 2) exhibited the greatest inhibitory activity followed by isolate 6 by *T. harzianum*. The result obtained from the greenhouse trial exhibited that the application of

Trichoderma spp. by spraying on the faba bean plants as bio fungicide in comparison to Bio-Zeid, 24 h before treatment with *B. fabae* considerably decrease the rate of disease severity after 14 days with a value which ranges from 3.0 to 4% in comparison to the control that had 8.7%. Also, isolate 2 from *T. album* exhibited the highest antagonistic effect of 3% followed by isolate 6 from *T. harzianum* then isolate 6 from *T. hamatum* and finally isolate 2 from *T. viride*, all having antagonistic values of 3.24, 3.30, and 3.40%, respectively. Finally, it was observed that the non-volatile and the volatile compounds secreted by the isolate 2 from *T. album* produced the maximum inhibitory effect on the growth of *B. fabae* followed by isolate 6 from *T. harzianum*.

Kowalska (2011) evaluated that the biocontrol effectiveness of strain T1 from *Trichoderma asperellum* isolated from the commercial product was evaluated in a field trial. It was stated that the biocontrol product possesses 5×10^8 of *Trichoderma* conidia per one gram was utilized in organic field strawberry. The application ration of 1:1 containing biocontrol-antagonist at a concentration of 10 g using a foliar sprayer during the growing season. It was discovered that the application of *Trichoderma* enhanced drastic improvement in the growth of the crop and reduced the occurrence of *B. cinerea* on harvested and stored fruits. Their study shows that strain from *Trichoderma* might be utilized for the post-harvest management and extension of the shelf life of fruits and vegetables for a period of 7 days after harvest. Moreover, it was established that the plant treated with *Trichoderma* exhibited an increase in the yield of the plant with 30.2% when compared to the control plant without any *Trichoderma*. Moreover, an increase was observed in the aerial part and the number of runners from the plant.

Zegeye et al. (2011) evaluated the biocontrol efficacy of two strains containing strain Bak150 from *Pseudomonas fluorescens* and strain ES1 from *Trichoderma viride* from the treatment of potato late blight pathogen caused by *Phytophthora infestans* under a greenhouse trial in an in vitro assay. The foliar application performed in the greenhouse consists of single utilization of *T. viride* and *P. fluorescens* only, the synergetic effect of the mixed culture, while Mancozeb was applied as a positive control. The pathogen was then applied 3 days after the application of the biocontrol agent. The result obtained from the in vitro assay shows that exhibited a radial growth inhibition against the tested pathogen from *P. infestans* with 36.7% and 88% by *T. viride and P. fluorescens*, respectively. Moreover, the affected leaf area by the pathogen was assessed weekly. It was observed that there was a drastic decrease in the diseases with the following values of *P. fluorescens* (AUDPC = 765.1), *T. viride* (AUDPC = 260) when compared to the control that had (AUDPC = 1045). In terms of effectiveness, *T. viride* exhibited more pronounced biocontrol effectiveness in comparison to *P. fluorescens* and mixed culture. Their study showed that the utilization of strain ES1 from *T. viride* could be explored for the prevention of late blight disease of potato.

Sundaramoorthy and Balabaskar (2013) assessed the biocontrol effectiveness of a local isolate *Trichoderma* species for its potential to enhance the enhancement of tomato crop (growth and yield) and toward the management of *Fusarium* wilt disease when performed during in vitro and in vivo environments. The pathogenic fungus that was isolated during this study was recognized as *Fusarium oxysporum* f. sp. *lycopersici.* During this study, 15 native *Trichoderma* antagonists were obtained from the rhizosphere of a healthy tomato plant from different geographical areas. Among all the screened isolates, it was discovered that strain ANR-1 from *T. harzianum* exhibited the highest biocontrol efficacy by showing the highest radial mycelia inhibition against the tested pathogen by 53% in comparison to the other biocontrol isolates. Moreover, it was discovered that the plant treated with strain ANR-1 from *T. harzianum* demonstrated the highest stimulatory effect on plant height with 73.62 cm and showed a dry weight of 288.38 g when compared to tomato plants without any treatment.

Singh and Singh (2009) wrote a comprehensive review on *Trichoderma* which has been identified as a mycoparasite that dwells in almost all agricultural soil with great biological control effectiveness against pathogenic fungi. They also highlighted that *Trichoderma* possesses several unique characteristics that make them a unique biological control agent like the capability to utilize several agricultural wastes for their proliferation, biodegradable, their capability to secret non-toxic and very potent and active metabolites which have been established to possess inhibitory effect against several plant pathogens and pests. Another interesting fact about *Trichoderma* is its capability to perform its inhibitory effect mode of action against the target organism without having a non-target effect on any other beneficial microorganisms. It has been reported that they possess the capability to enhance crop development, especially the capability to enlarge the root of agricultural crops. It was also stated that several researchers have identified almost 260 strains of *Trichoderma* obtained from different agro-ecological areas. Most of these strains have been documented with great biological control activity both at the laboratory and the field trial. It was also reported that most of the isolated *Trichoderma* could increase the number and sizes of deep roots that are available below the soil of some crops like ornamental plants, corn, fruit crops as well as increase their resistant to abiotic stress, especially drought. Moreover, it was further states that most of these *Trichoderma* species are efficient producers of ago-industrially important extracellular enzymes with diverse utility in various sectors (food, agriculture, and industries), especially for the production of cellulase enzyme used for the biodegradation of complex polysaccharides. Moreover, the application of protease produced by *Trichoderma* has been established to prevent the development of gray mold on the surface of the bean by inhibiting the germination of spore from this fungus as well as neutralizing destructive mold enzymes. Their application as biopesticides and biofungicides has been documented for the management of several diseases and pests. The authors also stated that more effort is now been added by several scientists to explore the biological activity of marine-derived *Trichoderma* species such as *Trichoderma reesie* which has been acclaimed with numerous biological activities and their capability to be utilized in different field trials for the management of different pathogens and pests mitigating against agricultural crops of greater interest.

Kuzmanovska et al. (2018) explored the ability of some *Trichoderma* species as a biotechnological weapon for the management of major pathogen that affects an increase in the growth of tomato plant (*Solanum lycopersicum* L.) from the Republic of Macedonia. The two biocontrol isolates were *T. harzianum* and *Trichoderma asperellum*, respectively, and their biological efficiency was tested against 18 genetically diverse *B. cinerea* isolates. The result obtained showed that all the *Trichoderma* species used during their study exhibited a very high biological control effectiveness against all the tested *B. cinerea* isolates. It was reported that *T. asperellum* exhibited mycelia growth inhibition that varies from 74.246 to 96.915% while *T. harzianum* showed values that vary from 71.072 to 95.889%. The conidial germination inhibitory effect varies from 76.932 to 95.107% and it was recorded by *T. asperellum* while *T. harzianum* exhibited values that vary from 76.933 to 93.658% against *B. cinerea* isolates. Their study showed that *T. harzianum and T. asperellum* could be utilized for the mass production of biological control of gray mold disease in tomato.

Onion (*Allium cepa* L.) has been highlighted as one of the most important vegetable crops in India, especially in the Maharashtra region, but their agricultural production is always affected by various pathogens. Jagtap and Suryawanshi (2015) isolated some biocontrol agents for their effectiveness toward the biological control of basal and onion rot as a result of an attack by *Fusarium oxysporum* f. sp. *cepae*. It has been reported to be one of the most dominant causal agents in the Nasik district, in India. which has been reported to be one of the most dominant causal agents in the Nasik district, in India. Some *Trichoderma* spp. like *T. koningii* and *Trichoderma viride* were evaluated along with several other prominent strains like *Pseudomonas fluorescens*, *Aspergillus niger*, *Penicillium expansum*, *Bacillus subtilis*, *A. flavus*, *Xanthomonos axonopodis*, *Curvularia lunata*, and *Alternaria alternate* which were all tested against *Fusarium oxysporum.* It was observed that all the tested biological control agents exhibited an inhibitory effect against the growth of the pathogen. Also, it was reported that *Trichoderma viride* showed the highest inhibitory value of dual culture technique followed by *Pseudomonas fluorescens* with a value of 12.85 against *Fusarium oxysporum.* They suggested that some of the modes of action that might be involved in the biocontrol of these pathogens might include hyperparasitism and competition, respectively. All these phenomena might be utilized for maintaining a balance between their relevance in the ecosystem and toward the prevention of various plant pathogens. Moreover, they stated that more considerable attention need to be given to the usage of biological control agents as biotechnological tools for the management of plant pathogen due to their various advantages which include low cost, improvement of soil fertility, and pathogen suppression.

Ghosh (2017) evaluated the biological control effectiveness of three soil fungi which include two *Trichoderma* (*T. harzianum and T. viride) and Beauveria bassiana* which were explored for the prevention of brinjal disease responsible for the *Phomopsis* fruit rot of brinjal as well as their capability to enhance the growth development of this crop. The efficacy of these biocontrol strains was compared with a synthetic fungicide Blitox 50. The result reveals that *Phomopsis vexans* together with *T. viride* injected fruits exhibited 20% disease index while *Phomopsis vexans* injected fruits exhibited 100% disease index followed by *Phomopsis vexans* with *T. harzianum* and *Phomopsis vaxans* with *B. bassiana* injected fruits exhibited 30 and 50% disease index, respectively. Moreover, it was observed that *T. viride* had the highest crop defense followed by *Trichoderma harzianum* and *B. bassiana*. The field trial was carried out for a period of 3 years from 2014 to 2016. It was revealed that percentage disease index values of 30.25 from *T. viride*, 35.00 from the combination of *T. harzianum* and Blitox 50 and untreated having a value of 28.75. Moreover, it was recorded that the plant treated with *T. viride* showed enhanced growth parameters (number of leaves, the dry weight, the height, number of leaf area) of brinjal plant from the treatment containing spore suspension of 1×10^7 CFU/mL when compared to the untreated without any spore suspension of *T. viride*. The authors advised that the application of *T. viride* spore suspension (10^7CFU/mL) might be an alternative replacement to the utilization of chemical fertilizer and fungicides during the period of seedling treatment but must be applied at four consecutive sprayings with an interval of 15 days after initiation of fruits.

Srivastava et al. (2015) wrote a comprehensive review of the probable modes of action utilized by the genus *Trichoderma* which portends them as a biocontrol agent for the prevention of phytopathogenic fungi affecting agricultural productivity. It has been reported that most of the strains from this genus can multiply asexually through the development of chlamydospore and cyanide while they multiply in the wild environment through the development of ascospores. Moreover, it was stated that most *Trichoderma* species possessed the capability to secrete various types of cell wall degrading enzymes that they used as a defensive mechanism toward the control of pathogens. These cell wall degrading enzymes are being regulated through the action of some certain genes that control their biological control action and they are referred to as biocontrol genes. Moreover, it was stated that most of these genes have been cloned for larger scale production while most of these genes have been documented for their utilization for the management of abiotic and biotic stress. Example of some of the mechanisms of these biocontrol agent utilized against phytopathogenic fungi include mycoparasitic, antibiosis, and competition for nutrients, and so on.

Awad et al. (2015) isolated and identified some novel fungal strains that could be utilized as a biocontrol agent against some soil-borne diseases. They screened 16 fungal isolates out of which the most active strain was obtained from the sugar beet rhizosphere. The best strain was later coded SRBP_ZSHSG1. The most active strain was later subjected to molecular characterization using 18S rRNA sequences. The blasting result carried out showed that strain SRBP_ZSHSG1 exhibited 100% similarity with *Trichoderma asperellum* which showed that they possess almost the same homology. It was also revealed that strain SRBP_ZSHSG1 possesses a very high inhibitor against the tested pathogen during the field trial. Furthermore, strain SRBP_ZSHSG1 was mass-produced using solid-state fermentation on rice straw (biostraw) and has been established to be able to produce some active metabolites with useful compounds. The result obtained from the gas chromatography–mass spectral analysis showed that the ethyl acetate extract possesses nine compounds containing four volatile alcohols and fatty acid esters.

Tančić et al. (2013) evaluated the effect of *Trichoderma* isolates isolated from different soil from different regions of Vojvodina, Serbia. The most potent biological control agents from this region were screened using dual culture assay against the pathogen *Sclerotinia sclerotiorum*. The result obtained showed that all the isolates possess the capability to colonize and exhibit high radial growth inhibition against the tested pathogens. The result obtained from the greenhouse also confirmed their biocontrol effectiveness of *Trichoderma* isolates which showed an enhancement in the growth of soybeans with an increase in all the tested parameters viz., root length and vigor index, root and shoot length, and enhanced germination.

Anthracnose disease has been highlighted as one of the most significant problems affecting an increase in the production of cayenne pepper (*Capsicum frutescens*), especially most cities that focus on the production of chilli pepper. Moreover, lots of economic losses are normally experienced which might be linked to the Anthracnose disease caused by numerous species of *Colletotrichum* including *C. acutatum*, *C. gloeosporiodes*, and *C. capsici.* Also, the incidence of diseases depends on several factors like weather and cultivar susceptibility. It has been observed that humid and warm weather conditions have a greater effect on the susceptibility of the cayenne pepper to anthracnose. Anthracnose has been discovered to be one of the diseases that is difficult to manage whenever the symptoms appear, especially when the environmental factors are in favor of the infection process.

In view of the aforementioned, Setiawati et al. (2016) utilized the combination of *Trichoderma* species and Azoxystrobin for the treatment of disease incidence and to increase the yield of four varieties of cayenne pepper. The result obtained showed that the synergetic effect of biological control agents and tolerant varieties exhibited the highest effectiveness for the prevention of anthracnose of cayenne pepper when compared to the synthetic control. The synergetic effect was preferred to chemical treatment because it is cheaper and cost-effective. The application of *T. harzianum* and Azoxystrobin showed a more inhibitory effect when compared to the combination of *T. koningii* and *T. viride* toward the reduction of the incidence of anthracnose on different fruit characteristics of cayenne pepper. Their study showed that the combination of *T. harzianum* and Azoxystrobin could be utilized for the management of anthracnose disease of *C. acutatum* because it resulted in an increase in the yield when compared to the other treatment.

12.4 Application of *Trichoderma* spp. as a Dependable Biotechnology Tool for Protection of Environment and Active Role They Play as a Bioremediation Agent

One of the utmost problems currently facing mankind worldwide is the problem of environmental pollution and the build-up of toxic substances. This has been a major challenge because of various hazards to the environment and human beings. The increase in the level of sensitization has led to enforcement of regulatory measures that could protect the environment from exploitation and future contamination. Therefore, there is a need to implement a sustainable technology that will mitigate against the future contamination of our environment. The application of microorganisms as a biotechnological tool for the bioremediation of polluted environment has been discovered as a cheaper and eco-friendly biotechnology technique. Several scientific reports have been documented regarding the utilization of enzymes from plant, bacteria, yeast, actinomycetes, and fungi for the bioremediation of toxic organic pollutants. Moreover, the application of fungi species such as Trichoderma has been identified in the biological removal of pollutants, treatment of effluent, and heavily polluted water containing toxic metallic ions. A typical example of such fungi is *Trichoderma* spp. because of the recent advances in their application as a biotechnological tool for reducing the level of toxicity of the pollutants using advance bioprocessing technology in order to obtaine useful and novel products or substances (Hasan 2016).

Blaszczyk et al. (2014) wrote a comprehensive review of the utilization of *Trichoderma* for the sustenance of a clean environment. They laid emphasis that *Trichoderma* spp. could utilize various modes of action in performing their role in the environment, especially how they colonize numerous ecological niches. They participate in the enhancement of crop development, their protection against pests and diseases. They are utilized in the bioremediation and biological plant defense as biofungicides. Moreover, they are utilized in various biological processes in the industry which include the production of useful metabolites, enzymes, antibiotics, and biofuel. They also emphasized that the application of genomics has made their biotechnological application most interesting by making their genome sequence more available. They advised that additional research needs to be carried out so as to establish the best effective methodology and the safety involved in the utilization of *Trichoderma* spp.

Nongmaithem et al. (2016) evaluated the potential of *Trichoderma* isolates for their capability to be utilized for the biosorption of nickel and cadmium. The authors screened 14 isolates in order to determine the best strain that could tolerate two heavy metals, nickel and cadmium. The result of the screening established that strains IBT-I, MT-4, and UBT-18 exhibited an enhanced tolerance of cadmium when compared to the other isolates. Moreover, it was observed that the biomass production improves up to a Ni concentration of 60 ppm in all the strains but reduced as the concentrations of nickel increased under nickel stress. It was reported that UBT-18 exhibited the maximum biomass when exposed to nickel-containing concentration of 150 ppm while the minimum concentration that will inhibit 50% of the growth was MIC50 from strain IBT-I. Also, strain IBT-II exhibited the highest biomass production and highest MIC50 value when subjected to cadmium stress among the cadmium-tolerant isolates. It was observed that the maximum percentage of nickel removal was noticed up to the concentration of 40 ppm as the biomass of *Trichoderma* isolates increased, as well as a rise in residual nickel and reduction in biomass production applied at a higher concentration during the submerged fermentation. Moreover, the rise in the concentration of cadmium led to a reduction in biomass production which demonstrated an enhancement in the value in the residual in the liquid fermentation. It was also observed that nickel and cadmium stress might be a factor that influences the level of sensitivity of the *Trichoderma* isolates to soil fungistasis while the strain isolates IBT-I and UBT-18 exhibited the highest fungistasis under cadmium and nickel stress.

Oladipo et al. (2018) isolated, screened, and evaluated the capacities of some fungal strains to tolerate exposure to different heavy metal concentrations of iron (Fe), arsenic (As), cadmium (Cd), arsenic (As), copper (Cu), and lead (Pb). These fungal strains were isolated from gemstone and gold mine site soils. They were later subjected to molecular characterization using internal transcribed spacers 1 and 2 (ITS 1 and ITS 2) regions. The various fungal strains were exposed to various concentrations containing (0–1000 Cu), (0–800 Fe), (0–100 Cd), (0–500 As), and (0–400 Pb) concentrations (mg kg^{-1}) were all amended into malt extract agar (MEA). The rate of the radial growths of the fungus was determined after incubating for a period of time that varies from 3 to 13 days. The fungal strains identified were *Trichoderma ghanense*, *Rhizopus microspores*, and *Fomitopsis meliae*. The values that range from 400 to 1000 mg kg^{-1} were observed from all the tested fungal strains when compared to the control that had tolerance index of >1. Moreover, it was observed that *Rhizopus microspores* and *Trichoderma ghanense* showed a greater potential to tolerate Cd and As concentrations when compared to the control which exhibited a tolerance index of >1 at 25 mg kg^{-1} Cd and 125 mg kg^{-1} As. It was noticed that these fungal strains exhibited high tolerance to metal concentrations greater than the global permissible limits recommended for contaminated soil. Their study exhibited that the fungal strain utilized during their study could be used as a biotechnological tool for the bioremediation of heavily polluted soil and heavy metal contaminated environments.

Isah et al. (2018) evaluated the biostimulatory influence of some fungal strains for the biodegradation of atrazine-contaminated soil. The physicochemical properties of the biostimulated soil were later assessed using standard protocol while the influence of the pre- and Post-biostimulation of fungal species were evaluated with the aid of enumeration techniques. Based on the microscopic and macroscopic identification the fungal strains identified were *Aspergillus niger* and *Trichoderma harzianum*. The level of biodegradation of atrazine-contaminated soil was carried out by evaluating the enzymes that are utilized in the degradation of atrazine-contaminated soil while the metabolomics of the metabolites secreted were assayed using GC-MS techniques. The result obtained showed that these fungal strains portend the capability to be used for the biostimulation and biodegradation of atrazine-contaminated soil.

It has been discovered that the amendment of soil with *Trichoderma* spp. has numerous capability to increase the development of important crops worldwide, especially rice (*Oryza sativa L.*) cultivated in acidic clay soil. This type of scenario is very peculiar in some regions in Cambodia where pH has been highlighted as a major factor preventing the uptake of nutrients from the soil. Hem and Pang (2017) evaluated the effect of *Trichoderma* as a biostimulant in order to determine its effect on the growth of rice (YRM70 variety) cultivated in strong acidic soil inside a glasshouse. The rice was cultivated in 80 pots having clay loam soil inoculated with different strains of *Trichoderma koningii*, *Trichoderma harzianum*, and *Trichoderma lignorum*. Growth parameters like chlorophyll content, root biomass, the number of tillers, the number of expanded leaves, root length, stem weight, total biomass, stem height, from the seedling of Zadoks' code GS12 to early tilling stage (GS14, 22) were determined. The result obtained shows that there was an increase in the growth parameters of the early tilling stage (GS14, 22), evaluated like the fresh weight of total biomass, chlorophyll content, leaf weight, and root mass. Their study showed that *Trichoderma* possessed a biostimulatory influence on crop improvement of rice by increasing the uptake of nutrient under very harsh soil pH conditions.

Fiorentino et al. (2013) tested the effect of strain A6 *Trichoderma harzianum* and compost fertilization amended with cadmium-polluted soil were performed in an open field condition in Southern Italy. The experiment was carried out so as to evaluate the influence of the treatment on the uptake of the soil N-cycle microflora and phytoextraction of heavy metals using the phytoextraction remediation technique. This was carried out using giant reed (*Arundo donax* L).

The result revealed that Cd concentration in the soil does not have any adverse effect on the giant reed biomass yield in the first growing season with an average of 12.8 Mg ha^{-1}. Moreover, it was established that the inoculation of *T. harzianum* and the compost fertilization enhanced the rate of translocation in the leaves and cadmium uptake. The result also established that *T. harzianum* could be utilized as a biotechnological tool that could be used for biomonitoring of soil quality. Their result shows that giant reed could be used for assisted-phytoremediation in the presence of strain A6 *Trichoderma harzianum* and compost fertilization.

The introduction of hexavalent chromium in the environment has been discovered as one of the reasons for the higher level of anthropogenic activities globally. The release of hexavalent chromium has led to the impairment of the reproductive, gastrointestinal, respiratory, and immunological systems. Therefore, the elimination of hexavalent chromium in the environment will minimize the exposure of various hazards to human beings and the environment. The application of microbial bioreduction of these heavy metals will go a long way toward the maintenance of the ecosystem and cost-effectiveness approaches for the chromate detoxification. Ray and Sur (2016) utilized *Trichoderma pseudokoningii* obtained from tannery effluent enriched soil near Kolkata for the transformation of hexavalent chromium by the chromium decreasing fungal used during their study. Their study shows that the fungal strain *Trichoderma pseudokoningii* could grow at a concentration of 1000 mg/L chromium, while the development of spore became flimsy as the level of the concentration was increased. The withdrawal of the hexavalent chromium was discovered to be feasible by bioreduction rather than the bioaccumulation or biosorption because no intracellular fraction bound Cr (VI) and the membrane was observed. It was observed that the pH of 7 enhanced the greatest extracellular chromium reduction when 0.09% peptone and 0.5% (w/v) pure dextrose were supplemented as a sole nitrogen and carbon sources. Moreover, it was observed that the concentration of 220 mg/L resulted in the maximum reduction of potassium dichromate Cr (VI) after 144 h of inoculation which could be linked to the stationary phase of growth of the tested strains. The rate of reduction might be linked to the availability of DDT and cysteine which might have enhanced the activity of chromium reductase enzyme possessing the thiol group at its active site. Moreover, the addition of feather and human hair in the culture medium enhanced the chromate reduction capability of *Trichoderma pseudokoningii*. The rate of the chromate reduction in the soil containing *Trichoderma pseudokoningii* was established using the data of atomic absorption spectroscopy which showed the significance of *Trichoderma pseudokoningii* in the bioremediation of contaminated soil.

Yao et al. (2015) utilized strain FS10-C for *Trichoderma reesei* for its capacity to bioremediate an aged polycyclic aromatic hydrocarbon (PAH)-contaminated and biological breakdown of benzo[a]pyrene (B[a]P). The result obtained showed that the amendment of the basal medium with glucose (10 g L^{-1}) as a co-metabolic substrate led to the removal of 54% of B[a]P (20 mg L^{-1}) after 12 days of incubation. Moreover, a 25% reduction was observed in the total polycyclic aromatic hydrocarbons concentration in the soil microcosms that were bioaugmented microcosms after 28 days. The degradation percentages of 3-, 4-, and 5(C6)-ring PAHs were 36%, 35%, and 25%, respectively. The augmented microcosms showed a better activity which includes increased average well color development, Shannon–Weaver index, fluorescein diacetate hydrolysis, Simpson index, and higher dehydrogenase. The result obtained by the authors from the principal component analysis revealed that the application of bioaugmentation affirmed the microbiological role of *Trichoderma reesei* FS10-C in the bioremediation of polycyclic aromatic hydrocarbons-contaminated soil. Their study shows that *T. reesei* FS10-C cold is utilized as a biotechnological tool for the bioaugmentation of polluted soil, especially polycyclic aromatic hydrocarbons-contaminated soil.

Siddiquee et al. (2013) utilized the application of some *Trichoderma* strains for the bioremediation of heavy metal from the environment. Some of the fungal strains evaluated include *T. virens*, *T. aureoviride*, and *T. harzianum.* The result revealed that strain T128 gave the maximum tolerance for Ni^{3+} and Pb^{2+} in a 1200 mg/L concentration. It was observed that the buildup and uptake capacity of *T. harzianum* was quantified by its ability to demonstrate maximum removal of Pb^{2+}, Cu^{2+}, and Ni^{3+} during the submerged fermentation when compared to other fungi while *T. virens* exhibited maximum uptake capacity and highest tolerance of metal which was recorded at 3.1789 g/g.

Cocaign et al. (2013) evaluated the effect of *T. virens* and *T. reesei* to tolerate the aromatic amines which have been recognized as a major class of pollutants comprising the extremely toxic pesticide residue 3,4-dichloroaniline (3,4-DCA). They cloned and evaluated the NATs from *T. virens* and *T. reesei*. This was evaluated in order to evaluate if the *N*-acetylation pathway enhances aromatic amines' tolerance in *Trichoderma* spp used in this study. *Trichoderma* spp. are diverse soil fungi that are highly impervious to many toxic compounds. The authors confirmed that NAT-independent conversion is solely (in *T. virens*) or principally (in *T. reesei*) accountable for the detected elimination of 3,4-DCA. Their study showed that *T. virens* as well as *T. reesei* to a lesser extent might be able to metabolize the metabolic pathway for the decontamination of aromatic amines aside from acetylation. Also, their study showed that functional and molecular characterization of aromatic amines biotransformation in *Trichoderma* spp. Their study showed that *Trichoderma* spp. could be utilized for the maintenance of a cleaner environment, especially contaminated soil containing aromatic amines-contaminated soil.

Teng et al. (2015) utilized strain FS10C *Trichoderma reesei* for the phytoremediation of Cd-contaminated soil that has been contaminated by hyperaccumulator *Sedum plumbizincicola* and the effect of the contamination was assessed on soil fertility. The ability of *Trichoderma reesei* FS10C to tolerate Cd was characterized while the potted experiment was performed to evaluate the growth and level of Cd uptake of *Sedum plumbizincicola* with or without *Trichoderma reesei* FS10C. The result obtained shows that *Trichoderma reesei* FS10C exhibited a very high Cd resistance around up to 300 mg L^{-1}. Also, 6–53% increment in the plant shoot biomass, Cd uptake by the shoots by10–53%, and dry weight of 16–61% were observed when linked to the control. The inoculation also stimulated the activities of soil microbial biomass and microbial activities like fluoresce diacetate hydrolysis activity, biomass C, and dehydrogenase activity. The result of the solid-state fermentation shows that *Trichoderma reesei* FS10C exhibited the highest potential to promote Cd uptake, microbial biomass, plant growth, nutrient release, and also enhanced the rate of *Trichoderma reesei* FS10C colonization on the solid substrates. Their study shows that solid fermentation powder of FS10-C could be utilized as an inoculating agent for *Trichoderma reesei* FS10C to improve soil fertility and high rate of phytoremediation effectiveness.

A typical example of anthropogenic activities which constitute to the high level of pollution is the release of waste from the mining industries which entails numerous heavy metals. The application of bioremediation, especially using fungi, can go along way toward the prevention of soil contamination due to the presence of these heavy metals. Tansengco et al. (2018) isolated, identified, and characterized some potential heavy metal-resistant fungi dwelling around the mine tailings in Itogon, Benguet. The isolation of these heavy metal-resistant fungi was performed using serial dilution and spread plate methods on potato dextrose agar amended with different heavy metal like nickel (Ni), chromium (Cr), copper (Cu), zinc (Zn), and lead (Pb). The result obtained from the isolation showed that 29 fungal isolates were present in the soil sampled assayed while 4 out of these strains were subjected to molecular characterization and they were identified as *T. gamsii*, *Trichoderma virens*, *T. saturnisporum*, and *T. harzianum*. It was observed that these fungi exhibited growth tolerance with a concentration that varies from 200 to 1000 ppm of heavy metal on potato dextrose agar in the following increasing order: *T. virens* > *T. harzianum* > *T. gamsii* > *T. saturnisporum*. The growth assessment carried out established that all the *Trichoderma* isolates possessed the capability to tolerate an enhanced level of Pb and Cr but the rate of tolerance to Ni, Cu, and Zn depends absolutely on the type of fungi species. It was detected that *T. virens* could remove 91–96% on different pH when the culture media was on shaking condition while *T. virens* demonstrated 70 and 63% decrease for Cu and Cr, at a neutral pH. Their study showed that *Trichoderma* isolates could be utilized for the treatment of biological wastewater treatment, especially in the mining industries.

12.5 Modes of Actions that Empower *Trichoderma* spp. to Perform Their Various Biotechnological Roles

Numerous *Trichoderma* spp. are normally found around the rhizosphere of numerous plants in different agro-ecological areas. Just like the situation with mycorrhizae, *Trichoderma* spp. also possess the capability to produce large quantities of hydrated polysaccharides (mono- and disaccharides) through the root-secreted mucigel layer. These polysaccharides attract *Trichoderma* spp. and help in stimulating their growth, especially the plant-derived sucrose which plays an active role in the stimulation of defense mechanism, encourages the colonization of root, and enhances the rate of photosynthesis, especially through the stomata available on the leaf surface (Vargas et al. 2009). Some important solute transporters including di/tripeptide transporter like permease or intracellular invertase system are utilized in the production of root exudates, especially from *Trichoderma* strains (Vizcaino et al. 2006; Vargas et al. 2009).

Moreover, some strains posses the capability to improve the development of plant growth and trigger their defense against pathogens that want to colonize the roots of a plant. The rate of colonization involves several steps including the capability of the fungi to identify and attach to the root, enter the plant, and tolerate toxic metabolites secreted by the plant as a reaction to an attack from pathogenic microorganisms. It has been observed that the attachment of *Trichoderma* to the root surface of the plant is normally facilitated by hydrophobins which are small hydrophobic proteins present in the outermost cell wall layer that surrounds the fungal cell surface as well as the presence of expansin-like proteins that play an active role in the active development of the cell wall. Specifically, *Trichoderma asperellum* has been documented to secret the class I hydrophobin TasHyd1 that enhances its attachment to the surface of the plant's root which position the hyphae tips to produce some useful compounds that play an active role in the plant defense (Viterbo and Chet 2006).

Furthermore, an expansin-like protein called swollenin TasSwo possessed a cellulose-binding domain that could identify cellulose and change the plant cell wall architecture enhancing the rate of their colonization of plant roots. Several cell wall-degrading enzymes are utilized in the colonization. A typical example of this is the endopolygalacturonase ThPG1 secreted by *Trichoderma harzianum* (Moran-Diez et al. 2009). Chacón et al. (2007) also observed that *Trichoderma* yeast-like cells were discovered to enhance the strengthening of cortical cell walls, plant epidermal, the introduction of newly synthesized numerous callose intrusions of cellulose. It has been discovered that plants could exhibit reaction against any fungi by producing several compounds which possessed antimicrobial properties. The capability of the fungi to colonize any plant depends on the fungal strain in tolerating them. Specifically, it has been observed that *Trichoderma* could resist any invading pathogen which might be linked to the existence of ABC transport systems which are important factors utilized by numerous *Trichoderma* fungal strains possessing biological control attributes against poisonous or antagonistic environment (Ruocco et al. 2009), quick dilapidation of the phenolic substances produced inside plant exuded from plants (Chen et al. 2011), the overpowering of phytoalexin fabrication exhibited on Lotus japonicas when colonized by *Trichoderma koningii* (Masunaka et al. 2011). Also, Harman et al. (2004) discovered that *T. harzianum* and *T. atroviride* have been established to produce a small secreted cysteine-rich protein which is known to be a homolog of the avirulence protein Avr4 obtained from *Cladosporium fulvum*. Stergiopoulos and de Wit (2009) also stated that the attachment of Avr4 to chitin could induce a defense mechanism that could protect *Trichoderma* from the vigorous action of chitinase produced by the plant.

Trichoderma-plant molecular signaling and plant induced effect could occur through the following stages: (1) Production of molecules in the roots apoplast inducers or through the induction of systemic resistance response; (2) Production of secondary roots and growth promotion attributes induced by indole-3-acetate acid; (3) Signaling strain/inoculum dependent, i.e., ethylene, salicylic acid, and jasmonic acid; and (4) Preparation of defense response upon abiotic and abiotic challenges.

12.6 Stimulation of Plant Resistance by *Trichoderma* spp.

It has been stated that plant possesses a defense system that can sense domains or motifs with preserved structural traits representative of different types of microorganisms which is not available in their host referred to as pathogen- or microbe-related molecular patterns. The sensitization of microbe-associated molecular patterns reactions elicited quickly and briefly. The quick microbe-associated molecular patterns reactions entail the movement of ion fluxes through the plasma membrane, the liberation of nitric oxide, reactive oxygen species, ethylene, deposition of callose, and fabrication of antimicrobial compounds. Some microbe-associated molecular patterns have been documented for their plant growth stimulating qualities. Examples of such include volatile compounds, lipopolysaccharides, antibiotics, biosurfactant, and flagellin which have been revealed to elicit systemic resistance.

Some active strains of *Trichoderma* have been documented to produce microbe-associated molecular patterns that have been reported among *Trichoderma* that have lots of biotechnological benefits. Brotman et al. (2008) reported that some *Trichoderma* protends the capability to produce some proteins that are used as an attachment to the root of the plant which could also play a significant role in microbe-associated molecular patterns. It was stated that Swollenin TasSwo could trigger a protective reaction in leaves and roots of cucumber in order to prevent the action of some pathogenic fungi and bacteria. Moran-Diez et al. (2009) stated that endopolygalacturonase ThPG1 could trigger a protective reaction in Arabidopsis the same as the induced systemic resistance stimulated by the plant growth-promoting rhizobacteria. Luo et al. (2010) showed that *Trichoderma peptaibols* utilized multiple defense signaling pathways as a protective mechanism against the action of tobacco mosaic virus. Engelberth et al. (2001) stated that 18mer peptaibols produced by *T. virens* could trigger systemic defenses in cucumber as a defensive mechanism against the action of *Pseudomonas syringae pv. Lachrymans* which is a leaf pathogen while Alamethicin, a 20mer peptaibol obtained from *T. viride* could trigger the production of salicylic acid and jasmonic acid production in lima bean.

12.7 Future Direction and Conclusion

This chapter has shown that *Trichoderma* spp. could be utilized as a biotechnological tool for the management of pests and diseases affecting agricultural crops. Especially they are normally applied as biological control agents against root-borne plant pathogens, soil-borne pathogenic microorganisms as well as for the pre-harvest and post-harvest management of agricultural commodities. It has been shown that the metabolites produced by *Trichoderma* spp. may act on the plant proteome, expressome, and metabolome through the action of the specific pathway that participates in the synthesis of the significant hormone, nutrient uptake, and protection against abiotic and biotic stress. The usage of biologically active components from two or more *Trichoderma* spp. may lead to the development of a more effective biopesticidal formulation which could mitigate all the various hazards relating to the utilization of synthetic pesticides. Moreover, it has been shown that *Trichoderma* spp. possessed the capability to be utilized as a bioremediation tool through the application of biotechnological techniques, thereby maintaining ecorestoration of a contaminated environment. Furthermore, the commercialization of the most effective strain needs to be encouraged because this has become the greatest challenge in the application of *Trichoderma* spp. as a biological control of diseases that might be liked to several factors like the same pathogen on different crops planted under different agro-ecosystems, incorporation with fungicides, disease control under different agro-climatic situations and over a number of years and the same pathogen on different crops plant under various agro-ecosystems. This has led to the discovery of potential isolates that may be effective under laboratory conditions but not effective under field conditions. There is a need for inter-institutional collaborations for large-scale assessment of these isolates. The most effective strain from these isolates might be developed into commercial products so as to mitigate all the challenges encountered during small-scale production.

References

Adedeji ARI, Odebode AC, Agbeniyi SO (2008) Bioassay of five *trichoderma* strains against *Phytophthora megakarya* (*Cacao* pod-rot) in Nigeria. Sci Res Essay 3(9):390–394
Adetunji CO, Adejumo IO (2017) Nutritional assessment of mycomeat produced from different agricultural substrates using wild and mutant strains from *Pleurotus sajor-caju* during solid state

Fermentation. Anim Feed Sci Technol 224:14–19. https://doi.org/10.1016/j.anifeedsci.2016.12.004

Adetunji CO, Adejumo IO (2018) Efficacy of crude and immobilized enzymes from *Bacillus licheniformis* for production of biodegraded feather meal and their assessment on chickens. Environ Technol Innov 11:116–124. https://doi.org/10.1016/j.eti.2018.05.002

Adetunji CO, Adejumo IO (2019) Potency of agricultural wastes in *Pleurotus sajor-caju* biotechnology for feeding broiler chicks. Int J Recycl Organ Waste Agric 8:37–45. https://doi.org/10.1007/s40093-018-0226-6

Adetunji CO, Oloke JK, Prasad G, Akpor OB (2017) Environmental influence of cultural medium on bioherbicidal activities of *Pseudomonas aeruginosa* C1501 on mono and dico weeds. Pol J Nat Sci 32(4):659–670

Adetunji CO, Adejumo IO, Afolabi IS, Adetunji JB, Ajisejiri ES (2018a) Prolonging the shelf-life of 'Agege Sweet' orange with chitosan-rhamnolipid coating. Hortic Environ Biotechnol 59 (5):687–697. https://doi.org/10.1007/s13580-018-0083-2

Adetunji CO, Oloke JK, Osemwegie OO (2018b) Environmental fate and effects of granular pesta formulation from strains of *Pseudomonas aeruginosa* C1501 and *Lasiodiplodia pseudotheobromae* C1136 on soil activity and weeds. Chemosphere 195:98–107

Adetunji CO, Oloke JK, Bello OM, Pradeep M, Jolly RS (2019a) Isolation, structural elucidation and bioherbicidal activity of an eco-friendly bioactive 2-(hydroxymethyl) phenol, from *Pseudomonas aeruginosa* (C1501) and its ecotoxicological evaluation on soil. Environ Technol Innov 13(2019):304–317. https://doi.org/10.1016/j.eti.2018.12.006

Adetunji CO, Afolabi IS, Adetunji JB (2019b) Effect of Rhamnolipid-*Aloe vera* gel edible coating on post-harvest control of rot and quality parameters of 'Agege Sweet' orange. Agric Nat Resour 53(2019):364–372

Adetunji CO, Oloke JK, Phazang P, Sarin NB (2020) Influence of eco-friendly phytotoxic metabolites from *Lasiodiplodia pseudotheobromae* C1136 on physiological, biochemical, and ultrastructural changes on tested weeds. Environ Sci Pollut Res Int 27(9):9919–9934. https://doi.org/10.1007/s11356-020-07677-9

Anand S, Jayarama R (2009) Biocontrol potential of *Trichoderma* Sp. against plant pathogens. Int J Agric Sci 1(2):30–39

Arotupin DJ, Ogunmolu FE (2012) Experimental investigations on the effects of Carbon and Nitrogen sources on concomitant Amylase and Polygalacturonase production by *Trichodermaviride* BITRS-1001 in submerged fermentation. Biotechnol Res Int 2012:904763, 8p. https://doi.org/10.1155/2012/904763

Awad HM, Hamed ER, Ghazi EA, El-gamal A, Shehata HS (2015) *Trichoderma asperellum* isolated from salinity soil using rice straw waste as biocontrol agent for cowpea plant pathogens. J Appl Pharm Sci 2(Supp. 2):091–098. https://doi.org/10.7324/JAPS.2015.58.S

Babu KN, Pallavi PN (2013) Isolation, identification and mass multiplication of *Trichoderma*—an important bio-control agent. Int J Pharm Life Sci 4(1):2320–2323

Barakat FM, Abada KA, Abou-Zeid NM, El-Gammal YHE (2014) Effect of volatile and non-volatile compounds of *Trichoderma* spp. on *Botrytis fabae* the causative agent of Faba bean chocolate spot. Special issue: role of combination between bioagents and solarization on management of crown-and stem-rot of egyptian clover. Am J Life Sci 2(6–2):11–18. https://doi.org/10.11648/j.ajls.s.2014020602.12

Begum M, Sariah M, Abidin MZ, Puteh A, Rahman MA (2008) Antagonistic potential of selected fungal and bacterial biocontrol agents against *Colletotrichum truncatum* of soybean seeds. Pertanika J Trop Agric Sci 31(1):45–53

Blaszczyk L, Siwulski M, Sobieralski K, Lisiecka J, Jedryczka M (2014) *Trichoderma* spp.—application and prospects for use in organic farming and industry. J Plant Protect Res 54 (4):309–317. https://doi.org/10.2478/jppr-2014-0047

Bogumił A, Paszt LS, Lisek A, Trzciński P, Harbuzov A (2013) Identification of new *Trichoderma* strains with antagonistic activity against *Botrytis cinerea*. Folia Hort 25(2):123–132. https://doi.org/10.2478/fhort-2013-0014

Brotman Y, Briff E, Viterbo A, Chet I (2008) Role of swollenin, an expansin-like protein from *Trichoderma*, in plant root colonization. Plant Physiol 147:779–789

Chacón MR, Rodríguez-Galán O, Benítez T, Sousa S, Rey M, Llobell A, Delgado-Jarana J (2007) Microscopic and transcriptome analyses of early colonization of tomato roots by *Trichoderma harzianum*. Int Microbiol 10:19–27

Chen LL, Yang X, Raza W, Li J, Liu Y, Qiu M, Zhang F, Shen Q (2011) *Trichoderma harzianum* SQR-T037 rapidly degrades allelochemicals in rhizospheres of continuously cropped cucumbers. Appl Microbiol Biotechnol 89:1653–1663

Cocaign A, Bui LC, Silar P, Tong LCH, Busi F, Lamouri A, Mougin C, Rodrigues-Lima F, Dupret JM, Dairoua J (2013) Biotransformation of *Trichoderma* spp. and their tolerance to aromatic Amines, a major class of pollutants. Appl Environ Microbiol 79(15):4719–4726

Ekefan EJ, Jama A, Gowen SR (2000) Potential of *Trichoderma harzianum* isolates in biocontrol of *Colletotrichum capsici* causing anthracnose of pepper (*Capsicum* spp.) in Nigeria. J Appl Biosci 20:1138–1145

Ekundayo EA, Ekundayo FO, Bamidele F (2016) Production, partial purification and optimization of a chitinase produced from *Trichoderma viride*, an isolate of maize cob. Mycosphere 7 (6):786–793. https://doi.org/10.5943/mycosphere/7/6/9

El-Mougy NS, Abdel-Kader MM, Aly MDE, Lashin SM (2012) Application of fungicides alternatives as seed treatment for controlling root rot of some vegetables in pot experiments. Adv Life Sci 2(3):57–64. https://doi.org/10.5923/j.als.20120203.03

Engelberth J, Koch T, Schüler G, Bachmann N, Rechtenbach J, Boland W (2001) Ion channel-forming alamethicin is a potent elicitor of volatile biosynthesis and tendril coiling. Cross talk between jasmonate and salicylate signaling in lima bean. Plant Physiol 125:369–377

Fajola AO, Alasoadura SO, Fajola AO, Alasoadura SO (1975) Antagonistic effects of *Trichoderma harzianum* on *Pythium aphanidermatum* causing the damping-off disease of tobacco in Nigeria. Mycopathologia 57(1):47–52

FAO (2017) The future of food and agriculture—trends and challenges. Rome. Foreign Agricultural Service/USDA. (2017). Office of Global Analysis, p 1

Faruq AN, Islam MT, Bhuiyan MZR, Mamun-ur-Rashid M, Amin MR, Hoque S (2014) Efficacy of soil application with *Trichoderma harzianum* T22 and some selected soil amendments on Fusarium wilt of eggplant (*Solanum melongena* L.). Appl Sci Rep 8(2):69–74. https://doi.org/10.15192/PSCP.ASR.2014.4.2.6974

Fiorentino N, Fagnano M, Adamo P, Impagliazzo A, Mori M, Pepe O, Ventorino V, Zoina A (2013) Assisted phytoextraction of heavy metals: compost and *Trichoderma* effects on giant reed (*Arundo donax* L.) uptake and soil N-cycle microflora. Ital J Agron 8(29):244–254

Gardener MBB, Fravel DR (2002) Biological control of plant pathogens: research, commercialization, and application in the USA. Plant Health Prog. https://doi.org/10.1094/PHP-2002-0510-01-RV

Ghosh SK (2017) Study of some antagonistic soil fungi for protection of fruit rot (*Phomopsis vexans*) and growth promotion of Brinjal. Int JAdv Res 5(7):485–494

Gutiérrez-Martínez P, Ramos-Guerrero A, Rodríguez-Pereida C, Coronado-Partida L, Angulo-Parra J, González-Estrada R (2018) Chitosan for postharvest disinfection of fruits and vegetables. In: Wasim Siddiqui M (ed) Postharvest disinfection of fruits and vegetables, 1st edn. Academic Press, Elsevier, pp 231–241. https://doi.org/10.1016/B978-0-12-812698-1.00012-1

Gwa VI, Nwankiti AO (2017) *In vitro* antagonistic potential of *Trichoderma harzianum* for biological control of *Fusarium moniliforme* isolated from *Dioscorea rotundata* tubers. Virologie 6:166. https://doi.org/10.4172/2161-0517.1000166

Harman GE (2000) Myths and dogmas of biocontrol: changes in perceptions derived from research on *Trichoderma harzianum* T-22. Plant Dis 84:377–393

Harman GE, Howell CR, Viterbo A, Chet I, Lorito M (2004) *Trichoderma* species—opportunistic, avirulent plant symbionts. Nat Rev Microbiol 2:43–56

Hasan S (2016) Potential of *Trichoderma* sp. in bioremediation: a Review. J Basic Appl Eng Res 3 (9):776–779

Hem R, Pang A (2017) Bio-stimulant effects of *Trichoderma spp.* on rice (*Oryza sativa L.*): an initial evaluation using a strongly acidic clay loam soil. Am Eur J Sustain Agric 11(5):42–48

Hermosa R, Viterbo A, Chet I, Monte E (2012) Plant-beneficial effects of *Trichoderma* and of its genes. Microbiology 158:17–25p

Howell C (2003) Mechanisms employed by *Trichoderma* species in the biological control of plant diseases: the history and evolution of current concepts. Plant Dis 87:4–10. https://doi.org/10.1094/PDIS.2003.87.1.4

Isah D, Milala AM, Ngala AL (2018) Biostimulation of some fungal strains for the degradation of Atrazine contaminated soil. Glob J Sci Front Res 18(2):33–37

Jagtap JD, Suryawanshi NS (2015) Potential of biocontrol agents against basal rot of onion caused by *Fusarium oxysporumf. sp. Cepae*. Int J Life Sci A5:65–69. Special Issue

Jonathan S G, Asemoloye M, Ahmad R, Olawuyi OJ, Adejoye D (2017) Response of a newly identified *Trichoderma harzianum* KY488466 to crude oil pollution and its expression of peroxidase genes. https://doi.org/10.2139/ssrn.3089090

Jonglaekha N, Vichittragoonthavorn K (2009) Success of using antagonistic fungi for control of plant pests in the Royal Project's area. J Agric Technol 5(1):65–73

Kowalska J (2011) Effects of *Trichoderma asperellum* [T1] on *Botrytis cinerea* [PERS.: FR.], growth and yield of organic strawberry. Acta Sci Pol Hortorum Cultus 10(4):107–114

Kuzmanovska B, Rusevski R, Jankulovska M, Oreshkovikj KB (2018) Antagonistic activity of *Trichoderma asperellum* and *Trichoderma harzianum* against genetically diverse *Botrytis cinerea* isolates. Chilean J Agric Res 78(3):391–399. https://doi.org/10.4067/S0718-58392018000300391

Luo Y, Zhang DD, Dong XW, Zhao PB, Chen LL, Song XY, Wang XJ, Chen XL, Shi M, Zhang YZ (2010) Antimicrobial peptaibols induce defense responses and systemic resistance in tobacco against tobacco mosaic virus. FEMS Microbiol Lett 313:120–126

Masunaka A, Hyakumachi M, Takenaka S (2011) Plant growth promoting fungus, *Trichoderma koningi* suppresses isoflavonoid phytoalexin vestitol production for colonization on/in the roots of *Lotus japonicus*. Microbes Environ 26:128–134

Medina-Cordova N, Rosales-Mendoza S, Hernández-Montiel L, Angulo C (2018) The potential use of *Debaryomyces hansenii* for the biological control of pathogenic fungi in food. Biol Control 121:216–222. https://doi.org/10.1016/j.biocontrol.2018.03.002

Moran-Diez E, Hermosa R, Ambrosino P, Cardoza RE, Gutiérrez S, Lorito M, Monte E (2009) The ThPG1 endopolygalacturonase is required for the *Trichoderma harzianum*—plant beneficial interaction. Mol Plant-Microbe Interact 22:1021–1031

Nongmaithem N, Roy A, Bhattacharya PM (2016) Screening of *Trichoderma* isolates for their potential of biosorption of nickel and cadmium. Braz J Microbiol 47(2016):305–313

Nunes C (2012) Biological control of postharvest diseases of fruit. Eur J Plant Pathol 133:181–196. https://doi.org/10.1007/s10658-011-9919-7

Oladipo OG, Awotoye OO, Olayinka A, Bezuidenhouta CC, Maboetaa MS (2018) Heavy metal tolerance traits of filamentous fungi isolated from gold and gemstone mining sites. Braz J Microbiol 49(2018):29–37

Peccatti A, Rovedder APM, Steffen GPK, Maldaner J, Missio EL, Witt CS, de Morais RM, Camargo B, Neuenschwander F, da Silva Júnior JC, Capitani LC, Dalcul LP (2019) Effect of *Trichoderma* spp. on the propagation of *Maytenus ilicifolia* Mart. ex Reissek. J Agric Sci 11 (3):435–442

Ray RR, Sur D (2016) Optimization of process parameters for effective bioremediation of chromium contaminated soil by *Trichoderma pseudokoningii*. Russ J Biol Res 8(2):69–79

Ricardo SA, Erick AE, Ramon SR, Tatiana CP, Veronica OA, Monica QF, Marcia CS, Anderson SS, Adriana COS, Adriano GC (2018) Physical hazards in dairy products: Incidence in a consumer complaint website in Brazil. Food Control 86:66–70

Ruocco M, Lanzuise S, Vinale F, Marra R, Turrà D, Woo SL, Lorito M (2009) Identification of a new biocontrol gene in *Trichoderma atroviride*: the role of an ABC transporter membrane pump in the interaction with different plant-pathogenic fungi. Mol Plant-Microbe Interact 22:291–301

Santos de Oliveira TA, Blum LEB, Duarte EAA, Luz EDMN (2018) Control of *Phytophthora palmivora* on postharvest papaya with *Trichoderma asperellum*, *T. virens*, *T. harzianum* and *T. longibrachiatum*. Biosci J 34(6):1513–1521

Sarrocco S, Matarese F, Moncini L, Pachetti G, Ritieni A, Moretti A, Vannacci G (2013) Biocontrol of fusarium head blight by spike application of *Trichoderma gamsii*. J Plant Pathol S1:19–S1.27

Senthil R, Prabakar K, Rajendran L, Karthikeyan G (2011) Efficacy of different biological control agents against major postharvest pathogens of grapes under room temperature storage conditions. Phytopathol Mediterr 50:55–65

Setiawati W, Hasyim A, Hudayya A (2016) The effect of fruit characteristics of cayenne pepper (*Capsicum frutescens*) and biocontrol agents (*Trichoderma* sp. and Azoxystrobin) on severity of anthracnose (*Colletotrichum acutatum*). AAB Bioflux 8(1):1–12

Sharma R, Singh D, Singh R (2009) Biological control of postharvest diseases of fruits and vegetables by microbial antagonists: A review. Biol Control 50:205–221. https://doi.org/10.1016/j.biocontrol.2009.05.001

Siddiquee S, Aishah SN, Azad SA, Shafawati SN, Naher L (2013) Tolerance and biosorption capacity of Zn^{2+}, Pb^{2+}, Ni^{3+} and Cu^{2+} by filamentous fungi (*Trichoderma harzianum*, *T. aureoviride* and *T. virens*). Adv Biosci Biotechnol 4:570–583. https://doi.org/10.4236/abb.2013.44075

Singh HB, Singh DP (2009) From biological control to bioactive metabolites: prospects with *Trichoderma* for safe human food. Pertanika J Trop Agric Sci 32(1):99–110

Singh A, Shukla N, Kabadwal B, Tewari A, Kumar J (2018) Review on plant-*Trichoderma*-pathogen interaction. Int J Curr Microbiol App Sci 7:2382–2397. https://doi.org/10.20546/ijcmas.2018.702.291

Srivastava M, Shahid M, Pandey S, Kumar V, Singh A, Trivedi S, Srivastava YK, Shivram (2015) *Trichoderma*: a scientific approach against soil borne pathogens. Afr J Microbiol Res 9 (50):2377–2384. https://doi.org/10.5897/AJMR2015.7788

Stergiopoulos I, de Wit PJ (2009) Fungal effector proteins. Annu Rev Phytopathol 47:233–263

Sundaramoorthy S, Balabaskar P (2013) Biocontrol efficacy of *Trichoderma* spp. against wilt of tomato caused by *Fusarium oxysporum* f. sp. *lycopersici*. J Appl Biol Biotechnol 1 (03):036–040. https://doi.org/10.7324/JABB.2013.1306

Tančić S, Skrobonja J, Lalošević M, Jevtić R, Vidić M (2013) Impact of *Trichoderma* spp. on soybean seed germination and potential antagonistic effect on *Sclerotinia sclerotiorum*. Pestic Phytomed (Belgrade) 28(3):181–185. https://doi.org/10.2298/PIF1303181T

Tansengco M, Tejano J, Coronado F, Gacho C, Barcelo J (2018) Heavy metal tolerance and removal capacity of *Trichoderma* species isolated from mine tailings in Itogon, Benguet. Environ Nat Resour J 16(1):39–57

Teng Y, Luo Y, Ma W, Zhu L, Ren W, Luo Y, Christie P, Li Z (2015) *Trichoderma reesei* FS10-C enhances phytoremediation of Cd-contaminated soil by *Sedum plumbizincicola* and associated soil microbial activities. Front Plant Sci 6:438. https://doi.org/10.3389/fpls.2015.00438

Terna TP, Odebode AC, Bem AA (2013) Growth suppression of some common post-harvest rot fungi by culture filtrates of a soil isolate of *Trichoderma viride*. IOSR J Environ Sci Toxicol Food Technol 3(3):90–96

Valenzuela N, Angel D, Ortiz D, Rosas R, García C, Santos M (2015) Biological control of anthracnose by postharvest application of *Trichoderma* spp. on maradol papaya fruit. Biol Control 91:88–93. https://doi.org/10.1016/j.biocontrol.2015.08.002

Vargas WA, Mandawe JC, Kenerley CM (2009) Plant-derived sucrose is a key element in the symbiotic association between *Trichoderma virens* and maize plants. Plant Physiol 151:792–808

Vinale F, Sivasithamparam K, Ghisalberti EL, Marra R, Woo SL, Lorito M (2008) *Trichoderma*-plant-pathogen interactions. Soil Biol Biochem 40:1–10

Viterbo A, Chet I (2006) TasHyd1, a new hydrophobin gene from the biocontrol agent *Trichoderma asperellum*, is involved in plant root colonization. Mol Plant Pathol 7:249–258

Vizcaino JA, Cardoza RE, Hauser M, Hermosa R, Rey M, Llobell A, Becker JM, Gutierrez S, Monte E (2006) ThPTR2, a di/tri-peptide transporter gene from *Trichoderma harzianum*. Fungal Genet Biol 43:234–246

Whipps JM (2001) Microbial interactions and biocontrol in the rhizosphere. J Exp Bot 52:487–511

Yao L, Teng Y, Luo Y, Christie P, Ma W, Liu F, Wu Y, Luo Y, Li Z (2015) Biodegradation of Polycyclic Aromatic Hydrocarbons (PAHs) by *Trichoderma reesei* FS10-C and effect of bioaugmentation on an Aged PAH-contaminated soil. Bioremediat J 19:9–17. https://doi.org/10.1080/10889868.2014.939137

Zegeye ED, Santhanam A, Gorfu D, Tessera M, Kassa B (2011) Biocontrol activity of *Trichoderma viride* and *Pseudomonas fluorescens* against *Phytophthora infestans* under greenhouse conditions. J Agric Technol 7(6):1589–1602

Chapter 13
Trichoderma as Biostimulant: Factors Responsible for Plant Growth Promotion

Nibha Gupta

Abstract *Trichoderma* has detonated as biostimulant and mycofungicide for improvement of economically important plants of different agriculture, forestry, horticulture sectors, in regard to their protection against abiotic and biotic stress as well as proper growth, development, and productivity. *Trichoderma* plays a vital role by enhancing and modifying the root surface so that plants can do better nutrient uptake and mobilize minerals fast. It can enhance the mineral content in the vicinity of the rhizosphere through solubilization of bound forms, significantly facilitating the plant growth by releasing growth hormones. It is evident that *Trichoderma* induces systemic resistance in plants against various pathogens with the help of various volatile and nonvolatile metabolites, siderohores, enzymes, antioxidants, and polysaccharides. On the one hand, the fungus creates rhizosphere competence, and on the other hand, efficiently eases the unfavorable effect of various environmental stress through antioxidant production and physiological modulation in plants. Recently, molecular and biochemical dialogs between *Trichoderma* and host plants have been studied thoroughly and envisaged the significance of gene–gene interaction corroborate with protein–protein interaction among them. Though the *Trichoderma* and genesis of its benefits have been studied, described, and cited comprehensively, the content of the chapter emphasizes the molecular, physiological, biochemical, and morphological interaction of *Trichoderma* and enlighten the compact and composed picture of its direct and indirect benefit to the host plants.

Keywords *Trichoderma* · Secondary metabolism · Antagonism · Biofungicide

N. Gupta (✉)
Plant Pathology and Microbiology Division, Regional Plant Resource Centre, Bhubaneswar, Odisha, India

C. Manoharachary et al. (eds.), Trichoderma: *Agricultural Applications and Beyond*, Soil Biology 61, https://doi.org/10.1007/978-3-030-54758-5_13

13.1 Introduction

The widely accepted definition of plant biostimulant is that the "plant biostimulant is applied to improve crop production and nutritional quality of agri food products. They are used in agricultural management practices aimed at reducing chemical input, increasing productivity and recovering natural equilibrium in agro ecosystem (EBIC 2013; du Jardin 2015). "The plant stimulants like organic and inorganic natural substances and beneficial microbes are being used for the growth promotion of economically important plants and proved sustainable and eco-friendly. A plant biostimulant is any substance or microorganism applied to plants with the aim to enhance nutrition efficiency, abiotic stress tolerance, and/or crop quality traits, regardless of its nutrients content. Commercial products containing mixtures of such substances and/or microorganisms are also designated as plant stimulants (du Jardin 2015). Plant biostimulants can be categorized into two categories, i.e., biochemical which includes humic and fulvic acids, protein hydrolysates and other N-containing compounds, seaweed extracts and botanicals, chitosan and other biopolymers, and inorganic compounds; whereas biological stimulants involve the role of obligate symbiont mycorrhiza, endosymbionts, and plant growth-promoting rhizobacteria (Colla et al. 2014; Deliopoulos et al. 2010; Hadwiger 2013; Halpern et al. 2015; Katiyar et al. 2015; Khan et al. 2009; Pilon-Smits et al. 2009; Shanmugaiah et al. 2009).

Trichoderma is free-living, green spored ascomycetes, opportunist, avirulent plant symbiont, ubiquitous habitants of soil, water, rhizosphere, and phyllosphere in tropical and temperate environment (Harman et al. 2004b; Howell 2003). It is more prevalent due to its attacking nature on other fungi. Though free living, they occasionally form endophytic associations with plant roots and may provide a range of benefits to their hosts (Colla et al. 2014; Cummings et al. 2016; Hermosa et al. 2012; Shoresh et al. 2010). The fungus is mycoparasite, antagonize other fungi, and utilize their nutrients. *Trichoderma* has gained importance as a microbial plant biostimulant in agriculture and horticulture due to its diverse applications as potential biological disease control agents, source of enzymes and health care medicines, and useful for bioremediation (Cristea et al. 2017; Woo et al. 2014). It is also present as one of the components in various biopesticides, biofertilizers, growth promoters, and biostimulants of commercial nature (Fiorentino et al. 2018; López-Bucio et al. 2015; Rao et al. 2016).

The prime factors behind plant growth promotion are categorized into three groups: (1) metabolite production (antibiotics, HCN, siderophores), (2) biostimulating phytohormone production (auxin, cytokinin, gibberaline), (3) biofertilizing potential through mineral solubilization or nitrogen fixation, and (4) bioprotection through bioactive secondary metabolites, antibiotics, siderophores (Hermosa et al. 2012; Puyam 2016; Van Loon 2007). The plant growth promotion is directly exhibited in terms of increased seed germination, in above- and below-ground plant parts, chlorophyll content and yield size and/or number of flowers and/or fruits (Mendoza-Mendoza et al. 2018). Indirectly, the modification of root

increase in absorption area, thereby improving the nutrient uptake and transport attributed to increase in biomass (Samolski et al. 2012). *Trichoderma* is endowed with these plant growth-promoting properties and significantly facilitates plant growth and development through numerous mechanisms including solubilization of soil nutrients increasing the nutrient efficacy and recycling releasing plant growth stimulatory agent and induced systemic resistance (Adams et al. 2007; Cai et al. 2013; Cornejo et al. 2009; Kapri and Tewari 2011; Li et al. 2015; Singh et al. 2015; Vinale et al. 2006).

The fungus is also very competent, interactive, and effective when co-inoculated with other beneficial microbes of agriculture importance (Colla et al. 2015; Kumar et al. 2009; Rani et al. 1998a, b; Singh and Kumar 2013; Vázquez et al. 2000; Zhuang et al. 2019). On account of plant growth promotion and development, *Trichoderma* is now being a major component of commercial biofertilizer products that contain microbial consortium beneficial for different crops for protecting seeds and strengthening seedlings, development of good root formation and establishment, and finally fully grown crop. *Trichoderma* is eco-friendly, leaves no chemical residue, reduces chemical fungicides, crop losses, and increases yields, is compatible with many crops and antagonistic toward many pathogenic fungi, cost-effective production and usage. It is evident that *Trichoderma* extends other indirect and multifarious benefits to different plant groups besides protection from plant pathogens (Sala et al. 2007; Singh et al. 2004). *Trichoderma* inhabits at the root and rhizosphere helps in decomposition and absorption of native organic materials. It can utilize wide-spectrum substrates and confer tough competition to other microbial rhizospheric associates. It elicits systemic resistance against pathogens. Root colonization by *Trichoderma* enhances plant root growth and nutrient and water uptake, increasing resistance to drought and crop productivity. The factors responsible for the intrinsic biological properties of *Trichoderma* that stimulate the positive effects on plant growth and development, enhancing their growth potential and nutritional uptake, fertilizer use efficiency, seed germination phenomenon, and stress tolerance are being reviewed here.

13.1.1 Modification in Rhizosphere and Roots

Rhizosphere is a composite system, acts as a plant–soil interface, is enriched with minerals, metabolites, gaseous compounds, and considered as a habitat of the microbiome, a variety of microorganisms belonging to nonsymbiotic and free-living, symbiotic, entophytic, parasitic, commensal group. Incidence of microbiome of different morphotypes, their biological properties and functions certainly depend upon the associated plant species, soil types and quality, ecological niche, and microhabitat. Externally added chemical fertilizers, pesticides, and biofertilizers also influence the biological role and functions of the existing microbiome either negatively or positively (Berendsen et al. 2012; Li et al. 2015; Fiorentino et al. 2018; Vázquez et al. 2000). It may affect the eukaryotic and prokaryotic population differently depending upon the host plants, habitat, and seasons under which applications are being carried out. It is known that varied crops and their metabolites are the prime components and can strongly affect soil microbial communities and dynamics (Larkin 2008).

Trichoderma is gaining importance due to its high activity in the edaphic, phyllosphere, and rhizosphere environment and has been very successfully used as mycofungicides, biofertilizers, and plant growth promoters. Their ability to colonize and grow in association with plant roots known as "Rhizosphere competence" is also one of the potential factors behind their role in plant growth promotion (Kaewchai et al. 2009). They are excellent competitors in the rhizosphere, have a capacity to modify the rhizosphere, are tolerant or resistant to soil stress or unfavorable conditions. They compete for the exudates produced by seedling, thereby restricting the growth of phytopathogens (Howell 2003). Root exudates of plants sometimes stimulate and attract *Trichoderma* and other microbial rhizosphere associates thereby helping in plant growth development and promotion under stress conditions also (Kandasamy et al. 2010; Lombardi et al. 2018). *Trichoderma* also plays an important role as decomposers, indirectly supporting root hair growth and allowing plants to take up more water and nutrient available distantly as their roots grow deeper into the soil.

Rhizospheric microflora are mostly influenced by root and soil composition and they establish beneficial interaction at the biochemical, physiological, or molecular level with co-microbiota of pathogenic nature or else affect the root architecture (Harman 2006; Hermosa et al. 2012). The carbon sources released through root exudates stimulate the growth and proliferation of several microbes including *Trichoderma* sp. which colonizes the root system and induces beneficial effects in plants (More et al. 2013; Fernandez et al. 2017; Vargas et al. 2009). Some *Trichoderma* strains do have rhizosphere competence and show a direct effect on plants, enhance their growth potential and nutrient uptake, fertilizer use efficiency, percentage and rate of seed germination, and stimulation of plant defense against biotic and abiotic stress (Shoresh et al. 2010).

Trichoderma is now considered as multi-tasked endophytic fungi of the host roots, as they are capable of residing in the root intracellular space, penetrating and colonizing the plant roots, especially intracellular space (Harman et al. 2004a; Harman 2011; Yedidia et al. 2001). The interaction of cysteine-rich cell wall protein is responsible for fungal adherence which has a vital role in lateral growth, air formation, and elongation. Such a phenomenon also imparts enhancing the root surface, indirectly helping in nutrient uptake and translocation in the shoots, thereby helping in enhancement in plant biomass and growth. As an example, *T. harzianum* is a most effective fungus and is able to colonize roots of most of the plant species, improve the rooting process, helpful in the establishment of plants in nursery conditions and thereby enhances the growth of several vegetables and floriculture crops (Chagas et al. 2017; MacKenzie and Starman 1995). Its inoculation influences the modification of root structure and stimulates the lateral root development in associated plants (Bjorkman et al. 1998; Cornejo et al. 2009). *Trichoderma* spp. produce and modulate hormonal signals in order to facilitate the colonization of roots' growth which, in turn, facilitates colonization by increasing the available surface area. The fungus produces auxins (indol 3 acetic ac (IAA), Indol-3-acetaldehyde (IAAld), Indol-3-ethanol (IEt.) (Casimiro et al. 2001; Reed et al. 1998). Manipulation of root system architecture (RSA) which involves the growth of lateral (LR) and adventitious root and root hairs (RH) formation is one important factor to regulate the effects of biotic and abiotic factors on plant growth and yield (Casimiro et al. 2003; Cornejo et al. 2014; López-Bucio et al. 2003, 2005). Such type of root growth-promoting behavior of *Trichoderma* has already been established under laboratory and field experiments done on various crop plants (Bal and Altintas 2006; Naseby et al. 2000; Yadav et al. 2009).

13.1.2 Bioaccumulation of Useful Metabolites in Rhizosphere System

Trichoderma spp. confer enhancement of growth and development of host plants and other biological associates. The fungus acts as a protective shield against adverse conditions like disease state, environmental conditions like high temperature, cold, drought, metals, acidic, salt, and alkaline conditions. These stress environments are managed by various metabolic processes and their product including enzymes, secondary metabolites, bioleaching, and mineral solubilization (Keller et al. 2005; Keller 2019; Manganiello et al. 2018).

Plant–microbe interaction is governed through communicating signals exerted by biomolecules (secondary metabolites like peptides, peptaibols, pyrones, siderophores, and volatile and nonvolatile metabolites) produced by rhizosphere inhabitants (Vinale et al. 2008a, b; Woo et al. 2014). *Trichoderma* also produces a variety of compound and metabolites which has a different function and potential application in different agriculture, biotechnology, and health care sectors (Singh et al. 2004). *Trichoderma* spp. produces over 250 metabolic products including cell wall degrading enzymes, peptides, secondary metabolites, and other proteins (Lombardi et al. 2018; Salwan et al. 2019; Sarrocco et al. 2009; Harman et al. 2004a; Sivasithamparam and Ghisalberti 1998; Vinale et al. 2009a, b, 2014). The plant growth-promoting effects are attributed to the role of *Trichoderma* alone and/or synergistic effect of other microbial associates and their induced metabolism which exhibited in the form of protection against plant pathogens, mineral solubilization capability, production of siderophores and secondary metabolites (Cornejo et al. 2014; Vieira et al. 2017). Besides plant growth-promoting activity, numerous evidence are available on the involvement of secondary metabolites in the antagonistic activity of *Trichoderma* against a considerable number of plant pathogens (Chet 1990; Kleifeld and Chet 1992; Inbar et al. 1994; Vinale et al. 2009a, b; Zeilinger et al. 2016).

13.1.3 Siderophores

Iron acts as a cofactor of numerous enzymes and an essential nutrient for the growth of plants and other microorganisms. In the aerobic environment (with oxygen and neutral pH), iron exists mainly as Fe3+ and tends to form insoluble ferric oxide, making it unavailable for root absorption and microbial growth (Miethke 2013). Fungal siderophores have been involved in transporting and storage of iron, competing for iron in natural soil, indirectly suppressing the plant pathogen by limiting the metabolism of iron to plants. *Trichoderma* secretes siderophore, an iron-chelating compound which binds with insoluble iron (FeIII) and converts to the soluble form (FeII) for plant absorption and inhibits the growth of plant pathogens by depriving them of iron sources (Howell 2003).

13.1.4 Volatile Compounds

Trichoderma produces volatile organic compounds (VoC) which are of low molecular mass, low boiling point, low polarity, and chemically these are hydrocarbon, aromatic, amine thiols and terpenes and now reported to mediate the plant growth and development in agricultural crops (Bitas et al. 2013; Hung et al. 2013; Junker and Tholl 2013; Korpi et al. 2009; Lee et al. 2015; Lee et al. 2016; Lemfack et al. 2014; Schulz and Dickschat 2007; Vinale et al. 2008b). The production of VoC is not only species-specific but also influenced by soil habitat, soil nutritional content, microbial composition, biomass, and environmental conditions (Insam and Seewald 2010; Lee et al. 2015; McNeal and Herbert 2009). Meena et al. (2017) reported the positive response of volatile compounds from *T. harzianum* for *Alternaria alternata*. As indicated, an auxin-like effect was observed in etiolated stems treated with harzianolide and 6-pentyl-a-pyrone, the major VOCs produced by different *Trichoderma* strains (Vinale et al. 2008a). This compound is important for multiple actions involving fungal mycelium growth inhibition, germination of spores, and pigmentation of plant pathogenic fungi (Salwan et al. 2019).

Many *Trichoderma* species are known as biofungicides and biofertilizers and helpful in crop growth enhancement. *Trichoderma* spp. are producers of many small metabolites having antimicrobial and anticancer properties (Cordovez et al. 2018; Tukhbatova et al. 2014). Nonvolatile metabolites from *Trichoderma* are summarized by Meng-Fei et al. (2019). He described 329 nonvolatile compounds from 20 known species and other unidentified species. Fungi produce a vast range of secondary metabolites and they are known for their capacity to secrete high levels of enzyme, antibiotics, vitamin, polysaccharide, and organic acids (Meyer 2008). Many reports are coming up on the antimicrobial compounds isolated from *Trichoderma* (Li et al. 2016). Zhang et al. (2019) reviewed novel and bioactive metabolites from endophytes including *Trichoderma* sp. They isolated two new isocoumarin and many other compounds having antibacterial activity.

13.1.5 Plant Growth Regulator

Fungi produce a variety of essential phytohormone and natural growth inducers like gibberellic acid and auxin which are crucial in maintaining normal growth and metabolic activity (Cornejo et al. 2009; Hermosa et al. 2012). Such fungi have a critical impact on the physiological status and adaptation of host plants that they colonize. IAA stimulates the higher production of longer roots with root hairs and root laterals which are finally involved in nutrient uptake. It also regulates the cell elongation and numbers which ultimately result in better growth and development. *Trichoderma* spp. are also reported to synthesize and produce IAA and exhibit plant growth promotion efficacy in many agricultural crops in field conditions (Guey et al. 2018; Kumar et al. 2017; França et al. 2017). The hormonal signal perceived by roots resultantly grow well, indirectly enhances nutrient and water uptake and ultimately plant growth. *Trichoderma* species, especially *T. virens* and *T. atroviride*, exhibited characteristic auxin-related phenotype that promoted the root growth, enhanced nutrient and water uptake, and finally increased biomass production (Kumar et al. 2017; Maria et al. 2017).

13.2 Alleviation of Abiotic Stress

13.2.1 Impact on Physiological Response of Plants

An alternative strategy to improve plant tolerance to stress is the use of plant growth-promoting microbes. *Trichoderma* species is a multitasker and rhizospheric salient biocomponent having beneficial effects on plant growth and enhancing resistance to both biotic and abiotic stress. They are known to produce different kinds of enzymes, elicit defense response, a fine metabolic regulation, thereby qualifying to combat the environmental changes and nutrient limitations (Mastouri et al. 2010; Schuster and Schmoll 2010; Singh et al. 2014).

The growth-promoting properties of *Trichoderma* inoculations on radish, pepper, cucumber, tomato, rice, wheat, etc. were demonstrated well (Baker et al. 1984; Chang et al. 1986; Harman 2000). It was thought to be due to increased root development and crop yield, the proliferation of secondary roots, and seedling biomass and foliar area. However, recent literature says it is due to the different physiological mechanisms responsible for the enhancement in plant growth (Doni et al. 2014). Application of *Trichoderma* increased photosynthetic rate, stomatal conductance, water use efficiency, transpiration, internal CO_2 content catalase and superoxide dismutase activities, proline content in treated plants grown in stress environment (Yasmeen and Siddiqui 2017). Mastouri et al. (2010) observed that the treatment of seed with *T. harzianum* accelerates seed germination, increases seedling vigor and ameliorates, water, osmotic, salinity, chilling and heat stress by inducing physiological protection in plants against oxidative damage. Ripa et al. (2019)

assessed the plant growth-promoting and abiotic stress tolerance property of wheat endophytic fungi including *Trichoderma* strains which exhibited salt, heavy metal and drought tolerance at a high level and also exhibited resistance to all tested antibiotic.

13.2.2 Nutritional Starvation

Competition for substrates is the most important factor for fungi as is competition for light in the case of the evolution of plants (Garrette 1956). Microbiome competition also causes nutritional starvation and ultimate defeat of weak associates (Benitez et al. 2004). In a similar way, the microorganisms growing in the vicinity of *Trichoderma* strains encounter the nutrient limitation and rhizospheric colonization. Root exudates and rhizosphere are rich sources of nutrients such as sugar, amino acids, iron, vitamins, organic acids, etc. Competition for carbon is an effective mode not only in *Trichoderma* but also in some other fungi such as strains of *Rhizoctonia solani* and *F. oxysporum* (Alabouvette et al. 2009; Sarrocco et al. 2009).

13.2.3 Salinity Tolerance

Salinity stress affects negatively on plant growth and causes ion toxicity, osmotic stress, oxidative stress, and nutrient deficiency which result in poor growth, reduction in yield, and nutritional deficiency (Chinnusamy et al. 2006). One of the phytohormone ethylenes and its direct precursor ACC is induced by salinity and many abiotic stressed imposed during host–pathogen interaction (Boller 1991; Gailīte et al. 2005). Indole acetic acid and ACC deaminase production by *Trichoderma* sp. was found to be an important factor behind enhanced tolerance toward salt stress when treated with wheat seedlings (Zhang et al. 2019). Besides GA and IAA, antioxidant compounds produced by these fungi especially *T. longibraciatum* are also known to alleviate the negative effects of salinity on many agricultural crops (Aban et al. 2017; Ahmad et al. 2010a, b; Mishra et al. 2015; Rawat et al. 2011). Application of *Trichoderma* in plants enhances the IAA levels reflected in the form of root development, enhanced level of abscisic acid, L proline, ascorbic acid and osmoprotective status, Na elimination through root exudates of plants under salt stress (Cornejo et al. 2014; Rawat et al. 2013). Stress tolerance is also induced due to the synthesis of phenol diacylglycerol, sterol esters, nonesterified fatty acid, and enzymatic antioxidants like SOD superoxide dismutase, catalase, peroxidase, ascorbate peroxidase glutathione reductase (Ahmad et al. 2015; Hashem et al. 2014).

Antioxidative defense mechanisms also play a vital role in mitigating salt stress in many plants. Prolonged salinity stress is responsible for oxidative stress that generates reactive oxygen species (ROS) deleterious to biological molecules (Ahmad et al.

2010a, b). *Trichoderma* induces resistance in host plants against NaCl stress through improved uptake of essential elements and modulation of osmolytes. Fu et al. (2017) studied the alleviation of the effect of *Trichoderma asperellum* on active oxygen production in maize seedling under saline-alkali stress condition. It has been reported that *Trichoderma harzianum*-inoculated plants restore the pigment content, enhances the proline content, plant growth, and development under stress conditions.

13.2.4 Drought Stress

Plant growth and development have also been affected by drought conditions. Plant growth-promoting microbes play a vital role in the alleviation of such stress in plants. Such microbial inoculants impart drought tolerance by producing various metabolites and hormones (Vurukonda et al. 2016). One of the responsible factors behind drought tolerance of plants under *Trichoderma* association is increased secondary metabolites and proline content. Under drought conditions, plant growth and physiological parameters decline as per the observation made on experimental tomato plants (AlwhibiMonaa et al. 2017). The *Trichoderma*-treated plants showed increased root and shoot growth and chlorophyll pigment under drought stress condition. Pectin and total protein content was also increased. An obvious increase in phenol and flavonoid content was observed. It also maintained a high level of growth regulators like indole acetic acid, indole butyric acid, and gibberellic acid under drought stress.

Trichoderma inoculations delayed the drought-induced physiological and biochemical changes in rice, wheat, and tomato (AlwhibiMonaa et al. 2017; Shukla et al. 2012, 2015; Rawat et al. 2016). The fungal treatment enhanced root growth, improved acquisition and storage of water in rice and phenolics, decreased stress-induced metabolites, delayed the stomatal conductance, net photosynthesis, proline, MDA and hydrogen peroxide content increase in phenolics. *Trichoderma* seed priming also reduces the accumulation of toxic reactive oxygen species (ROS) and resultantly root vigor enhances. The production of stress-related enzymes viz., superoxide dismutase (SOD), catalase (CAT), and peroxidase (POD), has been reported in rice under drought condition.

During *Trichoderma* and host plant association and interaction, the proteome and transcriptome of host plant change due to the fungal metabolite and colonization. Thus, the fungi reprogram plant gene expression resulting in alleviation of plant response to their environment (Bae et al. 2011). Alleviation of damage by reactive oxygen species (ROS) water use efficiency and secretion of phytohormonal analog are the three mechanisms employed by the fungi in enhancing plant growth under drought stress. It has been assumed that since the interaction between the plant and the fungus happens largely at the rhizosphere, such mechanism is probably connected to an increase in water absorption effectiveness due to the increased root capacity and hence increased water absorption (Mastouri et al. 2012).

13.2.5 Heat and Cold Tolerance

Low temperature is a major environmental factor limiting plant growth and development in high altitudes. In response to cold stress, plants regulate their physiological, biochemical, and molecular phenomenon like cell membrane permeability, photosynthesis, water absorption, and content and osmoregulation. *Trichoderma* also moderates the low-temperature stress in plants and efficiently alleviate the adverse effects of cold stress leading to enhancement in photosynthetic and growth rates (Ghorbanpour et al. 2018). Reduction in lipid peroxidation rate and electrolyte leakage and an increase in leaf water content and proline accumulation could also be observed as an effect of *Trichoderma* applications. Some *Trichoderma* spp. are isolated from glacial sites of the Indian Himalayan region and reported to be cold-tolerant antifungal strains (Ghildiyal and Pandey 2008). Such types of fungal inoculants as biological agents are useful for field applications in colder regions. Poosapati et al. (2014) studied high temperature-tolerant T isolate with antagonistic activity agent *Sclerotium rolfsii*. This strain was highly tolerant to heat showed at 52 C *T. asperellum*.

13.2.6 Metal Tolerance

Heavy metal contamination of soil and water has become an important environmental issue as it affects different microbiota drastically. Some filamentous fungi pave the way through bioremediation of heavy metal contamination. One of them is *Trichoderma* species which has shown tolerance to a range of toxicants and Cu, Cd, As, Zn heavy metal in vivo (Adams et al. 2007; Ezzi and Lynch 2005; Harman et al. 2004b; Hoseinzadeh et al. 2017; Karcprzak and Malina 2005; Maurya et al. 2019). Due to metal tolerance behavior, *Trichoderma* spp. became a dominant organism in some polluted environments and may play an important role in eco-friendly metal removal technology (Karcprzak and Malina 2005; Nongmaithem et al. 2016). *Trichoderma* cell wall revealed the presence of hydroxyl group and amide group that play a vital role in bioabsorption of heavy metals (Bishnoi et al. 2007). Such a metal tolerance trait of these fungal strains makes them effective cleaning agents of heavy metal polluted environments (Oladipo et al. 2018). Field application of these types of fungal strains has also exhibited a positive effect on translocation index and bioaccumulation factors besides enhancement in biomass and C, N. P, and solubility of heavy metal as compared to uninoculated plants (Nongmaithem et al. 2016). Babu et al. (2014) evaluated *Trichoderma virens*, a heavy metal-tolerant and plant growth-promoting fungus for remediation and bioenergy crop production on mining soil. The fungus tolerates heavy metal and reduces residual concentration in the soil thereby promoting phytostabilization in contaminated soil. The mycoremediation properties of *Trichoderma longibrachiatum* and its protective role for lead-induced oxidative stress in plants has also been studied (Devi et al. 2017). Bioremediation using efficient fungi like *Trichoderma virens*, *T. harzianum*, T. *saturnisporum*, and *T. gamsii* can help in eliminating heavy metal contaminants of wastewater in mining industries (Tansengco et al. 2018).

13.3 Enhancement in Mineral Solubilization and Uptake

Phosphorus is present in the soil in huge amounts but it a major plant growth-limiting nutrient because most of its amount is easily fixed in the soil in the form of insoluble phosphate. Other elements like Fe, Mn, Cu, and Zn which are very important in many physiological and metabolic processes are also not available in active forms. As a result, their deficiency affects the production, yield, and quality of agriculture production (Altomare et al. 1999; Lei and Zhang 2015; Lopaz et al. 2015). The mineral solubilization ability of *Trichoderma* is also one of the important biostimulating factors behind plant growth and development. Soil is a composite system of living and nonliving plethora of biological and nonbiological components including soluble and bound forms of different minerals (Rawat and Tewari 2005). Mineralization of different soluble and insoluble mineral is a dynamic process and greatly influenced by soil pH and extracellular secondary metabolites and enzymes which regulate the solubilization of minerals and uptake by plant system. It has been reported that *Trichoderma* solubilizes bound minerals through lowering the soil pH by releasing organic acid, gluconic acid, lactic acid, citric acid, tartaric acid, succinic acid, and fumaric acid extracellularly and allow the dissolution of phosphate as well as macro- and micronutrient, Fe, Mg, Mn, which are necessary for plant metabolism (Cao et al. 2008; Harman 2006). Besides, acidification of the surrounding media, *Trichoderma* solubilize minerals phytate, Fe2O3, CuO and metallic Zn through chelation by siderphores, reduce by ferric reduction, and hydrolysis by phytase (Li et al. 2015).

The mineral solubilization properties and activity of *Trichoderma* are species specific and environmentally regulated. *Trichoderma* produces organic acid to solubilize insoluble tricalcium phosphate at high pH stress whereas drought stress induces the production of alkaline phosphate enzymes. This beneficial activity of *Trichoderma* was evaluated and confirmed in many crop plants like rice, groundnut, tomato, etc. (Chagas et al. 2015; Singh et al. 2014; Shukla and Vyas 2014). Many species of *Trichoderma* are endowed with dual quality as hormone producer aided with mineral solubilizing potential makes them more useful mycopesticides for extensive commercial use in agriculture (Vinale et al. 2008b; Resende et al. 2014).

13.4 Enhancement in Plant Defense and Immune Stimulation

13.4.1 Mycoparasitism Related Metabolites

Trichoderma involves mycoparasitism for antagonistic behavior toward plant pathogens. The mycoparasitic event involves chemotropic growth, host recognition, coiling, and appressoria formation, secretion of hydrolytic enzymes like glucanases, chitinases and proteases, penetrations of the hyphae and lysis of the host cell (Harman et al. 2004a; Kumar et al. 2016). There are at least 20–30 genes, proteins, and other metabolites that are directly involved in this interaction. The functions of different glucanases and chitinases in the process of mycoparasitism are well studied from *Trichoderma* spp. using gene-for-gene experiments. Different types of *Trichoderma* produce mycoparasitin-related compounds. *T. harzianum* produces anthraquinone which enhances the number of coils. *Trichoderma atroviride*, *T. virens, T. reesei* produces ferricrocin a siderophors and key metabolite for iron chelation. There is a report on the inhibition of glucon biosynthesis by *T. longibrachiale*. Many *Trichoderma* species produce hydrolytic enzymes like glucanases, chitinases, endopolygalacturonase which hydrolase fungal cell wall (Daguerre et al. 2014).

13.4.2 Bioactive Metabolites

Trichoderma species are classified as microbial biological control agents “MBCA” (Woo et al. 2014). Numerous *Trichoderma* are successful MBCA of various plant pathogens. Initially, the biopesticidal properties of *Trichoderma* were considered as prime benefits, and eventually, these MBCAs are demonstrated to be effective biofertilizers, biostimulants, and bioenhancers of crop resistance to various biotic and abiotic stresses (Fontenelle et al. 2011). *Trichoderma* species are common in soil and root ecosystem, ubiquitous saprobes and have been tested as biological control agents against a wide range of pathogenic fungi like *Alternaria, Botrytis, Botryosphaeria, Dematophora, Fusarium, Lasiodiploidia, Rhizoctonia, Pythium, Phytophthora, Sclerotium*, and nematodes (Abdel Fattah et al. 2007; Manganiello et al. 2018; Singh et al. 2008). Various diseases controlled by *Trichoderma* spp. are sheath blight, bakanae, leaf blight, loose smut, wilt, root rot, ring rot, dieback, crown, black scurf, web blight of different crops like rice, wheat, chickpea, pigeon pea, apple, guava, chilli, tomato, potato, beans, etc. (Puyam 2016). Commercial formulation of *T. harzianum*, *T. polysporus, T koningii* is now available as brand names in aboard like Binab T, Plant Shield, Antagen, Promot plus, etc. In our country, most of the products are formulated from *T. viride* and *T. harzianum* on commercial productions like Antagen TV, Trichostar, Gliostar, Monitor, Birdene, Biofil, Ecofit, Trichoguard, Bicon, etc. (Puyam 2016).

Trichoderma is accredited with many biological control credentials like antibiosis, antagonisms, mycoparasitism, and induction of plant defense response. Rhizosphere interaction between plant and microbes involves communication between them through biomolecules synthesized inside and active extracellularly. The interactive host–microbe relationship establishes and is dependent upon their mutual molecular dialogs (Cornejo et al. 2014). Host plants have systemic acquire resistance or induced systemic resistance which is dependent upon the production of salicyclic acid, jasmonic acid, and ethylene (Meena et al. 2017; Yuan et al. 2019). The synthesis and production of signaling molecule like hydrogen peroxide, nitric oxide, and salicylic acid are activated by *Trichoderma* thereby inducing plant defense and mycoparasitism (Nawrocka et al. 2019). Such types of biocontrol activity is due to well-coordinated transcriptomic, proteomic, and metabolomic responses of plants in the presence of *Trichoderma* in its rhizosphere vicinity (Mukherjee et al. 2012). Production of phenolic compounds like hydroxyl benzoic acid, cinnamic acid, catechins, flavonols, flavones, flavanone also induces the systemic defense response (Nawrocka et al. 2019).

Trichoderma and its direct interaction with plant pathogens involve cell wall degrading substances including antibiotics (Benitez et al. 2004; Harman et al. 2004b; Kredics et al. 2001). *Trichoderma* produces a variety of antibiotics like trichokonins, glovinin, gliotoxin, viridian, pyrones, and reveal antibiosis against plant fungal pathogen (Howell 2003; Harman et al. 2004a). The beneficial interaction of *Trichoderma* with plants depends upon signal exchange among them and mediated by effector proteins known as hydrophobin that alter the host structure and help in the establishment of symbiotic relationship (Guzmán-G et al. 2017). To date 317 peptaibols are reported, and among them, 190 are synthesized by *Trichoderma* (Whitmore and Wallace 2004). These are characterized by the presence of unusual amino acid alpha aminoisobutyric acid isovalin, imino acid hydroxyproline (Chugh and Wallace 2001; Mukherjee et al. 2011). The production of cell wall degrading enzymes such as chitinase, cellulose protease, have a vital role in the inhibition of fungal pathogen and induced resistance of host plant system (El-Katathy et al. 2001; Gajera et al. 2012; Vinale et al. 2008b).

Trichoderma produces many antibiotics which have inhibitory action against many plant pathogens like *Rhizoctonia, Pythium, Gaeumannomyces, Candida, Penicillium, Aspergillus, Cryptococus, Sclerotium, Staphylococcus,* and *Mycena.* It is known that antimicrobial activity is species-specific and it produces specific metabolites against specific individual organisms. Besides antifungal properties, it produces protein inhibitors, antibacterial, antiviral, immunosuppressor compounds (Cornejo et al. 2014).

Trichoderma produces such types of compounds which alter the fungal growth of plant pathogen. Steroidal compounds viridian produced by *T. koningii, T. virens, T. viride* alter the spore germination of Botrytis, Colletotrichum, and Fusarium sp. Many *Trichoderma* sp. produces Trichothecene (Trichodermin) inhibiting the protein synthesis. *T. harzianum* produced by phenyl ethanol inhibits aflatoxin production by Aspergillus flavus. Disruption of cell wall cellulose is made by swollenin produced by *Trichoderma* (Andberg et al. 2015; Eibinger et al. 2016).

Fungal oligosaccharides are now focused on the biological management of crop diseases by elicitation of defense response (Boregowda et al. 2017). Crude oligosaccharide extracted from *Trichoderma* spp. enhanced the disease protection ability in pearl millet when they followed the seed priming process. Oligomers of chitin and glucan are fungal elicitors generated from the fungal cell wall and are measured as primary signals responsible for the initiation of plant resistance reactions. It is well known that several oligosaccharides of fungal cell wall components stimulate phytoalexin secretion and lignin and callose formation in plants (Kauss et al. 1989; Lattanzio et al. 2006).

13.5 Conclusions

Trichoderma resides mostly in soil and infrequently occurred as endophyte within host plants of agriculture, forestry, and horticulture importance. This fungus is also known as mycofungicide and endowed with many intrinsic properties like fast growth and development, inhibiting a broad spectrum of fungal disease, diversity of control mechanism, rhizosphere competence, tolerant or resistant to fungicides, stress tolerance, nutrient solubilization and mobilization and antagonism, etc. Such intrinsic nature of growth, biochemical, physiological, and metabolic behavior makes the fungus more beneficial for the growth and development of associated host plants. Factors responsible for the biostimulating characteristics of *Trichoderma* which includes morphological and microbial modification of host plants, bioaccumulation of metabolites, biotic and abiotic stress tolerance, nutrient solubilization, uptake and mobilization, biocontrol properties have been elaborated in detail. It is evident that the beneficial activity of *Trichoderma* is species-specific, and comprehensive search of this group of fungi from different ecological niche and agroclimatic zones is required as many more tropical regions remain to be unexplored in this regard.

References

Aban L, Barcelo RC, Oda EE, Reyes GA, Balangcod TD, Gutierrez RM, Hipol RM (2017) Production, phosphate Solubilisation and ACC Deaminase activity of root symbiotic fungi (RSF) from *Drynaria quercifolia* L Jomar. Bull Env Pharmacol Life Sci 6(5):18–23

Abdel Fattah MG, Shabana MY, Ismail EA, Rashad MY (2007) *Trichoderma harzianum*: a biocontrol agent against *Bipolaris oryzae*. Mycopathologia 164:81–89

Adams P, de Leij M, Lynch JM (2007) *Trichoderma harzianum* rifai 1295–22 mediates growth promotion of crack willow (*Salix fragilis*) saplings in both clean and metal-contaminated soil. Microb Ecol 54:306–313

Ahmad P, Hashem A, Abd-Allah EF, Alqarawi AA, John R, Egamberdieva D, Gucel S (2015) Role of *Trichoderma harzianum* in mitigating NaCl stress in Indian mustard (*Brassica juncea* L) through antioxidative defense system. Front Plant Sci 6:868

Ahmad P, Jaleel CA, Salem MA, Nabi G, Sharma S (2010a) Roles of enzymatic and non- enzymatic antioxidants in plants during abiotic stress. Crit Rev Biotechnol 30:161–175

Ahmad P, Jaleel CA, Sharma S (2010b) Antioxidative defense system, lipid peroxidation, proline metabolizing enzymes and biochemical activity in two genotypes of *Morus alba* L subjected to NaCl stress. Russ J Plant Physiol 57:509–517

Alabouvette C, Olivain C, Migheli Q, Steinberg C (2009) Microbiological control of soil-borne phytopathogenic fungi with special emphasis on wilt-inducing *Fusarium oxysporum*. New Phytol 184:529–544

Altomare C, Norvell WA, Björkman T, Harman GE (1999) Solubilization of phosphates and micronutrients by the plant-growth-promoting and biocontrol fungus *Trichoderma harzianum* Rifai 1295-22. Appl Environ Microbiol 65:2926–2933

AlwhibiMonaa S, Allah AHEF, Alqarawi AA, Sliman DWK, Wirth S, Gamberdieva DE (2017) Increase resistance of drought by *Trichoderma harzianum* fungal treatment correlates with increased secondary metabolites and proline content. J Inte Agri 16(8):1751–1757

Andberg M, Penttila M, Saloheimo M (2015) Swollenin from *Trichoderma reesei* exhibits hydrolytic activity against cellulosic substrates with features of both endoglucanases and cellobiohydrolases. Bio/Technology 181:105–113

Babu AG, Shim J, Bang KS, Shea PJ, Oh BT (2014) *Trichoderma virens* PDR-28: a heavy metal-tolerant and plant growth-promoting fungus for remediation and bioenergy crop production on mine tailing soil. J Environ Manag 132:129–134

Bae H, Roberts DP, Lim HS, Strem MD, Park SC, Ryu CM, Bailey BA (2011) Endophytic *Trichoderma* isolates from tropical environments delay disease onset and induce resistance against *Phytophthora capsici* in hot pepper using multiple mechanisms. Mol Plant-Microbe Interact 24(3):336–351

Baker R, Elad Y, Chet I (1984) The controlled experiment in the scientific method with special emphasis on biological control. Phytopathology 74:1019–1021

Bal U, Altintas S (2006) Effects of *Trichoderma harzianum* on the yield and fruit quality of tomato plants (*Lycopersicon esculentum*) grown in an unheated greenhouse. Aust J Exp Agric 46 (1):131–136

Benitez T, Rincon MA, Limon MC, Codon CA (2004) Biocontrol mechanisms of *Trichoderma* strains. Int Microbiool 7:249–260

Berendsen RL, Pieterse CMJ, Bakker PAHM (2012) The rhizosphere microbiome and plant health. Trends in Plant Sci 17(8):478–486

Bishnoi NR, Kumar R, Bishnoi K (2007) Biosorption of Cr (VI) with *Trichoderma viride* immobilized fungal biomass and cell free Ca – alginate beads. Indian J Exp Biol 45:657

Bitas V, Kim HS, Bennett JW, Kang S (2013) Sniffing on microbes: diverse roles of microbial volatile organic compounds in plant health. Mol Plant Microbe Interact J 4:835–843

Bjorkman T, Blanchard LM, Harman GE (1998) Growth enhancement of shrunken-2 sweet corn when colonized with *Trichoderma harzianum* 1295-22: effect of environmental stress. J Am Soc Hortic Sci 123:35–40

Boller T (1991) Ethylene in pathogenesis and disease resistance. In: Mattoo AK, Suttle JC (eds) The plant hormone ethylene. CRC Press, Boca Raton, FL, pp 293–314

Boregowda N, Puttaswamy H, Sripathy H, Nagaraja G (2017) *Trichoderma oligosaccharides* priming mediates resistance responses in pearl millet against downy mildew pathogen. J Applied Biol Biotechnol 5(2):97–103

Cai F, Yu G, Wang P, Wei Z, Fu L, Shen Q et al (2013) Harzianolide, a novel plant growth regulator and systemic resistance elicitor from *Trichoderma harzianum*. Plant Physiol Biochem 73:106–113

Cao L, Jiang M, Zeng Z, Du A, Tan H, Liu Y (2008) *Trichoderma* atroviride F6 improves phyto extraction efficiency of mustard (*Brassica juncea* (L) Coss var foliosa Bailey) in cd, Ni contaminated soils. Chemosphere 71(9):1769–1773

Casimiro I, Beeckman T, Graham N, Bhalerao R, Zhang H, Casero P, Sandberg G, Bennett MJ (2003) Dissecting *Arabidopsis* lateral root development. Trends Plant Sci 8:165–171

Casimiro I, Marchant A, Bhalerao RP, Swarup R, Graham N, Inzé D, Sandberg G, Casero PJ, Bennett M (2001) Auxin transport promotes Arabidopsis lateral root initiation. Plant Cell 13:843–852

Chagas LFB, Chagas AFJ, de Castro HG (2017) Phosphate solubilization capacity and indole acetic acid production by *Trichoderma* strains for biomass increase on basil and mint plants. Brazilian J Agriculture 92(2):176–185

Chagas LFB, Chagas Junior AF, de Carvalho MR, Miller LO, Colonia BSO (2015) Evaluation of the phosphate solubilization potential of *Trichoderma* strains (Trichoplus JCO) and effects on rice biomass. J Soil Sci Plant Nutr 15(3):794–804

Chang YC, Baker R, Kleifeld O, Chet I (1986) Increased growth of plants in the presence of the biological control agent *Trichoderma harzianum*. Plant Dis 70:145–148

Chet I (1990) Biological control of soil borne plant pathogens with fungal antagonists in combination with soil treatments. In: Hornby D (ed) Biological control of soil borne plant pathogens. CAB Intrnationals, Wallingford, pp 15–25

Chinnusamy V, Zhu J, Zhu JK (2006) Salt stress signaling and mechanisms of plant salt tolerance. Genet Eng (N Y) 27:141–177

Chugh JK, Wallace BA (2001) Peptaibols: models for ion channels. Biochem Soc Trans 29:565–570

Colla G, Rouphael Y, Canaguier R, Svecova E, Cardarelli M (2014) Biostimulantaction of a plant-derived protein hydrolysate produced through enzymatic hydrolysis. Front Plant Sci 5:1–6

Colla G, Rouphael Y, Di Mattia E, El-Nakhel C, Cardarelli M (2015) Co-inoculation of *Glomus intraradices* and *Trichoderma atroviride* acts as abiostimulant to promote growth, yield and nutrient uptake of vegetable crops. J Sci Food Agric 95:1706–1715

Cordovez V, Schop S, Hordijk K, Dupréde Boulois H, Coppens F, Hanssen I, Raaijmakers JM, Carrión VJ (2018) Priming of plant growth promotion by volatiles of root-associated *Microbacterium* spp. Appl Environ Microbiol 84:1865–1818

Cornejo HAC, Rodríguez LM, Cuevas RA, Bucio JL (2014) *Trichoderma* spp. improve growth of *Arabidopsis* seedlings under salt stress through enhanced root development, osmolite production, and Na+ elimination through root exudates. Mol Plant-Microbe Interact 27(6):503–514

Cornejo HAC, Rodriguez LM, Panagoz CC, Bucio JL (2009) *Trichoderma virens*, a plant beneficial fungus, enhances biomass production and promotes lateral root growth through an Auxin-dependent mechanism in *Arabidopsis*. Plant Physiol 149(3):1579–1592

Cristea VZ, Răut TE, Sesan BT, Oancea F (2017) Surface response optimization of submerged biomass production for a plant biostimulant *Trichoderma* strain. Scientific Bulletin Series F Biotechnologies 21:56–65

Cummings NJ, Ambrose A, Braithwaite M, Bisstt J, Roslan HA, Abdullah J, Stewart A, Agbayani FV, Steyaert J, Hill RA (2016) Diversity of root –endophytic *Trichoderma* from Malaysian borneo. Mycol Prog 15:50

Daguerre Y, Siegel K, Edel-Hermann V, Steinberg C (2014) Fungal proteins and genes associated with biocontrol mechanisms of soil-borne pathogens: a review. Fungal Biol Rev 28:97–125

Deliopoulos T, Kettlewell PS, Hare MC (2010) Fungal disease suppression byinorganic salts: a review. Crop Prot 29:1059–1075

Devi S, Sreenivasulu Y, Bhaskara Rao KV (2017) Protective role of *Trichoderma logibrachiatum* (WT2) on Lead induced oxidative stress in *Helianthus annus* L. Indian J Exp Biol 55:235–241

Doni F, Ishahak A, Radziah C, Zain CM, Mohtar W, Yusuff W (2014) Physiological and growth response of rice plants (*Oryza sativa* L) to *Trichoderma* spp. Inoculants ANB express 4:45

du Jardin P (2015) Plant biostimulants: definition, concept, main categories and regulation. Scientia Horticulture 196:3–14

EIBC (European Biostimulants Industry Council) (2013) Promoting the Biostimulant Industry and the Role of Plant Biostimulants in Making Agriculture More Sustainable Available online at: www biostimulants eu/

Eibinger M, Karin S, Sateelkow J, Thomas G, Ramoni J, Seiboth B, Plank H, Nidetzky B (2016) Functional characterization of the native swollenin from *Trichoderma reesei*: study of its possible role as C1 factor of enzymatic lignocellulose conversion. Biotechnol Biofuels 9(1):178

El-Katathy MH, Gudelj M, Robra KH, Elnaghy MA, Gubitz GM (2001) Characterization of chitinase and an endo B 1 3 glucanase from *Trichoderma harzianum* Rifai T 24 involved in control of the phytopahtogen Sclerotium rolfsii. Appl Microbiol Biotechnol 56(1–2):137–143

Ezzi MI, Lynch JM (2005) Biodegradation of cyanide by *Trichoderma* spp. and *Fusarium* spp. Enzym Microb Technol 36:849–854

Fernandez E, Trillas MI, Segarra G (2017) Increased rhizosphere populations of *Trichoderma* asperellum strain T34 caused by secretion pattern of root exudates in tomato plants inoculated with *Botrytis cinerea*. Plant Pathol 66(7):1110–1116

Fiorentino N, Ventorino V, Woo SL, Pepe O, De Rosa A, Gioia L, Romano I, Lombardi N, Napolitano M, Colla G, Rouphael Y (2018) *Trichoderma* based biostimulants modulate rhizosphere microbial populations and improve N uptake efficiency, yield, and nutritional quality of leafy vegetables. Front Plant Sci 9:743

Fontenelle ADB, Guzzo SD, Lucon CMM, Harakava R (2011) Growth promotion and induction of resistance in tomato plant against *Xanthomonas euvesicatoria* and *Alternaria solani* by *Trichoderma* spp. Crop Prot 30:1492–1500

França DVC, Kupper KC, Magri MMR, Gomes TM, Rossi F (2017) *Trichoderma* spp. isolates with potential of phosphate solubilization and growth promotion in cherry tomato. Pesquisa Agropecuaria Tropical 47(4):360–368

Fu J, Liu Z, Li Z, Wang Y, Yang K (2017) Alleviation of the effects of saline-alkaline stress on maize seedlings by regulation of active oxygen metabolism by *Trichoderma asperellum*. PLoS One 12(6)

Gailīte A, Samsone I, Ievinsh G (2005) Ethylene is involved in *Trichoderma*-induced resistance of bean plants against *Pseudomonas syringe*. Acta Univ Latv 691:59–67

Gajera HP, Bambharolia RP, Patel SV, Khatrani TJ, Goalkiya BA (2012) Antagonism of *Trichoderma* spp against *Macrophomina phaseolina*: evaluation of coiling and cell wall degrading enzymatic activities. J Plant Pathol Microb 3:7–10

Garrette SD (1956) Biology of root-infecting fungi. Cambridge university Press, Cambridge UK, p 293

Ghildiyal A, Pandey A (2008) Isolation of cold tolerant antifungal strains of *Trichoderma* sp. from glacial sites of Indian Himalayan region. Res J Microbiol 3:559–564

Ghorbanpour A, Salini A, Ali M, Ghanbary T, Pirdashti H, Dehestani A (2018) The effect of *Trichoderma* harzianum in mitigating low temperature stress in tomato (*Solanum lycopersicum* L) plants. Scientia Horticulurae 230:134–141

Guey N, Kumar K, Dangue A, Arama M (2018) Bioproduction of indol 3 acetic acid by *Trichoderma* strains isolated from agriculture field soils in Senegal. World J Pharmaceutical Res 7(17):817–825

Guzmán-G P, Duarte MIA, Delaye L, Estrella AH, Monfil VO (2017) Identification of effector-like proteins in *Trichoderma* spp. and role of a hydrophobin in the plant-fungus interaction and mycoparasitism. BMC Genet 18:16

Hadwiger LA (2013) Multiple effects of chitosan on plant systems: solid science orhype. Plant Sci 208:42–49

Halpern M, Bar-Tal A, Ofek M, Minz D, Muller T, Yermiyahu U (2015) The use of biostimulants for enhancing nutrient uptake. In: Sparks DL (ed) Advances in agronomy, vol 129. Elsevier, Boston, pp 141–174

Harman GE (2000) Myths and dogmas of biocontrol – changes in perceptions derived from research on *Trichoderma harzianum* T-22. Plant Dis 84:377–393

Harman GE (2006) Overview of mechanisms and uses of *Trichoderma* spp. Phytopathology 96:190–194

Harman GE (2011) Multifunctional fungal plant symbionts: new tools to enhance plant growth and productivity. New Phytol 189:647–649

Harman GE, Howell CR, Viterbo A, Chet I, Lorito M (2004a) *Trichoderma* species – opportunistic, avirulent plant symbionts. Nat Rev Microbiol 2:43–56

Harman GE, Lorito M, Lynch JM (2004b) Uses of *Trichoderma* spp.to remediate soil and water pollution. Adv Appl Microbiol 56:313–330

Hashem A, Allah AE, Alqarawi AA, Al Asma HA, Dilfuza E (2014) Alleviation of abiotic salt stress in *Ochradenus baccatus* (Del) by *Trichoderma hamatum* (Bonord) Bainier. J Plant Interact 9:10

Hermosa R, Viterbo A, Chet I, Monte E (2012) Plant-beneficial effects of *Trichoderma* and of its genes. Microbiology 158:17–25

Hoseinzadeh S, Shahabiv S, Aliloo AA (2017) Toxic metals accumulation in *Trichoderma asperellum* and *T.harzianum*. Microbiology 86(6):728–736

Howell CR (2003) Mechanisms employed by *Trichoderma* species in the biological control of plant diseases: the history and evolution of current concepts. Plant Dis 87:4–10

Hung R, Lee S, Bennett JW (2013) *Arabidopsis thaliana* as a model system for testing the effects of *Trichoderma* volatile organic compounds. Fungal Ecol 6:19–26

Inbar J, Abramsky M, Cohen D, Chet I (1994) Plant growth enhancement and disease control by *Trichoderma harzianum* in vegetable seedlings grown under commercial conditions. Eur J Plant Pathol 100:337–346

Insam H, Seewald SA (2010) Volatile organic compounds (VOCs) in soils. Biol Fertil Soils 46:199–213

Junker RR, Tholl D (2013) Volatile organic compound mediated interactions at the plant– microbe interface. J Chem Ecol 39:810–825

Kaewchai S, Soytong K, Hyde KD (2009) Mycofungicide and fungal biofertilizers. Fungal Divers 3:25–50

Kandasamy SK, Arasu VS, Kathiresan K (2010) Effect of *Trichoderma* on soil phosphate solubilization and growth improvement of *Avicennia marina*. Aquat Bot 41(3):787–795

Kapri A, Tewari L (2011) Phosphate solubilization potential and phosphatase activity of rhizospheric *Trichoderma* spp. Curr Microbiol 62(5):1521–1527

Karcprzak M, Malina G (2005) The tolerance and Zn2+, Ba2+ and Fe2+ accumulation by *Trichoderma atroviride* and *Mortierella exigua* isolated from contaminated soil. Can J Soil Sci 85:283–290

Katiyar D, Hemantaranjan A, Singh B (2015) Chitosan as a promising natural compound to enhance potential physiological responses in plant: a review. Indian J Plant Physiol 20:1–9

Kauss H, Jeblick W, Domard A (1989) The degrees of poylimerization and N-acetylation of chitosan determine its ability to elicite callose formation in suspension cells and protoplasts of *Catharanthus roseus*. Planta 1(178):385–392

Keller NP (2019) Fungal secondary metabolism: regulation, function and drug discovery. Nat Rev Microbiol 17:167–180

Keller NP, Turner G, Bennett JW (2005) Fungal secondary metabolism from biochemistry to genomics. Nat Rev Microbiol 3(12):937–947

Khan W, Rayirath UP, Subramanian S, Jithesh MN, Rayorath P, Hodges DM, Critchley AT, Craigie JS, Norrie J, Prithiviraj B (2009) Seaweed extracts as biostimulants of plant growth and development. J Plant Growth Regulation 28:386–399

Kleifeld O, Chet I (1992) *Trichoderma harzianum*-interaction with plants and effect on growth response. Plant Soil 144:267–272

Korpi A, Jarnberg J, Pasanen AL (2009) Microbial volatile organic compounds. Crit Rev Toxicol 139:93

Kredics L, Antal Z, Manczinger L, Nagy E (2001) Breeding of mycoparasitic *Trichoderma* strains for heavy metal resistance. Lett Appl Microbiol 2:112–116

Kumar A, Aggarwal A, Kaushish S (2009) Influence of arbuscular mycrooohizal fungi and *Trichoderma viride* ongrowth performance of *Salvia offficnalis* Linn. J Appl Nat Sci 1:13–17

Kumar NV, Rajam KS, Rani ME (2017) Plant growth promotion efficacy of Indole acetic acid (IAA) produced by a mangrove associated fungi-*Trichoderma virideVKF3*. Int J Curr Microbiol App Sci 6(11):2692–2701

Kumar SK, Yu C, Dou K, Wang M, Li Y, Chen J (2016) Synergistic effect of *Trichoderma*- derived antifungal metabolites and cell wall degrading enzymes on enhanced biocontrol of *Fusarium oxysporum* f sp *cucumerinum*. Biol Control 94:37–46

Larkin RP (2008) Relative effects of biological amendments and crop rotations on soil microbial communities and soilborne diseases of potato. Soil Biol Biochem 40:1341–1351

Lattanzio V, Lattanzio VMT, Cardinali A (2006) Role of phenolics in the resistance mechanisms of plants against fungal pathogens and insects. Phytochem Adv Res 6:23–67

Lee S, Hung R, Yap M, Bennett JW (2015) Age matters: the effects of volatile organic compounds emitted by *Trichoderma atroviride* on plant growth. Arch Microbiol 197:723–730

Lee S, Yap M, Behringer G, Hung R, Bennett JW (2016) Volatile organic compounds emitted by *Trichoderma* species mediate plant growth. Fungal Biology and Biotechnology 3:7

Lei Z, Zhang YQ (2015) Effects of phosphate solubilization and phytohormone production of *Trichoderma asperellum* Q1 on promoting cucumber growth under salt stress. J Integr Agric 14 (8):1588–1597

Lemfack MC, Nickel J, Dunkel M, Preissner R, Piechulla B (2014) VOC: a database of microbial volatiles. Nucleic Acids Res 42:744–752

Li R-X, Cai F, Pang G, Shen Q-R, Li R, Chen W (2015) Solubilisation of phosphate and micronutrients by *Trichoderma harzianum* and its relationship with the promotion of tomato plant growth. PLoS One 10(6):e0130081

Li W, Xu J, Li F, Xu L, Li C (2016) A new antifungal isocoumarin from the endophytic fungus *Trichoderma* sp 09 of *Myoporum bontioides* a gray. Phcog Mag 12(48):259–261

Lombardi N, Vitale S, Turrà D, Reverberi M, Fanelli C, Vinale F, Marra R, Ruocco M, Pascale A, D'Errico G, Woo SL, Lorito M (2018) Root exudates of stressed plants stimulate and attract *Trichoderma* soil fungi. Mol Plant Microbe 31(10):982–994

Lopaz MG, Avilez M, Dalgado A (2015) Plant uptake of phosphorus from sparingly available as P sources as affected by *Trichoderma* asperellum T −34. Agriculture Food Sci 24:249–260

López-Bucio J, Cruz-Ramírez A, Herrera-Estrella L (2003) The role of nutrient availability in regulating root architecture. Curr Opin Plant Biol 6:280–287

López-Bucio J, Cruz-Ramírez A, Pérez-Torres A, Ramírez-Pimentel JG, Sánchez-Calderón L, Herrera-Estrella L (2005) Root architecture. In: Turnbull C (ed) Plant architecture and its manipulation. Blackwell Annual Review Series, Oxford, pp 181–206

López-Bucio J, Pelagio-Flores R, Herrera-Estrella A (2015) *Trichoderma* as biostimulant: exploiting the multilevel properties of a plant beneficial fungus. Sci Hortic 196:109–123

MacKenzie AJ, Starman TW (1995) Enhanced root and shoot growth of *Chrysanthemum* cuttings propagated with the fungus *Trichoderma harzianum*. HortScience 30(3):496–498

Manganiello G, Sacco A, Ercolano MR, Vinale F, Lanzuise S, Pascale A et al (2018) Modulation of tomato response to *Rhizoctonia solani* by *Trichoderma harzianum* and its secondary metabolite harzianic acid. Front Microbiol 9:1966

Maria FNJ, Steyaert JM, Salazar Badillo FB, Nguyen DV, Rostas M, Baithwaite M, De Souza JT, Bremont JFJ, Ohkura M, Stewart A, Mendoza AM (2017) Environmental growth conditions of *Trichoderma* spp. affects Indole acetic acid derivatives, volatile organic compounds, and plant growth promotion. Front. Plant Sci 8:102

Mastouri F, Björkman T, Harman GE (2010) Seed treatment with *Trichoderma harzianum* alleviatesbiotic, abiotic, and physiological stresses in germinating seeds and seedlings. Phytopathology 100(11):1213–1221

Mastouri F, Bjorkman T, Harman GE (2012) *Trichoderma harzianum* enhances antioxidant defense of tomato seedlings and resistance to water deficit. Mol Plant-Microbe Interact 25:1264–1271

Maurya S, Rashk-E-Eram NSK, Choudhary JS, Kumar S (2019) Heavy metals scavenging potential of *Trichoderma asperellum* and *Hypocrea nigricans* isolated from acid soil of Jharkhand. Indian J Microbiol 59(1):27–38

McNeal L, Herbert B (2009) Volatile organic metabolites as indicators of soil microbial activity and community composition shifts. Soil Sci Soc Am J 73:579–588

Meena M, Prashant S, Zehra A, Dubey MK, Upadhyay RS (2017) Antagonistic assessment of *Trichoderma* spp. by producing volatile and non-volatile compounds against different fungal pathogens. Arch Phytopathol Plant Protect 50(13–14):629–648

Mendoza-Mendoza A, Zaid R, Lawry R, Hermosa R, Monte E, Horwitz BA et al (2018) Molecular dialogues between *Trichoderma* and roots: role of the fungal secretome. Fungal Biol Rev 32:62–85

Meng-Fei L, Guo-Hong L, Zhang K-Q (2019) Non-volatile metabolites from *Trichoderma* spp. Meta 9(58):1–24

Meyer V (2008) Genetic engineering of filamentous fungi-Progress, obstacles and future trends. Biotechnol Adv 26:177–185

Miethke M (2013) Molecular strategies of microbial iron assimilation: from high affinity complexes to cofactor assembly systems. Metallomics 5:15–28

Mishra S, Singh A, Keswani C, Saxena A, Sarma BK, Singh HB (2015) Harnessing plant-microbe interactions for enhanced protection against phytopathogens. In: Arora NK (ed) Plant microbes Symbiosis: applied facets. Springer, India, pp 111–125

More A, Giacobbe S, Faraco V (2013) Regulation of cellulase and hemi-cellulase gene expression in fungi. Curr Genomics 14:230–249

Mukherjee M, Mukherjee PK, Horwitz BA, Zachow C, Berg G, Zeilinger S (2012) *Trichoderma* plant pathogen interactions: advances in genetics of biological control. Indian J Microbiol 52 (4):522–529

Mukherjee PK, Wiest A, Ruiz N, Keightley A, Diez M, McCluskey K, Pouchus YF, Kenerley CM (2011) Two classes of new peptaibols are synthesized by a single non ribosomal peptide synthetase of *Trichoderma virens*. J Biol Chem 286:4544–4554

Naseby DC, Pascual JA, Lynch JM (2000) Effect of biocontrol strains of *Trichoderma* on plant growth, *Pythium ultimum* populations, soil microbial communities and soil enzyme activities. J Appl Microbiol 88:161–169

Nawrocka J, Gromek A, Małolepsza U (2019) Nitric oxide as a beneficial signaling molecule in *Trichoderma atroviride* TRS25-induced systemic defense responses of cucumber plants against *Rhizoctonia solani*. Front Plant Sci 10:421

Nongmaithem N, Roy A, Bhattacharya PM (2016) Screening of *Trichoderma* isolates for their potential of biosorption of nickel and cadmium. Braz J Microbiol 47(2):305–313

Oladipo OG, Awotoye OO, Olayinka A, Bezuidenhout CC, Maboeta MS (2018) Heavy metal tolerance traits of filamentous fungi isolated from gold and gemstone mining sites. Braz J Microbiol 49(1):29–37

Pilon-Smits EAH, Quinn CF, Tapken W, Malagoli M, Schiavon M (2009) Physiological functions of beneficial elements. Curr Opin Plant Biol 12:267–274

Poosapati S, Ravulapalli PD, Tippirishetty N, Vishwanathaswamy DK, Chunduri S (2014) Selection of high temperature and salinity tolerant *Trichoderma* isolates with antagonistic activity against *Sclerotium rolfsii*. Springerplus 3:641

Puyam A (2016) Advent of *Trichoderma* as a bio-control agent- a review.Journal of. Applied Natural Sci 8(2):1100–1109

Rani P, Agarwal A, Mehrotra RS (1998b) Growth responses in *Acacia nilotica* inoculated with VAM fungi (*Glomus fasciculatum*) *Rhizobium* sp. and *Trichoderma harzianum*. J mycopath Res 36:13–16

Rani P, Aggarwal A, Mehrotra RS (1998a) Establishment of nursery technology through *G. mosseae*, *Rhizobium* sp. and *Trichoderma harzianum* on better biomass yield of *Prosopis cineraria* Linn. Proc Nat Acad Sci Sec B 68:301–305

Rao KLNM, Siva RK, Ravisankarc H (2016) Cultural conditions on the production of extracellular enzymes by *Trichoderma* isolates from tobacco rhizosphere. Braz J Microbiol 47:25–32

Rawat L, Bisht TS, Kukreti A, Prasad M (2016) Bioprospecting drought tolerant *Trichoderma harzianum* isolates promote growth and delay the onset of drought responses in wheat (*Triticum aestivum* L). Molecular Soil Biol 7(4):1–15

Rawat L, Singh Y, Shukla N, Kumar J (2011) Alleviation of the adverse effects of salinity stress in wheat (*Triticum aestivum* L) by seed biopriming with salinity tolerant isolates of *Trichoderma harzianum*. Plant Soil 347:387–400

Rawat L, Singh Y, Shukla N, Kumar J (2013) Salinity tolerant *Trichoderma harzianum* reinforces NaCl tolerance and reduces population dynamics of *Fusarium oxysporum* f sp *ciceri* in chickpea (*Cicer arietinum* L) under salt stress conditions. Arch Phytopathol Plant Protect 46 (12):1442–1467

Rawat R, Tewari L (2005) Effect of abiotic stress on phosphate solubilization by biocontrol fungus *Trichoderma* sp. Can J Microbiol 51(3):217–222

Reed RC, Brady SR, Muday G (1998) Inhibition of auxin movement from the shoot into the root inhibits lateral root development in *Arabidopsis*. Plant Physiol 118:1369–1378

Resende MP, Jakoby ICMC, Dos Santos LCR, Soares MA, Pereira FD, Souchie EL, Silva FG (2014) Phosphate solubilization and phytohormone production by endophytic and rhizosphere *Trichoderma* isolates of *Guanandi* (*Calophyllum Brasiliense Cambess*). Afr J Microbiol Res 8:2616–2623

Ripa FA, Cao W, Tong S, Sun J (2019) Assessment of plant growth promoting and abiotic stress tolerance properties of wheat endophytic fungi. Bio Med Res Int 6105865:12

Sala E, Burzi PL, Galletti SM, Cerato C (2007) Multiple effects of *Trichoderma* spp. applied to sugar beet towards soil-borne pathogens. Bulletin-OILB/SROP 30(6–1):199–202

Salwan R, Rialch N, Sharma V (2019) Bioactive volatile metabolites of *Trichoderma*: an overview. In: Singh HB, Keswani C, Reddy M, Sansinenea E, García-Estrada C (eds) Secondary metabolites of plant growth promoting Rhizomicroorganisms. Springer, Singapore, pp 87–111

Samolski I, Rincón AM, Pinzón LM, Viterbo A, Monte E (2012) The qid74 gene from *Trichoderma harzianum* has a role in root architecture and plant biofertilization. Microbiology 158:129–138

Sarrocco S, Guidi L, Fambrini S, DesI-Innocenti E, Vannacci G (2009) Competition for cellulose exploitation between *Rhizoctonia solani* and two *Trichoderma* isolated in the decomposition of wheat straw. J Plant Pathol 91:331–338

Schulz S, Dickschat JS (2007) Bacterial volatiles: the smell of small organisms. Nat Prod Rep 24:814–842

Schuster A, Schmoll M (2010) Biology and biotechnology of *Trichoderma*. Appl Microbiol Biotechnol 87(3):787–799

Shanmugaiah V, Balasubramanian N, Gomathinayagam S, Manoharan PT, Rajendran A (2009) Effect of single application of *Trichoderma viride* and *Pseudomonas fluorescens* on growth promotion in cotton plants. Afr J Agric Res 4:1220–1225

Shoresh M, Harman GE, Mastouri F (2010) Induced systemic resistance and plant responses to fungal biocontrol agents. Annu Rev Phytopathol 48:21–43

Shukla N, Awasthi RP, Rawat L, Kumar J (2012) Biochemical and physiological responses of rice (Oryza sativa L) as influenced by *Trichoderma harzianum* under drought stress. Plant Physiol Biochem 54:78–88

Shukla N, Awasthi RP, Rawat L, Kumar J (2015) Seed biopriming with drought tolerant isolates of *Trichoderma harzianum* promote growth and drought tolerance in *Triticum aestivum*. Ann Appl Biol 166(2):171–182

Shukla RM, Vyas RV (2014) Effects of phosphate solubilization and phytohormone production of *Trichoderma asperellum* Q1 on promoting cucumber growth under salt stress. Int J Agriculture, Environ Biotechnol 7(4):705–710

Singh V, Joshi BB, Awasthi SK, Srivastava SN (2008) Eco-friendly management of red rot disease of sugarcane with *Trichoderma* strains. Sugar Tech 10:158–161

Singh PK, Kumar V (2013) Differential biocontrol and rhizosphere competence ability in strains of *Trichoderma harzianum*. J Agricultural Tech 8:2245–2257

Singh R, Shelke G, Kumar A, Jha P (2015) Biochemistry and genetics of ACC deaminase: a weapon to "stress ethylene" produced in plants. Front Microbiol 6(937):1–14

Singh SP, Singh HB, Singh DK, Rakshit A (2014) *Trichoderma* mediated enhancement of nutrient uptake and reduction in incidence of *Rhizoctonia solani* in tomato. Egypt J Biol 16:29–38

Singh VS, Zaidi NW, Joshi D, Khan T, John D, Bajpai A (2004) *Trichoderma*: a microbe with multifaceted activity. Annu Rev Plant Pathol 3:33–75

Sivasithamparam K, Ghisalberti EL (1998) Secondary metabolism in *Trichoderma* and *Gliocladium*. In: Kubicek CP, Harman GE (eds) *Trichoderma* and *Gliocladium*, basic biology, taxonomy and genetics. Taylor and Francis Ltd., London, pp 139–191

Tansengco M, Tejano J, Coronado F, Gacho C, Barcelo J (2018) Heavy metal tolerance and removal capacity of *Trichoderma* species isolated from mine tailings in Itogon, Benguet. Environ Natural Res J 16(1):39–57

Tukhbatova RI, Fattakhova AN, Alimova FK (2014) Anticancer properties of *Trichoderma asperellum* 302 from buried soils. Tsitologiia 56(6):450–452

Van Loon LC (2007) Plant responses to plant growth-promoting rhizobacteria. Eur J Plant Pathol 119(3):243–254

Vargas WA, Mandawe JC, Kenerley CM (2009) Plant-derived sucrose is a key element in the symbiotic association between *Trichoderma virens* and maize plants. Plant Physiol 151:792–808

Vázquez MM, César S, Azcón R, Barea JM (2000) Interactions between arbuscular mycorrhizal fungi and other microbial inoculants (*Azospirillum*, *Pseudomonas*, *Trichoderma*) and their effects on microbial population and enzyme activities in the rhizosphere of maize plants. Appl Soil Ecol 15:261–272

Vieira D, França C, Kupper KC, Magri MMR, Gomes TM, Rossi F (2017) *Trichoderma* spp. isolates with potential of phosphate solubilization and growth promotion in cherry tomato. Pesq Agropec Trop, Goiânia 47(4):360–368

Vinale F, Flematti G, Sivasithamparam K, Lorito M, Marra R, Skelton BW, Ghisalberti EL (2009b) Harzianic acid, an antifungal and plant growth promoting metabolite from *Trichoderma harzianum*. J Nat Prod 72:2032–2035

Vinale F, Ghisalberti EL, Sivasithamparam K, Marra R, Ritieni A, Ferracane R, Woo S, Lorito M (2009a) Factors affecting the production of *Trichoderma harzianum* secondary metabolites during the interaction with different plant pathogens. Lett Appl Microbiol 48:705–711

Vinale F, Marra R, Ruocco M (2014) *Trichoderma* secondary metabolites active on plants and fungal pathogens. Open Mycology J 8:127–139

Vinale F, Marra R, Scala F, Ghisalberti EL, Lorito M, Sivasithamparam K (2006) Major secondary metabolites produced by two commercial *Trichoderma* strains active against different phytopathogens. Lett Appl Microbio 43:143–148

Vinale F, Sivasithamparam K, Ghisalberti EL, Marra R, Barbetti MJ, Li H, Lorito M (2008a) A novel role for *Trichoderma* secondary metabolites in the interactions with plants. Physiol Mol Plant Pathol 72(1–3):80–86

Vinale F, Sivasithamparam K, Ghisalberti EL, Marra R, Woo SL, Lorito M (2008b) *Trichoderma* plant pathogen interactions. Soil Biol Biochem 40:1–10

Vurukonda SSKP, Vardharajula S, Shrivastava M, Ali SZ (2016) Enhancement of drought stress tolerance in crops by pant growth promoting rhizobacteria. Microbiol Res 184:13–24

Whitmore L, Wallace BA (2004) The Peptaibol database: a database for sequences and structures of naturally occurring peptaibols. Nucleic Acids Res 32(Database issue):D593–D594

Woo SL, Ruocco M, Vinale F, Nigro M, Marra R, Lombardi N, Pascale A, Lanzuise S, Manganiello G, Lorito M (2014) *Trichoderma*-based products and their widespread use in agriculture. Open Mycology J 8:71–126

Yadav RL, Shukla SK, Suman A, Singh PN (2009) *Trichoderma* inoculation and trash management: effects on soil microbial biomass, soil respiration, nutrient uptake and yield of sugarcane under subtropical conditions. Biol Fertil Soils 45:461–468

Yasmeen R, Siddiqui ZS (2017) Physiological responses of crop plants against *Trichoderma harzianum* in saline environment. Acta Bot Croat 76(2):154–162

Yedidia I, Shrivasta AK, Kapulnik Y, Chet I (2001) Effect of *Trichoderma harzianum* on micro-element concentration and increased growth of cucumber plants. Plant Soil 2:235–242

Yuan M, Yuanyuan H, Weina GE, Zhenhua J, Song S, Zhang L, Huang Y (2019) Involvement of jasmonic acid, ethylene and salicylic acid signaling pathways behind the systemic resistance induced by *Trichoderma longibrachiatum* H9 in cucumber. BMC Genomics 20:144

Zeilinger S, Gruber S, Bansal R, Mukherjee PK (2016) Secondary metabolism in *Trichoderma*: chemistry meets genomics. Fungal Biol Rev 30:74–90

Zhang S, Gan Y, Xu B (2019) Mechanisms of the IAA and ACC-deaminase producing strain of *Trichoderma longibrachiatum* T6 in enhancing wheat seedling tolerance to NaCl stress. BMC Plant Biol 19:22

Zhuang D, Tao T, Wang L, Zhao Y, Huang H, Zhang D, Liu M, Wang Z, Han J (2019) Bioprospecting of novel and bioactive metabolites from Endophytic fungi isolated from rubber tree *Ficus elastica* leaves. J Microbiol Biotechnol 29(5):731–738

Chapter 14
Trichoderma Proteome: Multifunctional Role in Plant Defense

Akansha Jain and Sampa Das

Abstract *Trichoderma* spp., endophytic plant symbionts have long been recognized as biocontrol agents against plant-pathogenic fungi and for improving growth and yield. These fungi are ubiquitous in the soil and are being successfully exploited and commercialized as biofungicides against a broad range of phytopathogenic fungi such as *Rhizoctonia solani*, *Pythium ultimum*, and *Botrytis cinerea*. Signaling during plant–pathogen interaction has always been an important topic in phytopathology for many years, whereas recent studies are more focused to understand the communication processes involved in plant–nonpathogenic microorganisms interaction both of bi-partite and tri-partite mode, especially for improving plant yield or inducing systemic resistance. The improvement of *Trichoderma* species as biocontrol agents requires extensive studies to have a complete repertoire of proteins involved in mycoparasitism, antibiosis as well as other components. This chapter highlights the proteins associated with the biocontrol mechanism of *Trichoderma* spp. and future prospects in addressing the gap to accelerate agricultural use of these fungi.

Keywords Biocontrol · Disease response · Enzymes · Mycoparisitism · Proteins · *Trichoderma*

14.1 Introduction

Trichoderma spp. (teleomorph *Hypocrea*), the most common saprophytic fungi in the rhizosphere, are widely used in agriculture with more than 60% of the registered biofungicides worldwide being *Trichoderma*-based (Samuels 1996; Verma et al. 2007). These fungi have the ability to attack a large number of aerial and soilborne plant pathogens by mycoparasitism (killing of one fungus by other), enzymatic lysis, antibiotic production, and competition for niche and nutrients (Chet 1987; Jain et al. 2012). They indirectly protect plants by inducing systemic resistance (ISR) and

A. Jain (✉) · S. Das
Division of Plant Biology, Bose Institute, Kolkata, West Bengal, India

C. Manoharachary et al. (eds.), Trichoderma: *Agricultural Applications and Beyond*, Soil Biology 61, https://doi.org/10.1007/978-3-030-54758-5_14

enhancing plant growth (Shoresh et al. 2005; Singh et al. 2014a; Jain et al. 2015a). Due to increasing concerns surrounding food and environmental safety, biocontrol agents have received increased attention. The biocontrol agents are being used more and more in combination with or without agrochemicals (Martinez-Medina et al. 2014; Jain et al. 2013a, b; Jain and Das 2016). *Trichoderma* spp. colonize plant roots and elicitate defense response in the upper parts of plants, thereby enhancing plants' resistance to foliar pathogens. The mycoparisitic strains of *Trichoderma* are reported to have a high abundance of such genes in the genome (Lin et al. 2012; Atanasova et al. 2013) with an expression of over 60% of the encoding transcripts during interaction of *T. virens* and *T. atroviride* with *Rhizoctonia* (Atanasova et al. 2013).

Plant–microbe interaction requires extensive communication and involvement of a large number of signaling molecules playing an essential role in plant defense response. Microbes are able to detect the plant-derived molecules, and plants are able to recognize microbe-derived molecules and thus they strategize defense responses accordingly. The initial molecular dialog is established between the two coordinates of the cellular processes, thus determining the final result of their relationship, ranging from parasitism to mutualism (Pozo et al. 2004; Djonović et al. 2006b). The presence of *Trichoderma* in the rhizosphere induces coordinated changes at transcriptomic, proteomic, and metabolomic responses in the plant. Proteomics studies have already been started in *Trichoderma harzianum* (Grinyer et al. 2005) and *Trichoderma atroviride* (Grinyer et al. 2004) and have provided insights to understand mechanism involving biological control of pathogenic fungi. The ability of *Trichoderma* spp. to sense a pathogenic fungal host has been reported and regulatory sequences in the promoter region of mycoparasitism linked genes and other important elements in the signaling transduction pathways have been identified (Cortes et al. 1998; Mendoza-Mendoza et al. 2003; Zeilinger et al. 2005).

14.1.1 Mechanism of Action of Trichoderma

Trichoderma strains are used as BCAs in agriculture largely for their abilities to act as biocontrol agents against plant-pathogenic fungi. The principal mechanism of action includes mycoparasitism, production of hydrolases (Gruber and Seidl-Seiboth 2012), antibiotics (Vinale et al. 2014), and competition for space and nutrients. Recent findings indicate that *Trichoderma* alone or in combination with other beneficial microbes reprograms plants' defense response by induction of ISR and systemic resistance and improved plant growth and yield (Harman 2011; Jain et al. 2014a, b; Singh et al. 2014b). For example, strains of *T. harzianum* TNHU27 promote plant growth and boost immune responses, seed germination, and enhance the population of other beneficial microbes in the presence and absence of plant pathogen, *Sclerotinia sclerotiorum* in pea (Jain et al. 2012; Jain et al. 2013b). Induction of systemic resistance involves complex signaling of jasmonic acid/ ethylene-induced systemic resistance and/or salicylic acid pathway in systemic acquired resistance in plant–*Trichoderma* interactions (Shoresh et al. 2010; Jain et al. 2015b). The tripartite interaction between *Trichoderma*, host plant, and pathogen is known to induce oxidative burst along with changes induced at proteome level of host plants (Segarra et al. 2007; Shoresh and Harman 2008; Jain et al. 2015c; Singh et al. 2013; Pelagio-Flores et al. 2017). Peptaibiotics and peptaibols are a class of linear peptides synthesized by *Trichoderma*, and more than 300 of such have been described to date. These compounds exhibit antimicrobial activity and are referred to as "antibiotic peptides."

14.1.2 Proteome of Trichoderma

Even though ISR has been confirmed as a mechanism of bioprotection by *Trichoderma* spp., the underlying molecular mechanisms involved remain largely unknown. Several previous reports at the proteomic level on different strains of *T. reesei, T. harzianum, T. atroviride* total protein extracts or their interaction with the pathogen and host plants are available to improve our understanding of the agricultural importance of this ascomycete fungus (Moran-Diez et al. 2009; Jain et al. 2015c) (Table 14.1). Recent studies have reported some specific secreted proteins to have a target in the host plant (Plett et al. 2014; Lamdan et al. 2015). A large array of signaling molecules or microbial elicitors have been characterized and are known to play a role in initiating plant defense responses (Nimchuk et al. 2003).

In this context, the newly emerging proteomics techniques involving global analysis of proteins have largely contributed to our understanding of the role of these proteins in plant defense response as a whole. Two-dimensional electrophoresis (2DE) and two-dimensional fluorescence difference gel electrophoresis (2D-DIGE) and non-gel-based (multidimensional liquid chromatography) protein separation techniques are widely used in fungal proteomics (Bhadauria et al. 2007). These separation techniques when coupled with various mass spectrometry (MS) technologies are the most important tools for protein identification (Figeys et al. 2001). Matrix-assisted laser desorption ionization (MALDI) and electrospray ionization (ESI) have assisted in the volatilization and ionization of large peptides and proteins (Yates III 1998; Godovac-Zimmermann and Brown 2001; Mann et al. 2001). ESI can be coupled to separation techniques such as LC and high-performance liquid chromatography (HPLC), allowing high throughput analysis of peptide or protein mixtures (Ducret et al. 1998; Gatlin et al. 1998). iTRAQ uses isobaric reagents to label primary amines in proteins allowing multiplexing of up to eight samples in one MS experiment (Noire et al. 2011; Evans et al. 2012). Stable isotope labeling by amino acid in cell cultures has also been introduced for high-throughput quantitative proteomics (Mann et al. 2001). Peptide mass fingerprints (PMF) are used for protein identification by analyzing the tryptic fragments via the MASCOT (http://www.matrixscience) search engine using the NCBI protein database.

Trichoderma secretes complex mixtures of hydrolytic enzymes to degrade the host cell wall (mycoparasitism). In a previous study, *T. harzianum* CECT 2413

Table 14.1 List of a proteins of *Trichoderma* characterized for their role in biocontrol

Protein involved	Reference	Biocontrol function
Chitinases Endochitinases, chitobiosidase, N-acetyl-β-D-glucosaminidase	Yang et al. 2009; Lopez-Mondejar et al. 2009; Ihrmark et al. 2010; Gruber et al. 2011; Xie et al. 2015; Singh et al. 2014a, b; Jain et al. 2014a, b; Jain et al. 2015a, b; de Lima et al. 2017	• Breakdown of pathogen's cell wall through mycoparasitism. • As elicitors. • Induction of systemic resistance against biotic stress. • Root colonization.
Glucanases-Exo-β-D-(1,3/4/6)-glucanases, Endo-β-D-(1,3/4/6)-glucanases	de la Cruz and Llobell 1999; Kim et al. 2002; Teresa et al. 2003; Nobe et al. 2004; Mittler et al. 2004 Djonović et al. 2006a, b; Singh et al. 2014a, b; Jain et al. 2015c; Jain and Das 2016; de Lima et al. 2017	
Protease	Chet et al. 1998; Hanson and Howell 2004; Suarez et al. 2007; de Lima et al. 2017; Jain et al. 2015c; 2013b; Wu et al. 2017	
Amylase	de Azevedo et al. 2000; Jain et al. 2015c)	
Cellulase and xylanase	Calderon et al. 1993; Martinez et al. 2001; Ron et al. 2000; Reithner et al. 2011	
Expansin-like protein	Georgelis et al. 2014; Lamdan et al. 2015	• For endophyic lifestyle.
Glycoside hydrolase	Atanasova et al. 2013; Lamdan et al. 2015; de Lima et al. 2017	• Hydrolase activity, degrade pathogenic fungi's cell wall.
L-amino acid oxidase	Yang et al. 2009; Yang et al. 2011	• Oxidoreductase activity, leads to pathogen's hyphal lysis (apoptosis-like response).
SSCPs (Sm-1 and Ep1–1)	Seidl et al. 2006; Djonović et al. 2006a, b; 2007a, b; Vargas et al. 2008; Lamdan et al. 2015; Gomes et al. 2015; Ramada et al. 2016; Salas-Marina et al. 2015	• In initial plant-fungus signalling. • Elicitation of defense responses.

extracellular proteome was analyzed in the presence of different fungal cell walls. Significant differences were detected in 2DE maps, depending on the use of specific components of cell walls (Suárez et al. 2005). The study reported a novel aspartic protease (P6281: MW 33 and pI 4.3) that was identified to be induced by fungal cell walls with implications in mycoparasitism. Xylanase and peptaibols like alamethicin and trichovirin II are also produced by *Trichoderma* spp. and are known to elicit an immune response in plants (Leitgeb et al. 2007; Viterbo et al. 2007; Luo et al. 2010; Druzhinina et al. 2011).

T. harzianum T1A secretome was also investigated against deactivated mycelium of *Guignardia citricarpa* (causal agent of Citrus black spot in citrus plants) containing growth medium and compared to a control condition (de Lima et al. 2017). Out of 65 differentially expressed proteins identified by mass spectrometry,

54% of the total proteins represented glycoside hydrolases, an L-amino acid oxidase, a serine protease, and Epl 1 protein, which were not only associated to mycoparasitism, but also to plant defense elicitation and biological control.

14.1.3 Proteins Related in Primary Metabolism

Glycoside hydrolase (GH) is reported for having a central role in mycoparasitism (Atanasova et al. 2013). These enzymes are known to degrade fungal cell wall, in turn providing additional defensive benefit to *T. harzianum* colonized plants. They are associated both with metabolism and mycoparasitism processes. *T. harzianum* T1A expressed these enzymes both in control and *G. citricarpa* deactivated mycelium treated media, with higher expression in control media (de Lima et al. 2017). The results suggest that the expression of GH is more linked to the primary metabolism than to mycoparasitism. In a recent report, 17 significantly differential upregulated expansin-like protein required for endophytic lifestyle has also been reported in *T. asperellum*–cucumber and *T.virens*–maize root secretome interaction (Georgelis et al. 2014; Lamdan et al. 2015). Similarly, β-1,6-glucanase (B9VQ17) was also detected in the control medium indicating its role in primary metabolism (de Lima et al. 2017). In a similar study involving *T. harzianum* ETS 323, L-amino acid oxidase (LAAO) and two endochitinases were uniquely induced in the media that contained deactivated *B. cinerea* mycelia as the sole carbon source (Yang et al. 2009). Yang et al. (2011) using in vitro assays demonstrated that *T. harzianum* ETS 323 LAAO had an antagonistic effect against *R. solani* and a stimulatory effect on hyphal density and sporulation in *Trichoderma* itself.

14.2 Proteins Related in Biocontrol

Cellular changes occurring on the induction of systemic resistance include generation of reactive oxygen species, accumulation of polyphenols and phytoalexins, and synthesis of pathogenesis-related (PR) proteins such as chitinases and glucanases (Mittler et al. 2004; Jain et al. 2012; Jain et al. 2014b). Some of the cysteine-rich small proteins are proposed to play an important role in signaling, specificity, recognition, and adhesion of symbiotic fungi to their host plants (Tagu et al. 2002; Wosten 2001). *Trichoderma* spp. have three MAPK cascades comprising MAPKKK, MAPKK, and MAPK (Schmoll 2008) and are involved in mycoparasitism and biocontrol processes (Reithner et al. 2007; Kumar et al. 2010)

Trichoderma spp. is capable of producing extracellular enzymes viz., chitinases, β-glucosidases, mannosidases, phosphatases, and proteases required for fungal cell wall degradation (Chet et al. 1998). Xylanases are also known to be involved in mycoparasitism by *Trichoderma* when induced by deactivated mycelium of *R. solani* (Tseng et al. 2008). The reduction of pathogen growth was positively correlated with in vitro *Trichoderma* challenge. Similarly, the secretome of *T. harzianum* in medium containing mycelium of pathogens showed the role of endo-1,4-β xylanase in the biological control of pathogenic fungi (Reithner et al. 2011). GH was found to be expressed at a higher level in *T. virens*–maize root–*C. heterostrophus* interaction in comparison to only *T. virens* secretome analyses (Lamdan et al. 2015). Ten of these were predicted to be specific for plant cell wall degradation. Recently, Mukherjee et al. (2012) identified a PKS/NRPS hybrid enzyme in maize with involvement in defense responses. Transmembrane protein G-coupled to receptor Gpr1 is involved in sensing the fungal prey in *T. atroviride* and is responsible for its mycoparasitic ability (Omann et al. 2012)

The β-1,3-exoglucanase and endochitinase are reported in the secretome of *T. harzianum* and are able to degrade fungal cell walls (Bolar et al. 2000; Harman 2000; de Lima et al. 2017). Several other enzymes such as β-1,3-glucanases, β-1,6-glucanases, chitinases, proteases, and xylanases are also secreted by *T. harzianum* ETS 323 when grown in the presence of the cell wall of *B. cineria* (Yang et al. 2009). The study indicated that the cell wall of *B. cinerea* is the main target of *T. harzianum* ETS 323 in the biocontrol mechanism. Also, in vitro assays on L-AAO indicated its role in inhibited *B. cinerea* hyphal growth along with cytosolic vacuolization in the hyphae that led to hyphal lysis (Cheng et al. 2012). Th-L-AAO also showed disease resistance against *B. cinerea* on postharvest apple fruit and tobacco leaves. This study further reported the role of L-AAO in an apoptosis-like response, including the generation of reactive oxygen species, indicating that Th-L-AAO triggers programmed cell death in *B. cinerea*.

14.3 Proteins Involved in Elicitation of Plant Defense

Trichoderma spp. not only directly fights against pathogens but also promotes growth and induces resistance. Serine endopeptidase, an extracellular protease, has been reported to induce defense responses by *T. virens* (Hanson and Howell 2004) and biocontrol process in *Trichoderma* sp. (Flores et al. 1997; Pozo et al. 2004). Proteins such as cellulase and xylanase have been reported as proteinaceous elicitors in *Trichoderma* spp. because they induce a hypersensitive response, expression of PR proteins, and phytoalexins in different plant species (Calderon et al. 1993; Martinez et al. 2001; Ron et al. 2000).

Serine proteases are also reported to be induced during and before the contact with the prey in different *Trichoderma* species (Suarez et al. 2007; de Lima et al. 2017). Pozo et al. (2004) proposed that the expression of proteolytic enzymes during the starting of mycoparasitism leads to the formation of nitrogenous metabolites, derived from the pathogenic fungi. These nitrogenous metabolites interact with the nitrogen sensors of *Trichoderma*. Proteases provide easier penetration of the prey fungal tissue by the degradation of the protein links of the outer layer of the host and/or use these proteins for their nutrition.

Initial studies on *Trichoderma* secreted proteins reported an abundant secreted protein, belonging to the cerato-platanins which are proteins belonging to the fungal family of secreted elicitors and toxins. Proteins, Sm-1 (in *T. virens*)/Epl-1 (in *T. atroviride*) (Seidl et al. 2006; Djonović et al. 2006a, b; 2007a, b; Vargas et al. 2008) from the same family have been shown to ameliorate ISR. Both these proteins also belong to a larger class of fungal proteins defined as SSCPs or SSPs [small, secreted (cysteine-rich) proteins] (Rep 2005; Stergiopoulos and de Wit 2009; de Wit et al. 2009). SSCPs are proposed to change their abundance in response to association with plant roots and may function in the fungal–plant molecular dialog (Lamdan et al. 2015). Epl-1 has been reported to elicitate defense responses in plants (Gomes et al. 2015; Ramada et al. 2016; Salas-Marina et al. 2015). In fact, Gomes et al. (2015) proposed that the absence of Epl-1 protein may affect the expression of all mycoparasitism genes analyzed and that Epl-1 might act as a recognition molecule to identify its own and/or host hyphae, therefore avoiding self-degradation. Lamdan et al. (2015) suggested the role of SSCPs as negative effectors reducing the defense levels in the maize plants and may be important for the fine-tuning of ISR by *T. virens*.

Trichoderma spp. are widely used as biocontrol agents against fungal phytopathogens. The mechanisms of their biocontrol action involve mycoparasitism, antibiosis, enhancing plant growth, and induction of plant resistance (Elad 1996; Perazzolli et al. 2008; Cheng et al. 2012; Singh et al. 2013; Ram et al. 2015). Exploration of proteins involved in mycoparasitism, elicitation, and induction of defense responses is an active field of research. Secreted proteins are central to the molecular interaction taking place between *Trichoderma* and their host plants. Recent studies are focusing more on the molecular basis for mutualistic interactions between beneficial microbes and plants for nutrient acquisition and disease management.

Immense progress has been made in the field of *Trichoderma* proteomics in the past few years. The use of high-throughput techniques has led to a rapid increase in the availability of transcriptomics data of *Trichoderma.* The availability of microarrays, next-generation DNA sequencing, RNA-seq, and genome annotation along with data at proteome level will generate insight picture into the transcriptome response of plant–*Trichoderma* and pathogen interaction. However, our knowledge is still incomplete and requires technical upgradation. For example, the detection of low-abundance proteins require from one to several million molecules per cell is needed. On the other hand, post-translational modifications make analysis challenging and difficult for designing and application. Further studies are needed to integrate raw information from all omics approaches to integrate variations of active molecular components caused by the translational response of BCAs in interaction with plants and phytopathogens (Sharma et al. 2017). The studies at mRNA level supported with proteome abundance data will help in the accurate presentation of actively engaged mRNAs in translation involving *Trichoderma*–plant–pathogens.

14.4 Conclusion

Trichoderma spp. are extensively researched for both agricultural benefits as well as for studying plant–microbe interaction. Even though there have been several studies on their interaction with plants and pathogenic microbes, a complete understanding of the mechanisms is lacking. The proteome is dynamic, reflecting the particular conditions to which *Trichoderm*a is exposed, for example, different proteins are expressed by the same strain in the presence of different pathogens. Integration of transcriptome, translatome, and proteome studies can provide a better state-of-the-art understanding of these adaptive responses during biocontrol interaction. Omic studies under conditions of mycoparasitism and plant–*Trichoderma* interaction would help in identifying novel proteins involved in the interactions. Detailed understanding will allow possible engineering to tailor strains for enhanced biocontrol potential and other biotechnological applications.

Acknowledgments AJ is thankful to SERB, Department of Science and Technology, New Delhi, India, for providing funding under DST FAST Track Young Scientist Scheme (YSS/2015/000773). SD acknowledges the Indian National Science Academy for her Senior Scientist program.

References

Atanasova L, Crom SL, Gruber S, Coulpier F, Seidl-Seiboth V, Kubicek CP et al (2013) Comparative transcriptomics reveals different strategies of *Trichoderma* mycoparasitism. BMC Genomics 14:121

Bhadauria V, Zhao WS, Xia L, Wang Y, Zhang JH, Yang LJ, AnKong L, Peng YL (2007) Advances in fungal proteomics. Microbiol Res 162:193–200

Bolar JP, Norelli JL, Wong KW, Hayes CK, Harman GE, Aldwinckle HS (2000) Expression of endochitinase from *Trichoderma harzianum* in transgenic apple increases resistance to apple scab and reduces vigor. Phytopathology 90:72–77

Calderon AA, Zapata JM, Munoz R, Pedreno MA, Barcelo AR (1993) Resveratrol production as a part of the hypersensitive-like response of grapevine cells to an elicitor from *Trichoderma-viride*. New Phytol 124:455–463

Cheng CH, Yang CA, Peng KC (2012) Antagonism of *Trichoderma harzianum* ETS 323 on *Botrytis cinerea* mycelium in culture conditions. Phytopathology 102:1054–1063

Chet I (1987) *Trichoderma*—application, mode of action and potential as a biocontrol agent of soilborne plant pathogenic fungi. In: Chet I (ed) Innovative approaches to plant disease control. Wiley, New York, pp 137–160

Chet I, Benhamou N, Haran S (1998) Mycoparasitism and lytic enzymes. In: Harman GE, Kubicek CP (eds) *Trichoderma* and *Gliocladium* – enzymes, biological control and commercial applications. Taylor and Francis Ltd, London, pp 153–172

Cortes C, Gutierrez A, Olmedo V, Inbar J, Chet I, Herrera-Estrella A (1998) The expression of genes involved in parasitism by *Trichoderma harzianum* is triggered by a diffusible factor. Mol Gen Genet 260:218–225

de Azevedo AM, De Marco JL, Felix CR (2000) Characterization of an amylase produced by a Trichoderma harzianum isolate with antagonistic activity against Crinipellis perniciosa, the causal agent of witches' broom of cocoa. FEMS Microbiol Lett 188(2):171–175 de la Cruz J, Llobell A (1999) Purification and properties of a basic endo-β-1,6-glucanase (BGN16.1) from the antagonistic fungus Trichoderma harzianum. FEBS J 265 145–151

de Lima FB, Félix C, Osório N, Alves A, Vitorino R, Domingues P, Rute T, da Silva R, Esteves AC (2017) *Trichoderma harzianum* T1A constitutively secretes proteins involved in the biological control of *Guignardia citricarpa*. Biol Control 106:99–109

de Wit PJ, Mehrabi R, Van den Burg HA, Stergiopoulos I (2009) Fungal effector proteins: past, present, and future. Mol Plant Pathol 10:735–747

Djonović S, Pozo MJ, Dangott LJ, Howell CR, Kenerley CM (2006a) Sm1, a proteinaceous elicitor secreted by the biocontrol fungus *Trichoderma virens* induces plant defense responses and systemic resistance. Mol Plant-Microbe Interact 19:838–853

Djonović S, Pozo MJ, Kenerley CM (2006b) Tvbgn3, a beta-1,6- glucanase from the biocontrol fungus *Trichoderma virens*, is involved in mycoparasitism and control of *Pythium ultimum*. Appl Environ Microbiol 72:7661–7670

Djonović S, Vargas WA, Kolomiets MV, Horndeski M, Wiest A, Kenerley CM (2007b) A proteinaceous elicitor Sm1 from the beneficial fungus *Trichoderma virens* is required for induced systemic resistance in maize. Plant Physiol 145:875–889

Djonović S, Vittone G, Mendoza-Herrera A, Kenerley CM (2007a) Enhanced biocontrol activity of *Trichoderma virens* transformants constitutively coexpressing beta-1,3- and beta-1,6-glucanase genes. Mol Plant Pathol 8:469–480

Druzhinina IS, Seidl-Seiboth V, Herrera-Estrella A, Horwitz BA, Kenerley CM, Monte E, Mukherjee PK, Zeilinger S, Grigoriev IV, Kubicek CP (2011) Trichoderma—the genomics of opportunistic success. Nat Rev Microbiol 9:749–759

Ducret A, van Oostveen I, Eng JK, Yates JR III, Aebersold A (1998) High throughput protein characterization by automated reverse-phase chromatography/electrospray tandem mass spectrometry. Protein Sci 7:706–719

Elad Y (1996) Mechanism involved in the biological control of *Botrytis cinerea* incited diseases. Eur J Plant Pathol 102:719–732

Evans C, Noirel J, Ow SY, Salim M, Pereira-Medrano AG, Couto N, Pandhal J, Smith D, Pham TK, Karunakaran E, Zou X, Biggs CA, Wright PC (2012) An insight into iTRAQ: where do we stand now? Anal Bioanal Chem 404:1011–1027

Figeys D, Linda D, McBroom LD, Moran MF (2001) Mass spectrometry for the study of protein–protein interactions. Methods 24:230–239

Flores A, Chet I, Herrera-Estrella A (1997) Improved biocontrol activity of *Trichoderma harzianum* by over-expression of the proteinase-encoding gene prb1. Curr Genet 31:30–37

Gatlin CL, Kleemann GR, Hays LG, Link AJ, Yates JR (1998) Protein identification at the low femtomole level from silver stained gels using a new fritless electrospray interface for liquid chromatographymicrospray and nanospray mass spectrometry. Anal Biochem 263:93–101

Georgelis N, Nikolaidis N, Cosgrove DJ (2014) Biochemical analysis of expansin-like proteins from microbes. Carbohydr Polym 100:17–23

Godovac-Zimmermann J, Brown LR (2001) Perspectives for mass spectrometry and functional proteomics. Mass Spectrom Rev 20:1–57

Gomes EV, Costa Mdo N, de Paula RG, de Azevedo RR, de Silva FL, Noronha EF, Ulhoa CJ, Monteiro VN, Cardoza RE, Gutierrez S, Silva RN (2015) The Cerato-Platanin protein Epl-1 from *Trichoderma harzianum* is involved in mycoparasitism, plant resistance induction and self cell wall protection. Sci Rep 5:17998

Grinyer J, Hunt S, Mckay M, Hervert BR, Nevalainen H (2005) Proteomic response of the biological control fungus *Trichoderma atroviride* to growth on the cell walls of *Rhizoctonia solani*. Curr Genet 47:381–388

Grinyer J, Mckay M, Nevalainen H, Hervert BR (2004) Fungal proteomics: initial mapping of biological control strain *Trichoderma harzianum*. Curr Genet 45:163–169

Gruber S, Kubicek CP, Seidl-seiboth V (2011) Differential regulation of orthologous chitinase genes in mycoparasitic Trichoderma species. Appl Environ Microbiol 77:7217–7226

Gruber S, Seidl-Seiboth V (2012) Self-versus non-self: fungal cell wall degradation in *Trichoderma*. Microbiol 158:26–34

Hanson LE, Howell CR (2004) Elicitors of plant defense responses from biocontrol strains of *Trichoderma virens*. Phytopathology 94:171–176

Harman GE (2000) Myths and dogmas of biocontrol-changes in perceptions derived from research on *Trichoderma harzianum* T-22. Plant Dis 84:377–393

Harman GE (2011) Multifunctional fungal plant symbionts: new tools to enhance plant growth and productivity. New Phytol 189:647–649

Ihrmark K, Asmail N, Ubhayasekera W, Melin P, Stenlid J, Karlsson M (2010) Comparative molecular evolution of Trichoderma chitinases in response to mycoparasitic interactions. Evol Bioinforma 6:1–26

Jain A, Das, S (2016) In-sight to the interaction between plants and associated fluorescent *Pseudomonas* spp. Int J Agronomy, Hindawi Publications, Article ID 4269010

Jain A, Singh A, Sarma BK, Singh HB (2012) Microbial consortium mediated reprogramming of defense network in pea to enhance tolerance against *Sclerotinia sclerotiorum*. J Appl Microbiol 112(3):537–550

Jain A, Singh A, Singh BN, Singh S, Upadhyay RS, Sarma BK, Singh HB (2013a) Biotic stress management in agricultural crops using microbial consortium. In: Maheshwari DK (ed) Bacteria in agrobiology: disease management. Springer-Verlag, Berlin Heidelberg, pp 427–448

Jain A, Singh A, Singh S, Singh HB (2013b) Microbial consortium-induced changes in oxidative stress markers in pea plants challenged with *Sclerotinia sclerotiorum*. J Plant Growth Regul 32:388–398

Jain A, Singh A, Chaudhary A, Singh S, Singh HB (2014a) Modulation of nutritional and antioxidant potential of seeds and pericarp of pea pods treated with microbial consortium. Food Res Int 64:275–282

Jain A, Singh S, Singh S, Sarma BK, Singh HB (2014b) Bicontrol agents mediated suppression of oxalic acid induced programmed cell death during *Sclerotinia sclerotiorum*-pea interaction. J Basic Microbiol Special Issue: Signaling 55(5):601–606

Jain A, Singh A, Singh S, Singh HB (2015a) Biological management of *Sclerotinia sclerotiorum* in pea using plant growth promoting microbial consortium. J Basic Microbiol 55(8):961–972

Jain A, Singh A, Singh S, Singh HB (2015b) Phenols enhancement effect of microbial consortium in pea plants restrains *Sclerotinia sclerotiorum*. Biol Control 89:23–32

Jain A, Singh A, Singh S, Singh V, Singh HB (2015c) Comparative proteomics analysis in pea treated with microbial consortium of beneficial microbes reveals changes in protein network to enhance resistance against *Sclerotinia sclerotiorum*. J Plant Physiol. https://doi.org/10.1016/j.jplph.2015.05.004

Kim DJ, Baek JM, Uribe P, Kenerley CM, Cook DR (2002) Cloning and characterization of multiple glycosyl hydrolase genes from Trichoderma virens. Curr Genet 40 374–384

Kumar A, Scher K, Mukherjee M, Pardovitz-Kedmi E, Sible GV, Singh US, Kale SP, Mukherjee PK, Horwitz BA (2010) Overlapping and distinct functions of two *Trichoderma virens* MAP kinases in cell-wall integrity, antagonistic properties and repression of conidiation. Biochem Biophys Res Commun 398:765–770

Lamdan NL, Shalaby S, Ziv T, Kenerley CM, Horwitz BA (2015) Secretome of *Trichoderma* interacting with maize roots: role in induced systemic resistance. Mol Cell Proteomics 14 (4):1054–1063

Leitgeb B, Szekeres A, Manczinger L, Vagvolgyl C, Kredics L (2007) The history of alamethicin: a review of most extensively studied peptaibol. Chem Biodivers 4:1027–1051

Lin YR, Lo CT, Liu SY, Peng KC (2012) Involvement of pachybasin and emodin in self-regulation of *Trichoderma harzianum* mycoparasitic coiling. J Agric Food Chem 60:2123–2128

Lopez-Mondejar R, Catalano V, Kubicek CP, Seidl V (2009) The β-N-acetylglucosaminidases NAG1 and NAG2 are essential for growth of Trichoderma atroviride on chitin. FEBS J 276:5137–5148

Luo Y, Zhang DD, Dong XW, Zhao PB, Chen LL, Song XY, Wang XJ, Chen XL, Shi M, Zhang YZ (2010) Antimicrobial peptaibols induce defense responses and systemic resistance in tobacco against tobacco mosaic virus. FEMS Micro Lett 313:120–126

Mann M, Hendrickson RC, Pandey A (2001) Analysis of proteins and proteomes by mass spectrometry. Annu Rev Biochem 70:437–473

Martinez C, Blanc F, LeClaire E, Besnard O, Nicole M, Baccou JC (2001) Salicylic acid and ethylene pathways are differentially activated in melon cotyledons by active or heat-denatured cellulase from *Trichoderma longibrachiatum*. Plant Physiol 127:334–344

Martinez-Medina A, Del Mar AM, Pascual JA, Van Wees SC (2014) Phytohormone profiles induced by *Trichoderma* isolates correspond with their biocontrol and plant growth-promoting activity on melon plants. J Chem Ecol 40:804–815

Mendoza-Mendoza A, Pozo MJ, Grzegorski D, Martinez P, Garcia JM, Olmedo-Monfil V, Cortes C, Kenerley C, Herrera-Estrella A (2003) Enhanced biocontrol activity of *Trichoderma* through inactivation of a mitogen-activated protein kinase. Proc Natl Acad Sci U S A 100:15965–15970

Mittler R, Vanderauwera S, Gollery M, Van Breusegem F (2004) Reactive oxygen gene network of plants. Trends Plant Sci 9(10):490–498

Moran-Diez E, Hermosa R, Ambrosino P, Cardoza RE, Gutierrez S, Lorito M, Monte E (2009) The ThPG1 endopolygalacturonase is required for the *Trichoderma harzianum*-plant beneficial interaction. Mol Plant-Microbe Interact 22:1021–1031

Mukherjee PK, Buensanteai N, Moran-Diez ME, Druzhinina IS, Kenerley CM (2012) Functional analysis of non-ribosomal peptide synthetases (NRPSs) in *Trichoderma virens* reveals a polyketide synthase (PKS)/NRPS hybrid enzyme involved in induced systemic resistance response in maize. Microbiol 158:155–165

Nimchuk Z, Eulgem T, Holt BF 3rd, Dangl JL (2003) Recognition and response in the plant immune system. Annu Rev Genet 37:579–609

Nobe R, Sakakibara Y, Ogawa K, Suiko M (2004) Cloning and expression of a novel Trichoderma viride laminarinase AI gene (lamAI). Biosci Biotechnol Biochem 68:2111–2119

Noire J, Evans C, Salim M, Mukherjee J, Ow SY, Pandhal J, Pham TK, Biggs CA, Wright PC (2011) Methods in quantitative proteomics: setting iTRAQ on the right track. Curr Prot 8:17–30

Omann MR, Lehner S, Escobar Rodriguez C, Brunner K, Zeilinger S (2012) The seven-transmembrane receptor Gpr1 governs processes relevant for the antagonistic interaction of *Trichoderma atroviride* with its host. Microbiol 158:107–118

Pelagio-Flores R, Esparza-Reynoso S, Garnica-Vergara A, López-Bucio J, Herrera-Estrella A (2017) *Trichoderma*-induced acidification is an early trigger for changes in *Arabidopsis* root growth and determines fungal phytostimulation. Front Plant Sci 17:822

Perazzolli M, Dagostin S, Ferrari A, Elad Y, Pertot I (2008) Induction of systemic resistance against *Plasmopara viticola* in grapevine by *Trichoderma harzianum* T39 and benzothiadiazole. Biol Control 47:228–234

Plett JM, Daguerre Y, Wittulsky S, Vayssieres A, Deveau A, Melton SJ, Kohler A, Morrell-Falvey JL, Brun A, Veneault-Fourrey C, Martin F (2014) Effector MiSSP7 of the mutualistic fungus *Laccaria bicolor* stabilizes the Populus JAZ6 protein and represses jasmonic acid (JA) responsive genes. Proc Natl Acad Sci U S A 111:8299–8304

Pozo MJ, Baek JM, Garcia JM, Kenerley CM (2004) Functional analysis of Tvsp1, a serine protease-encoding gene in the biocontrol agent *Trichoderma virens*. Fungal Genet Biol 41:336–348

Ram RM, Jain A, Singh A, Singh HB (2015) Biological management of *Sclerotinia* rot of bean through enhanced host defense responses triggered by *Pseudomonas* and *Trichoderma* species. J Pure Appl Microbiol 9(1):523–532

Ramada MH, Steindorff AS, Bloch C Jr, Ulhoa CJ (2016) Secretome analysis of the mycoparasitic fungus *Trichoderma harzianu*m ALL 42 cultivated in different media supplemented with *Fusarium solani* cell wall or glucose. Proteomics 16:477–490

Reithner B, Ibarra-Laclette E, Mach RL, Herrera-Estrella A (2011) Identification of mycoparasitism-related genes in *Trichoderma atroviride*. Appl Environ Microbiol 77:4361–4370

Reithner B, Schuhmacher R, Stoppacher N, Pucher M, Brunner K, Zeilinger S (2007) Signaling via the *Trichoderma atroviride* mitogen-activated protein kinase Tmk1 differentially affects mycoparasitism and plant protection. Fungal Genet Biol 44:1123–1133

Rep M (2005) Small proteins of plant-pathogenic fungi secreted during host colonization. FEMS Microbiol Lett 253:19–27

Ron M, Kantety R, Martin GB, Avidan N, Eshed Y, Zamir D, Avni A (2000) High-resolution linkage analysis and physical characterization of the EIX-responding locus in tomato. Theor Appl Genet 100:184–189

Salas-Marina MA, Isordia-Jasso MI, Islas-Osuna MA, Delgado-Sanchez P, Jimenez-Bremont JF, Rodriguez-Kessler M, Rosales-Saavedra MT, Herrera-Estrella A, Casas-Flores S (2015) The Epl1 and Sm1 proteins from *Trichoderma atroviride* and *Trichoderma virens* differentially modulate systemic disease resistance against different life style pathogens in *Solanum lycopersicum*. Front Plant Sci 6:77

Samuels GJ (1996) *Trichoderma*: a review of biology and systematics of the genus. Mycol Res 100:923–935

Schmoll M (2008) The information highways of a biotechnological workhorse—signal transduction in *Hypocrea jecorina*. BMC Genomics 9:430

Segarra G, Casanova E, Bellido D, Odena MA, Oliveira E, Trillas I (2007) Proteome, salicylic acid, and jasmonic acid changes in cucumber plants inoculated with *Trichoderma asperellum* strain T34. Proteomics 7:3943–3952

Seidl V, Marchetti M, Schandl R, Allmaier G, Kubicek CP (2006) Epl1, the major secreted protein of *Hypocrea atroviridis* on glucose, is a member of a strongly conserved protein family comprising plant defense response elicitors. FEBS J 273:4346–4359

Sharma V, Salwan R, Sharma PN, Gulati A (2017) Integrated translatome and proteome: approach for accurate portraying of widespread multifunctional aspects of *Trichoderma*. Front Microbiol. https://doi.org/10.3389/fmicb.2017.01602

Shoresh M, Harman GE (2008) Genome-wide identification, expression and chromosomal location of the genes encoding chitinolytic enzymes in *Zea mays*. Mol Gen Genomics 280(2):173–178

Shoresh M, Harman GE, Mastouri F (2010) Induced systemic resistance and plant responses to fungal biocontrol agents. Annu Rev Phytopathol 48:21–43

Shoresh M, Yedidia I, Chet I (2005) Involvement of jasmonic acid/ethylene signaling pathway in the systemic resistance induced in cucumber by *Trichoderma asperellum* T-203. Phytopathology 95:76–84

Singh A, Jain A, Sarma BK, Upadhyay RS, Singh HB (2013) Rhizosphere microbes facilitate redox homeostasis in *Cicer arietinum* against biotic stress. Ann Appl Biol 163:33–46

Singh A, Jain A, Sarma BK, Upadhyay RS, Singh HB (2014a) Rhizosphere competent microbial consortium mediates rapid changes in phenolic profiles in chickpea during *Sclerotium rolfsii* infection. Microbiol Res 169:353–360

Singh A, Jain A, Sarma BK, Upadhyay RS, Singh HB (2014b) Beneficial compatible microbes enhance antioxidants in chickpea edible parts through synergistic interactions. LWT- Food Sci Technol 56:390–397

Stergiopoulos I, de Wit PJ (2009) Fungal effector proteins. Annu Rev Phytopathol 47:233–263

Suárez MB, Sanz LM, Chamorro I, Rey M, González FJ, Llobell A, Figeys M (2005) Proteomic analysis of secreted proteins from *Trichoderma harzianum*: identification of a fungal cell wall-induced aspartic protease. Fung Genet Biol 42:924–934

Suarez MB, Vizcaino JA, Llobell A, Monte E (2007) Characterization of genes encoding novel peptidases in the biocontrol fungus *Trichoderma harzianum* CECT 2413 using the TrichoEST functional genomics approach. Curr Genet 51:331–342

Tagu D, Lapeyrie F, Martin F (2002) The ectomycorrhizal symbiosis: genetics and development. Plant Soil 244:97–105

Teresa M, Bara F, Lima AL, Ulhoa CJ (2003) Purification and characterization of an exo-β-1,3-glucanase produced by Trichoderma asperellum. FEMS Microbiol Lett 219:81–85

Tseng SC, Liu SY, Yang HH, Lo CT, Peng KC (2008) Proteomic study of biocontrol mechanisms of *Trichoderma harzianum* ETS 323 in response to *Rhizoctonia solani*. J Agric Food Chem 56:6914–6922

Vargas WA, Djonović S, Sukno SA, Kenerley CM (2008) Dimerization controls the activity of fungal elicitors that trigger systemic resistance in plants. J Biol Chem 283:19804–19815

Verma M, Brar SK, Tyagi RD, Surampalli RY, Val'ero JR (2007) Antagonistic fungi, *Trichoderma* spp.: panoply of biological control. Biochem Eng J 37:1–20

Vinale F, Sivasithamparam K, Ghisalberti EL, Woo SL, Nigro M, Marra R et al (2014) *Trichoderma* secondary metabolites active on plants and fungal pathogens. Open Mycol J 8:127–139

Viterbo A, Wiest A, Brotman Y, Chet I, Kenerley CM (2007) The 18mer peptaibols from *Trichoderma virens* elicit plant defence responses. Mol Plant Pathol 8:737–746

Wosten HAB (2001) Hydrophobins: multipurpose proteins. Annu Rev Microbiol 55:625–646

Wu Q, Sun R, Ni M, Yu J, Li Y, Yu C, Dou K, Ren J, Chen J (2017) Identification of a novel fungus, Trichoderma asperellum GDFS1009, and comprehensive evaluation of its biocontrol efficacy. PLoS One 12(6):e0179957

Xie BB, Li D, Shi WL, Qin QL, Wang X, Rong JC, Sun CY, Huang F, Zhang XY, Dong XW, Chen XL, Zhou BC, Zhang YZ, Song XY (2015) Deep RNA sequencing reveals a high frequency of alternative splicing events in the fungus Trichoderma longibrachiatum. BMC Genomics 16:54

Yang CA, Cheng CH, Lo CT, Liu SY, Lee JW, Peng KC (2011) A novel L-amino acid oxidase from *Trichoderma harzianum* ETS 323 associated with antagonism of *Rhizoctonia solani*. J Agric Food Chem 59(9):4519–4526

Yang HH, Yang SL, Peng KC, Lo CT, Liu SY (2009) Induced proteome of *Trichoderma harzianum* by *Botrytis cinerea*. Mycol Res 113:924–932

Yates JR III (1998) Mass spectrometry and the age of the proteome. J Mass Spectrom 33:1–19

Zeilinger S, Reithner B, Scala V, Peissl I, Lorito M, Mach RL (2005) Signal transduction by Tga3, a novel g protein subunit alpha of *Trichoderma atroviride*. Appl Environ Microbiol 71:1591–1597

Chapter 15
Strategies of Biotechnological Innovations Using *Trichoderma*

Kandasamy Saravanakumar, Sabarathinam Shanmugam, Anjana Hari, Zeba Usmani, Davoodbasha Mubarak Ali, Kandasamy Kathiresan, Yevgen Karpichev, Bhim Pratap Singh, Susana Rodriguez-Couto, Myeong-Hyeon Wang, and Vijai Kumar Gupta

Abstract Naturally occurring *Trichoderma* species and their recombinant strains are powerful scaffolds for high production of valuable biochemicals and metabolites of interest. This fungal group lives in diversified ecological niches. Its metabolic products such as carbohydrate-active enzymes (CAZymes), proteins, polysaccharides, lipids, and peptides are undergoing a variety of industrial and biological utilization. Recently, the range of omics-based molecular approaches has facilitated the successful screening and genetic modifications in the gene cluster or gene

---

K. Saravanakumar · M.-H. Wang
Department of Medical Biotechnology, College of Biomedical Sciences, Kangwon National University, Chuncheon, Gangwon, Republic of Korea

S. Shanmugam · A. Hari
Bioprocess Laboratory, Department of Microbial Biotechnology, Bharathiar University, Coimbatore, Tamil Nadu, India

Z. Usmani · Y. Karpichev
Department of Chemistry and Biotechnology, Tallinn University of Technology, Tallinn, Estonia

D. M. Ali
National Repository for Microalgae and Cyanobacteria – Freshwater (DBT, Govt. of India), Department of Microbiology, Bharathidasan University, Tiruchirappalli, Tamil Nadu, India

K. Kathiresan
Centre of Advanced Study in Marine Biology, Faculty of Marine Sciences, Annamalai University, Parangipettai, Tamil Nadu, India

B. P. Singh
Department of Agriculture and Environmental Sciences, National Institute of Food Technology Management and Entrepreneurship (NIFTEM), Kundli, Haryana, India

S. Rodriguez-Couto
Ikerbasque, Basque Foundation for Science, Bilbao, Spain

V. K. Gupta (✉)
Biorefining and Advanced Materials Research Center, Scotland's Rural College (SRUC), Edinburgh, UK

C. Manoharachary et al. (eds.), Trichoderma: *Agricultural Applications and Beyond*, Soil Biology 61, https://doi.org/10.1007/978-3-030-54758-5_15

transfer in the fungal species. The present review discusses improved bioprocessing potential, a novel method to tailor omics on the intelligent selection of *Trichoderma* strains with new and functional traits, followed by advanced system biology and bioengineering for heterologous protein production and epigenetic manipulation in a biomolecular cluster of the fungal strains.

Keywords *Trichoderma* · Omic · System biology · Bioactive metabolites · Heterologous proteins

15.1 Introduction

Filamentous fungi are well known as cell factories to extensively produce therapeutically important enzymes and antibiotics, proteins, and cephalosporins (Cho et al. 2014; Hamad 2010). The fungal genes of *Trichoderma* species are of great biotechnological interest and wide industrial importance for the production of organic acids, metabolites, and enzymes. Genetic engineering techniques based on RNA and DNA manipulations are developed in industrially important fungi (Meyer 2008), particularly, *Trichoderma* comprising a bundle of filamentous fungi, isolated from various habitats (Brotman et al. 2010) with nifty biological activity in agriculture (Kiriga et al. 2018; Zaidi et al. 2017; Mendoza-Mendoza et al. 2018), cellulosic biofuels (Gupta et al. 2016; Srivastava et al. 2018; Saravanakumar et al. 2015), biomedical (Saravanakumar et al. 2015), bioremediation (Zhao et al. 2018; Saravanakumar and Kathiresan 2015), and nanotechnology (Saravanakumar et al. 2015; Saravanakumar and Wang 2018; Guilger et al. 2017; Omran et al. 2018; Menon et al. 2019). Research on *Trichoderma* started in 1794 (Persoon 1794), and its characteristics, identification, and biological activities have been extensively investigated. Morphological keys and molecular markers have also been generated for the appropriate identification of *Trichoderma* isolates (Samuels 2006).

Several *Trichoderma* species that can produce mycotoxins were assessed for toxicity and pathogenicity for industrial applications (Schuster and Schmoll 2010). Enzymes from *Trichoderma* species have been widely used in food, feed, textile, biofuels, and heterologous protein cell factories for centuries (Blaszczyk et al. 2014; Mukherjee et al. 2013; Chavez-Guerrero et al. 2019). Diversified industrially important molecules including simple organic to complex metabolites were reported to form *Trichoderma* sp. (for the review see (Contreras-Cornejo et al. 2016). Some chemical products were commercialized from *Trichoderma* cell factories such as cellulase from *T. reesei* for the benefit of the paper industry; cellobiohydrolase from *T. viride* and *T. reesei* for textile, pulp, and paper industries; pectin lyases from *T. reesei* for food industry; xylanases from *T. reesei* and *T. konignii* for textile, pulp, and paper industries; and hydrophobin from *T. reesei* for tissue engineering (Meyer 2008). Some *Trichoderma* strains were genetically modified as cell factories in biotechnology for enhanced production of heterologous proteins and metabolites for various industrial applications (Kidwai and Nehra 2017). Thus, this review comprehensively reports the current status and innovations in genetic tools for engineering of *Trichoderma* species toward enhanced biotechnological applications.

15.2 *Trichoderma* Genera as Improved Bioprocessing Cell Factories

Fungi are well-known producers of plant cell wall-degrading enzymes and, thus, there is a growing interest for them in the biofuel industry. *Trichoderma* strains are energetic producers of secretome proteins and carbohydrates active enzymes (CAZymes) comprising cellulase, β-glucosidase, endoglucanase, exoglucanase, xylanase, pectinase, amylase, glucoamylase, glucose, isomerase, protease, β-glucanase, phytase, lipase, phospholipase, lysophospholipase, but the amount of the enzyme production from the naturally occurring strains is low for industrial exploitation. Despite all this information about enzyme production, very little is known about the secretory pathway in *Trichoderma*.

In order to enhance enzyme production, several traditional strategies are applied such as modification in fermentation media constituents and strain improvement through mutations to reach the required level for industrial applications. Thus, these attempts have attained a high-scale production of 100 g/L of protein that comprises up to 60% of cellobiohydrolase I and 20% of cellobiohydrolase II (Schuster and Schmoll 2010).

Potential applications of *Trichoderma* strains have recently been demonstrated and it is now indistinct that hundreds of separate functions are involved in the processes of enzyme production (Monte 2001). Some of these protein functions have been identified and expressed in *Trichoderma* spp. and offer great assurance toward producing higher levels of valuable enzymes for various biotechnological processes (Table 15.1).

Biotechnology behind gene expression can help to develop efficient *Trichoderma* cell microbial factories for various bioprocess applications (Gusakov 2011). Strains have been improved by random mutagenesis, resulting in substantially improved enzyme/protein functions (Tolan and Foody 1999). A *Trichoderma* strain, *T. reesei* Rut C-30 is one of the very high yielding wild strains of many commercially used strains currently.

Trichoderma strains can be cultured using economical sources such as agricultural wastes and plant materials and do not require any amino acids or vitamins as constituents for the fermentation medium. This is due to the ability of *Trichoderma* strains to produce cellulase complex enzymes for the conversion of cellulosic waste (agricultural waste: straw, husk, wood, corn cob, leaves, and biogas; food processing waste, municipal waste, etc.) into fermentable sugars (Li et al. 2016) followed by the generation of biofuels using yeast: an eco-friendly and economically important method for the production of petroleum products (Saravanakumar and Kathiresan 2014).

Table 15.1 *Trichoderma* spp. and their respective enzyme functions in bioenergy and biorefinery sectors (Source: Gupta et al. 2014)

S. no.	*Trichoderma* spp.	Protein function	Productivity level	Reference
1	*T. reesei*	Cellobiohydrolase II	Not quantified	Chen et al. (1987)
2	*T. harzianum*	Alkaline proteinase	26.4 units	Geremia et al. (1994))
3	*T. reesei*	β-D-glucoside glucohydrolase	Not quantified	Mach et al. (1994)
4	*T. koningii*	Cellulose1,4- β-cellobiosidase/Cellobiohydrolase I	90–100 mg/L	Wey et al. (1994)
5	*T. reesei*	Endoglucanase III	Not quantified	Saloheimo et al. (1988)
6	*T. reesei*	Endo-1-4- β-glucanase V	Not quantified	(Saloheimo et al. 1994)
7	*T. harzianum*	Glucan endo-1,6- β-glucosidase	Not quantified	(Lora et al. 1995)
8	*T. harzianum*	Endochitinase	3 mg/L	(Draborg et al. 1996)
9	*T. harzianum*	Endo-1,3(4)- β-glucanase	Not quantified	(de la Cruz et al. 1992)
10	*T. reesei*	Cellobiohydrase II	0.04 U/10^8 conidia	(Stangl et al. 1993)
11	*T. reesei*	Endoglucanase I	Not quantified	(Penttila et al. 1987)
12	*T. koningii*	Arabinofuranosidase/β-xylosidase	4.5%	(Huang et al. 1991)
13	*T. reesei*	endoxylanaseII	3700 nkat/ml	(Saarelainen et al. 1993)
14	*T. viride*	1,4- β-D-glucan cellobiohydrolase	65%	(Cheng et al. 1990)
15	*T. reesei*	Endo- β-1,-4-xylanase I (XYN1)	100 U/mg	(Torronen et al. 1992)
16	*T. reesei*	Endo- β-1,-4-xylanase II (XYNII)	1600 U/mg	(Torronen et al. 1992)
17	*T. reesei*	β −1,3-endoglucanase	1.6 nkat mg^{-1}	(El-Katatny et al. 2000)
18	*T. reesei*	Chitinase	16.6 nkat mg^{-1}	(El-Katatny et al. 2000)
19	*T. reesei*	β −1,3-endoglucanase	4.3 nkatal mg^{-1}	(Lorito et al. 1994)
20	*T. reesei*	β −1,3-exo glucanase	Not quantified	(Cohen-Kupiec et al. 1999)
21	*T. reesei*	β −1,3-exo glucanase	10 mmol/g/min	(Ramot et al. 2000)
22	*T. longibrachiatum*	β-1,4-endoglucanase	45.7 (mU mg^{-1} dry weight)	(Migheli et al. 1998)
23	*T. reesei*	Cellobiohydrolase	Not quantified	(Shoemaker et al. 1983)
24	*T. reesei*	Cellobiohydrolase	Not quantified	(Teeri et al. 1987)
25	*T. reesei*	Endo-1,4-glucanase	Not quantified	(Penttila et al. 1987)
26	*T. reesei*	Endo-1,4-glucanase	Not quantified	(Saloheimo et al. 1988)

27	*T. reesei*	Endo-1,4-glucanase	15 (mU/mL)	(Okada et al. 1998)
28	*T. reesei*	Endo-1,4-glucanase	Not quantified	(Saloheimo et al. 1997)
29	*T. reesei*	Endo-1,4-glucanase	Not quantified	(Saloheimo et al. 1994)
30	*T. reesei*	β -Glucosidase	Not quantified	(Barnett et al. 1991)
31	*T. reesei*	β -Glucosidase	23.9 $U.mg^{-1}$	(Takashima et al. 1999)
32	*T. reesei*	Cellulase	2.2 U/mg of myc.	(Nogawa et al. 2001)
33	*T. reesei*	Xylanase	Not quantified	(Saloheimo et al. 2002)
34	*T. reesei*	β-Xylosidase	16.3 nkat/mL	(Margolles-Clark, et al. 1996)
35	*T. reesei*	Acetyl xylan esterase	Not quantified	(Foreman et al. 2003)
36	*T. reesei*	Arabinofuranosidase	*4* nkat/mL	(Margolles-Clark et al. 1996)
37	*T. reesei*	Mannanase	*Not quantified*	(Stalbrand et al. 1995)
38	*T. reesei*	α-Galactosidase	*0.1* nkat/mL	(Margolles-Clark et al. 1996)
39	*T. reesei*	α-Galactosidase	*0.082 U/mg protein*	(Kubicek 1987)
40	*T. reesei*	Cellulases (CMCase, CBH, BGL),	230 IU/g of cellulose	(Chahal 1985)
41	*T. reesei*	Hemicellulase (xylanase)	25 ± 5 U/mg of protein	(Kurzatkowski et al. 1996)
42	*T. harzianum*	Cellulases (CMCase, CBH), β-1,3-glucanases	1.43 IU/mL; 2.40 IU/mL	(Khan et al. 2007)
43	*T. virens*	Endochitinase activity	2.91 ± 0.15 units/mg	(Kim et al. 2002)
44	*T. virens*	N-acetylglucosaminidase Activity	10.31 ± 0.24 units/mg	(Kim et al. 2002)
45	*T. reesei*	B-glucosidase I	63 IU/mL	(Nakazawa et al. 2012)
46	*T. reesei*	Protease	0.1 mg/L	(Bali 2013)
47	*T. estonicum*	Protease	41.54 IU/mL	(Saravanakumar et al. 2013)
48	*T. reesei*	Protease	30.25 IU/mL	(Zhang et al. 2014)
49	*T. asperellum*	Protease	9.52 U/mL	(Doua et al. 2014)
50	*T. reesei*	Protease	23.4 mg/mL	(Landowski et al. 2015)
51	*T. harzianum*	Protease	321.8 U/mL	(Deng et al. 2018)

Since recent years, *Trichoderma* strains are used as an excellent expression host organism for the production of heterologous proteins. The heterologous expression is the transformation of novel genes from one organism to another host organism to achieve higher yields of the scaffold. Moreover, the gene transfer approaches involving amplification of targeted genes, promoter system generation by cloning using the PCR approaches, characterization, and expression of the genes in appropriate hosts can enhance the production level of targeted enzymes or metabolites (Punt et al. 2002). Among *Trichoderma* strains, *T. reesei* (anamorph *Hypocrea jecorina*) is considered to be one of the best cell factories and the most credible host for genetically engineered strains through molecular tools along with extraordinary heterologous protein secretion potential (Nakari-Setala et al. 2009; Seidl and Seiboth 2010), higher compared to other strains and host adaptability with specific production of the hydrolyzing enzymes such as cellulases and hemicellulases (Xia et al. 2017) which requires additional improvements through molecular modifications via gene editing or transformations. In *Trichoderma,* transcriptomic and genomic information has uncovered the secretion pathway (Wu et al. 2017). But little is known about the new molecular strategies facilitating the rapid progress in amplifying the useful genes and annihilating the unfavorable ones toward enhancing the targeted expression of the heterologous genes (Punt et al. 2002; Wu et al. 2017). Several engineered strains are available such as hyper cellulose, cellulase knockout strains, and protease deficient strains.

Recent reports on higher heterologous protein expression through the modification of the promoter or host genes by genetic engineering tools using *Trichoderma* as a model organism of host or donor of respected genes are gathered in Table 15.2. Cellulase is the major industrial enzyme used in the conversion of cellulosic waste materials into bioethanol or other chemical products (Liu et al. 2017a, b). Cellulases can be classified as cellobiohydrolase I, cellobiohydrolase II, β-glucosidase, and endoglucanase. These enzymes are extensively involved in the conversion of cellulosic waste at different stages (Gupta et al. 2016; Liu et al. 2017a, b). β- glucosidase (βgls) belongs to the member of the cellulase enzyme complex widely used in various applications and active against several substrates such as glycolipids, flavonoids, glycoceramides, cellobiose, and glucosides (Florindo et al. 2018). The gene manipulation or strain constriction in *Pichia pastoris* using the βgls (bgl1) from *T. viride* by PCR overlapping approach has resulted in increased production of the βgls toward scarification of gentiooligosaccharides from the gentiobiose (Wang et al. 2018a, b). *T. harzianum* has been recently reported to produce a higher level of βgls (3.55BGU/ml) than the commercial strain *T. reessi* RUT C-30 (Souza et al. 2018). Many strategies are reported to improve the yield of the βgls in order to reduce the biorefinery. One of such approaches is using a mixed culture. The combination of the genetically engineered *T. reesei* with *A. niger* was reported to have significantly increased the production of βgls (Zhao et al. 2018).

The gene AnGOD of *A. niger* when expressed in *T. reesei*, there was an overexpression of heterogonous protein genes (*snc 1*) related to the secretion pathway (Wu et al. 2017). The overexpression of the Trvib-1 encoding putative transcription factor in *T. reessi* RUT C-30 has resulted in about 40% enhanced

Table 15.2 *Trichoderma* as cell donor or recipient in genetically engineered strains for enhanced bioprocessing potencies

Gene	Donor	Expressing host	Activity	References
bgl1	*T. viride*	*Pichia pastoris*	Scarification of gentiooligosaccharides	(Florindo et al. 2018)
rP6281	*T. harzianum*	*P. pastoris*	Antifungal activity	(Deng et al. 2018)
Tv-ECH1	*T. virens*	*P. pastoris*	Enzyme activity	(Bubwinkel et al. 2018)
chit42	*T. harzianum*	*Daucus carota L*	Antifungal activity	(Ojaghian et al. 2018)
AnGOD	*A. niger*	*T. reesei*	–	(Wu et al. 2017)
Cel12A (EG III)	*Trichoderma reesei*	*L. Lactis subsp. lactisMG1363*	Extracellular activity of heterologous protein	(Liu et al. 2017a, b)
Cel12A (EG III)	*T. reesei*	*E. coli DH5α*	Extracellular activity of heterologous protein	(Liu et al. 2017a, b)
Cel5A (EG II)	*T. reesei*	*E. coli Rosetta-gami B (DE3) pLacI*	Hydrolysis of CMC	(Nakazawa et al. 2008)
Cel12A (EG III)	*T. reesei*	*E. coli Rosetta blue (DE3) pLacI*	Hydrolysis of CMC	(Nakazawa et al. 2008)
Lip	*A. niger*	*T. reesei*	–	(Qin et al. 2012)

production of cellulase and protein secretion for the efficient conversion of lignocellulose waste (Zhang et al. 2018). The aspartic protease P6281 from the *T. harzianumc,* when transformed and expressed in *P. pastoris* using recombinant technology produced more quantity of enzyme against gray mold disease (Deng et al. 2018). Furthermore, recent reports on the transformation of *Trichoderma* genes expressed in other host organisms are listed in Table 15.2.

15.3 Novel Omics Approaches to Tailor *Trichoderma* for Biotech Potential

Structural and functional diversity and interactions within the community as well as with the hosts are essential issues in understanding microbial ecosystems. Furthermore, they are a basis for its biotechnological exploitation. Microorganisms are known to be important drivers for the functioning of the earth's ecosystems. The study of single protein or transcripts is unable to correlate the whole biological pathways. But, omics-based study through the systematic biological approach can provide a comprehensive set of genomic, transcriptomes, translatomes, metabolomics, and proteomic data, which facilitate to understand the cellular function pathways in fungi. *Trichoderma* strains have unique characteristics for tailoring themselves for enhanced productivities. Additionally, there are several novel omics

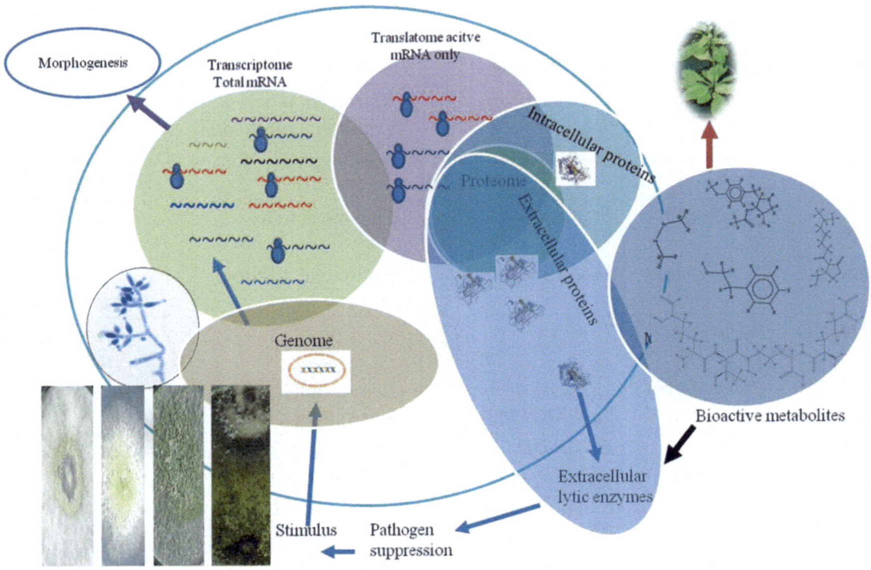

Fig. 15.1 Omics approaches in *Trichoderma*. (Source: *Sharma, Salwan, Sharma & Gulati*, 2018)

approaches proposed to improve the higher yield by molecular methods. The omics approaches facilitate the study of *Trichoderma* strains physiology through integrated genomic, transcriptomes, translatomes, and proteomic approaches (Fig. 15.1) (Sharma et al. 2018).

A number of integrated molecular studies were carried out to understand the genomic and transcriptomic behavior of *Trichoderma* strains at the omics level. The whole-genome sequencing of *Trichoderma* strains enabled the design of targeted molecular approaches (see reviews of Lorito et al. (2010)). All the mutant strains are derived from the strain QM6a, which is a native reference strain of *T. reesei* isolated from the Natick Army Research Laboratory (Bischof et al. 2016). Within the cycle of elements, fungi are the dominant players in the degradation of organic matter. Thereby, the genus *Trichoderma* plays a crucial role. In plant–host interactions, they mediate nutrient delivery, activation of the immune system, and tolerance against various stressors. Due to their versatile capabilities, *Trichoderma* communities essentially contribute to a healthy environment and also represent a resource for biomolecules. Bae et al. (2016) investigated the metabolites extracted from 128 *Trichoderma* isolates against seven *Phytophthora* species. Among them, ethyl acetate extract of *Trichoderma* sp. strain KACC 40557 inhibited *P. capsici* growth via changes in plant hormone levels and induction of defense-related genes in leaf tissues. Further, the utilization of Trichoderma biofertilizer increased the antifungal compounds which relatively decrease the presence of *Ophiosphaerella* and simultaneously improved the abundance of plant-available phosphorus by increasing Archaeorhizomyces in grassland biomass (Zhang et al. 2018). The molecular tools

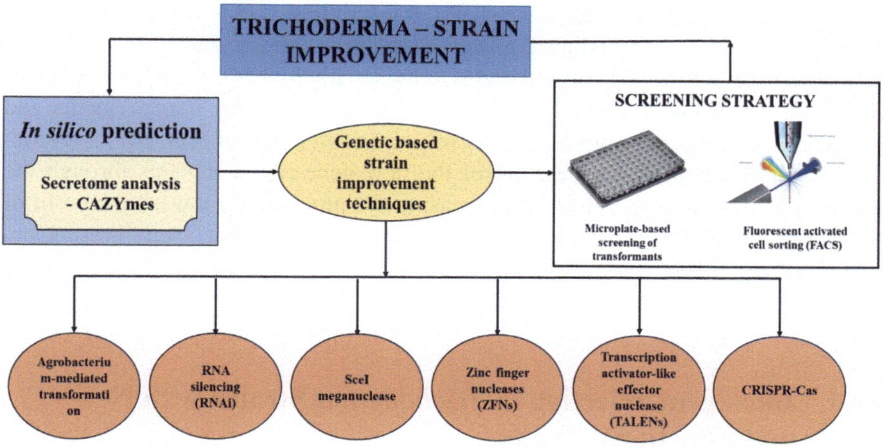

Fig. 15.2 Tools for *Trichoderma* strain improvement

available for working with *T. reesei* include mutagenesis, gene deletion, promoters, multiplication of strains, optimization of carriers and codons (Bischof et al. 2016; Glass et al. 2013; Tisch et al. 2017) for the regulation of secondary metabolism and gene regulations and it signaling pathways which provide novel approaches for strain improvement (Fig. 15.2) (Tisch et al. 2017). Moreover, it shows the major prerequisites for the successful cultivation of *Trichoderma* and discusses the potential of metagenomic strategies for biotechnological applications.

During the early decades of molecular research, about 20% of the cellulase production was increased through mutagenesis in the native strain of QM6a, and finally, the mutant was named as RUT-C30 (Mandels et al. 1971). Later, many researchers applied several molecular approaches for editing the genes in host strains resulting in a high industrial level yield of biotechnological products. In the early 1990s, transformations techniques were widely used for genetic engineering of *T. reesei* (Penttila et al. 1987; Gruber et al. 1990). During this period, *T. reesei* was used as the first host for heterogonous expression of the mammalian proteins, i.e., calf chymosin through cbh1 signals (Harkki et al. 1989). *Hypocrea jecorina* is a sexual form of *T. reesei* (Kuhls et al. 1996) whose sexual development is described elsewhere (Seidl et al. 2009) which has led to the subsequent link of mutations in MAPK scaffold genes regulations (Linke et al. 2015). The omics-based sexual crossing of the strains has revealed the CAzymes gene regulations and metabolic pathways associated with gene expression. The generation of a DNA microarray library-based cDNA of *T. reesei* corresponding to the 5000 transcripts of its genome and in a combination of wide genome sequencing of the original strain QM6a has enabled broad-scale applications (Martinez et al. 2008). Furthermore, omics studies on *Trichoderma* species include comparative omics analysis, gene silencing approaches, hyperproductive strains discovered through nucleocytoplasmic transport, vascular proteins, mRNA turnover and gene-editing CRISPR/Cas9 system (Liu et al. 2015; Bischof and Seiboth 2015; Le Crom et al. 2009). The regulation of the catabolite expression mediates the C_2H_2 transcription factor CRE1 and CRE2 through the targeting signaling pathway-based gene editing and is favorable for enhanced digestion of polycarbohydrates to monocarbohydrates using its CAZymes (Schmoll and Kubicek 2003). The genetic engineering-based improvements and generation of *T. reesei* strains are extensively reviewed (Bischof et al. 2016). Overall genetic modification and generation of the new gene-edited mutants through the omics approach, for synergistically increasing the productivity toward the profit of the biotechnology-based cell factory is an emerging trend in genetic engineering and bioprocessing.

15.4 Intelligent Selection of *Trichoderma* Strains with New and Functional Traits

Trichoderma is one of the most common fungi that can be isolated from different ecological niches viz., free soil, xenobiotics-infested soil, dead wood, indoor building walls, kerosene tanks in aircraft (Papavizas 1985), and various biotrophic associations ranging from rhizosphere colonization and endophytism to facultative *Trichoderma* pathogenesis (Druzhinina et al. 2011). To adapt to the above-mentioned environments, *Trichoderma* needs to produce a wide range of secretory proteins to break down complex organic polymers into simpler forms for its growth and metabolism. Among *Trichoderma* species, *T. reesei* has gained interest owing to its extracellular protein secretion machinery, endogenous host protein modification mechanism, and post-translation modification machinery similar to that of mammals, and hence it acts as a potential cell factory for different heterologous proteins (Peterson and Nevalainen 2012).

Further, *T. reesei* has achieved GRAS (Generally Recognized as Safe) status by the US Food and Drug Administration (FDA). The generation of the myriad of mutants from *T. reesei* facilitated efficient utilization in biotechnological industries for the production of cellulases and hemicellulases (Fig. 15.3) that were applied for food and feed, textile, and particularly, for biofuel production (Kubicek 2013). Druzhinina et al. (2011) reviewed the identification of novel traits of *Trichoderma* by the composition and properties of secretome via in silico prediction. 747 proteins were secreted by *T. reesei* (as of 2012) through the plasma membrane, in which CAZymes, small secreted cysteine-rich proteins (SSCPs) and unknown proteins, alone account for 60% of all secretome (Fig. 15.3). The presence of SSCPs and unknown orphan proteins which accounts for 50% of secretome holds the key factor for controlling the recognition of macromolecules/partner organism which is engaged in the degradation of complex organic polymers like lignin. Furthermore, proteomic analysis combined with available genomic sequence and bioinformatic tools of *T. reesei* makes it possible to identify novel enzyme homologs that may be involved in biomass conversion.

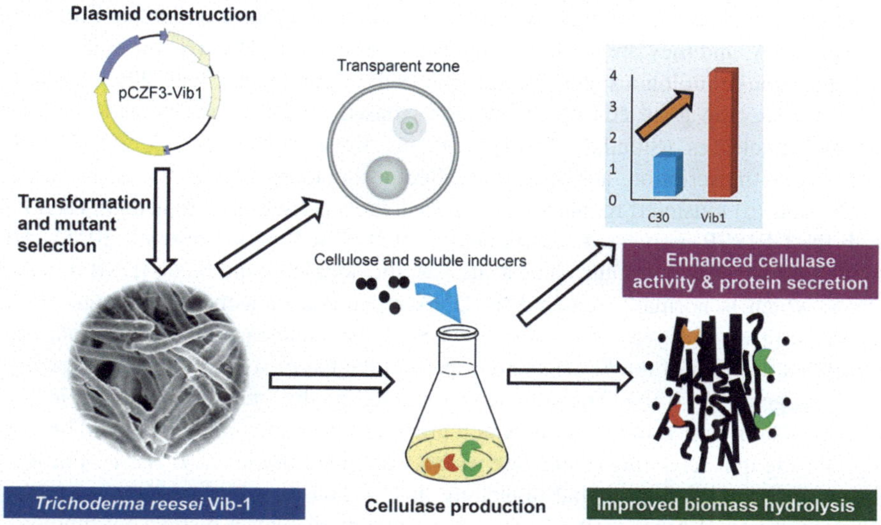

Fig. 15.3 *Trichoderma* as a cell factory for enzymes and protein production

A lot of genes are involved in the secretion of various CAZymes, but comprehensive information about those genes involved in the cellulolytic machinery in *T. reesei* is not available so far. Therefore, after comparison with the cellulase genes and their regulatory elements from different species or genera, techniques like inter−/intraspecies recombination via protoplast fusion could be utilized for the generation of better traits. When protoplast fusion is combined with mutation techniques, it is possible to generate recombinants with genetic improvement (Dillon et al. 2008).

Classical strain improvement methods involved in the development of mutants via chemical and physical mutagens have generated improved cellulolytic potential; however, the genetic basis for strain improvement remains unknown (Fujii et al. 2010) and also the characterization of the mutations at the molecular level is tedious (Gehring et al. 2000). Insertional mutagenesis via *Agrobacterium*-mediated transformation has proven to be one of the most powerful techniques for genetic modification in fungi which leads to single-copy T-DNA integration into the genomes, and this has high efficiency (Zhong et al. 2011). It also offers a decisive advantage over chemical and physical mutagenesis as the mutated genes are tagged by the inserted element, which can then be used to identify the disrupted genes/flanking sequences (Jeong et al. 2005). Zhong et al. (2011) have conducted T-DNA insertional mutagenesis for the strain improvement of *T. reesei* with improved cellulase production using *Agrobacterium tumefaciens*-mediated transformation (AMT) containing T-DNA binary vector pBI-hph. The three generated mutants TA-32, TB-87, and TE-6 can produce higher cellulase activities with improved ability to hydrolyze cellulose substrates.

In eukaryotic cells, there are two most prominent mechanisms for repairing the broken DNA, and they are the homologous recombination (HR) (Heyer et al. 2010) and the non-homologous end-joining method (Davis and Chen 2013). Major sequence-specific gene editing technologies based on these mechanisms include (a) meganucleases (Stoddard 2014), (b) zinc finger nucleases (ZFNs) (Carroll 2011), (c) transcription activator-like effector nuclease (TALENs) (Liu et al. 2015), and (d) clustered regularly interspaced short palindromic repeats (CRISPR) with the CRISPR-associated (Cas) nuclease (CRISPR-Cas9) (Liu et al. 2015). All these four nuclease types introduce a site-specific double-strand break (DSBs) in the DNA, which is normally repaired by DNA repair mechanisms. Ouedraogo et al. (2016) utilized the yeast *S. cerevisiae* I-SceI meganuclease to demonstrate the combined deletion of tku70 gene and improved the targeted integration frequencies in the range of 90–100%. The advantage rendered by this gene target system is that I-SceI cuts only once in the genome at a predetermined site and is well suitable for high throughput screening of enzyme variants or gene libraries in *T. reesei*. Further, three genes involved in hyphal branching in *T. reesei* are successfully silenced by RNA silencing (RNAi) method where the mutants display a distinct morphology from that of the parents, which proves that RNAi is an efficient tool for gene manipulation in *T. reesei*. Wang et al. (2018a, b) developed a copper-responsive RNAi for the inhibition and derepression of cellulase genes driven by the tcu1 promoter. The copper-responsive RNAi system acted as a toggle switch which can be turned on or off based on the presence/absence of copper in the medium, and not on the nutritional states. Thus, it acted as a powerful tool for characterizing the target gene of interest in *T. reesei* by changing the nutritional state of the medium. Liu et al. (2015) have successfully implemented CRSIPR/Cas9 system for efficient genome editing in *T. reesei*. In that work, *T. reesei* codon-optimized Cas9 gene linked to GFP was employed for targeting single and multiple genes simultaneously and was able to generate site-specific mutation via homologous recombination. However, the CRISPR/Cas9 system employed in *T. reesei* may have the possibility of generating off-target effects (Liu et al. 2015). Recently Liu et al. (2017a, b) utilized the transcription activator-like effectors (TALEs) for the efficient and reliable gene editing in *T. reesei*. This technique circumvents the requirement of a 500-bp homology arm for the recombination event, and t displays a higher frequency of gene deletion devoid of off-target effects.

T. reesei strains were improved via genetic modification for increasing the concentration of cellulase enzyme yield from cheap renewable substrates. Typically, the new recombinants obtained through genetic engineering were isolated using a low-to-medium throughput agar plate method (Bodie et al. 1994) or on larger volume cultivation shake methods. However, these techniques often limit the number of strains that can be screened or selected from a large number of transformants. This problem can be overcome by the novel microplate-based screening strategy in which the screening process is miniaturized, which allows the parallel assessment of a large number of fungal strains (Stefano et al. 2010). In this method, *Trichoderma* strains are cultivated in microcrystalline cellulose using 24-well plates in medium with agar. After the incubation period, the supernatant is obtained by rapid centrifugation and evaluated for cellulase activity. This method addresses the major hindrance in the screening programs and maybe automated using robots, in which thousand to hundreds and thousands of variants can be screened and easily integrated with high-throughput enzyme assays.

The identification of recombinants with highly branched mycelium hampers the selection and screening of improved *T. reesei* traits. Bradner and Nevalainen (2003) utilized the flow cytometric analysis for the sorting of the metabolically active germinating spores which were able to pass through the nozzle of a cell sorter. Notably, this technique was upgraded by the incorporation of the green fluorescent protein gene from *Renilla reniformis* as a reporter protein coupled with high-speed fluorescent activated cell sorting (FACS), which allowed the high-speed sorting of mutant strains coupled with enzyme activity/expression. This technique offers the feasibility to screen 10^9 mutants per day for the isolation of strains with improved enzyme expression (Throndset et al. 2010). Through the above-mentioned techniques, it is now possible to generate new variants of *T. reesei* mutant with novel traits which can be utilized for the generation of higher concentration of CAZymes and related regulatory enzymes for various downstream application processes.

15.5 Advance Systems Biology and Bioengineering Approaches for Heterologous Protein and Bioactive Production

Fungal cells communicate with the external environment through the secretion of different molecules such as proteins. The protein secretion pathway is a complex mechanism in charge to deliver proteins from the intracellular space to the extracellular compartment. This protein transporting pathway is quite conserved in eukaryotic cells and has been thoroughly studied in the baker yeast and some filamentous fungi. Some organisms belonging to the *Trichoderma* genus are able to establish the plant and mycoparasitic interactions, and this is in part achieved by the controlled secretion of proteins with different biological functions. Despite several variations to the tools of the trade, surprisingly little progress has been made over the last 20 years in terms of the yields of heterologous gene products produced in fungi. While recent approaches, including genome sequencing and transcriptional and proteomic studies, have provided some leads for further development, there seems to be additional physiological factors that would need to be addressed in order to better understand and overcome the bottlenecks of heterologous proteins production in *T. reesei*. *Trichoderma* strains are the producer of primary, secondary extracellular metabolites and enzymes of industrial and medicinal importance. The main constituents for expressing heterologous gene products in *T. reesei* are the strong inducible cellobiohydrolase 1 (*cbh1*) promoter, high protein-secreting mutant strains, and the heterologous protein typically fused to an endogenous well-secreted carrier protein. Kiesenhofer et al. (2017) investigated the configuration effect of cis-elements on the promoter strength of *cbh1* and hemicellulase (*xyn1*) in *T.reesei* which impacts the transcriptional regulation of industrially relevant enzymes. Based on the study, it was evident that the inverted repeat and the distance of cis-element Xyr1-binding sites (*XBS*) increased the promoter strength of *cbh1* and allow the induction of enzymes by exposure toward lignocellulosic biomass.

In order to increase the bioactive production, a variety of the novel omics approaches were applied to improve the natural strains through genetic manipulations using high-throughput targeted gene engineering tools: split-marker system (Derntl et al. 2015), non-homologous end joining (Catalano et al. 2011), RNA interference (Kuck and Hoff 2010), CRISPR/cas9 system (Liu et al. 2015), and gene transformations (Qian et al. 2016), adjusting the signaling transduction pathways (Tisch et al. 2011). Protoplasts for transformations (Penttila et al. 1987), *Agrobacterium*-mediated transformations (de Groot et al. 1998), electroporation transformations (Schuster et al. 2012), and biolistic transformations (Lorito et al. 1993). Several methods have been developed and currently used in the replacement of the targeted genes to increase the functional characteristics of fungi. The available whole-genome sequences data facilitate the intelligent selection of the appropriate method based on the targeted gene regions. Among the methods, the slit marker system is extensively used for targeted gene replacement (Goswami 2012).

15.6 Chemical Epigenetic Manipulation of *Trichoderma* Biomolecule Cluster

All the above-mentioned genetic engineering approaches involve manipulations by either overexpression or repression of one or more genes for the enhanced production of secondary metabolites or extracellular enzymes by modifying the genome at the DNA level. However, a cell is recently known to undergo reversible heritable changes in gene expression (active vs inactive state) in the absence of changes in DNA sequence. The phenomenon of changes in phenotype expression without a genotype modification is termed as epigenetics, which occurs via regulation of DNA methylation, chromatin remodeling by histone modification and RNA interference. Hence, engineering of epigenetic regulations and their influence on product formation can be considered as a potential new tool to promote the productivity of industrially important *T. ressei* (Aghcheh and Kubicek 2015). Recent developments in the epigenetic engineering are reviewed by Aghcheh and Kubicek (2015) and Druzhinina and Kubicek (2017). In DNA methylation, transcription is regulated by the covalent addition of a methyl group at the 5-C of cytosine ring resulting in 5-methylcytosine (5-mC) which extends into the major groove of DNA. However, reports on the occurrence and importance of DNA methylation in fungi are still scarce. DNA methylation studies in *Neurospora crassa* clearly demonstrates that the genome defense system repeat-induced point (RIP) mutation is the potent signal for the involvement of de novo methylation (Tamaru and Selker 2003). Further, genomic evidence confirms that RIP occurs in the life history of several fungi including *Trichoderma* (Kubicek et al. 2011). However, the regulatory role of DNA methylation in *Trichoderma* still remains elusive. In addition to DNA methylation, epigenetic regulation by chromatin remodeling through transcriptional activators has been established in *Trichoderma* (Fig. 15.4).

The genes that code for the plant biomass-degrading enzymes (PBDE) in *T. reesei* are often found in the biomolecular clusters along with the genes involved in the production of secondary metabolites. Hence, the effect of chromatin packing on metabolic processes has been investigated with respect to secondary metabolism (Gacek and Strauss 2012). Gupta et al. (2016) reviewed chromatin remodeling by transcriptional activators and highlighted it as an emerging method for the improvement of the extracellular enzymes in *T. reesei*. During cellulase production, the chromatin packing around the cellulase-encoding genes cbh1 and cbh2 opens, which is absent in the xylanase regulator 1 (XYR1) mutant strain. Further, carbon catabolite repressor (CRE1) is also involved in chromatin remodeling. With CRE-1 in mutant condition, the *T. reesei* strain exhibits the loss of positioned nucleosomes within the coding regions or promoters of cbh1 and cbh2 which are also involved in the chromatin remodeling (Zeilinger et al. 2003; Ries et al. 2014). Notably, the truncated version of CRE1 (CRE1–96) acts on the promoters and induces chromatin opening (Mello-de-Sousa et al. 2014). Also, CRE1–96 is likely to contribute to chromatin remodeling by regulating the expression of snf2/htf1, a putative helicase (Ponting et al. 2009) that participates in an ATP-dependent chromatin remodeling complex (Mello-de-Sousa et al. 2014). The cellulase and hemicellulase production in *T. reesei* can also be modulated by the manipulation of fungal genes—Velvet-LaeA/LAE1 complex—which is also involved in chromatin modification (Seiboth et al. 2012; Liu et al. 2016).

During the last decades, investigations on the epigenetic regulation of gene expression by the non-coding RNAs (ncRNAs) at a transcriptional and post-transcriptional level in eukaryotes has gained much interest among the researchers. The ncRNAs are functional RNA molecules that are transcribed from DNA; yet are not translated into the functional proteins (Ponting et al. 2009). Based on the length of nucleotides, they are differentiated into short ncRNAs [sncRNAs] (less than 30 nts) and long ncRNAs [lncRNAs] (more than 200 nts), in which the latter comprise a major group of functional ncRNA. lncRNAs have been identified in all eukaryotes, in which they are engaged in the formation of double-stranded RNA, transcriptional interference, and chromatin remodeling for regulating the gene expression. In fungi, lncRNAs appear to be involved in the regulation of mating and meiosis, cell aging, carbon metabolism, circadian rhythm, and plant pathogenesis (Donaldson and Saville 2012). The formation of CAZYmes is controlled at the level of RNA transcriptional interference in *A. niger* and *T. reesei* (Delmas et al. 2012; Ries et al. 2013). However, there is no direct evidence cited for the involvement of the lncRNAs in the regulation of CAZYmes production. A recent report by a researcher has uncovered the regulatory impact of a lncRNA, HAX1 on PBDE in *T. reesei* (Till et al. 2018). In that study, a different version of HAX1 was characterized in terms of RNA length on the cellulase expression from different strains viz.,

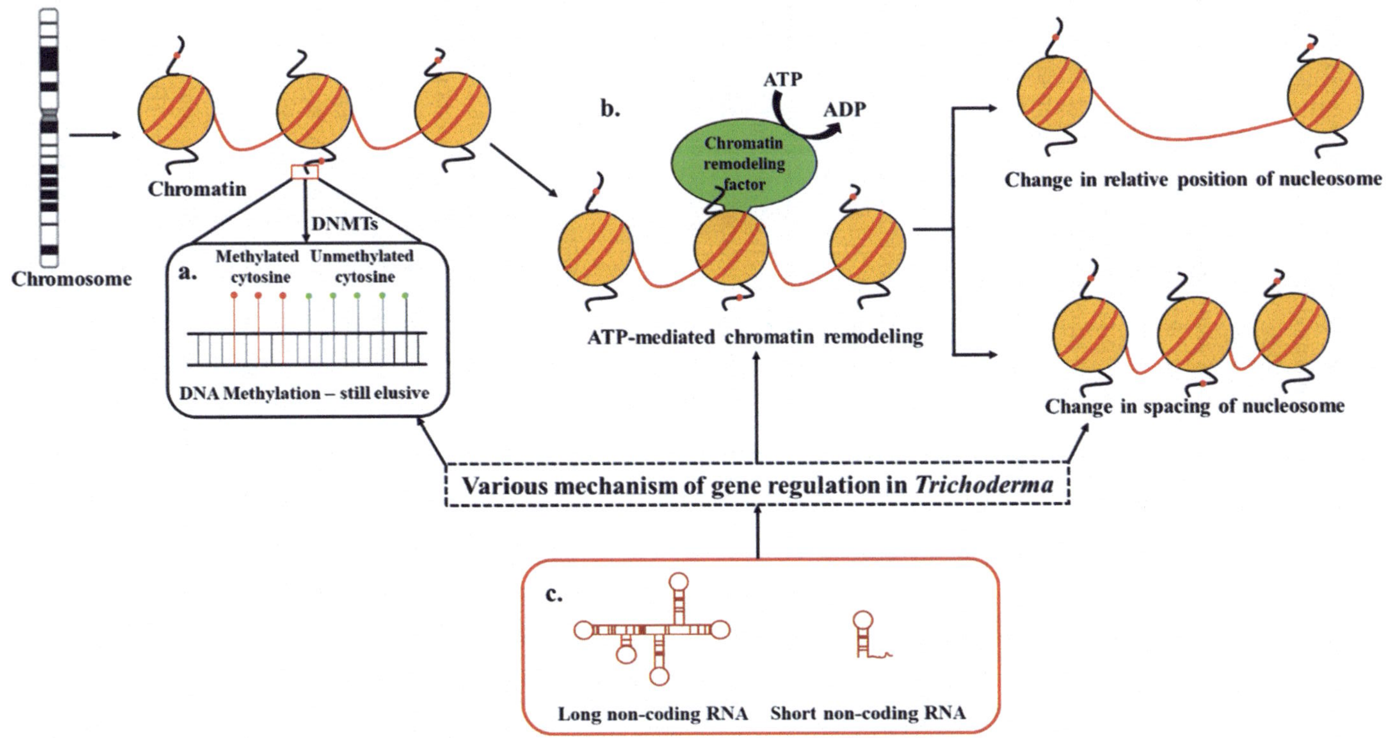

Fig. 15.4 Gene regulation mechanism in *Trichoderma*

QM6a, QM9414, and Rut-C30. The longer variants of HAX1 occur in the hyper cellulolytic strains QM9414 and Rut-C30 and are potentially involved in the PBDE expression. With the longer variant of HAX1, the overexpression has regained the function of hax1 disrupted strain and has also promoted hyper cellulose activity in QM6a. Thus, the lncRNA HAX1 can be considered as a potential target for the regulation of gene expression in *T. reesei* which will eventually improve the PBDE in these strains. The application of epigenetic principles for fungal strain improvement is achieved only in a small number of cases. However, the identification of new potential targets for histone modification and RNA interference has shown that this is an area of research with immense potential for *Trichoderma* strain improvement.

15.7 Conclusion and Future Prospects

By unique properties, *Trichoderma* strains are industrially valuable fungi for various applications such as agricultural, bioremediation, biofuels, enzymes generation, biofuels, and therapeutic agents. Recently, genetic engineering of *Trichoderma* strains has received much attention due to their flexibility in large-scale cultivation with high yield and low cost. Furthermore, *Trichoderma* strains are used as donors or receptors for heterologous expression of genes for enhanced and powerful generation of the recombinant proteins with versatile functions.

Moreover, the knowledge on omics of *Trichoderma* facilitates the successful generation of genetically engineered strains through removing, replacing/inserting by gene editing or gene manipulation, transformations, microinjection, restriction enzyme-mediated integration (REMI), transposon-arrayed gene knock out (TAGKO), *Agrobacterium tumefaciens*-mediated transformation (ATMT), RNA Interference (RNAi) (Jiang et al. 2013). The zinc-finger nucleases (ZEN) and transcription activator-like effector nucleases (TALENs) (Wood et al. 2011) and RNA-guided CRISPR-Cas9 nuclease system (Zheng et al. 2017) are used to identify the functional genes (Jiang et al. 2013). However, the present review confines to general omics of *Trichoderma* and several methods on the generation of engineered strains. It is important to choose the appropriate method and strategies based on different hosts or donors of fungal strains for the enhanced homolog expression of the targeted genes. Based on the existing knowledge on the development of novel approach with economic viability, highly efficient strains can be developed for industrial utilization of *Trichoderma* as a cell factory.

Acknowledgments This work was supported by Korea Research Fellowship Program through the National Research Foundation of Korea (NRF) funded by the Ministry of Science ICT (2017H1D3A1A01052610). This work was partially supported by Brain Korea 21 PLUS, UGC, New Delhi for the BSR Faculty Fellowship (KK); EU seventh Framework Programme for research, technological development and demonstration activities under grant agreement No. 621364 (TUTIC-Green) (ZU, YK) and Mohammed VI Polytechnic University (UM6P), Morocco (VKG).

References

Aghcheh RK, Kubicek CP (2015) Epigenetics as an emerging tool for improvement of fungal strains used in biotechnology. Appl Microbiol Biot 99(15):6167–6181. https://doi.org/10.1007/s00253-015-6763-2

Bae SJ, Mohanta TK, Chung JY, Ryu M, Park G, Shim S, Hong SB, Dong HS, Bae DW, Bae I, Jookim JJ, Bae H (2016) *Trichoderma* metabolites as biological control agents against Phytophthorapathogens. Biol Control 92:128–138

Bali H (2013) A thesis submitted in fulfilment of the requirements for the degree of master of philosophy, Department of Chemistry and Biomolecular Sciences, Macquarie University, Australia. Macquarie University, Sydney, Australia

Barnett C, Berka R, Fowler T (1991) Cloning and amplification of the gene encoding an extracellular/3-glucosidase from *Trichoderma reesei*: evidence for improved rates of saccharification of cellulosic substrates. Nat Biotechnol 9:562–567

Bischof R, Seiboth B (2015) Systems biology of carbohydrate active enzyme production in *Trichoderma reesei*. Advances in Enzymatic Conversion of Biomass to Biofuels:6–19

Bischof RH, Ramoni J, Seiboth B (2016) Cellulases and beyond: the first 70 years of the enzyme producer *Trichoderma reesei*. Microb Cell Factories 15(1):106

Blaszczyk L, Siwulski M, Sobieralski K, Lisiecka J, Jedryczka M (2014) Trichoderma spp. – application and prospects for use in organic farming and industry. J Plant Prot Res 54 (4):309–317

Bodie EA, Armstrong GL, Dunn-Coleman NS (1994) Strain improvement of chymosin-producing strains of *Aspergillus niger* var. awamori using parasexual recombination. Enzyme Microb Technol 16(5):376–382

Bradner JR, Nevalainen KMH (2003) Metabolic activity in filamentous fungi can be analysed by flow cytometry. J Microbiol Methods 54(2):193–201

Brotman Y, Kapuganti JG, Viterbo A (2010) Trichoderma. Curr Biol 20(9):R390–R391

Bubwinkel F, Goni O, Cord-Landwehr S, Connell SO, Moerschbacher BM (2018) Endochitinase 1 (Tv-ECH1) from *Trichoderma virens* has high subsite specificities for acetylated units when acting on chitosans. Int J Biol Macromol:114

Carroll D (2011) Genome engineering with zinc-finger nucleases. Genetics 188(4):773–782. https://doi.org/10.1534/genetics.111.131433

Catalano V, Vergara M, Hauzenberger JR, Seiboth B, Sarrocco S, Vannacci G, Kubicek CP, Seidl-Seiboth V (2011) Use of a non-homologous end-joining-deficient strain (delta-ku70) of the biocontrol fungus *Trichoderma virens* to investigate the function of the laccase gene lcc1 in sclerotia degradation. Curr Genet 57(1):13–23. https://doi.org/10.1007/s00294-010-0322-2

Chahal DS (1985) Solid-state fermentation with *Trichoderma reesei* for Cellulase production. Appl Environ Microbiol 49(1):205–210. https://doi.org/10.1128/aem.49.1.205-210.1985

Chavez-Guerrero L, Silva-Mendoza J, Sepulveda-Guzman S, Medina-Aguirre NA, Vazquez-Rodriguez S, Cantu-Cardenas ME, Garcia-Gomez NA (2019) Enzymatic hydrolysis of cellulose nanoplatelets as a source of sugars with the concomitant production of cellulose nanofibrils. Carbohyd Polym 210(15):85–91. https://doi.org/10.1016/j.carbpol.2019.01.055

Chen CM, Gritzali M, Stafford DW (1987) Nucleotide sequence and deduced primary structure of cellobiohydrolase II from *Trichoderma reesei*. Nat Biotechnol 5:274–278. https://doi.org/10.1038/nbt0387-274

Cheng C, Tsukagoshi N, Udaka S (1990) Transformation of Trichoderma viride using the Neurospora crassa pyr4 gene and its use in the expression of a taka-amylase a gene from Aspergillus oryzae. Curr Genet 18:453. https://doi.org/10.1007/BF00309916

Cho H, Uehara T, Bernhardt TG (2014) Beta-lactam antibiotics induce a lethal malfunctioning of the bacterial Cell Wall synthesis machinery. Cell 159(6):1300–1311. https://doi.org/10.1016/j.cell.2014.11.017

Cohen-Kupiec R, Broglie K, Friesem D, Broglie R, Chet I (1999) Molecular characterisation of a novel a1,3-exoglucanase related to mycoparasitism of *Trichoderma harzianum*. Gene 226:147–154. https://doi.org/10.1016/S0378-1119(98)00583-6

Contreras-Cornejo HA, Macias-Rodriguez L, del-Val E, Larsen J (2016) Ecological functions of *Trichoderma* spp. and their secondary metabolites in the rhizosphere: interactions with plants. FEMS Microbiol Ecol 92(4):fiw036-fiw036. https://doi.org/10.1093/femsec/fiw0

Davis AJ, Chen DJ (2013) DNA double strand break repair via non-homologous end-joining. Transl Cancer Res 2(3):130–143. https://doi.org/10.3978/j.issn.2218-676X.2013.04.02

de Groot MJA, Bundock P, Hooykaas PJ, Beijersbergen AG (1998) *Agrobacterium tumefaciens*-mediated transformation of filamentous fungi. Nat Biotechnol 16(11):839–1074. https://doi.org/10.1038/nbt0998-839

De la Cruz J, Hidalgo-Gallego A, Lora JM, Benitez T, Pintor-Toro JA, Llobell A (1992) Isolation and characterization of three chitinases from *Trichoderma harzianum*. Eur J Biochem 206:859–867. https://doi.org/10.1111/j.1432-1033.1992.tb16994.x

Delmas S, Pullan ST, Gaddipati S, Kokolski M, Malla S, Blythe MJ, Ibbett R, Campbell M, Liddell S, Aboobaker A, Tucker GA, Archer DB (2012) Uncovering the genome-wide transcriptional responses of the filamentous *Fungus Aspergillus niger* to lignocellulose using RNA sequencing. PLoS Genet 8(8):e1002875. https://doi.org/10.1371/journal.pgen.1002875

Deng JJ, Huang WQ, Li ZW, Lu DL, Zhang Y, Luo XC (2018) Biocontrol activity of recombinant aspartic protease from *Trichoderma harzianum* against pathogenic fungi. Enzym Microb Technol 12:35–42. https://doi.org/10.1016/j.enzmictec.2018.02.002

Derntl C, Kiesenhofer DP, Mach RL, Mach-Aigne A (2015) Novel strategies for genomic manipulation of *Trichoderma reesei* with the purpose of strain engineering. Appl Environ Microbiol 81(18):6314–6323. https://doi.org/10.1128/AEM.01545-15

Dillon A, Camassola M, Henriques JAP, Fungaro MH, Azevedo ACS, Velho TAF, Echeverrigaray S (2008) Generation of recombinants strains to cellulases production by protoplast fusion between Penicillium echinulatum and Trichoderma harzianum. Enzyme Microbial Technol 43 (6):403–409. https://doi.org/10.1016/j.enzmictec.2008.07.009

Donaldson ME, Saville BJ (2012) Natural antisense transcripts in fungi. Mol Microbiol 85 (3):405–417. https://doi.org/10.1111/j.1365-2958.2012

Doua K, Wanga Z, Zhang R, Wanga N, Fana H, Diaoa G, Liua Z (2014) Cloning and characteristic analysis of a novel aspartic protease gene Asp55 from *Trichoderma asperellum* ACCC30536. Microbiol Res 169(12):915–923. https://doi.org/10.1016/j.micres.2014.04.006

Draborg H, Christgau S, Halkier T, Rasmussen G, Dalboge H, Kaupinnen S (1996) Secretion of an enzymatically active *Trichoderma harzianum* endochitinase by *Saccharomyces cerevisiae*. Curr Genet 29(4):404–409. https://doi.org/10.1007/BF02208622

Druzhinina SI, Kubicek CP (2017) Genetic engineering of *Trichoderma reesei* cellulases and their production. Microbial Biotechnol 10(6):1485–1499. https://doi.org/10.1111/1751-7915

Druzhinina IS, Seidl-Seiboth V, Herrera-Estrella A, Horwitz BA, Kenerly CM, Monte E, Mukherjee PK, Zeilinger S, Grigoriev IV, Kubicek CP (2011) *Trichoderma*: the genomics of opportunistic success. Nat Rev Microbiol 9(10):749–759. https://doi.org/10.1038/nrmicro2637

El-Katatny MH, Somitsch W, Robra KH, El-Katatny MS, Gubitz GM (2000) Production of chitinase and β-1,3-glucanase by *Trichoderma harzianum* for control of the phytopathogenic fungus, *Sclerotium rolfsii*. Food Technol Biotechnol 38:173–180

Florindo RN, Souza VP, Mutti HS, Camillo C, Manzine LR, Marana SR, Polikarpov I, Nascimento AS (2018) Structural insights into β-glucosidase transglycosylation based on biochemical, structural and computational analysis of two GH1 enzymes from *Trichoderma harzianum*. New Biotechnol 40:218–227. https://doi.org/10.1016/j.nbt.2017.08.012

Foreman PK, Brown D, Dankmeyer L, Dean R, Diener S, Dunn-Coleman NS, Goedegebuur F, Houfek TD, England GJ, Kelley AS, Meerman HJ, Mitchell T, Mitchinson C, Olivares HA, Teunissen PJ, Yao J, Ward M (2003) Transcriptional regulation of biomass-degrading enzymes in the filamentous fungus *Trichoderma reesei*. J Biol Chem 278(34):31988–31997. https://doi.org/10.1074/jbc.M304750200

Fujii T, Murakami K, Sawayama S (2010) Cellulase Hyperproducing mutants derived from the fungus *Trichoderma reesei* QM9414 produced large amounts of Cellulase at the enzymatic and transcriptional levels. Biosci Biotechnol Biochem 74(2):419–422. https://doi.org/10.1271/bbb.90655

Gacek A, Strauss J (2012) The chromatin code of fungal secondary metabolite gene clusters. Appl Microbiol Biotechnol 95(6):1389–1404. https://doi.org/10.1007/s00253-012-4208-8

Gehring AM, Nodwell JR, Beverley SM, Losick R (2000) Genomewide insertional mutagenesis in *Streptomyces coelicolor* reveals additional genes involved in morphological differentiation. PNAS 97(17):9642–9647. https://doi.org/10.1073/pnas.170059797

Geremia RA, Goldman GH, Jacobs D, Vila SB, Ardiles W, Vanmontagu M, Herrerra-Esterella A (1994) Molecular characterization of proteinase encoding gene, prb1, related to Mycoparasitism by *Trichoderma harzianum*. Mol Microbiol 8:603–613. https://doi.org/10.1111/j.1365-2958.1993.tb01604.x

Glass NL, Schmoll M, Cate JH, Coradetti S (2013) Plant Cell Wall deconstruction by Ascomycete fungi. Ann Rev Microbiol 67(1):477–498. https://doi.org/10.1146/annurev-micro-092611-150044

Goswami RS (2012) Targeted gene replacement in fungi using a Split-marker *Approach*. In: Bolton MD, Thomma BPHJ (eds) Plant fungal pathogens: methods and protocols. Humana Press, Totowa, NJ, pp 255–269

Gruber F, Visser J, Kubicek CP, Graaff LH (1990) The development of a heterologous transformation system for the cellulolytic fungus *Trichoderma reesei* based on a pyrG-negative mutant strain. Curr Genet 18(1):71–76. https://doi.org/10.1007/BF00321118

Guilger M, Pasquoto-Stigliani T, Bilesky-Jose N, Grillo R, Abhilash PC, Fraceto LF, de Lima R (2017) Biogenic silver nanoparticles based on trichoderma harzianum: synthesis, characterization, toxicity evaluation and biological activity. Sci Rep 7:44421. https://doi.org/10.1038/srep44421

Gupta VK, Schmoll M, Herrera-Estrella A, Upadhyay R, Druzhinina I, Tuohy M (2014). Biotechnology and Biology of *Trichoderma*. ISBN: 9780444595942, Elsevier, p. 650

Gupta VK, Steindorff AS, de Paula RG, Silva-Rocha R, Mach-Aigner AR, Mach RL, Silva RN (2016) The post-genomic era of *Trichoderma reesei*: What's next? Trends Biotechnol 34 (12):970–982. https://doi.org/10.1016/j.tibtech.2016.06.003

Gusakov (2011) Alternatives to *Trichoderma reesei* in biofuel production. Trends Biotechnol 29 (9):419–425. https://doi.org/10.1016/j.tibtech.2011.04.004

Hamad B (2010) The antibiotics market. Nat Rev Drug Discov 9(9):675–676. https://doi.org/10.1038/nrd3267

Harkki A, Uusitalo J, Bailey M, Pentill M, Knowles J (1989) A novel fungal expression system: secretion of active calf Chymosin from the filamentous fungus *Trichoderma Reesei*. Nat Biotechnol 7:596–603. https://doi.org/10.1038/nbt0689-596

Heyer WD, Ehmsen KT, Liu J (2010) Regulation of homologous recombination in eukaryotes. Ann Rev Genet 44(1):113–139. https://doi.org/10.1146/annurev-genet-051710-150955

Huang L, Hseu TH, Wey TT (1991) Purification and characterization of an endoxylanase from *Trichoderma koningii* G-39. Biochem J 278:329–333. PMCID:PMC1151344

Jeong S, Yeo S, Yi S (2005) The effect of filler particle size on the antibacterial compounded properties of polymers/silver fibres. J Mater Sci 40:5407–5411. https://doi.org/10.1007/s10853-005-4339-8

Jiang D, Zhu W, Wang Y, Sun C, Zhang KQ, Yang J (2013) Molecular tools for functional genomics in filamentous fungi: recent advances and new strategies. Biotechnol Adv 31 (8):1562–1574. https://doi.org/10.1016/j.biotechadv.2013.08.005

Khan SA, Mulvaney RL, Ellsworth TR, Boast CW (2007) The myth of nitrogen fertilizer for soil organic sequestration. J Environ Qual 36:1821–1832. https://doi.org/10.2134/jeq2007.0099

Kidwai MK, Nehra M (2017) Biotechnological applications of *Trichoderma* species for environmental and food security. In: Gahlawat SK et al (eds) Plant biotechnology*: Recent Advancements and Developments*. Springer, Singapore, pp 125–156

Kiesenhofer DP, Mach RL, Mach-Aigner AR (2017) Influence of cis element arrangement on promoter strenghth in *Trichoderma reesei*. Appl Environ Microbiol 84(1):e01742–e01717. https://doi.org/10.1128/AEM.01742-17

Kim D-J, Bae J-M, Uribe P, Kenerley CM, Cook DR (2002) Cloning and characterization of multiple glycosyl hydrolase genes from *Tricoderma virens*. Curr Genet 40:374–384. https://doi.org/10.1007/s00294-001-0267-6

Kiriga AW, Haukeland S, Kariuki GM, Coyne DL, Beek NV (2018) Effect of *Trichoderma* spp. and *Purpureocillium lilacinum* on *Meloidogyne javanica* in commercial pineapple production in Kenya. Biol Control 119:27–32. https://doi.org/10.1016/j.biocontrol.2018.01.005

Kubicek CP (1987) Involvement of a conidial endoglucanase and a plasma-membrane-bound p-glucosidase in the induction of endoglucanase synthesis by cellulose in *Trichoderma reesei*. J Gen Microbiol 133:1481–1487. https://doi.org/10.1099/00221287-133-6-1481

Kubicek CP et al (2011) Comparative genome sequence analysis underscores mycoparasitism as the ancestral life style of *Trichoderma*. Genome Biol 12(4):R40. https://doi.org/10.1186/gb-2011-12-4-r40

Kubicek CP (2013) Systems biological approaches towards understanding cellulase production by *Trichoderma reesei*. J Biotechnol 163(2):133–142. https://doi.org/10.1016/j.jbiotec.2012.05.020

Kuck U, Hoff B (2010) New tools for the genetic manipulation of filamentous fungi. Appl Microbiol Biotechnol 86(1):51–62. https://doi.org/10.1007/s00253-009-2416-7

Kuhls K, Lieckfeldt E, Samuels GJ, Kovacs W, Meyer W, Pertini O, Gams W, Borner T, Kubicek CP (1996) Molecular evidence that the asexual industrial fungus *Trichoderma reesei* is a clonal derivative of the ascomycete *Hypocrea jecorina*. PNAS 93(15):7755–7760. https://doi.org/10.1073/pnas.93.15.7755

Kurzatkowski W, Torronen A, Filipek J, Mach R, Herzog P, Sovka S, Kubicek CP (1996) Glucose induced secretion of *Trichoderma reesei* xylanases. Appl Environ Microbiol 62:2859–2865. PMCID:PMC168071

Landowski CP et al (2015) Enabling low cost biopharmaceuticals: a systematic approach to delete proteases from a well-known protein production host *Trichoderma reesei*. PLoS One 10(8): e0134723. https://doi.org/10.1371/journal.pone.0134723.eCollection2015

Le Crom S et al (2009) Tracking the roots of cellulase hyperproduction by the fungus *Trichoderma reesei* using massively parallel DNA sequencing. PNAS 106(38):16151–16156. https://doi.org/10.1073/pnas.0905848106

Li C, Lin F, Li Y, Wei W, Wang H, Qin L, Zhou Z, Li B, Wu F, Chen Z (2016) A β-glucosidase hyper-production *Trichoderma reesei* mutant reveals a potential role of cel3D in cellulase production. Micro Cell Fact 15(1):151. https://doi.org/10.1186/s12934-016-0550-3

Lora JM, De la Cruz J, Llobell A, Benitez T, Pintor-Tora JA (1995) Molecular characterization and heterologous expression of an endo-beta-1,6-glucanase gene from the mycoparasitic fungus *Trichoderma harzianum*. Mol Gen Genet 247:639–645. https://doi.org/10.1007/BF00290356

Lorito M, Hayes CK, Di Pietro A, Woo SL, Harman GE (1994) Purification of chitinolytic and glucanolytic enzymes from *Trichoderma barxianurn* and their synergistic activity against *Botrytis cinerea*. Phytopathology 84

Lorito M, Woo SL, Harman GE, Monte E (2010) Translational research on *Trichoderma*: from 'Omics to the field. Annu Rev Phytopathol 48(1):395–417. https://doi.org/10.1146/annurev-phyto-073009-114314

Lorito M, Hayes CK, Di Pietro A, Harman GE (1993) Biolistic transformation of *Trichoderma harzianum* and *Gliocladium virens* using plasmid and genomic DNA. Curr Genet 24 (4):349–356. https://doi.org/10.1007/BF00336788

Linke R, Thallinger GG, Haarmann T, Eidner J, Schreiter M, Patrick L, Seiboth B, Kubicek CP (2015) Restoration of female fertility in *Trichoderma reesei* QM6a provides the basis for inbreeding in this industrial cellulase producing fungus. Biotechnol Biofuels 8(1):155. https://doi.org/10.1186/s13068-015-0311-2

Liu R, Chen L, Jiang Y, Zhou Z, Zou G (2015) Efficient genome editing in filamentous fungus *Trichoderma reesei* using the CRISPR/Cas9 system. Cell Discov 1:15007. https://doi.org/10.1038/celldisc.2015.7

Liu Q, Dong Y, Wang F, Jiang B, Wang M, Fang X (2016) Regulation of cellulase expression, sporulation, and morphogenesis by velvet family proteins in *Trichoderma reesei*. Appl Microbiol Biotechnol 100(2):769–779. https://doi.org/10.1007/s00253-015-7059-2

Liu Q, Dong Y, Wang F, Jiang B, Wang M, Fang X (2017a) Solution for promoting egl3 gene of *Trichoderma reesei high-efficiency secretory expression in Escherichia coli and Lactococcus lactis*. Process Biochem 62:135–143. https://www.cheric.org/research/tech/periodicals/doi.php?art_seq=1607235

Liu P, Wang W, Wei D (2017b) Use of transcription activator-like effector for efficient gene modification and transcription in the filamentous fungus *Trichoderma reesei*. J Ind Microbiol Biotechnol 44(9):1367–1373. https://doi.org/10.1007/s10295-017-1963-7

Mach L, Mort JS, Glossl I (1994) Noncovalent complexes between the lysosomal proteinase cathepsin B and its propeptide account for stable, extracellular, high molecular mass forms of the enzyme. J Biol Chem 269:13030–13035

Mandels M, Weber J, Parizek R (1971) Enhanced Cellulase production by a mutant of *Trichoderma viride*. J Appl Microbiol 21(1):152–154. PMCID:PMC377137

Margolles-Clark E, Hayes CK, Harman G, Pentilla M (1996) Improved production of *Trichoderma harzianum* endochitinase by expression in *Trichoderma reesei*. Appl Environ Microbiol 62 (6):2145–2151

Martinez D, Berka RM, Brettin TS (2008) Genome sequencing and analysis of the biomass-degrading fungus *Trichoderma reesei* (syn. *Hypocrea jecorina*). Nature Biotechnol 26:553–560. https://doi.org/10.1038/nbt1403

Mello-de-Sousa TM, Gorsche R, Rassinger A, Pocas-Fonseca M-J, Mach RL, Mach-Aigner AR (2014) A truncated form of the carbon catabolite repressor 1 increases cellulase production in *Trichoderma reesei*. Biotechnol Biofuels 7(1):129. https://doi.org/10.1186/s13068-014-0129-3

Mendoza-Mendoza A, Zaid R, Lawry R, Hermosa R, Monte E, Horwitz BA, Mukherjee PK (2018) Molecular dialogues between Trichoderma and roots: role of the fungal secretome. Fungal Biol Rev 32(2):62–85. https://doi.org/10.1016/j.fbr.2017.12.001

Menon S, Agarwal H, Kumar VS, Rajeshkumar S (2019) Biomemetic synthesis of selenium nanoparticles and its biomedical applications. Green Synthesis, Characterization and Applica tions of Nanoparticles. Micro and Nano Technologies:165–197. https://doi.org/10.1016/B978-0-08-102579-6.00008-3

Meyer V (2008) Genetic engineering of filamentous fungi - Progress, obstacles and future trends. Biotechnol Adv 26(2):177–185. https://doi.org/10.1016/j.biotechadv.2007.12.001

Migheli Q, Gonzalez-Candelas L, Dealessi L, Camponogara A, Ramon-Vidal D (1998) Transformants of *Trichoderma longibrachiatum* overexpressing the beta-1,4-endoglucanase gene egl1 show enhanced biocontrol of *Pythium ultimum* on cucumber. Phytopathology 88:673–677. https://doi.org/10.1094/PHYTO.1998.88.7.673

Monte (2001) Understanding *Trichoderma*: between biotechnology and microbial ecology. Interl Microbiol 4(1):1–4. https://doi.org/10.1007/s101230100001

Mukherjee PK, Horwitz B, Herrera-Estrella A, Schmoll M, Kenerley C (2013) *Trichoderma* research in the genome era. Annu Rev Phytopathol 51(1):105–129. https://doi.org/10.1146/annurev-phyto-082712-102353

Nakazawa H, Kawai T, Ida N, Shida Y, Kobayashi Y, Okada H, Tani S, Sumitani J, Kawaguchi T, Morikawa Y, Ogasawara W (2012) Construction of a recombinant *Trichoderma reesei* strain expressing *Aspergillus* aculeatus β-glucosidase I for efficient biomass conversion. Biotechnol Bioeng 109:92–99. https://doi.org/10.1002/bit.23296

Nakazawa H, Okada K, Kobayashi R, Kubota T, Onodera T, Ochiai N, Omata N, Ogaswara W, Okada H, Morikawa Y (2008) Characterization of the catalytic domains of *Trichoderma reesei* endoglucanase I, II, and III, expressed in *Escherichia coli*. Appl Microb Biotechnol 81 (4):681–689. https://doi.org/10.1007/s00253-008-1667-z

Nakari-Setala T, Paloheimo M, Kallio J, Vehmannpera J, Pentilla M, Saloheimo M (2009) Genetic modification of carbon Catabolite repression in *Trichoderma reesei* for improved protein production. J Appl Environ Microbiol 75(14):4853–4860. https://doi.org/10.1128/AEM.00282-09

Nogawa M, Goto M, Okada H, Morikawa Y (2001) L-Sorbose induces cellulase gene transcription in the cellulolytic fungus *Trichoderma reesei*. Curr Genet 38(6):329–334. PMID:11270575

Papavizas GC (1985) *Trichoderma* and Gliocladium: biology, ecology, and potential for biocontrol. Annu Rev Phytopathol 23(1):23–54. https://doi.org/10.1146/annurev.py.23.090185.000323

Penttila M, Nevalainen H, Ratto M, Salminen E, Knowles J (1987) A versatile transformation system for the cellulolytic filamentous fungus *Trichoderma reesei*. Gene 61(2):155–164. https://doi.org/10.1016/0378-1119(87)90110-7

Persoon CH (1794) Disposita methodica fungorum. Romer's Neues Mag Bot 1:81–128

Peterson R, Nevalainen H (2012) *Trichoderma reesei* RUT-C30-thirty years of strain improvement. Microbiol 158(1):58–68. https://doi.org/10.1099/mic.0.054031-0

Punt PJ, Van Biezen N, Conesa A, Albers A, Mangnus J, Hondel VC (2002) Filamentous fungi as cell factories for heterologous protein production. Trends Biotechnol 20(5):200–206. https://doi.org/10.1016/S0167-7799(02)01933-9

Omran BA, Nassar HN, Younis SA, Fatthallah NA, Hamdy A, El-Shatoury EH, El-Gendy N s (2018) Physiochemical properties of *Trichoderma longibrachiatum* DSMZ 16517-synthesized silver nanoparticles for the mitigation of halotolerant sulphate-reducing bacteria. J Appl Microbiol 126(1):138–154. https://doi.org/10.1111/jam.14102

Okada H, Sekiya T, Yokoyama K, Tohda H, Kumagai H, Morikawa Y (1998) Efficient secretion of *Trichoderma reesei* cellobiohydrolase II in Schizosaccharomyces pombe and characterization of its products. Appl Microbiol Biotechnol 49(3):301–308. https://doi.org/10.1007/s002530051173

Ojaghian S, Wang L, Xie G-L, Zhang J-Z (2018) Increased resistance against storage rot in transgenic carrots expressing chitinase chit42 from *Trichoderma harzianum*. Sci Hortic 234:81–86. https://doi.org/10.1016/j.scienta.2018.02.025

Ouedraogo JP, Arenthorst M, Nikolaev I, Barends S, Ram AF (2016) I-SceI enzyme mediated integration (SEMI) for fast and efficient gene targeting in *Trichoderma reesei*. J Biotechnol 222:25–28. https://doi.org/10.1016/j.jbiotec.2016.02.012

Ponting CP, Oliver PL, Reik W (2009) Evolution and functions of long noncoding RNAs. Cell 136 (4):629–641. https://doi.org/10.1016/j.cell.2009.02.006

Ramot O, Cohen-Kupiec R, Chet I (2000) Regulation of β-1,3- glucanase by carbon starvation in the mycoparasite *T. harzianum*. Mycol Res 104:415–420. https://doi.org/10.1017/S0953756299001471

Ries L, Pullan ST, Delmas S, Malla S, Blythe MJ, Archer DB (2013) Genome-wide transcriptional response of *Trichoderma reesei* to lignocellulose using RNA sequencing and comparison with *Aspergillus niger*. BMC Genomics 14(1):541. https://doi.org/10.1186/1471-2164-14-541

Ries L, Belshaw N, Ilmen M, Pentilla ME, Archer DB (2014) The role of CRE1 in nucleosome positioning within the cbh1 promoter and coding regions of *Trichoderma reesei*. Appl Microbiol Biotechnol 98(2):749–762. https://doi.org/10.1007/s00253-013-5354-3

Saarelainen R, Paloheimo M, Fagerstrom R, Suominen PL, Nevalainen KMH (1993) Cloning, sequencing and enhanced expression of the Trichoderma reesei endoxylanase II (pI 9) gene xln2. Mol Gen Genet 241(5–6):497–503. https://doi.org/10.1007/bf0027989

Saloheimo M, Lehtovaara P, Penttila M, Teeri TT, Stahlberg J, Johansson G, Pettersson G, Claeyssens M, Tomme P, Knowles JKC (1988) EG III, a new endoglucanase from *T. reesei*: the characterization of both gene and enzyme. Gene 63:11–21. https://doi.org/10.1016/0378-1119(88)90541-0

Saloheimo A, Henrissat B, Hoffren AM, Teleman O, Pentilla M (1994) A novel, small endoglucanase gene *egl5* from *Trichoderma reesei* isolated by expression in yeast. Mol Microbiol 13:219–228. https://doi.org/10.1111/j.1365-2958.1994.tb00417.x

Saloheimo M, Nakari-Setala T, Tenkanen M, Pentilla M (1997) cDNA cloning of a *Trichoderma reesei* Cellulase and demonstration of Endoglucanase activity by expression in yeast. Eur J Biochem 249(2):584–591. https://doi.org/10.1111/j.1432-1033.1997.00584.x

Saloheimo M, Paloheimo M, Hakola S, Pere J, Swanson B, Nyyssonen E, Bhatia A, Ward M, Penttila M (2002) Swollenin, a *Trichoderma reesei* protein with sequence similarity to the plant expansins, exhibits disruption activity on cellulosic materials. Eur J Biochem 269:4202–4211. https://doi.org/10.1046/j.1432-1033.2002.03095.x

Samuels GJ (2006) *Trichoderma*: systematics, the sexual state, and ecology. Phytopathology 96 (2):195–206. https://doi.org/10.1094/PHYTO-96-0195

Saravanakumar K, Shanmuga Arasu V, Kathiresan K (2013) Effect of *Trichoderma* on soil phosphate solubilization and growth improvement of *Avicennia marina*. Aquat Bot 104:101–105. https://doi.org/10.1016/j.aquabot.2012.09.001

Saravanakumar K, Kathiresan K (2015) Bioremoval of lead and iron from sewage water by mangrove-derived *Hypocrea lixii*. Int Environ Sci Technol 12(10):3341–3350. https://doi.org/10.1007/s13762-014-0703-z

Saravanakumar K, Vivek R, Boopathy S, Yaqian N, Kathiresan LK, Chen J (2015) Anticancer potential of bioactive 16-methylheptadecanoic acid methyl ester derived from marine *Trichoderma*. J Appl Biomed 13(3):199–212. https://doi.org/10.1016/j.jab.2015.04.001

Saravanakumar K, Kathiresan K (2014) Bioconversion of lignocellulosic waste to bioethanol by *Trichoderma* and yeast fermentation. Biotech 4(5):493–499. https://doi.org/10.1007/s13205-013-0179-4

Saravanakumar K, Wang MH (2018) *Trichoderma* based synthesis of anti-pathogenic silver nanoparticles and their characterization, antioxidant and cytotoxicity properties. Microb Pathog 114:269–273. https://doi.org/10.1016/j.micpath.2017.12.005

Schmoll M, Kubicek CP (2003) Regulation of *Trichoderma cellulase* formation: lessons in molecular biology from an industrial fungus. Acta Microbiol Imm H 50(2–3):125–145. https://doi.org/10.1556/AMicr.50.2003.2-3.3

Schuster A, Schmoll M (2010) Biology and biotechnology of *Trichoderma*. Appl Microbiol Biotechnol 87(3):787–799

Schuster A, Bruno KS, Collett JR, Baker SE, Seiboth B, Kubicek CP, Schmoll M (2012) A versatile toolkit for high throughput functional genomics with *Trichoderma reesei*. Biotechnol Biofuels 5 (1):1. https://doi.org/10.1186/1754-6834-5-1

Seiboth B, Karimi RA, Phatale PA, Linke R, Hartl L, Sauer DG, Smith KM, Baker SE, Freitag M, Kubicek CP (2012) The putative protein methyltransferase LAE1 controls cellulase gene expression in *Trichoderma reesei*. Mol Microbiol 84(6):110–1164. https://doi.org/10.1111/j.1365-2958.2012.08083.x

Seidl V, Seibel C, Kubicek CP, Schmoll M (2009) Sexual development in the industrial workhorse *Trichoderma reesei*. PNAS 106(33):13909–13914. https://doi.org/10.1073/pnas.0904936106

Seidl V, Seiboth B (2010) *Trichoderma reesei*: genetic approaches to improving strain efficiency. Biofuels 1(2):343–354. https://doi.org/10.4155/bfs.10.1

Sharma V, Salwan R, Sharma PN, Gulati A (2018) Integrated translatome and proteome: approach for accurate portraying of widespread multifunctional aspects of *Trichoderma*. Front Microbiol 8:1602. ISSN: 1664-302X. https://doi.org/10.3389/fmicb.2017.01602

Shoemaker S, Watt K, Tsitovsky G, Cox R (1983) Characterization and properties of cellulases purified from *Trichoderma Reesei* strain L27. Nat Biotechnol 1:687–690. https://doi.org/10.1038/nbt1083-687

Souza MFD, Silva ASAD, Bon EAS (2018) A novel Trichoderma harzianum strain from the Amazon Forest with high cellulolytic capacity. Biocatal Agric Biotechnol 14:183–188. https://doi.org/10.1016/j.bcab.2018.03.008

Srivastava N, Srivastava M, Mishra PK, Gupta VK, Molina G, Rodriguez-Couto S, Ambepu M, Ramteke PW (2018) Applications of fungal cellulases in biofuel production: advances and limitations. J Renew Sustain Energy 82:2379–2386. https://doi.org/10.1016/j.rser.2017.08.074

Stangl H, Gruber F, Kubicek CP (1993) Characterization of the *Trichoderma reesei* cbh2 promoter. Curr Genet 23(2):115–122. https://doi.org/10.1007/BF00352009

Stalbrand H, Saloheimo A, Vehmaanpera J, Henrissat B, Penttila M (1995) Cloning and expression in *Saccharomyces cerervisiae* of a *Trichoderma reesei* B-mannanase gene containing a cellulose binding domain. Appl Environ Microbiol 61:1090–1097

Stefano C, Galletti S, Burzi PL, Cerato C (2010) A novel microplate-based screening strategy to assess the cellulolytic potential of *Trichoderma* strains. Biotechnol Bioeng 107(3):461–468. https://doi.org/10.1002/bit.22816

Stoddard BL (2014) Homing endonucleases from mobile group I introns: discovery to genome engineering. Mob DNA 5:7. https://doi.org/10.1186/1759-8753-5-7

Takashima S, Nakamura A, Hidaka M, Uozumi HMT (1999) Molecular cloning and expression of the novel fungal β-Glucosidase genes from *Humicola grisea* and *Trichoderma reesei*. J Biochem 125(4):728–736

Tamaru H, Selker EU (2003) Synthesis of signals for De novo DNA methylation in *Neurospora crassa*. Mol Cell Biol 23(7):2379–2394. https://doi.org/10.1128/mcb.23.7.2379-2394.2003

Teeri TT, Kumar V, Lehtovaara P, Knowles JKC (1987) Construction of cDNA libraries by blunt end ligation: high frequency cloning of long cDNAs from filamentous fungi. Anal Biochem 164:60–67. https://doi.org/10.1016/0003-2697(87)90367-8

Throndset W, Kim S, Bower B, Lantz S, Kelemen B, Pepsin M, Chow N, Mitchinson C, Ward M (2010) Flow cytometric sorting of the filamentous fungus *Trichoderma reesei* for improved strains. Enzym Microb Technol 47(7):335–341. https://doi.org/10.1016/j.enzmictec.2010.09.003

Till P, Pucher ME, Mach RL, Mach-Aigner AR (2018) A long noncoding RNA promotes cellulase expression in *Trichoderma reesei*. Biotechnol Biofuels 11(1):78. https://doi.org/10.1186/s13068-018-1081-4

Tisch D, Kubicek CP, Schmoll M (2011) New insights into the mechanism of light modulated signaling by heterotrimeric G-proteins: ENVOY acts on gna1 and gna3 and adjusts cAMP levels in *Trichoderma reesei (Hypocrea jecorina)*. Fungal Genet Biol 48(6):631–640. https://doi.org/10.1016/j.fgb.2010.12.009

Tisch D et al (2017) Omics analyses of *Trichoderma reesei* CBS999.97 and QM6a indicate the relevance of female fertility to carbohydrate-active enzyme and transporter levels. Appl Environ Microbiol 83(22)

Tolan JS, Foody B (1999) Cellulase from submerged fermentation. In: Tsao GT et al (eds) Recent Progress in bioconversion of Lignocellulosics. Advances in biochemical engineering/biotechnology, vol 65. Springer, Berlin, Heidelberg. https://doi.org/10.1007/3-540-49194-5_3

Torronen A, Mach RL, Messner R, Gonzalez R, Kalkkinen N, Harkki A, Kubicek CP (1992) The two major xylanases from *Trichoderma reesei*: characterization of both enzymes and genes. Nat Biotechnol 10:1461–1465. https://doi.org/10.1038/nbt1192-1461

Qian Y, Zhong L, Hou Y, Qu Y, Zhong Y (2016) Characterization and strain improvement of a Hypercellulytic variant, *Trichoderma reesei* SN1, by genetic engineering for optimized Cellulase production in biomass conversion improvement. Front Microbiol 7:1349. https://doi.org/10.3389/fmicb.2016.01349

Qin LN, Cai FR, Dong XR, Huang ZB, Tao Y, Huang JZ, Dong ZY (2012) Improved production of heterologous lipase in *Trichoderma reesei* by RNAi mediated gene silencing of an endogenic highly expressed gene. Bioresour Technol 109:116–122. https://doi.org/10.1016/j.biortech.2012.01.013

Xia J, He A, Li R, Zhang Y, Xu J, Liu X, Xu J (2017) Enzymatic activity and protein expression of cellulase from rice straw produced by *Trichoderma reesei* in the presence of oxygen vectors. Ind Crop Prod 109:654–660. https://doi.org/10.1016/j.indcrop.2017.09.017

Wang F, Wu J, Chen S (2018a) Preparation of gentiooligosaccharides using *Trichoderma viride β-glucosidase*. Food Chem 248:340–345. https://doi.org/10.1016/j.foodchem.2017

Wang L, Zheng F, Zhang W, Zhong Y, Chen G, Meng X, Liu W (2018b) A copper-controlled RNA interference system for reversible silencing of target genes in *Trichoderma reesei*. Biotechnol Biofuels 11:33. https://doi.org/10.1186/s13068-018-1038-7

Wey TT, Hseu TH, Huang L (1994) Molecular cloning and sequence analysis of the cellobiohydrolase I gene from *Trichoderma koningii* G-39. Curr Microbiol 28:31–39. https://doi.org/10.1007/BF01575983

Wood AJ, Lo TW, Zeitler B, Pickle CS, Ralston EJ, Lee AH, Amora R, Miller JC, Leung E, Meng X, Zhang L, Rebar EJ, Gregory PD, Urnov FD, Meyer BJ (2011) Targeted genome editing across species using ZFNs and TALENs. Sci 333(6040):307. https://doi.org/10.1126/science.1207773

Wu Y, Sun X, Xue X, Luo H, Yao B, Xie X, Su X (2017) Overexpressing key component genes of the secretion pathway for enhanced secretion of an *Aspergillus niger* glucose oxidase in *Trichoderma reesei*. Enzyme Microb Technol 106:83–87. https://doi.org/10.1016/j.enzmictec.2017.07.007

Zaidi NW, Singh D, Kumar S, Sangle U, Nityanand Singh R, Sachitanand Prasad R, Singh S, Singh S, Yadav AK, Singh A, Waza SA, Singh US (2017) *Trichoderma harzianum* improves the performance of stress-tolerant rice varieties in rainfed ecologies of Bihar, India. Fields Crop Res 220:97–104. https://doi.org/10.1016/j.fcr.2017.05.003

Zeilinger S, Schmoll M, Pail M, Mach RL (2003) Nucleosome transactions on the Hypocrea jecorina (Trichoderma reesei) cellulase promoter cbh2 associated with cellulase induction. Mol Gen Genomics 270(1):46–55. https://doi.org/10.1007/s00438-003-0895-2

Zhao J, Zeng S, Xia Y, Xia L (2018) Expression of a thermotolerant laccase from Pycnoporus sanguineus in *Trichoderma reesei* and its application in the degradation of bisphenol a. J Biosci Bioeng 125(4):371–376. https://doi.org/10.1016/j.jbiosc.2017.11.010

Zhang G, Zhu Y, Wei D, Wang W (2014) Enhanced production of heterologous proteins by the filamentous fungus *Trichoderma reesei* via disruption of the alkaline serine protease SPW combined with a pH control strategy. Plasmid 71:16–22. https://doi.org/10.1016/j.plasmid.2014.01.001

Zhang F, Zhao X, Bai F (2018) Improvement of cellulase production in *Trichoderma reesei* rut-C30 by overexpression of a novel regulatory gene Trvib-1. Bioresour Technol 247:676–683. https://doi.org/10.1016/j.biortech.2017.09.126

Zheng YM, Lin FL, Gao H, Zou G, Zhang JW, Wang GQ, Chen GD, Zhou ZH, Yao XS, Hu D (2017) Development of a versatile and conventional technique for gene disruption in filamentous fungi based on CRISPR-Cas9 technology. Sci Rep 7(1):9250. https://doi.org/10.1038/s41598-017-10052-3

Zhong Y, Yu H, Wang X, Lu Y, Wang T (2011) Towards a novel efficient T-DNA-based mutagenesis and screening system using green fluorescent protein as a vital reporter in the industrially important fungus *Trichoderma reesei*. Mol Biol Rep 38(6):4145–4151. https://doi.org/10.1007/s11033-010-0534-z

Chapter 16
Trichoderma spp.: Expanding Potential beyond Agriculture

Ratul Moni Ram, Anukool Vaishnav, and Harikesh Bahadur Singh

Abstract *Trichoderma* is a genetically diverse group of fungi present in different ecological niches with multiple capabilities. Most of the *Trichoderma* spp. are reported as plant growth promoters and efficient biocontrol agents against various biotic and biotic stresses. Besides that genus *Trichoderma* is also utilized for bioremediation of heavy metal contamination, pesticide residue degradation, and industrial purposes for food, beverages, nanoparticles, and pharmaceuticals. These fungal species produce a vast variety of extracellular enzymes including cellulase, which play a key role in the degradation of complex polysaccharides and other organic compounds. The application of these enzymes into industries has been an economically and environmentally sustainable approach for producing high-quality products. As *Trichoderma* genomic sequences are now available in the public domain, it can be explored to search its wider applicability in the scientific arena. This chapter presents an overview of the application of *Trichoderma* beyond the agriculture areas like food industries, pharmaceuticals, beverages, bioremediation, and nanotechnology.

Keywords Bioremediation · Heavy metals · Enzymes · Nanotechnology · *Trichoderma*

R. M. Ram
Department of Plant Pathology, A.N.D.U.A & T, Ayodhya, India

A. Vaishnav
Department of Biotechnology, GLA University, Mathura, Uttar Pradesh, India

H. B. Singh (✉)
Department of Biotechnology, GLA University, Mathura, Uttar Pradesh, India

Somvanshi Research Foundation, Lucknow, India

C. Manoharachary et al. (eds.), Trichoderma: *Agricultural Applications and Beyond*, Soil Biology 61, https://doi.org/10.1007/978-3-030-54758-5_16

16.1 Introduction

Trichoderma species are cosmopolitan in nature and are dominant members of different ecosystems in diverse geographical locations (Kubicek et al. 2008). The presence of *Trichoderma* spp. in nature is governed by various factors including substrate availability, climatic conditions along with intricate ecological interactions (Hoyos-Carvajal and Bissett 2011). The cosmopolitan nature of these fungi is a result of their high reproducibility, metabolic diversity, and competitive capabilities of its different strains in nature (Cardoso Lopes et al. 2012).

The genus *Trichoderma* is a member of phylum Ascomycetes, class Sordariomycetes, order Hypocreales, and family Hypocreaceae. The history of these fungi rewinds to 1794 as Persoon in the same year coined the term "*Trichoderma*." In 1865, famous French scientists Tulasne brothers assigned *Hypocrea rufa* as the teleomorph of *Trichoderma viride* Pers. (Tulasne and Tulasne 1865). Their illustrations were remarked as guidelines of identification for this holomorph in that era. Prior to 1969, only a few species were included in genus *Trichoderma*. Later on, with the passage of time, a few genera comprising many species were merged with *Trichoderma*, thereby expanding the genus dimension. The primary concept behind addition was on the basis of the production of hairs and colorless conidia or/and species producing green conidia. The introduction of advanced tools especially the sequence analysis of internal transcribed spacer (ITS) region, restriction fragment length polymorphisms, random amplified polymorphic DNA, and chromosome have come into light which brought a revolution in morphology-based taxonomy of *Trichoderma*. Kindermann et al. (1998) was the pioneer in molecular identification of species by analyzing the internal transcribed region 1 (ITS1) sequence of the rRNA coding region. This technique has brought a revolution for the identification of a particular strain of any species. The development of newer molecular methods, including the fungal oligonucleotide barcode for identification of *Trichoderma* species and its holomorph *Hypocrea*, i.e., *TrichO*Key version 1.0 (Druzhinina et al. 2006), has played a potent role in the description of newly evolved species.

Trichoderma spp. are omnipresent in nature. They reside freely in soil as saprophytes owing to their ability to decompose organic matter. They thrive to grow on other fungi, and in turn, colonize the rhizosphere zone. In addition to the rhizospheric region, many species are reported to be phyllospheric in nature. These fungi produce various secondary metabolites in abundance, which have multifarious applications (Sivasithamparam and Ghisalbarti 1998; Singh et al. 2019a, b). These fungi are very popular in agriculture as biocontrol agents against the most notorious plant pathogens (Singh et al. 2011). Few species are regarded as excellent antagonists and plant growth promoters (Harman et al. 2004). The antagonistic potential of these fungi can be attributed to their secondary anti-microbial metabolites, high metabolic rates, and physiological conformation (Mukhopadhyay et al. 1992; Singh et al. 2019a, b). Antibiosis, mycoparasitism, competition for nutrients, and induction of plant defense system are their prime mechanisms to confront antagonism (Singh et al. 2011; Ram and Singh 2017).

Apart from their tremendous potential in the agricultural sector, *Trichoderma* spp. has wide applications in industries and pharmaceuticals. They produce various hydrolytic enzymes rendering them applicable in industry (Mach and Zeilinger 2003). In the industrial sector, they are widely used in the production of commercial enzymes namely lipase, proteases, cellulases, hemicellulases, and ß-1,3- glucanase (Singh 2014; Keswani et al. 2014). A list of *Trichoderma* producing enzymes and their uses in different industrial purposes are summarized in Table 16.1. Various kinds of drugs and cosmetic products are being manufactured using different species of *Trichoderma.* All these properties have rendered *Trichoderma* spp. a superstar in agricultural as well as industrial sector (Fig. 16.1).

However, in contrast to their enormous beneficial effects, the genus also comprises several species posing threat to humans, flora, and fauna (Kraus et al. 2004; Park et al. 2006; Kredics et al. 2006). These species have emerged as a potent human pathogen along with pathogens of insects and invertebrates viz. *T. longibrachiatum, T. brevicompactum, T .atroviride, T. citrinoviride*, etc. (Kuhls et al. 1999; Kredics et al. 2003; Druzhinina et al. 2011). Apart from causing human ailments, they are also the prime agents of green mold disease of mushrooms. *T. aggressivum, T. pleutorum,* and *T. pleuroticola* are the main species causing green mold disease (Samuels et al. 2002; Hatvani et al. 2007).

16.2 Application in Beverage Industry

A number of *Trichoderma* spp. has been widely used as a source of enzymes or secondary metabolites having ample applications in food/beverage industrial processes. *Trichoderma* spp. possess tremendous potential in wine and brewing industries, can be used as a direct source of enzymatic blends, or can be combined with fungal genes applied for the transformation of industrial yeast strains and barley cultivars. In beverages viz. beer and wine industries, flavor, texture, and aroma are prime quality parameters for producers and consumers (Styger et al. 2011). Many *Trichoderma* metabolites are applied to enhance desirable properties and to enhance the commercial value of the product. *Trichoderma* produces a great number of extracellular enzymes, many of which have a pivotal role in biotechnology. *T. reesei, T. harzianum, T. viride, T. atroviride, T. virens, T. lignorum,* and *T. longibrachiatum* are the best known (Lorito et al. 2010).

Recombinant yeast strains; produced from genes encoding endo-β-1,4-glucanases and xylanases, which is used to prepare free-flow wine for Ruby Cabernet, have multiple colors, concentration and constancy in 6-month aging Pinot Noir, Ruby Cabernet and scotch (Perez-Gonzalez et al. 1993). Moreover, these cell wall degrading enzymes are also used in the wine making process to enhance juice yield, taste, filterability, and clarification. In addition, it also facilitates the release

Table 16.1 List of different enzymes, their mode of action, and applications produced by *Trichoderma* spp.

Enzyme group	Enzymes	Mode of action	Application	Reference
Cellulases	Cellobiohydrolose	Breakdown the cellulose to cellobiose from the free chain end	Mainly applied in food and detergent industry	Chakraborty et al. (2019)
	Endo glucanase	Digest the amorphous regions of cellulose		
	β-D-glucosidase	Degrade small soluble oligosaccharides and cellobiose to glucose.		
Mannases	Acetyl mannan esterases	Hydrolyzes mannan yielding mannotriose and mannobiose	Employed inwashing powders for removal of food-base stains	Nevalainen (2017); Arisan-Atac et al. (1993)
	β-Mannosidases,			
	β-Glucosidases,			
	Endo-1,4-β-mannanases			
Pectinases	Polygalacturonase	Breaks the glycosidic bonds of the long chains of galacturonic acid residues in pectic substances	Macerating enzymes in fruit juice production, fruit juice clarification, wine production, and treatment of softwoods.	Rebello et al. (2019)
	Pectin methylesterase			
	Pectate lyase			
	Pectin lyase			
A-L-Arabinofaranosidase	A-Galactosides	Catalyzes cleavage of terminal α-galactose residues from α-*O* galactosides	Modification ofwood-derived materials, digestion of guar gum. Also use inmedicine	Florencio et al. (2016)
Polyphenol oxidases containing copper containing enzyme	Laccases	Catalyze oxidation of various compounds viz. carbohydrates,unsaturated fatty acids, phenolic, based compounds and thiol-containingproteins	Used in food industry for the production of cost-effective and healthy foods. Also used in biosensor	Patel et al. (2019)
Mutanase	a-1,3-glucan 3 glucanohydrolases	Degradation of mutan to glucose	Used in toothpasteto prevent accumulation of the polysaccharide mutan in dental plaque	Wiater et al. (2012)

Xylanases	Endo-1,4-β-xylanases	Hydrolysis of xylan to xylose	Feed additive in human and livestockfarming. Also used in paper-making and biofuel production process	Marques et al. (2018)
	β-Xylosidase			
	Acetylxylan esterases	Deacetylation of partiallyacetylated 4-*O*-methyl-D-glucuronoxylan		

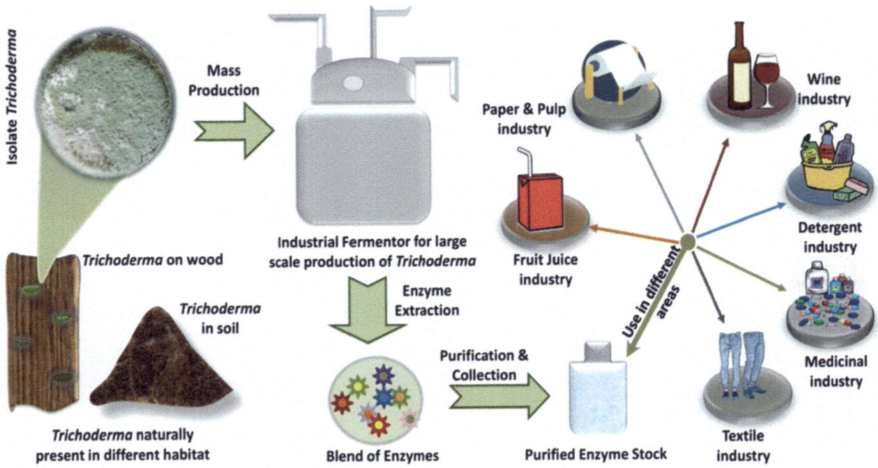

Fig. 16.1 A schematic representation for the application of *Trichoderma* producing enzymes in different industrial sectors

and solubilization of phenolic compounds extracted from seeds, skin, and flesh of grape berry.

β-glucan content in the final end product is a major hindrance for beer manufacturers in the beer industry. It causes adverse effects on appearance, causing considerable qualitative loss to producers. The genes encoding fungal thermotolerant cellulolytic enzymes are regarded as key factors to attain barley varieties having lower β-glucan content. The genes encoding cellulolytic enzymes are usually responsible for such activities (Nuutila et al. 1999). In an experiment, egl1 gene from *T. reesei* was used to develop "Golden Promise" and "Kymppi" varieties, which exhibited an increased amount of EGI content in their seeds and reduced content of soluble β-glucan.

Apart from the beneficial aspects, *Trichoderma* spp. has also been found to be responsible for unpleasant properties of wine, such as cork taint and musty off-odors (Coque et al. 2003). Molds are considered to be the prime microbes liable for cork taint, with the fungal microbes on cork including *Trichoderma* spp., *Rhizoctonia* spp., *Mucor* spp., *Cladosporium* spp., *Fusarium* spp., *Aspergillus* spp., *Penicillium* spp., *Verticillium* spp., *Monilia* spp., *Acremonium* spp., *Paecilomyces* spp., *Chrysonilia* spp., and *Mortierella* spp. (Prak et al. 2007). Among these, *T. longibrachiatum* and *T. viride* altogether with *Fusarium* spp. are most efficient in producing 2,4,6-trichloroanisole; the principal component involved in cork taint (Alvarez-Rodriguez et al. 2002; Coque et al. 2003).

16.3 Applications in Food, Flavor and Aroma Industry

Apart from playing a significant role in agriculture, the potential of *Trichoderma* secondary metabolites was also exhibited in food, flavor, and confectionaries (Blumenthal 2004; Nevalainen et al. 1994). Approximately 20 known species of *Trichoderma* are known to produce enzymes. The various enzymes produced by *Trichoderma* spp. are used for enhancing viz., brewing process (β-glucanases), in juice production (cellulases, hemicellulases, pectinases,) or as feed additives to livestock (xylanases) and pet food. Cell wall-degrading enzymes produced by *T. harzianum* acts as food preservatives due to their antifungal nature (Fuglsang et al. 1995), but they have not gained widespread application. Cellulases are usually involved in baking, malting, and fermentation (Galante et al. 1998; Schuster and Schmoll 2010). Similarly, mutanase is admixed in toothpaste in order to prevent the accumulation of the polysaccharide mutan in dental plaque (Wiater et al. 2005).

In an experiment conducted by Bayitse et al. (2015), fermentation of cassava residues with *T. pseudokoningii* for 12 days leads to enhanced protein content by 48.2% when urea act as a source of nutrient and the substrate moisture was adjusted to 70% w/v. It facilitated the feed manufacturers for better utilization of plant proteins for the production of economical products in animal husbandry.

However, in addition to enzymes, the secondary metabolites of *Trichoderma* spp. also contribute as food additives. 6-pentyl-2H-pyran-2-one, a volatile compound imparts coconut aroma in viable *Trichoderma* cultures and used as a flavoring agent in confectionaries (Claydon et al. 1987). The production of 6-pentyl-2H-pyran-2-one is enhanced by solid-state fermentation with bagasse and desiccated coconut (Sarhy-Bagnon et al. 2000). Similarly, 1-octen-3-ol, 3-octanone, and 3-octanol add flavor and aroma in mushroom (Combet et al. 2006).

Methyl ketone produced by *T. viride* acts as a flavoring agent particularly for cheese and fruits and has gained widespread popularity (Patton 1950; Janssens et al. 1992; Hagedorn and Kaphammer 1994). *T. reesei* is used extensively in the cheese industry. The application of such products in the food and confectionaries sector has become a prime feature, rendering *Trichoderma* spp. a prolific fungus with large-scale industrial applications. Past reports have documented the role of *Trichoderma* exhibiting biosecurity attributes owing to their bioactive metabolites and enhancing antioxidants stream (Singh and Singh 2009; Singh et al. 2010). With growing interest and thrust laid down in this area, the possible role of *Trichoderma* secondary metabolites in different sectors has been thoroughly explored.

16.4 Application in Pharmaceutical Industry

The secondary metabolites produced by *Trichoderma* strains exhibit wide application potential in the medical sector. The secondary metabolites of *Trichoderma* have antifungal (Ande et al. 2008), antibacterial (Cheng et al. 2011), antiviral (Lu et al.

2002), antiprotozoal (Ciscotto et al. 2009), and anticancer activities (Liu et al. 2009). However, marine *Trichoderma* species are popular in the pharmaceutical industry. Marine *T. virens* isolated from ascidians and green algae are efficient producers of Trichodermamides-A and B having potent pharmaceutical value (Garo et al. 2003). The genomic study of *Trichoderma* has highlighted its widespread property within the genus (Druzhinina et al. 2011).

Few *Trichoderma* strains isolated from marine sponges and seaweeds possess significant antimicrobial activity against human pathogens (Masuma et al. 2001). Trichodermatides A–D (1–4) from *T. reesei* exhibits strong cytotoxicity against A375-S2 human melanoma cells (Sun et al. 2008). Similarly, Trichodenone A, B, and C, obtained from marine *T. harzianum* OUPSN115 display cytotoxicity against leukemia P388 cells (Thakur et al. 2003). *T. longibrachiatum* isolated from mussels are producers of fatty acids (Ruiz et al. 2007). Marine *Trichoderma* also reported to produce 16-methylheptadecanoic acid methyl ester (HDA) and 9,12-octadecadienoic acid (ODA) induces ROS-dependent programmed cell death (PCD) in oncogenic cells and thus shows its utility to be a promising chemotherapeutic agent for treating cancers (Saravanakumar et al. 2015).

T. viride plays a crucial role in antibiotic production by converting gliotoxin and viridin (Dennis and Webster 1971), to alpha-pyrones (Keszler et al. 2000). Based on the fact that fungal extracts provide evidence to develop anticancer drugs, a study was conducted for evaluation of the anticancer activity of an Endolichenic fungi (ELF), *T. harzianum,* (strain No: MF029755) extract against NCI-H292 lung cancer cells (Sinthujah et al. 2017). Similarly, L-Lysine a-oxidase (LO) enzyme isolated from *Trichoderma* cf. *aureoviride* Rifai has a characteristic cytotoxicity against the tumor cell lines and helps to eliminate the malady to a certain extent (Pokrovsky et al. 2013).

16.5 *Trichoderma* in Bioremediation

Disposal of heavy metal and organic contaminants in soil and water cause environmental pollution. These pollutants cause negative effects on flora and fauna of polluted sites and looses the quality of products. These pollutants in soil restrict plant growth and their productivity. The toxicity of pollutants cause a negative effect on the metabolic processes of plants like nitrogen fixation, nitrogen reduction, irregularities on enzyme synthesis, etc. (Nwuche and Ugoji 2008). Further accumulation of pollutants in soil and water enters the food chain and causes problems in human health. The degradation of these pollutants is necessary for waste management and land/water clearance (Xiezhi et al. 2005). Bioremediation is an approach to transform toxic metal compounds into non-hazardous substances through microbial interventions. This process is based on the attack of microbial enzymes on immobilizing waste materials. Microbes have the ability to link metal ions on their cell walls and use them as food source material. Thus, these microbes increase chemical and physical properties of soil including nutrient contents and their absorption in plants consequently improve plant growth, yield, and productivity. In addition, the presence of such microbes in the rhizosphere intensifies the phytoremediation process by enhancing phytostimulation. *Trichoderma* spp. are widely used in the bioremediation process. This group of fungi produces a diverse range of organic acids, which decrease pH in soil and allow the dissolution of complex compounds in their available forms for plants (Hasan 2016).

Trichoderma spp. have the potency to tolerate many agrochemicals, heavy metals, and pesticides residues, thus it is used in integrated pest management (Cao et al. 2008). The resistant ability of this fungal genus attracts to explore genetic material to employ in the bioremediation of toxic pollutants and other biotechnological applications. Some reports are available in which *Trichoderma* spp. reported to degrade chemical pesticides such as Pentachlorophenol (PCP) (Sing et al. 2014). In addition, *Trichoderma* isolates have been also reported to tolerate benzene compounds and crude oils (Argumedo-Delira et al. 2012).

A plethora of studies are available on *Trichoderma* mediated bioremediation and plant growth promotion under contaminated soil. *Trichoderma* spp. absorb various ions and remove them from soils (Errasquın and Vazquez 2003; Yazdani et al. 2009; Zeng et al. 2010; Srivastava et al. 2011). In a study, *T. viride* was able to remove cadmium and lead from polluted water and suggested as an economical and eco-friendly approach for treating effluents charged with toxic metallic ions (Sahu et al. 2012). Another study demonstrated the arsenic removal ability of *T. asperellum* and *T. viride* from liquid media through biovolatilization (Srivastava et al. 2011). Adams et al. (2007) have documented the activity of *T. atroviridae* and *T. harzianum* in modulating the uptake efficiency of metal ions (Zn, Ni, and Cd) in *Brassica juncea* and *Salix fragilis* plants in contaminated soil. In another study, *T. harzianum* strains were found to detoxify potassium cyanide compound and augment root growth of *Pteris vittata* fern in contaminated soil (Lynch and Moffat 2005). In a consortium study, *T. harzianum* and AM fungi altogether in *Eucalyptus globulus* plant exhibited increase tolerance and accumulation of aluminum and arsenic in soil (Arriagada et al. 2009).

In organic pollutants, cellulose and hemicellulose compounds covered a large portion. Hemicellulose is heterogeneous polymers of sugars and sugar acids. Most of the hemicelluloses are comprised of xylans and glucomannans and require endoxylanases and endomannanases enzymes for complete degradation (Adav and Sze 2014). Hemicellulose is more complex than cellulose due to several side-chain residues and need multi-enzyme action for complete hydrolysis. *Trichoderma reesei* has been reported to produce xylanases (endoxylanase), arabinofuranosidases, β-xylosidases, β-glucuronidase (GH79), acetyl xylan esterase, acetyl esterase to degrade cellulose, sugar beet pulp, sawdust, and corn stover (Olsson et al. 2003; Chundawat et al. 2011).

Various *Trichoderma* spp. such as *T. harzianum, T. atroviride,* and *T. longibrachiatum* have been isolated and characterized to degrade cellulose and hemicellulose compounds and further utilize them as carbon and energy source (Holker et al. 2002; Chakroun et al. 2010). These fungal strains degrade lignocellulosic biomass by producing antioxidative enzymes including catalase (CAT), peroxidase (POD), glutathione S-transferase glyoxalase, glutathione reductase (GR), and laccase enzymes (Holker et al. 2002; Adav et al. 2012). Laccase enzyme play role in lignin degradation and have commercial applications in different industries. In addition, peptidases, chitinases, phosphatase, transport proteins, and hypothetical proteins have been also reported in lignocellulosic hydrolysis (Adav et al. 2011). However, their role in degradation is not described properly.

16.6 *Trichoderma* in Nanotechnology

In recent times, the biological synthesis of nanoparticles is safer than hazardous material production, gaining attention for industrial purposes (Vahabi and Dorcheh 2014). Different microorganisms including fungi, yeast, bacteria, and algae are used as biological systems for the synthesis of nanoparticles (Das and Marsili 2011). The biosynthetically produced nanomaterials have unique biological activity and characteristics like size, dispersity, and mechanical properties (Gade et al. 2010). These nanoparticles have been used successfully in bioremediation of contaminated environment. In addition, these nanoparticles also have wide applications in pharmaceuticals, cosmetics, agriculture, and electronic industries.

Synthesis of nanoparticles from fungi recognized as mycosynthesis or myconanotechnology are emerging potential for interdisciplinary sciences (Rai et al. 2009). Different *Trichoderma* species *T. viride, T. harzianum, T. reesei,* and *T. asperellum* are key players in the biosynthesis of silver nanoparticles (Mukherjee et al. 2008; Fayaz et al. 2009; Vahabi et al. 2011). Most of the fungi reported for the biosynthesis of nanoparticles are pathogenic; *Trichoderma* is the only non-pathogenic fungi used for nanoparticle synthesis. *Trichoderma* originating nanoparticles are used for the preservation of fruits and vegetables and as a biosensor for the detection of microbial specific gene sequences (Siddiquee et al. 2011). These fungi produce extracellular enzymes and metabolites for the reduction of metal compounds. *T. reesei* is the best candidate among other *Trichoderma* spp. produce extracellular enzymes up to 100 g/lit and reduce toxic metal ions to nontoxic nanoparticles (Oksanen et al. 2000; Vahabi et al. 2011). Due to the higher expression of extracellular enzymes in *T. reesei,* they have a high biotechnological and industrial application. In addition, reductase enzymes such as nitrate reductase, naphthoquinones, and anthraquinones are also responsible for the metal reduction process by *T. ressei* (Baker and Tatum 1998; Bell et al. 2003). Biosynthesis of nanoparticles outside the cells has advantages in point of view of safety, economy, and technical inputs (Vahabi et al. 2011).

16.7 In Vitro Study on Nitrile Degrading Ability of *Trichoderma* Spp.: A Case Study

Nitriles are cyano group compounds (RC ≡ N) that are pervasive in the environment as their role is an excessive industrial arena. The main sources of nitrile contamination in agriculture soil are (1) contaminated water from chemical industries and (2) the use of agrochemicals. The nitrile group containing herbicides and pesticides such as dichlobenil, bromoxynil, ioxynil, buctril, and chlorothalonil, etc. are widely used in Indian states namely Uttar Pradesh followed by Punjab, Haryana, and Maharashtra (Abhilash and Singh 2009). The residues of these chemicals in soil caused pollution due to their recalcitrant nature and toxicity. Soil microbes have the ability to utilize nitrile compounds as carbon and nitrogen source. Three nitrile hydrolyzing enzymes, i.e., [nitrile hydratase (NHase), nitrilase, and amidase] participate in nitrile degradation through microbes. The presence of nitrile degrading microbes in plant soil suggests their explicit roles in nitrogen utilization, catabolism of cyanogenic glycosides and glucosinolates, detoxification of nitriles, and cyanides along with the synthesis of various plant hormones (Howden and Preston 2009). In this context, a plant beneficial fungi *T. harzianum* BHU P4 (MH730446—previously isolated from agriculture field) was tested for nitrile degrading ability on different nitrile compounds including acetonitrile, acrylonitrile, butyronitrile, benzonitrile, and indole-3-acetonitrile in the range of 10–100 mM individually. The in vitro plate assay was performed with minimal salt agar medium supplemented with nitrile substrates (10–100 mM) as a whole carbon and nitrogen source. Bromophenol blue dye was added for zone visualization. A 5 mm of fungal disk was placed on the center of petriplate containing autoclaved medium. Petriplates were incubated at 30 ± 2 °C for 72 hours. A clear orange zone was visualized on each nitrile substrate which indicates acid production in the medium due to the degradation of nitrile compounds (Fig. 16.2). *T. harzianum* BHU P4 showed tolerance up to 40 mM acetonitrile, 20 mM acrylonitrile, 60 mM butyronitrile, 30 mM benzonitrile, and 50 mM indole-3-acetonitrile. Further study will be conducted to analyze the role of *T. harzianum* BHU P4 in plant interaction and bioremediation of nitrile contaminated soil.

16.8 Conclusion and Future Prospects

Trichoderma spp. earns a reputation in the scientific community on the basis of their multifarious applications. These fungi have gained widespread popularity owing to their biocontrol and plant growth promotion attributes, but apart from agricultural applications, these fungi are also admired in other sectors such as food, beverage, pharmaceuticals, bioremediation, nanotechnology, etc. Different *Trichoderma* species such as *T. harzianum, T. reesei, T. viride, T. atroviride,* and *T. asperellum* have been reported for the biosynthesis of silver nanoparticles. Apart from that

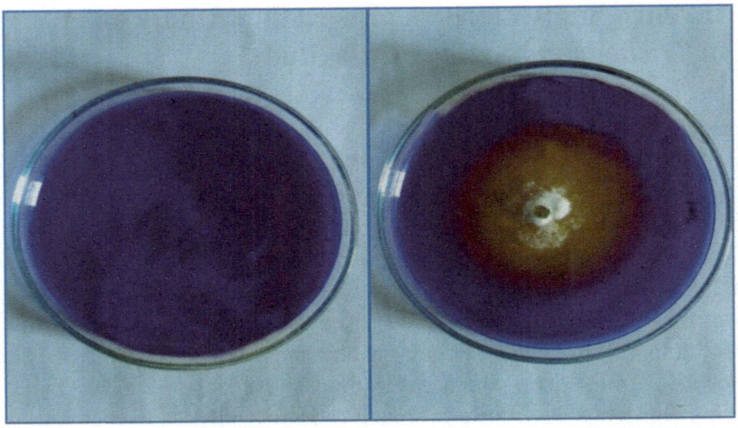

Fig. 16.2 Nitrile degrading ability of *T. harzianum* BHU P4 on minimal media supplemented with 40 mM acetonitrile as nitrogen and carbon source

Trichoderma spp. are also used in xenobiotics for mitigating metal toxicity problems. The different secondary metabolites of *Trichoderma* are highly used in imparting flavor and aroma to various food products. The anti-cancerous and anti-aging property of *Trichoderma* spp. is also highly appreciated. So it is high time to explore other benefits of this "wonder fungi." Thus, further studies and research attention is needed to identify potential isolates of *Trichoderma* with multifarious applications in order to reap their beneficial aspects. Advanced molecular and biotechnological tools may be advocated in order to develop new strains.

Acknowledgments AV is grateful to SERB-NPDF (PDF/2017/000689) for providing financial assistance.

References

Abhilash PC, Singh N (2009) Pesticide use and application: an Indian scenario. J Hazard Mater 165:1–12

Adams P, De-Leij FAAM, Lynch JM (2007) *Trichoderma harzianum* Rifai 1295-22 mediates growth promotion of crack willow (*Salix fragilis*) saplings in both clean and metal-contaminated soil. Microb Ecol 54:306–313

Adav SS, Chao LT, Sze SK (2012) Quantitative secretomic analysis of *Trichoderma reesei* strains reveals enzymatic composition for lignocellulosic biomass degradation. Mol Cell Proteomics 11:M111.012419

Adav SS, Ravindran A, Chao LT, Tan L, Singh S, Sze SK (2011) Proteomic analysis of pH and strains dependent protein secretion of *Trichoderma reesei*. J Proteome Res 10:4579–4596

Adav SS, Sze SK (2014) *Trichoderma* secretome: an overview. In: biotechnology and biology of *Trichoderma*. Elsevier, pp 103–114

Alvarez-Rodriguez ML, Lopez-Ocana L, Lopez-Coronado JM, Rodriguez E, Martinez MJ, Larriba G, Coque JJR (2002) Cork taint of wines: role of the filamentous fungi isolated from cork in the formation of 2,4,6-trichloroanisole by O methylation of n 2,4,6-trichlorophenol. Appl Environ Microbiol 68:5860–5869

Ande SR, Fussi H, Knauer H, Murkovic M, Ghisla S, Frohlich KU, Macheroux P (2008) Induction of apoptosis in yeast by L-amino acid oxidase from the Malayan pit viper *Calloselasma rhodostoma*. Yeast 25:349–357

Argumedo-Delira R, Alarcón A, Ferrera-Cerrato R, Almaraz JJ, Peña-Cabriales JJ (2012) Tolerance and growth of 11 *Trichoderma* strains to crude oil, naphthalene, phenanthrene and benzo [a] pyrene. J Env Manage 95:S291–S299

Arisan-Atac I, Hodits R, Kristufek D, Kubicek CP (1993) Purification, and characterization of a β-mannanase of Trichoderma reesei C-30. Appl Microbiol Biotechnol 39(1):58–62

Arriagada C, Aranda E, Sampedro I, Garcia-Romera I, Ocampo JA (2009) Contribution of the saprobic fungi *Trametes versicolor* and *Trichoderma harzianum* and the arbuscular mycorrhizal fungi *Glomus deserticola* and *G. claroideum* to arsenic tolerance of *Eucalyptus globulus*. Bioresour Technol 100:6250–6257

Baker RA, Tatum JH (1998) Novel anthraquinones from stationary cultures of *Fusarium oxysporum*. J Ferment Bioeng 85:359–361

Bayitse, R., Hou, X., Laryea, G., & Bjerre, A. B. (2015). Protein enrichment of cassava residue using Trichodermapseudokoningii (ATCC 26801). AMB Express,5(1), 80.

Bell AA, Wheeler MH, Liu J, Stipanovic RD, Puckhaber LS, Orta H (2003) United States department of agriculture-agricultural research service studies on polyketide toxins of *Fusarium oxysporum* f. sp. *vasinfectum*: potential targets for disease control. Pest Manag Sci 59:736–747

Blumenthal CZ (2004) Production of toxic metabolites in *Aspergillus niger*, *Aspergillus oryzae*, and *Trichoderma reesei*: justification of mycotoxin testing in food grade enzyme preparations derived from the three fungi. Regul Toxicol Pharmacol 39(2):214–228

Cao L, Jiang M, Zeng Z, Du A, Tan H, Liu Y (2008) *Trichoderma atroviride* F6 improves phytoextraction efficiency of mustard (*Brassica juncea* (L.) Coss. Var. foliosa bailey) in cd, Ni contaminated soils. Chemosphere 71:1769–1773

Cardoso Lopes FA, Steindorff AS, Geraldine AM, Brandao RS, Monteiro VN, Junior ML, Guedes Coelho AS, Ulhoa CJ, Silva RN (2012) Biochemical and metabolic profiles of *Trichoderma* strains isolated from common bean crops in the Brazilian Cerrado, and potential antagonism against *Sclerotinia sclerotiorum*. Fungal Biol 116:815–824

Chakraborty S, Yadav G, Saini JK, Kuhad RC (2019) Comparative Study of Cellulase Production Using Submerged and Solid-State Fermentation. In New and Future Developments in Microbial Biotechnology and Bioengineering (pp. 99113). Elsevier

Chakroun H, Mechichi T, Martinez MJ, Dhouib A, Sayadi S (2010) Purification and characterization of a novel laccase from the ascomycete *Trichoderma atroviride*: application on bioremediation of phenolic compounds. Process Biochem 45:507–513

Cheng H, Yang CA, Liu SY, Lo CT, Huang HC, Liao FC, Peng KC (2011) Cloning of a novel L-amino acid oxidase from *Trichoderma harzianum* ETS 323 and bioactivity analysis of overexpressed L-amino acid oxidase. J Agric Food Chem 59:9142–9149

Chundawat SPS, Lipton MS, Purvine SO, Uppugundla N, Gao D, Balan V, Dale BE (2011) Proteomics based compositional analysis of complex cellulase -hemicellulase mixtures. J Proteome Res 10:4365–4372

Ciscotto P, Machado RA, de Avila EA, Coelho J, Oliveira CG, Farais LM, de Carvalho MA, Maria WS, Sanchez EF, Borges A, Chavez-Olortegui C (2009) Antigenic microbial and parasitic properties of an L-amino acid oxidase isolated from *Bothrops jararaca* snake venom. Toxicon 53:330–341

Claydon N, Allan M, Hanson JR, Avent AG (1987) Antifungal alkyl pyrones of *Trichoderma harzianum*. Trans Br Mycol Soc 88:503–513

Combet E, Henderson J, Eastwood DC, Burton KS (2006) Eight-carbon volatiles in mushrooms and fungi: properties, analysis, and biosynthesis. Mycoscience 47:317–326

Coque JJ, Alvarez-Rodríguez ML, Larriba G (2003) Characterization of an inducible chlorophenol *O*-methyltransferase from *Trichoderma longibrachiatum* involved in the formation of chloroanisoles and determination of its role in cork taint of wines. Appl Environ Microbiol 69:5089–5095

Das SK, Marsili E (2011) Bioinspired metal nanoparticle: synthesis, properties and application. Nanomaterials: 253–274

Dennis C, Webster J (1971) Antagonistic properties of species groups of *Trichoderma*: III. Hyphal interactions. Trans Br Mycol Soc 57:363–369

Druzhinina IS, Kopchinskiy AG, Kubicek CP (2006) The first 100 *Trichoderma* species characterized by molecular data. Mycoscience 47:55–64

Druzhinina IS, Seidl-Seiboth V, Herrera-Estrella A, Horwitz BA, Kenerley CM, Monte E, Mukherjee PK, Zeilinger S, Grigoriev IV, Kubicek CP (2011) *Trichoderma*: the genomics of opportunistic success. Nat Rev Microbiol 16:749–759

Errasquın EL, Vazquez C (2003) Tolerance and uptake of heavy metals by *Trichoderma atroviride* isolated from sludge. Chemosphere 50:137–143

Fayaz AM, Balaji K, Girilal M, Kalaichelvan PT, Venketesan R (2009) Mycobased synthesis of silver nanoparticles and their incorporation into sodium alginate films for vegetable and fruit preservation. J Agric Food Chem 57:6246–6252

Florencio C, Cunha FM, Badino AC, Farinas CS, Ximenes E, Ladisch MR (2016) Secretome data from *Trichoderma reesei* and *Aspergillus niger* cultivated in submerged and sequential fermentation methods. Data Brief 8:588–598

Fuglsang CC, Johansen C, Christgau S, Adler Nissen J (1995) Antimicrobial enzymes: applications and future potential in the food industry. Trends Food Sci Technol 6(12):390–396

Gade A, Ingle A, Whiteley C, Rai M (2010) Mycogenic metal nanoparticles: progress and applications. Biotechnol Lett 32:593–600

Galante Y, De Conti A, Monteverdi R (1998) Application of *Trichoderma* enzymes in the food and feed industry. In: Harman G, Kubicek C (eds) *Trichoderma* and *Gliocladium*, enzymes, biological control and commercial applications, vol 2, pp 327–342

Garo E, Starks CM, Jensen PR, Fenical W, Lobkovsky E, Clardy J (2003) Trichodermamides a and B, cytotoxic modified dipeptides from the marine derived fungus *Trichoderma virens*. J Nat Pro 66:423–426

Hagedorn S, Kaphammer B (1994) Microbial biocatalysis in the generation of flavor and fragrance chemicals. Annu Rev Microbiol 48:773–800

Harman GE, Howell CR, Viterbo A, Chet I, Lorito M (2004) *Trichoderma* species-opportunistic, avirulent plant symbionts. Nat Rev Microbiol 2:43–56

Hasan S (2016) Potential of *Trichoderma* sp. in bioremediation: a review. J Basic Appl Eng Res 3:776–779

Hatvani L, Antal Z, Manczinger L, Szekeres A, Druzhinina IS, Kubicek CP, Nagy A, Nagy E, Vagvolgyi C, Kredics L (2007) Green mold diseases of *Agaricus* and *Pleurotus* spp. are caused by related but phylogenetically different *Trichoderma* species. Phytopathology 97:532–537

Holker U, Dohse J, Höfer M (2002) Extracellular laccases in ascomycetes *Trichoderma atroviride* and *Trichoderma harzianum*. Folia Microbiol 47:423–427

Howden AJ, Preston GM (2009) Nitrilase enzymes and their role in plant–microbe interactions. Microb Biotechnol 2(4):441–451

Hoyos-Carvajal L, Bissett J (2011) Biodiversity of *Trichoderma* in neotropics. In: Grillo, O, Venora G. (Eds.), The dynamical processes of biodiversity – case studies of evolution and spatial distribution, Intech, pp. 303–320

Janssens L, de Pooter HL, Vandamme EJ, Schamp NM (1992) Production of flavours by microorganisms. Process Biochem 27:195–215

Keswani C, Mishra S, Sarma BK, Singh SP, Singh HB (2014) Unraveling the efficient applications of secondary metabolites of various *Trichoderma* spp. Appl Microbiol Biotechnol 98 (2):533–544

Keszler A, Forgacs E, Kotal L, Vizcaino JA, Monte E, Garcia-Acha I (2000) Separation and identification of volatile components in the fermentation broth of *Trichoderma atroviride* by solid phase extraction and gas chromatography-mass spectroscopy. J Chromatograph Sci 38:421–424

Kindermann J, El-Ayouti Y, Samuels GJ, Kubicek CP (1998) Phylogeny of the genus *Trichoderma* based on sequence analysis of the internal transcribed spacer region 1 of the rDNA cluster. Fungal Genet Biol 24:298–309

Kraus G, Druzhinina I, Bissett J, Prillinger HJ, Szakacs G, Gams W, Kubicek CP (2004) *Trichoderma brevicompactum* sp. nov. Mycologia 96:1059–1073

Kredics L, Antal Z, Doczi I, Manczinger L, Kevei F, Nagy E (2003) Clinical importance of the genus *Trichoderma.* A review. Acta Microbiol Immunol Hung 50:105–117

Kredics L, Diczi I, Antal Z, Bartyik K, Molnar EG, Manczinger L, Hatvani L, Vagvolgyi C, Nagy E (2006) Emergence of the filamental fungus opportunist *Trichoderma longibrachiatum* in Hungary. Acta Microbiol Immunol Hung 53(3):305

Kubicek CP, Komon-Zelazowska M, Druzhinina IS (2008) Fungal genus *Hypocrea/Trichoderma*: from barcodes to biodiversity. J Zhejiang Univ Sci B 9:753–763

Kuhls K, Lieckfeldt E, Borner T, Gueho E (1999) Molecular re-identification of human pathogenic *Trichoderma* isolates as *Trichoderma longibrachiatum* and *Trichoderma citrinoviride.* Med Mycol 37:25–33

Liu SY, Lo CT, Shibu MA, Leu YL, Jen BY, Peng KC (2009) Study on the anthraquinones separated from the cultivation of *Trichoderma harzianum* strain Th-R16 and their biological activity. J Agric Food Chem 57:7288–7292

Lorito M, Woo SL, Harman GE, Monte E (2010) Translational research on *Trichoderma*: from 'omics to the field. Annu Rev Phytopathol 48:395–417

Lu QM, Wei Q, Jin Y, Wei JF, Wang WY, Xiong YL (2002) L-amino acid oxidase from *Trimeresurus jerdonii* snake venom: purification, characterization, platelet aggregation-inducing and antibacterial effects. J Nat Toxins 11:345–352

Lynch JM, Moffat AJ (2005) Bioremediation—prospects for the future application of innovative applied biological research. Ann Appl Biol 146:217–221

Mach R, Zeilinger S (2003). Regulation of gene expression in industrial fungi: Trichoderma. Appl Microbiol Biotechnol 60(5):515–522

Marques SF, Minafra CS, Cafe MB, Stringhini JH, Ulhoa CJ (2018) Production and characterization of a *Trichoderma harzianum* multienzyme complex and its application in broiler Chicks' diets. Curr Biotechnol 7(1):26–33

Masuma R, Yamaguchi Y, Noumi M, Omura S, Namikosh M (2001) Effect of sea water concentration on hyphal growth and antimicrobial metabolite production in marine fungi. Mycoscience 42:455–459

Mukherjee P, Roy M, Mandal BP, Dey GK, Mukherjee PK, Ghatak J, Tyagi AK, PKale S (2008) Green synthesis of highly stabilized nanocrystalline silver particles by a non-pathogenic and agriculturally important fungus *T. asperellum.* Nanotechnology 19:075103–075110

Mukhopadhyay AN, Shrestha SM, Mukherjee PK (1992) Biological seed treatment for control of soilborne plant pathogens FAO. Plant Prot Bull 40:21–30

Nevalainen KH (2017). The molecular biology of *Trichoderma* and its application to the expression of both homologous and heterologous genes. In Molecular Industrial Mycology (pp. 129–148). Routledge

Nevalainen H, Suominen P, Taimisto K (1994) On the safety of *Trichoderma-reesei.* J Biotechnol 37(3):193–200

Nuutila AM, Ritala A, Skadsen RW, Mannonen L, Kauppinen V (1999) Expression of fungal thermo tolerant endo-1,4-beta-glucanase in transgenic barley seeds during germination. Plant Mol Biol 41:777–783

Nwuche CO, Ugoji EO (2008) Effect of heavy metal pollution on the soil microbial activity. J Environ Sci 5:409–414

Oksanen T, Pere J, Paavilainen L, Buchert J, Viikari L (2000) Treatment of recycled Kraft pulps with *Trichoderma reesei* hemicellulases and cellulases. J Biotechnol 78:39–44

Olsson L, Christensen TMIE, Hansen KP, Palmqvist EA (2003) Influence of the carbon source on production of cellulases, hemicellulases and pectinases by *Trichoderma reesei* rut C-30. Enzym Microb Technol 33:612–619

Park MS, Bae KS, Yu SH (2006) Two new species of *Trichoderma* associated with green mold of oyster mushroom cultivation in Korea. Mycobiol 34:111–113

Patel N, Shahane S, Majumdar R, Mishra U (2019) Mode of action, properties, production, and application of Laccase: a review. Recent Pat Biotechnol 13(1):19–32

Patton S (1950) The methyl ketones of blue cheese and their relation to its flavor. J Dairy Sci 33:680–684

Perez-Gonzalez JA, Gonzalez R, Querol A, Sendra J, Ramon D (1993) Construction of a recombinant wine yeast strain expressing β-(1,4)-endoglucanase and its use in microvinification processes. Appl Environ Microbiol 59:2801–2806

Persoon CH (1794) Disposita methodica fungorum. Romer's Neues Mag Bot 1:81–128

Pokrovsky VS, Treshalina HM, Lukasheva EV, Sedakova LA, Medentzev AG, Arinbasarova AY, Berezov TT (2013) Enzymatic properties and anticancer activity of L-lysine α-oxidase from *Trichoderma* cf. *aureoviride* Rifai BKMF-4268D. Anti-Cancer Drugs 24(8):846–851

Prak S, Gunata Z, Guiraud JP, Schorr-Galindo S (2007) Fungal strains isolated from cork stoppers and the formation of 2,4,6-trichloroanisole involved in the cork taint of wine. Food Microbiol 24:271–280

Rai M, Yadav P, Bridge P, Gade A (2009) Myconanotechnology: a new and emerging science. In: Rai B (ed) Applied mycology. CABI publication, UK, pp 258–267

Ram RM, Singh HB (2017) *Trichoderma* spp: Nature's gift to mankind. In: Plant systematics & Biotechnology: Challenges and opportunities. Today and tomorrow's printers and publishers, 133–141

Rebello S, Aneesh EM, Sindhu R, Binod P, Pandey A, Gnansounou E (2019) Enzyme Catalysis: a Workforce to Productivity of Textile Industry. A handbook on high value fermentation products, Volume 2: Human Welfare, 49

Ruiz N, Dubois N, Wielgosz C, Robiou PT, Berge EP, Pouchus YF, Barnathan G (2007) Lipid content and fatty acid composition of a marine-derived *Trichoderma longibrachiatum* strain cultured by agar surface and submerged fermentations. Process Biochem 42:676–680

Sahu A, Manna MC, Mandal A, Rao SA, Thakur J (2012) Exploring bioaccumulation efficacy of *Trichoderma viride*: an alternative bioremediation of cadmium and Lead. Natl Acad Sci Lett 35:299–302

Samuels GJ, Dodd SL, Gams W, Castlebury LA, Petrini O (2002) *Trichoderma* species associated with the green mold epidemic of commercially grown *Agaricus bisporus*. Mycol 94:146–170

Saravanakumar K, Vivek R, Boopathy NS, Yaqian L, Kathiresan K, Chen J (2015) Anticancer potential of bioactive 16-methylheptadecanoic acid methyl ester derived from marine *Trichoderma*. J Appl Biomed 13(3):199–212

Sarhy-Bagnon V, Lozano P, Saucedo Castañeda G, Roussos S (2000) Production of 6-pentyl-α-pyrone by *Trichoderma harzianum* in liquid and solid state cultures. Process Biochem 36:103–109

Schuster A, Schmoll M (2010) Biology and biotechnology of *Trichoderma*. Appl Microbiol Biotechnol 87(3):787–799

Siddiquee S, Yusof NA, Salleh AB, Tan SG, Abu Bakar F (2011) Electrochemical DNA biosensor for the detection of *Trichoderma harzianum* based on a gold electrode modified with a composite membrane made from an ionic liquid, ZnO nanoparticles and chitosan, and by using acridine orange as a redox indicator. Microchim Acta 172:357–363

Sing NN, Zulkharnain A, Roslan HA, Assim Z, Husaini A (2014) Bioremediation of PCP by *Trichoderma* and *Cunninghamella* strains isolated from sawdust. Braz Arch Biol Technol 57 (6):811–820

Singh HB (2014) Management of plant pathogens with microorganisms. Proceedings of National Academy of Science 80:443–454

Singh HB, Singh DP (2009) From biological control to bioactive metabolites: prospects with *Trichoderma* for safe human food. Pertanika Jn Trop Agric Sci 32:99–110

Singh P, Singh J, Rajput RS, Vaishnav A, Ray S, Singh RK, Singh HB (2019a) Exploration of multitrait antagonistic microbes against *Fusarium oxysporum* f. sp. *lycopersici*. J Appl Nat Sci 11:503–510

Singh P, Singh J, Rajput RS, Vaishnav A, Ray S, Singh RK, Singh HB (2019b) *Trichoderma* mediated seed biopriming augments antioxidant and phenylpropanoid activities in tomato plant against *Sclerotium rolfsii*. J Pharma Phytochem 8:2641–2647

Singh HB, Singh BN, Singh SP, Nautiyal CS (2010) Solid-state cultivation of *Trichoderma harzianum* NBRI-1055 for modulating natural antioxidants in soybean seed matrix. Bioresour Technol 101:6444–6453

Singh BN, Singh A, Singh SP, Singh HB (2011) *Trichoderma harzianum* mediated reprogramming of oxidative stress response in root apoplast of sunflower enhances defense against *Rhizoctonia solani*. European J Plant Pathol 131:121–134

Sinthujah S, Samarakoon SR, Tennakoon KH, Attanayake RN, Weerakoon G, Gunasegara DS, Paranagama PA (2017) Anticancer activity of *Trichoderma harzianum* extract against NCI-H292 lung cancer cells. International Research Symposium on Pure and Applied Sciences, 2017 Faculty of Science, University of Kelaniya, Sri Lanka

Sivasithamparam K, Ghisalbarti E (1998) Secondary metabolism. In: Harman GE, Kubicek CP (eds) *Trichoderma* and *Gliocladium*, Basic biology, taxonomy and genetics, vol 1. Taylor and Francis London, UK, pp 139–191

Srivastava PK, Vaish A, Dwivedi S, Chakrabarty D, Singh N, Tripathi RD (2011) Biological removal of arsenic pollution bysoil fungi. Sci Total Environ 409:2430–2442

Styger G, Prior B, Bauer FF (2011) Wine flavor and aroma. Ind Microbiol Biotechnol 38:1145–1159

Sun Y, Tian L, Huang J, Ma HY, Zheng Z, Yasukawa K, Pei YH (2008) Novel polyketides from the marine-derived fungus *Trichoderma reesei*. Oncol Lett 10(3):393–396

Thakur NL, Hentschel U, Krasko A, Pabel CT, Anr AC, Muller WEG (2003) Antibacterial activity of the sponge *Suberites domuncula* and its primmorphs: potential basis for epibacterial chemical defense. Aquat Microbial Ecol 31:77–83

Tulasne LR, Tulasne C (1865) Selecta fungorum carpologia. Jussu, Paris

Vahabi K, Dorcheh SK (2014) Biosynthesis of silver nano-particles by Trichoderma and its medical applications. In: biotechnology and biology of *Trichoderma*. Elsevier: 393-404

Vahabi K, Mansoori GA, Karimi S (2011) Biosynthesis of silver nanoparticles by fungus *Trichoderma reesei* (a route for large-scale production of AgNPs). Insciences J 1:65–79

Wiater A, Pleszczynska M, Szczodrak J, Janusz G (2012) Comparative studies on the induction of *Trichoderma harzianum* mutanase by α-(1→3)-glucan-rich fruiting bodies and mycelia of *Laetiporus sulphureus*. Int J Mol Sci 13(8):9584–9598

Wiater A, Szczodrak J, Pleszczynska M (2005) Optimization of conditions for the efficient production of mutan in streptococcal cultures and post-culture liquids. Acta Biol Hung 56 (1–2):137–150

Xiezhi Y, Jieming C, Ming HM (2005) Earthworm-mycorrhiza interaction on cd uptake and growth of ryegrass. Soil Biol Biochem 37:195–201

Yazdani M, Yap CK, Abdullah F, Tan SG (2009) *Trichoderma atroviride* as a bioremediator of cu pollution: an *in vitro* study. Toxicol Environ Chem 91:1305–1314

Zeng X, Su S, Jiang X, Li L, Bai L, Zhang Y (2010) Capability of pentavalent arsenic bioaccumulation and biovolatilization of three fungal strains under laboratory conditions. Clean: Soil, Air, Water 38:238–241

The manufacturer's authorised representative in the EU is Springer Nature Customer Service Centre GmbH, Europaplatz 3, 69115 Heidelberg, Germany. If you have any concerns regarding our products, please contact ProductSafety@springernature.com

Printed and bound by CPI Group (UK) Ltd, Croydon, CR0 4YY
30/09/2026
02225456-0019